56 Springer Series in Chemical Physics
Edited by J. Peter Toennies

Springer Series in Chemical Physics
Editors: Vitalii I. Goldanskii   Fritz P. Schäfer   J. Peter Toennies

Managing Editor: H.K.V. Lotsch

Volume 40   **High-Resolution Spectroscopy of Transient Molecules**
By E. Hirota

Volume 41   **High Resolution Spectral Atlas of Nitrogen Dioxide 559–597 nm**
By K. Uehara and H. Sasada

Volume 42   **Antennas and Reaction Centers of Photosynthetic Bacteria**
Structure, Interactions, and Dynamics
Editor: M.E. Michel-Beyerle

Volume 43   **The Atom-Atom Potential Method.** Applications to Organic
Molecular Solids
By A.J. Pertsin and A.I. Kitaigorodsky

Volume 44   **Secondary Ion Mass Spectrometry SIMS V**
Editors: A. Benninghoven, R.J. Colton, D.S. Simons, and H.W. Werner

Volume 45   **Thermotropic Liquid Crystals, Fundamentals**
By G. Vertogen and W.H. de Jeu

Volume 46   **Ultrafast Phenomena V**
Editors: G.R. Fleming and A.E. Siegman

Volume 47   **Complex Chemical Reaction Systems**
Mathematical Modelling and Simulation
Editors: J. Warnatz and W. Jäger

Volume 48   **Ultrafast Phenomena VI**
Editors: T. Yajima, K. Yoshihara, C.B. Harris, and S. Shionoya

Volume 49   **Vibronic Interactions in Molecules and Crystals**
By I.B. Bersuker and V.Z. Polinger

Volume 50   **Molecular and Laser Spectroscopy**
By Zu-Geng Wang and Hui-Rong Xia

Volume 51   **Space-Time Organization in Marcomolecular Fluids**
Editors: F. Tanaka, M. Doi, and T. Ohta

Volume 52   **Clusters of Atoms and Molecules.** Theory, Experiment,
and Clusters of Atoms
Editor: H. Haberland

Volume 53   **Ultrafast Phenomena VII**
Editors: C.B. Harris, E.P. Ippen, G.A. Mourou, and A.H. Zewail

Volume 54   **Physics of Ion Impact Phenomena**
Editor: D. Mathur

Volume 55   **Ultrafast Phenomena VIII**
Editors: J.-L. Martin, A. Migus, G.A. Mourou, and A.H. Zewail

Volume 56   **Clusters of Atoms and Molecules II.** Solvation and Chemistry of
Free Clusters, and Embedded, Supported and Compressed Clusters
Editor: H. Haberland

Volume 1–39 are listed on the back inside cover

Hellmut Haberland (Ed.)

Clusters of Atoms and Molecules II

Solvation and Chemistry of Free Clusters,
and Embedded, Supported and Compressed Clusters

With 174 Figures and 15 Tables

Springer-Verlag

Berlin Heidelberg New York
London Paris Tokyo
Hong Kong Barcelona
Budapest

Professor Dr. Hellmut Haberland
Albert-Ludwigs-Universität
Fakultät für Physik
Hermann-Herder Strasse 3
D-79104 Freiburg, Germany

Series Editors

Professor Dr. Fritz Peter Schäfer
Max-Planck-Institut
für Biophysikalische Chemie
D-37073 Göttingen-Nikolausberg, Germany

Professor Vitalii I. Goldanskii
Institute of Chemical Physics
Academy of Sciences
Kosygin Street 4
Moscow, 117334, USSR

Professor Dr. J. Peter Toennies
Max-Planck-Institut
für Strömungsforschung
Böttingerstrasse 6–8
D-37073 Göttingen, Germany

Managing Editor: Dr. Helmut K.V. Lotsch
Springer-Verlag, Tiergartenstrasse 17,
D-69121 Heidelberg, Germany

ISBN-13:978-3-642-84987-9   e-ISBN-13:978-3-642-84985-5
DOI: 10.1007/978-3-642-84985-5

Library of Congress Cataloging-in-Publication Data. (Revised for vol. 2) Clusters of atoms and molecules. (Springer series in chemical physics; v. 52, 56) Includes bibliographical references and indexes. Contents: [1] Theory, experiment, and clusters of atoms – [2] Solvation and chemistry of free clusters, and embedded, supported, and compressed clusters. 1. Microclusters. 2. Metal crystals. 3. Chemistry, Physical and theoretical. I. Haberland, Hellmut, 1939– . I. Series. QC173.4.M48C55 1993   546.3   93-28934 ISBN-13:978-3-642-84987-9

Typesetting: Macmillan India Ltd., Bangalore 25
SPIN: 10122951   54/3140/SPS – 5 4 3 2 1 0 – Printed on acid-free paper

Preface

Cluster science studies the transition from atomic, and molecular physics or chemistry to the science and technology of condensed matter. Two main topics from this large field will be emphasized in this second volume of *Atomic and Molecular Clusters*. After an Introduction, Chap. 2 deals mainly with molecular clusters, how they react to positive or negative charges (Sect. 2.1 to 2.5), how they decompose and how they can be charged (Sect. 2.6 and 2.7), and how one can do chemistry with them (2.8 and 2.9).

Clusters in contact with a macroscopic medium are treated in Chap. 3. It is from this domain that one can expect possible new applications of cluster science. The optical spectra of silver clusters in a dielectric medium are discussed in Sect. 3.1. Their properties have since long been used unknowingly to stain glass windows. Large clusters floating in an ambient pressure gas are called aerosols (Sect. 3.2). Their properties can be used to monitor air pollution. Development of a photographic film is due to supported silver clusters in a liquid environment (Sect. 3.3). Large semiconductor clusters, also called "quantum dots", have novel optical and electronic properties (Sect. 3.4). The optical properties of large clusters, in general, are reviewed in Sect. 3.5, and properties of clusters supported on clean surfaces are discussed in Sect. 3.6. Large clusters can be collected under clean conditions by letting them stick to a cold surface in vacuum. A "nanocrystalline material" is formed if the clusters are collected and compressed. This macroscopic material is thus composed of compressed clusters. It has many unique physical properties not found in the normal bulk state.

Each part has been written by expert in the field. The intention was to start out as simply as possible, then to progress to more difficult problems and to end with the latest results of the field. My sincere thanks go to all contributors and Dr. H. Lotsch, who have all helped to realise this book.

Freiburg, February 1994   Hellmut Haberland

Contents

1 Introduction
H. Haberland 1
1.1 Historical Application of Metal Clusters 2
1.2 Outlook 3
Reference 3

2 Solvation, Chemistry, and Charging of Free Clusters
2.1 Solvated Atoms in Polar Solvents
C.P. Schulz and I.V. Hertel (With 8 Figures) 7
    2.1.1 Introduction 7
    2.1.2 Experimental 8
    2.1.3 Ionisation Potentials 10
    2.1.4 Further Spectroscopic Studies 16
    References 18
2.2 IR Spectroscopy of Solvated Molecules
F.G. Amar, S. Goyal, D.J. Levandier, L. Perera, and G. Scoles
(With 12 Figures) 19
    2.2.1 Introduction 19
    2.2.2 Solvation Studies 21
    2.2.3 Computer Simulations 28
    2.2.4 Complex Forming Reactions on the Surface and
          in the Interior of a Cluster 37
    2.2.5 Recent Experimental Results 40
    References 42
2.3 IR Spectroscopy of Hydrogen Bonded Charged Clusters
M.W. Crofton, J.M. Price, and Y.T. Lee (With 10 Figures) . . . 44
    2.3.1 General Background and Motivation 44
    2.3.2 Some Previous Studies 47
    2.3.3 Consequence Spectroscopy of Charged Clusters 49
    2.3.4 Basic Principles of the Experiment 49
    2.3.5 Typical Results and Interpretations 54
    2.3.6 Concluding Remarks 69
    References 74
2.4 Solvated Cluster Ions
A.W. Castleman, Jr. (With 17 Figures) 77
    2.4.1 Introduction 77

2.4.2   Experimental Techniques   79
2.4.3   Reactions Influenced by Solvation   87
2.4.4   Solvation Phenomena   111
2.4.5   Photodissociation Experiments of Solvated Cluster Ions   122
2.4.6   Recent Developments   127
References   127

2.5   Solvated Electron Clusters
      H. Haberland and K.H. Bowen (With 12 Figures)   134
2.5.1   Introduction   134
2.5.2   Case Studies   137
2.5.3   Outlook   152
2.5.4   Recent Developments   152
References   152

2.6   Internal Reactions and Metastable Dissociations
      after Ionization of van der Waals Clusters
      T.D. Märk and O. Echt (With 17 Figures)   154
2.6.1   Introduction   154
2.6.2   Ionization Mechanisms and Processes   155
2.6.3   Ionization Efficiency   157
2.6.4   Post Collision Internal Reactions   159
2.6.5   Metastable Dissociations   161
2.6.6   Recent Developments   178
References   178

2.7   Multiply Charged Clusters
      O. Echt and T.D. Märk (With 17 Figures)   183
2.7.1   Introduction   183
2.7.2   Formation of Multiply Charged Clusters   185
2.7.3   Stability and Fragmentation of Multiply Charged Clusters   201
2.7.4   Outlook   216
2.7.5   Recent Developments   216
References   217

2.8   Chemistry with Neutral Metal Clusters
      S.J. Riley (With 7 Figures)   221
2.8.1   Introduction – Clusters and Heterogeneous Catalysis . .   221
2.8.2   Experimental   222
2.8.3   Adsorbate Uptake – The Path to Coverage   223
2.8.4   Kinetics – Strong Cluster Size Dependence and
        the Approach to Bulk   225
2.8.5   Equilibrium – The Thermodynamics
        of Adsorbate Binding   227
2.8.6   Saturated Compositions – The Number and Nature
        of Adsorption Sites   229
2.8.7   Chemical Probes of Metal Cluster Structure   231
2.8.8   Chemistry on Clusters – Adsorbate Decomposition . . .   235

2.8.9   Future Prospects 238
2.8.10 Recent Developments 239
References 239
2.9   Chemistry with Cluster Ions
S.L. Anderson (With 7 Figures) 241
2.9.1   Introduction 241
2.9.2   Experimental Methods 243
2.9.3   Comparison of Ion and Neutral Cluster Reactivity ... 247
2.9.4   Boron Cluster Ions: A Case Study with Ion Beams ... 248
2.9.5   Chemistry Studies with ICR Methods 254
2.9.6   Chemical Identification of Isomers 255
2.9.7   Future Directions 257
2.9.8   Recent Developments 258
References 258

3   **Embedded, Supported, and Compressed Clusters**
3.1   Optical Properties of Silver Clusters in Dielectric Matrices
K.-P. Charlé and W. Schulze (With 3 Figures) 263
3.1.1   Introduction 263
3.1.2   Optical Absorption of Colloids 263
3.1.3   Size Effect and the Influence of the Matrix 266
References 271
3.2   Aerosols, Large Clusters in Gas Suspensions
H. Burtscher and H.C. Siegmann (With 11 Figures) 272
3.2.1   Introduction 272
3.2.2   Some Tools of Aerosol Science 273
3.2.3   Diffusion Charging of Particles 274
3.2.4   Photoelectric Charging of Particles 275
3.2.5   The Photoelectric Yield of Small Metals Particles ... 277
3.2.6   Adsorption of Gas Molecules at the Surface of Particles 281
3.2.7   Applications of Photoelectron Emission from Particles . 284
3.2.8   X-Ray Absorption of Particles 286
3.2.9   Conclusions 287
3.2.10 Recent Developments 288
References 288
3.3   Metal Clusters in a Liquid Environment
Photographic Development
J. Belloni, J. Amblard, J.L. Marignier and M. Mostafavi
(With 4 Figures) 290
3.3.1   Introduction 290
3.3.2   Synthesis 291
3.3.3   Physical Properties 295
3.3.4   Chemical Properties 298
3.3.5   Photographic Development 306

   3.3.6 Conclusion 308
   3.3.7 Recent Developments 309
   References 309
3.4 Larger Semiconductor Clusters ("Quantum Dots")
   L. Brus (With 4 Figures) 312
   3.4.1 Introduction 312
   3.4.2 Elementary Theory 313
   3.4.3 Summary 319
   References 320
3.5 Electromagnetic Excitations of Large Clusters
   U. Kreibig and M. Quinten (With 19 Figures) 321
   3.5.1 Introduction 321
   3.5.2 Extinction of Radiation by Single Particles 327
   3.5.3 Discussion of Optical Properties of Isolated Clusters . . 330
   3.5.4 Intrinsic Optical Cluster Size Effects 338
   3.5.5 Optical Properties of Cluster Matter 342
   3.5.6 Appendix: Computer Program of the Mie Theory 348
   References 358
3.6 Supported Clusters
   S.B. DiCenzo and G.K. Wertheim (With 10 Figures) 361
   3.6.1 Introduction 361
   3.6.2 Preparation of Supported Clusters 362
   3.6.3 Cluster Growth 364
   3.6.4 Band Structure of Clusters 365
   3.6.5 Core Electron Spectra 367
   3.6.6 Other Types of Measurement 376
   3.6.7 Metal-Insulator Transition 378
   3.6.8 Prospects for the Future 380
   3.6.9 Conclusion 382
   References 382
3.7 Nanocrystalline Materials
   R. Birringer and H. Gleiter (With 16 Figures) 384
   3.7.1 Introduction 384
   3.7.2 Basic Ideas 384
   3.7.3 Preparation and Characterization 387
   3.7.4 Structural Studies 390
   3.7.5 Properties 395
   3.7.6 Multiphase Nanocrystalline Materials 401
   3.7.7 Prospects 402
   References 403

Subject Index 405

Subject Index of Volume 52 409

Contributors

F.G. Amar
Department of Chemistry, University of Maine, Orono, ME 04469, USA

J. Amblard
Laoratoire de Physico-Chimie des Rayonnements, URA 75 du CNRS, Université Paris-Sud, Bât. 350., F-91405 Orsay Cedex, France

S.L. Anderson
Department of Chemistry, State University of New York at Stony Brook, Stony Brook, NY 11794–3400, USA

J. Belloni
Laoratoire de Physico-Chimie des Rayonnements, URA 75 du CNRS, Université Paris-Sud, Bât. 350., F-91405 Orsay Cedex, France

R. Birringer
Werkstoff-Wissenschaften, University, D-66123 Saarbrücken, Germany

K.H. Bowen
Department of Chemistry, Johns Hopkins University, Baltimore, MD 21218, USA

L. Brus
AT&T Bell Laboratories, Murray Hill, New Jersey, NJ 07974, USA

H. Burtscher
Laboratory for Solid Physics, ETH-Zurich, CH-8093 Zurich, Switzerland

A.W. Castleman, Jr
Department of Chemistry, The Pennsylvania State University, University Park, PA 16802, USA

K.-P. Charlé
Fritz-Haber-Institut der Max-Planck-Gesellschaft, Faradayweg 4–6, D-14195 Berlin, Germany

M.W. Crofton
The Aerospace Corporation, P.O. Box 92957/M5-754, Los Angeles,
CA 90009, USA

S.B. DiCenzo
Department of Physics, University of Colorado, Colorado Springs,
CO 80933, USA

O. Echt
Department of Physics, University of New Hampshire, Durham,
NH 03824-3568, USA

H. Gleiter
Werkstoff-Wissenschaften, University, D-66123 Saarbrücken, Germany

H. Haberland
Albert-Ludwigs-Universität, Fakultät für Physik, Hermann Herderstr. 3,
D-79104 Freiburg, Germany

I.V. Hertel
Max-Born-Institut, Rudower Chaussee 6, Gebande 19.29, D-12474 Berlin,
Germany

U. Kreibig
1. Physikalisches Institut der Rheinisch Westfälischen Technischen Hochschule, Sommerfeldstrasse, Turm 28, D-52056 Aachen, Germany

Y.T. Lee
Department of Chemistry, University of California, Berkeley, and Materials
and Chemical Sciences Division, Lawrence Berkeley Laboratory, CA 94720,
USA and Institute of Atomic and Molecular Sciences, Academica Sinica,
P.O. Box 23-166, Taipei, Taiwan

D.J. Levandier
Department of Chemistry, University of Rochester, Rochester, NY 14627,
USA

J.L. Mariginer
Laoratoire de Physico-Chimie des Rayonnements, URA 75 du CNRS,
Université Paris-Sud, Bât. 350., F-91405 Orsay Cedex, France

T.D. Märk
Institut für Ionenphysik, Leopold-Franzens-Universität, Technikerstr. 25,
A-6020 Innsbruck, Austria

M. Mostafavi
Laoratoire de Physico-Chimie des Rayonnements, URA 75 du CNRS, Université Paris-Sud, Bât. 350., F-91405 Orsay Cedex, France

L. Perera
Department of Chemistry, University of Maine, Orono, ME 04469, USA

J.M. Price
SRI International, 333 Ravenswood, Menlo Park, CA 94025-3493, USA

M. Quinten
1. Physikalisches Institut der Rheinisch Westfälischen Technischen Hochschule, Sommerfeldstrasse, Turm 28, D-52056 Aachen, Germany

S.J. Riley
Chemistry Division, Argonne National Laboratory, Argonne, IL, USA

C.P. Schulz
Max-Born-Institut, Rudower Chaussee 6, Gebande 19.29, D-12474 Berlin, Germany

W. Schulze
Fritz-Haber-Institut der Max-Planck-Gesellschaft, Faradayweg 4–6, D-14195 Berlin, Germany

G. Scoles
Department of Chemistry, Princeton University, Princeton, NJ 08540, USA

H.C. Siegmann
Laboratory for Solid Physics, ETH-Zurich, CH-8093 Zurich, Switzerland

G.K. Wertheim
AT&T Bell Laboratories, Murray Hill, NJ 07974, USA

Contents of Volume 52

1 Introduction
R.S. Berry and H. Haberland (With 2 Figures) 1
1.1 What are Clusters? 2
1.2 What Makes Clusters Interesting? 5
1.3 How Does One Make Clusters? 6
1.4 Experiments with Clusters 7
1.5 Experiments Not Possible Today 8
1.6 Cluster, Tantalizing Subjects for Theoretical Studies 9
1.7 Clusters Make New Kinds of Materials 10
1.8 New Chemistry 11
1.9 Outlook 12

2 Theoretical Concepts 13
2.1 Quantum Chemistry of Clusters
V. Bonačič-Koutecký, P. Fantucci, and J. Koutecký
(With 9 Figures) 15
2.1.1 Introduction 16
2.1.2 Quantum Mechanical Background 30
2.1.3 Ground State Properties of Metal Clusters 37
2.1.4 Excited States of Alkali Metal Clusters
and their Spectroscopical Properties 37
2.1.5 Conclusions 46
References 48
2.2 Tight-Binding and Hückel Models of Molecular Clusters
D.A. Jelski, T.F. George, and J.M. Vienneau (With 8 Figures). 50
2.2.1 Introduction 50
2.2.2 Quantum Chemistry Background 51
2.2.3 Application to Clusters 54
2.2.4 TB Model Applied to Silicon Clusters 55
2.2.5 Applications of the Hückel Model 63
2.2.6 Conclusions 65
References 65
2.3 Density Functional Calculations for Clusters
R.O. Jones (With 5 Figures) 67
2.3.1 Introduction 67

2.3.2   Calculating Structures   68
2.3.3   Application to Clusters of Group VIa Elements   74
2.3.4   Structure of Phosphorus Clusters, P_2 to P_{10}   79
2.3.5   Models Based on Electron Gas Calculations   80
2.3.6   Concluding Remarks   83
References   84
2.4   Transition from van der Waals to Metallic Bonding
in Clusters
G.M. Pastor and K.H. Bennemann (With 17 Figures)   86
2.4.1   Introduction   86
2.4.2   Theory for the Electronic Properties of Divalent-
Metal Clusters   90
2.4.3   Properties of Hg_n-Clusters as a Function of Cluster
Size: The Transition from van der Waals
to Covalent to Metallic Bonding   95
2.4.4   Summary and Outlook   104
Appendix A: Slave-Boson Approach to Electron Correlations
in Small Clusters   107
Appendix B: On the Size Dependence of the Ionization Energy
of Small Clusters   110
References   113
2.5   Analytic Cluster Models and Interpolation Formulae
for Cluster Properties
H. Müller, H.-G. Fritsche, and L. Skala (With 13 Figures) . . .   114
2.5.1   Introduction   114
2.5.2   Special Role of the Analytic Cluster Model (ACM) . . .   114
2.5.3   Quantum Chemical Analytic Cluster Model (QACM) .   116
2.5.4   Topological Analytic Cluster Model (TACM)   126
2.5.5   Theoretical Background of Interpolation Formulae . . .   133
2.5.6   Concluding Remarks   138
References   138
2.6   Shell Structure in Atoms, Nuclei and in Metals Clusters
S. Bjørnholm (With 9 Figures)   141
2.6.1   Quantum Shells in Spherical Fermion Systems   141
2.6.2   Nuclear Shell Structure and Deformations   146
2.6.3   Shells and Supershells in Large Fermion Systems . . .   153
2.6.4   Further Reading   161
2.6.5   Recent Developments   161
References   162
2.7   Introduction to Statistical Reaction Rate Theories
M.F. Jarrold (With 8 Figures)   163
2.7.1   Introduction   163
2.7.2   RRK Theory   163
2.7.3   RRKM Theory and the Transition State   165
2.7.4   Phase Space Theory   169
2.7.5   Product Kinetic Energy Distributions   176

2.7.6   Evaporative Cooling 181
2.7.7   Determining Cluster Dissociation Energies 182
2.7.8   Problems Associated with the Application of Statistical
        Theories to Clusters 183
2.7.9   Summary 185
References 186

2.8   Melting and Freezing of Clusters: How They Happen and What
      They Mean
      R.S. Berry (With 8 Figures) 187
2.8.1   Introduction: The "Phases" of Clusters 188
2.8.2   Theoretical Basis 189
2.8.3   Simulations and Experiments 195
2.8.4   Implications for Bulk Matter 202
References 204

3   Experimental Methods
    H. Haberland (With 34 Figures) 205
3.1   Sources 207
3.1.1   Supersonic Jets 208
3.1.2   Gas Aggregation 223
3.1.3   Surface Erosion Sources 225
3.1.4   Pick-up Sources 227
3.2   Detection of Cluster Ions 229
3.3   Electron Diffraction 231
3.4   Methods for the Production of (Nearly) Mass Selected
      Neutral Cluster Beams
      U. Buck 232
3.4.1   Scattering from Atomic Beams 233
3.4.2   Re-Neutralization of Ions 237
3.4.3   Summary 239
3.5   Mass Spectrometers 239
3.6   Optical Spectroscopy 243
3.7   Infrared Spectroscopy
      G. Scoles 245
3.8   Photo Electron Spectroscopy 247
3.9   Recent Developments 249
References 250

4   Across the Periodic Table
4.1   Alkali Clusters
      C. Bréchignac (With 17 Figures) 255
4.1.1   Introduction 255
4.1.2   Ionization of s^1 Clusters 255
4.1.3   Stability of s^1 Clusters 263
4.1.4   Optical Response of s^1 Clusters 278
References 286

4.2 Clusters of s^2p^1 Metals and Semiconductors
    M.F. Jarrold (With 12 Figures)   288
    4.2.1 Introduction   288
    4.2.2 Boron Clusters   288
    4.2.3 Aluminum Clusters   293
    4.2.4 Gallium, Indium, and Thallium Clusters   309
    4.2.5 Conclusions   311
    References   312
4.3 Transition Metal Clusters: Physical Properties
    M.F. Jarrold (With 6 Figures)   315
    4.3.1 Introduction   315
    4.3.2 Electronic Configuration and Bonding   315
    4.3.3 Mass Spectra and Magic Numbers   317
    4.3.4 Ionization Energies   317
    4.3.5 Dissociation and Dissociation Energies   320
    4.3.6 Magnetic Properties   322
    4.3.7 Optical Spectroscopy   323
    4.3.8 Electron Affinities and Photoelectron Spectroscopy ...   325
    4.3.9 Geometric Structure   327
    4.3.10 Summary   328
    References   328
4.4 Carbon Clusters
    E.E.B. Campbell (With 18 Figures)   331
    4.4.1 Introduction   331
    4.4.2 The Small Clusters   333
    4.4.3 The Fullerenes: C_n with $n \geq 24$ 339
    4.4.4 The "Buckyball" Era   349
    4.4.5 Recent Developments   353
    References   353
4.5 Oxides and Halides of Alkali Metals
    T.P. Martin (With 9 Figures)   357
    4.5.1 Introduction   357
    4.5.2 Interatomic Forces   358
    4.5.3 Neutral Clusters at Zero Temperature   359
    4.5.4 Charged Clusters at Zero Temperature   362
    4.5.5 Catchment Area and Free Energy   363
    4.5.6 Atomic Vibrations   366
    4.5.7 Photoionization of Cs–O Clusters   368
    4.5.8 Recent Developments   372
    References   372
4.6 Rare Gas Clusters
    H. Haberland (With 12 Figures)   374
    4.6.1 Neutral Rare Gas Clusters in the Ground
           Electronic State   377
    4.6.2 Potentials for Excited and Ionized Rare Gas Dimers and
           Clusters   379

   4.6.3 Experiments with Neutral Rare Gas Clusters 381
   4.6.4 Experiments with Positively Charged Rare Gas
     Clusters 389
   4.6.5 Experiments with Negatively Charged Rare Gas
     Clusters 394
   4.6.6 Summary 394
   References 394
 4.7 Neutral Molecular Clusters
   U. Buck (With 10 Figures) 396
   4.7.1 Introduction 396
   4.7.2 Structure Calculations 397
   4.7.3 Electronic Spectroscopy 401
   4.7.4 Vibrational Spectroscopy 404
   4.7.5 Infrared Photodissociation of Size Selected Clusters ... 407
   4.7.6 Summary 413
   4.7.7 Recent Developments 415
   References 416

Subject Index 419

1 Introduction

H. Haberland

In the first part 1. of this two-volume treatment on *Clusters of Atoms and Molecules* the emphasis was mainly on theory, experimental details and clusters of atoms studied in vacuum. In this second volume the broad applicability to many aspects of physics, chemistry, and technology is discussed in detail. The wide variety of the subjects treated have been grouped under two main headings.

First, it is well known from macroscopic physics and chemistry that an electric charge in a dielectric introduces novel effects and can trigger new chemistry. These effects are studied here as a function of cluster size for molecular clusters in vacuum. For example a beam of molecular clusters is made and its charging is studied. Closely related are questions of internal reactions, ejection of atoms on a small or large time scale. Chemistry with neutral or charged clusters is also directly related to these questions.

Second, the properties of clusters change, but often not very dramatic, if they are deposited on a surface. In this case one speaks of supported clusters. Or clusters can be embedded into a solid, a liquid, or a gas. In the latter case one speaks of an aerosol. It is from this area that technological applications of clusters have come and will be coming. In the following an example is given, highlighting these ideas.

1.1 Historical Application of Metal Clusters

One can dissolve metal containing chemicals in molten glass. If the temperature is high enough the metal atoms diffuse through the molten glass and will stick together if they meet by chance. By carefully controlling the time and amount of heating, clusters of changing size can be grown, as discussed in more detail in the contribution by *Kreibig* and *Quinten* in this volume. Silver stains blue, gold induces a deep red colour, and so on. This technique leads to some of the beautiful shining colours of the medieval church windows. The old glass makers did of course not know what they were doing. The technique had been found by accident and refined by trial and error.

The technological use of metal clusters can be traced to far more ancient times. Thousands of years ago people learned to put a hard shining surface on their pottery. And they quickly discovered how to stain it. Up to the present day, clusters of metal atoms are important for the hue and shade of the colour.

Only after *Mie* published his theory in 1907, it is known how this colouring can be explained on a rational basis.

1.2   Outlook

This historical excursion shows how cluster properties can change with clusters size. The "silvery" bulk metal Ag induces a blue colour once the clusters become sufficiently small. Thus the cluster size can be used as a new variable to influence optical, electronic, and chemical properties. It is this new variability which has already lead to many a surprising technological application. And more can be expected.

Reference

1. Clusters of Atoms and Molecules, ed. H. Haberland, Springer Verlag, Heidelberg 1993, Springer Series in Chemical Physics, Vol. 52

2 Solvation, Chemistry, and Charging of Free Clusters

2.1 Solvated Atoms in Polar Solvents

C.P. Schulz and *I.V. Hertel*

2.1.1 Introduction

The study of the development of solvation from small complexes to the bulk (or liquid) phase is a challenging subject in cluster physics. The solvation of molecules, ions and electrons by various solvents will be discussed in subsequent chapters. Here we will concentrate on the interaction of neutral alkali metal atoms with polar solvents.

When introducing sodium into liquid ammonia an intensely blue colored solution is produced. It is fairly well established, that this observation is due to the separation of the valence electron from the sodium atom thus forming a so-called "solvated electron" and a sodium cation. It has been found experimentally [1, 2] and confirmed theoretically [3, 4], that the spectroscopic properties, e.g. photoemission and photoabsorption, of these solutions are governed by the interaction of the solvated electron with the solvent molecules and are (nearly) independent of the solvated alkali metal. The recent observation of isolated negatively charged ammonia and water clusters [5–7] and the corresponding neutral clusters enclosing one alkali atom [8–13] have focused new attention on this fundamental process.

By studying the spectroscopic properties of these complexes from small micro clusters to the bulk one may hope to gain a detailed understanding of the solvation process. At first sight the most interesting question is the number of solvent molecules needed to introduce a spontaneous dissociation of the metal atom into a localized excess electron and a positive ion. The energetics of this process is very important since in the gas phase approximately 5 eV are required for this charge separation. Directly related with these questions is the understanding of the interaction of the metal atom and the solvent molecules and the formation of solvation shells. Furthermore there is presently an intensive discussion under which conditions the excess electron is bound to the surface or in the bulk of the cluster [14, 15].

Only a small number of systems have been studied in a free cluster beam so far. Among those $Na(NH_3)_n$ and $Na(H_2O)_n$ are the systems where most experimental [9, 11, 12] and theoretical [16–19, 30, 31] data are available. Thus, we will concentrate on these systems.

2.1.2   Experimental

The preparation of the alkali atom solvent complexes needs a special experimental technique, because of the high reactivity of the alkali atom and the polar solvent. To form these clusters in a free jet a special type of "pickup source" has been developed [9]. For the formation of $Na(H_2O)_n$ clusters a carrier gas, typically argon at a stagnation pressure of 1 to 8 bar, is bubbled through water, whereas neat ammonia is used to form $Na(NH_3)_n$. The gas is expanded through a pulsed nozzle and interacts with a cloud of sodium atoms, which is produced by an effusive sodium oven located as close as possible in front of the nozzle orifice. Figure 1 illustrates this nozzle expansion schematically by two snapshots. The sodium density in the interactions region is about 10^{12} to 10^{14} cm^{-3}. The sodium atoms are "picked-up" by the expanding mixture of carrier gas and molecular clusters, which are formed near the nozzle orifice (Fig. 1a). Due to the high number of collisions the temperature of the beam decreases further and mixed clusters are formed (Fig. 1b). It is essential to operate this source in a

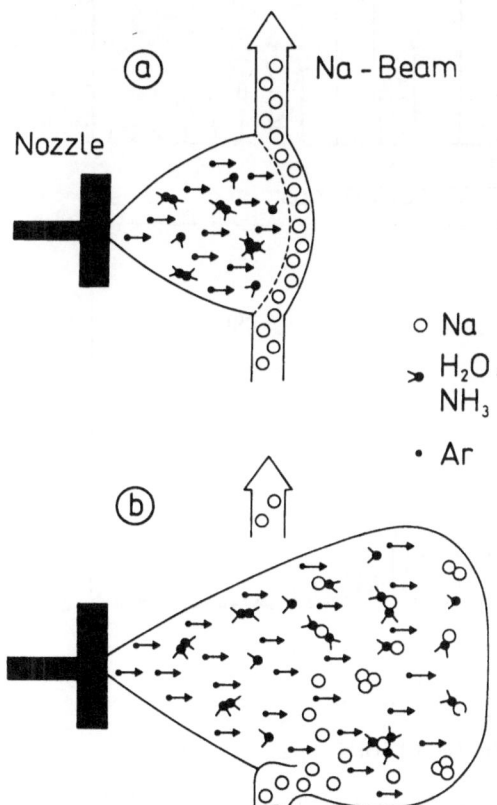

Fig. 1. Operation of the "pick-up" source demonstrated schematically by two snapshots: **(a)** Immediately after opening of the pulsed nozzle the expanded NH_3 or Ar/H_2O mixture reaches a cloud of Na atoms in front of the nozzle. The mixed clusters are formed and cooled due to the high collision rate. A few μs later **(b)** the formed cluster and the remaining Na atoms are carried with the gas flow. Additional atoms from the sodium beam are scattered at the boundary of gas pulse and cannot contribute to the cluster formation

pulsed mode in order to allow the sodium cloud at the beam axis to be replenished before the next gas pulse arrives. A simple consideration of the pressure difference between the gas beam and the sodium beam leads to the conclusion, that a continuous operation of this source is not possible: The local pressure of the gas beam close to the orifice is usually in the mbar range, while the sodium beam has a local pressure of only 10^{-3} mbar ($n \approx 10^{-14}$ cm^{-3}) at

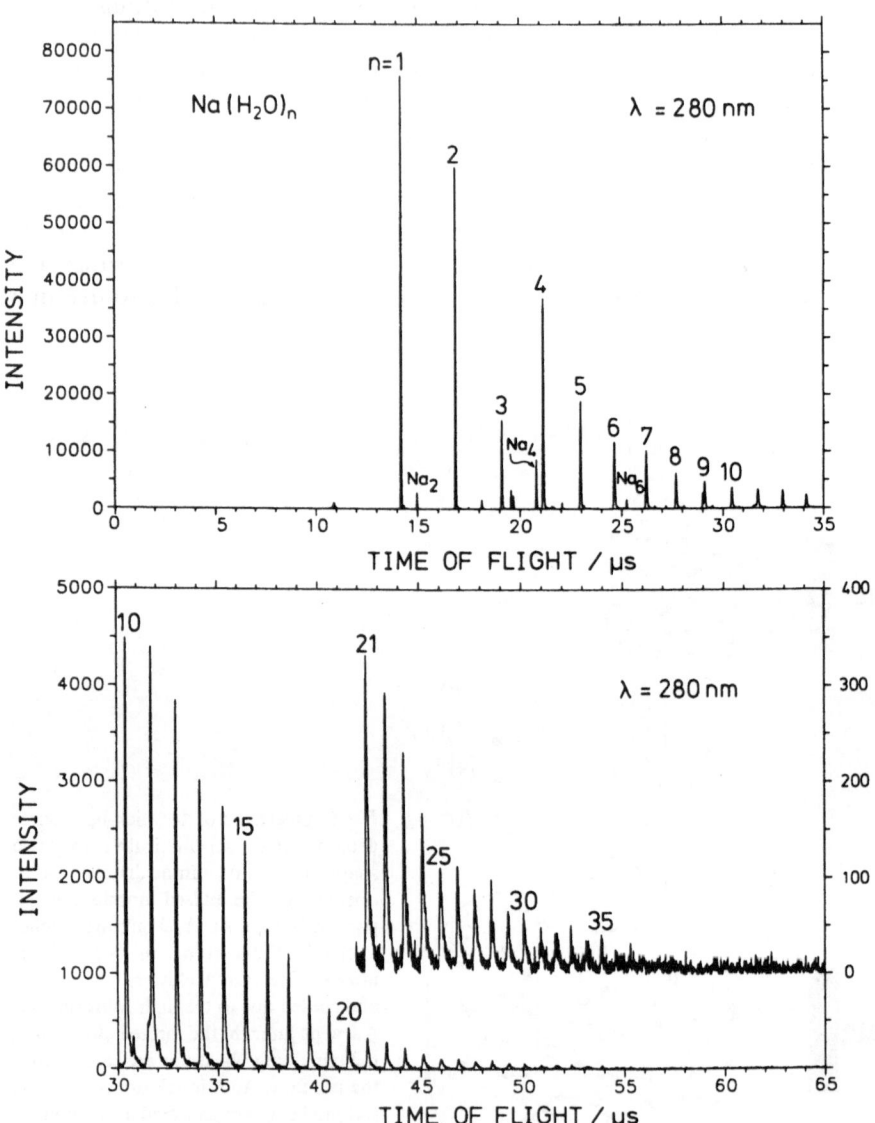

Fig. 2. TOF mass spectrum of Na(H$_2$O)$_n^+$ produced in the pick-up source with 2.5 bar of argon, mixed with 32 mbar of water by bubbling the gas through the liquid at room temperature. Ionisation was achieved with a laser pulse at the wavelength of 280 nm and a fluence of 3.5 mJ cm^{-2} (from [9])

the interaction region. Thus, in continuous operation all Na atoms would be scattered at the outer boundary of the gas beam and never reach it's centre.

After the cluster formation the beam is shaped by a skimmer and enters the detection region. There the clusters are ionised by a pulsed laser beam. The ions are mass analysed by a standard time of flight arrangement. A typical $Na(H_2O)_n$ mass spectrum is shown in Fig. 2. The ionising laser wavelength is 280 nm and its fluence 3.5 mJ cm^{-2}. Sodium water aggregates up to $n = 35$ are observed. A small signal of pure sodium clusters is also present in the spectrum. These are created in the primary sodium jet and picked up by the pulsed gas beam. Similar spectra have been observed for $Na(NH_3)_n$ clusters [12].

2.1.3 Ionisation Potentials

The most interesting question in the context of the atom interaction with the polar solvent is the delocalisation of the (atomic) valence electron as more and more polar molecules are bond to the atom. For alkali atoms solvated in liquid ammonia, traditionally two competing models were used to describe the experimental observations such as photoemission of electrons and photo-absorption: (a) the cavity model in which the electron is separated from the atomic ion and both are screened individually by the surrounding dielectric medium [3]. Alternatively (b), an electron trap is assumed to be formed by the field of cation embedded in a shell of ammonia molecules [20, 21]. By measuring the ionisation potential as a function of the cluster size one in principle can follow how the liquid phase is built up from its constituents.

Usually, ionisation thresholds are measured by scanning the laser wavelength and registering the mass resolved ion signal for all cluster sizes of interest. To interpret the photo ion efficiency (PIE) spectra obtained in this manner one has to make sure that one photon only has been absorbed by the cluster. This is done, usually, by studying the dependence of the cluster ion yield on the laser intensity which should be linear. Even so, in general saturation and geometry effects cannot be excluded completely. A second problem in the interpretation of PIE spectra arises when clusters have different structures and equilibrium distances in their neutral and ionised form.

Fortunately, in the case of the sodium water complex theory predicts nearly identical structures for neutral [17] and the ionic [22] $Na(H_2O)$, so that the interpretation of the ionisation threshold is straight forward. This is illustrated schematically by the potential energy diagram Fig. 3, where the molecular potential for an Na–solvent complex is displayed. Since the ionisation potential of the Na atom is 5.14 eV and the ionic complex is bound significantly stronger than the neutral we expect the ionisation potential of the complex to be substantially less than 5.14 eV. An energy range very conveniently accessible to single photon laser ionisation (frequency doubled pulsed dye lasers). This is indeed verified experimentally as documented by Fig. 4. Shown is the $Na(H_2O)$

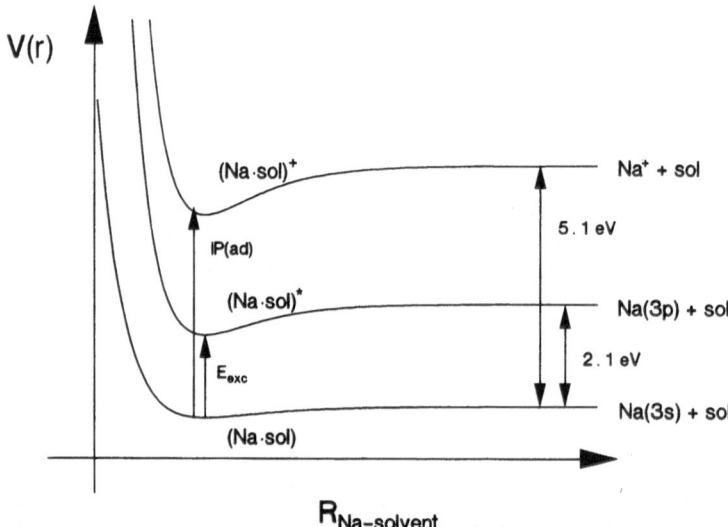

Fig. 3. Schematic potential energy diagram for Na–sol (sol = solvent: NH_3 or H_2O) as a function of the Na $\cdots$ solvent distance. The adiabatic ionisation potential (= IP (ad)) for these complexes is lower than asymptotic value of 5.1 eV due to difference in binding energy for the ionic complex and the neutral ground state. For the same reasons the transition energy (E_{exc}) to the first electronically excited state is expected to shift towards lower energy compared to the 2.1 eV for the Na(3p) excitation

PIE spectrum in an energy range close to the ionisation threshold. The steep rise of the ion signal at 35320 cm^{-1} is attributed to a vertical transition at the ionisation threshold. 305 cm^{-1} above the threshold a second step is seen, which we identify as being due to ionisation into the first vibrationally excited mode of Na(H$_2$O)$^+$. The high resolution obtained here allows one to study field ionisation of excited Rydberg states in the presence of the extraction field. Thus, a shift of the ionisation threshold is observed when comparing the PIE spectrum in a continuous extraction field (Fig. 4, lower part) and pulsed electric field which was delayed by 1 μs with respect to the laser pulse (Fig. 4, upper part). A detailed discussion of this phenomenon can be found in [9]. The PIE spectrum of Na(NH$_3$) shows a similar structure and dependence of the extraction field [12]. Thus, these complexes with only one solvent molecule bound to the sodium atom show a strictly molecular behaviour. The sodium valence electron is centered around the Na$^+$ core and the dipole molecules are aligned in the field of the ion (Na–OH$_2$, Na–NH$_3$). Interestingly, theoretical calculations show that for these neutral complexes even a small electron transfer from the molecule to the sodium atom occurs (Lewis base type binding) [17].

For complexes with more solvent molecules the photoionisation threshold decreases but also the shape of the ionisation threshold changes. The ion yield rises more or less slowly with increasing photon energy. Figure 5 gives an example Na(NH$_3$)$_n$, n = 2 to 7, for ease of comparison with identically spaced

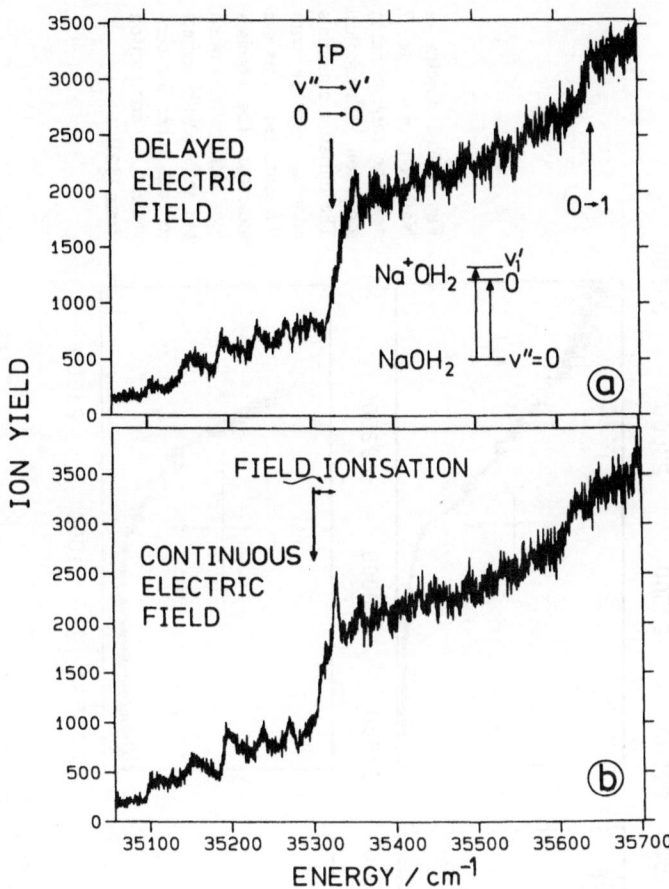

Fig. 4. Ion yield (PIE) spectrum for Na(H$_2$O)$^+$ near the ionisation threshold. The ions are accelerated either by an electric field that is switched on only 1 µs after the laser pulse (*upper part*) or by a static field (*lower part*). The shift in the ionisation threshold is due to a reduction of the ionisation threshold by Stark effect and thus ionisation of Rydberg molecules excited in the laser field (from [9])

abscissa but shifted in absolute energy with respect to each other. From the different shapes of the ion yield curves near the threshold some information on the structure of these clusters can be extracted. Na(NH$_3$)$_2$ e.g. shows a particular smooth rise of the ion signal. From theoretical calculations [23, 24] it is known, that the neutral Na(NH$_3$)$_2$ is bent while the ion is linear, the sodium atom being in the middle. Thus, the Franck–Condon factors for ionisation are small at the adiabatic threshold. In addition it can be assumed, that the H$_3$N ··· Na ··· NH$_3$ bending mode has a low frequency and will be excited significantly due to the finite temperature in the beam. Both effects give rise to the observed smooth increase of the ion signal near threshold. For larger clusters ab initio calculations are not yet available. However, the distinct onset of the ion signal in the

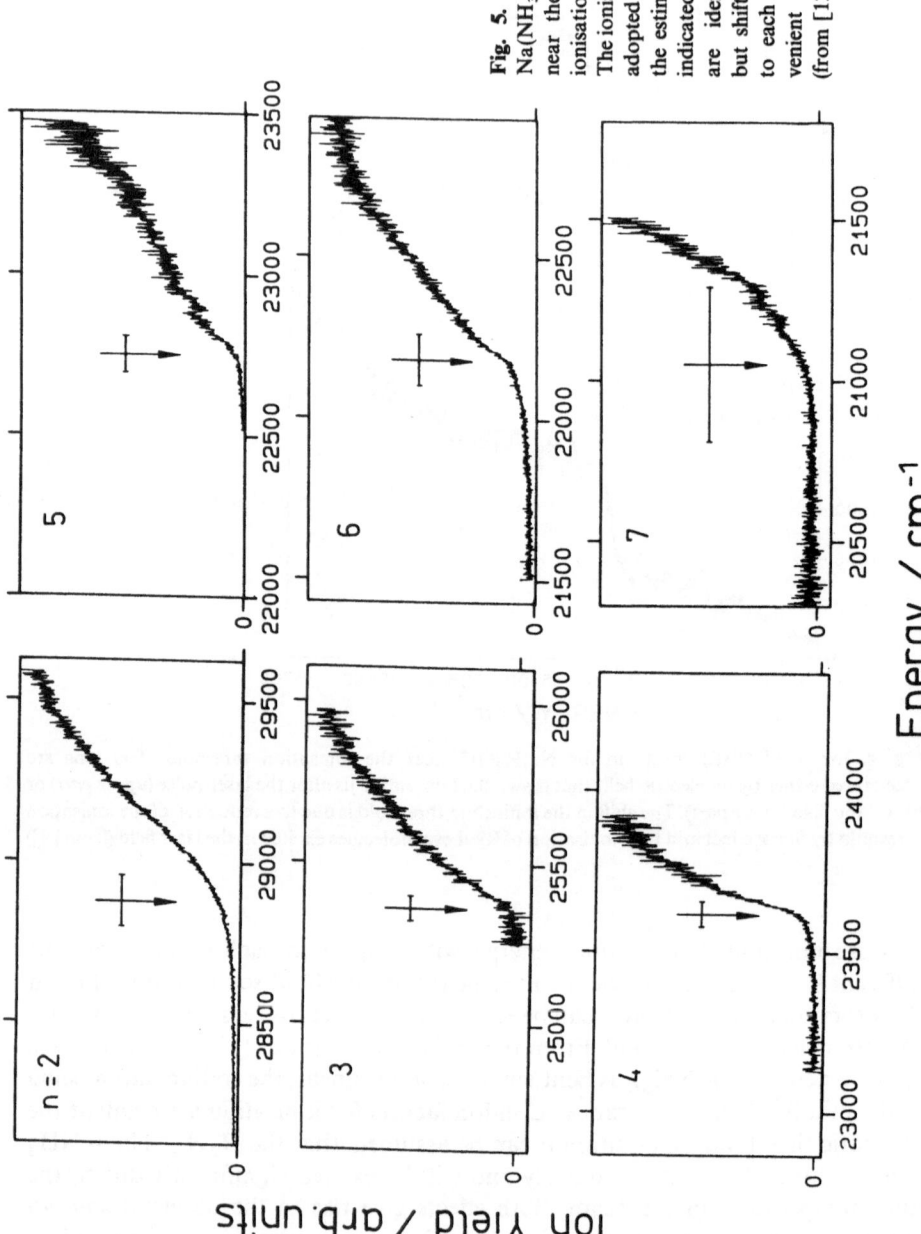

Fig. 5. PIE spectra of Na(NH$_3$)$_n$ $n = 2$ to 7 near the corresponding ionisation thresholds. The ionisation potentials adopted together with the estimated errors are indicated. The abscissa are identically spaced but shifted with respect to each other for convenient comparison (from [12])

Energy / cm^{-1}

Ion Yield / arb. units

observed PIE spectra for $n = 3$ to 6 suggests that these clusters are more rigid, i.e. only a small fraction of the clusters is found in a vibrationally excited mode. For larger clusters ($n \geq 7$) a smooth rise of the ion yield is observed again. This might be explained by assuming a first solvation shell to be filled with $6NH_3$ molecules surrounding the Na atom core. An increasing manifold of isomeric structures is expected when adding extra NH_3 molecules outside this first solvation shell. The measured ion yield represents an average over the all different structures and the corresponding Franck–Condon factors with the ion continuum.

As a general trend the ionisation potential of $Na(NH_3)_n$ decreases more or less monotonically with the cluster size n. Figure 6 gives a graphic representation of the measured ionisation potentials of $Na(NH_3)_n$ and $Na(H_2O)_n$ as a function of $(n + 1)^{-1/3}$, the latter being roughly inverse proportional to the cluster radius R_c including the Na^+ core. One immediately notes the different behaviour of H_2O and NH_3 as solvent. The ionisation potential for $Na(H_2O)_n$ decreases with the cluster size up to $n = 4$ and then stays constant at 3.17(5) eV for all larger cluster ($n \leq 23$). *Fuke* and *coworkers* [13] have observed the same features for solvated Cs atoms as shown in Fig. 7: a constant IP of 3.12(6) eV for $Cs(H_2O)_n$ $n \geq 4$ and deceasing IP for $Cs(NH_3)_n$ n up to 31 in excellent agreement with the solvated Na data. The extrapolation of the measured ionisation potentials to the bulk value coincides with photoemission data for the liquids namely for NH_3 with the experimental determined value of 1.45 eV [1]

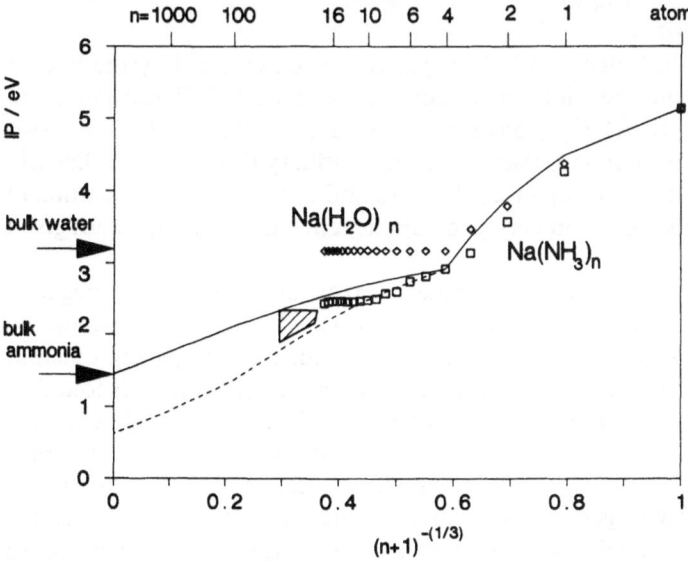

Fig. 6. Experimental ionisation potentials of $Na(H_2O)_n$ (*diamonds*) and $Na(NH_3)_n$ (*squares*) vs. $(n + 1)^{-1/3}$, the shaded area giving an upper bound for $20 < n < 35$. Experimental errors correspond to the size of the symbols. The *dashed line* represents the calculated IP's using the model potential for dielectric screening; the *solid line* for $n \geq 5$ includes the self-consistent-field screening potential

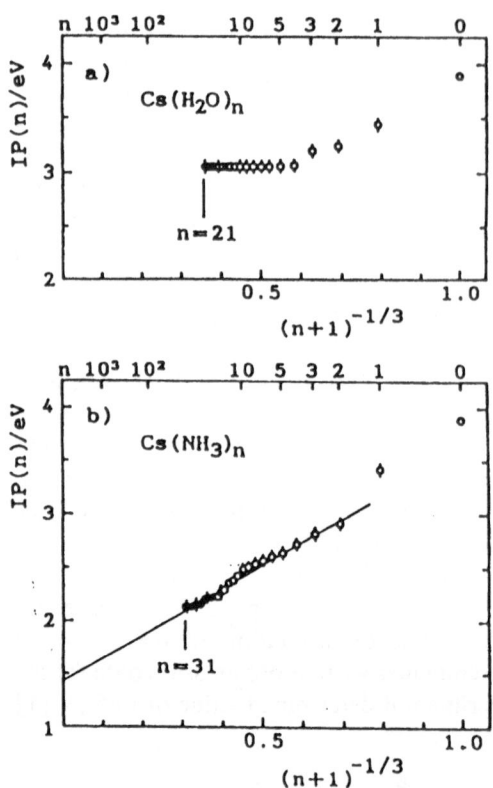

Fig. 7. Ionisation potentials of $Cs(H_2O)_n$ (*upper part*) and $Cs(NH_3)_n$ (*lower part*) plotted vs. $(n+1)^{-1/3}$. The *solid line* represents a least square fit for $n \geq 2$ with the intercept at the bulk value of 1.41 eV (from [13])

and for H_2O with the value of 3.2 eV which can be determined by the known photoconductivity and the width of the conduction band [7]. Vertical electron affinities of $(H_2O)_n^-$ and $(NH_3)_n^-$ have been measured by *Lee* et al. [7, see also Chapter 2.5]. They have found, that the electron affinity (EA) increases linearly with the cluster radius (see Chapter 2.5, Figs. 10 and 11) and the extrapolation of these data to the bulk limit is in very good agreement with the bulk values given above.

A very remarkable similarity of these electron affinities for the negative clusters and our ionisation potentials for the solvated alkali atoms should be noted. When comparing the two sets of data obtained in completely different experiments by different groups of researchers an empirical relation is found to hold for both solvents: $\Delta E = IP - EA = 5.9 n^{-1/3}$. This energy difference ΔE can be compared with the potential energy e^2/R_p of an electron in a spherical electric field at the radius R_p which gives $R_p \approx 1.3 R_{ws} n^{1/3}$. Interestingly this radius is only slightly larger than the cluster radius $R_c = R_{ws} n^{-1/3}$. Thus, the energy difference ΔE might be viewed at as the energy needed to move an electron from the surface of the cluster to infinity.

To obtain a more quantitative concept for the evolution of the ionisation potential with the cluster size we have applied a simple model by which the binding energy for the valence electron is calculated in the field of the Na ion core *and* its dielectric surrounding. The result of such a calculation for $Na(HN_3)_n$ for different n are also shown in Fig. 6 (dashed line) [11]. In this model the dielectric screening has been constructed by two different regions: The first filled solvation shell around the Na^+ core and the dielectric medium out side this first solvation shell up to the cluster radius. The only free parameter in this model is the surface charge of the oriented dipoles in the first solvation shell. This parameter has been adjusted in such a way, that the experimental value of the IP of $Na(NH_3)_4$ is reproduced by the calculation. The solid line in Fig. 6 represents the result of a more consequent calculation which also includes the influence of the electron itself on the orientation of the dielectric medium outside the first solvation shell as suggested by *Jortner* [21]. The experimental IP's for $Na(NH_3)_n$ and the result of the refined model (solid line in Fig. 6) up to $n = 10$ are in very good agreement, especially when considering the fact that only one fitting parameter has been used. A clear deviation is observed for larger clusters and the bulk value is not reproduced at all. It may be concluded from these results, that the one centre model is only valid for smaller clusters, while for larger clusters a complete charge separation appears to occur. In the case of $Na(H_2O)_n$, where the parameters are very similar to that of $Na(NH_3)_n$, the model calculation fails to reproduce the experimental data for $n \geq 5$.

A number of calculations on negatively charges $(NH_3)_n$ and $(H_2O)_n$ clusters are available [e.g. 14, 15]. In accord with the experimental findings [5, 6] these calculations have shown that for $(H_2O)_n^-$ and $(NH_3)_n^-$ a minimum number of molecules are necessary to bind an excess electron (≈ 8 for H_2O and ≈ 32 for NH_3). Furthermore these calculation revealed that for small clusters the electron can only be bound in a surface state while for larger clusters internal localisation (solvation) is energetically favoured. On the basis of these findings *Jortner* [25] has proposed, that solvated alkali atoms the constant IP beginning at $n = 4$ for H_2O as solvent and the flat region between $n = 10$ and $n \approx 16$ for NH_3 indicates the formation of two charged centers: a Na^+ ion core and an electron in a surface state. The further decreasing IP for larger clusters of $Na(NH_3)_n$ may be produced by a third region where a mixture of surface and interior electron states are formed.

2.1.4 Further Spectroscopic Studies

All the above speculations of the different states of charge separation during the solvation process have to be supported by rigorous quantum calculations which are for larger clusters presently not available. In addition more experimental data are needed, e.g. on the excited stated of the solvated atom clusters, to obtain a more detailed understanding of the solvation process on a microscopic scale. Particularly the energy shift of the first excited state as a function of cluster size is of great interest. First results for $Na(NH_3)$ are shown in Fig. 8 [26]. These data were obtained by resonant two photon ionisation, where a variable photon

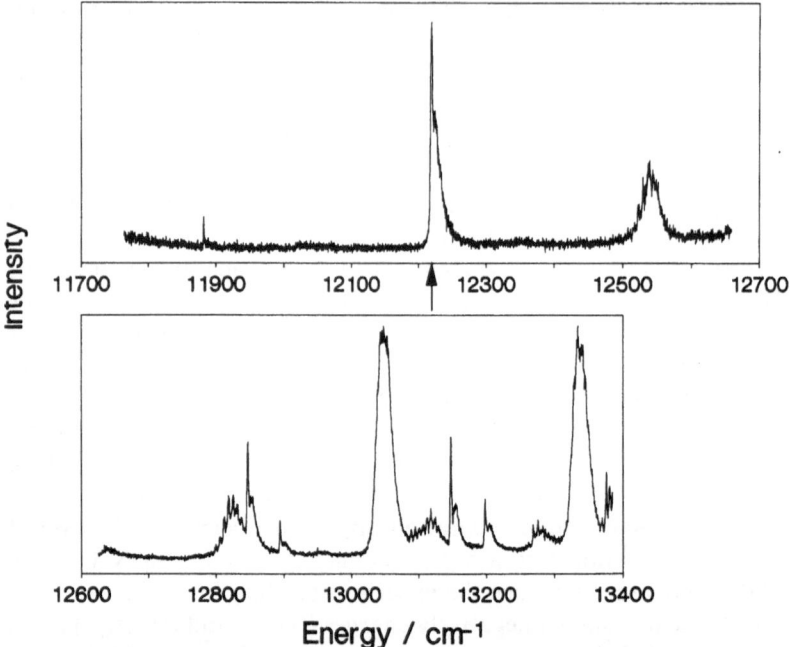

Fig. 8. Resonant two photon ionisation of Na(NH$_3$). The energy scale corresponds to the energy of the first photon, which have been scanned. The second photon has a fixed energy of 24 009 cm^{-1}. The arrow indicates a tentative assignment of the 0–0 transition between the ground A$_1$ and the excited E state

excites the Na(NH$_3$) into the first excited state (see Fig. 3) and a second photon at a fixed energy (24 009 cm^{-1} = 2.98 eV in this case) ionises the excited clusters. A rich structure is observed in the shown energy regions. We have tentatively assigned the line at 12 220 cm^{-1} (= 1.52 eV) as the 0–0 transition, since no other intense line have been observed below this energy. This assignment is in good agreement with a SCF calculation, which gave 12 200 cm^{-1} for the lowest excited state [24]. The large energy shift of about 0.59 eV compared to the first excited state of the naked Na atom indicates an increase of the binding energy between the Na atom and the ammonia molecule with respect to the ground state. The lines above this energy clearly originates from additional vibrational excitation of six possible modes of Na(NH$_3$) and combination of these modes. By comparing with Na(NH$_3$) ion one expects that the Na $\cdots$ NH$_3$ stretching mode should have the lowest frequency ($\approx$ 300 cm^{-1}). Nevertheless, the corresponding bending mode which is known to be very soft in electronic ground state (< 30 cm^{-1}) may be in the same energy range for the excited state. Thus, for a detailed assignment further experimental and theoretical data are necessary.

All the studies discussed above have to be extended in various direction: the search for the excited states for larger cluster represent a great challenge for both theorists and experimentalists. With the now available solid-state femto second lasers it should be also possible to look at the electron solvation in real time. First experiments in liquid water have been made [27, 28]. Furthermore the studies have to be extended to other elements with more than one valence electron. Up to now only small complexes of $Hg(NH_3)_n$ have been observed [29]. It can be concluded, that the investigation of solvation effects represents a fascinating facet of cluster research and will be an active field in the near future.

References

1. J. Häsing: Ann. Phys. 5. Folge **37**, 509 (1940)
2. H. Aulich, B. Baron, P. Delahay, R. Lugo: J. Chem. Phys. **58**, 4439 (1973)
3. see e.g. D.A. Copeland, N.R. Kestner, J. Jortner: J. Chem. Phys. **53**, 1189 (1970)
4. see e.g. W.L. Jolly: *Metal-Ammonia Solutions*, Stroudsburg, Pennsylvania: Dowden, Hutchinson, and Ross, Inc. 1972; *Electrons in Fluids*, ed. by J. Jortner, N.R. Kestner, Springer, Berlin (1973); J.C. Thompson: *Electrons in Liquid Ammonia*, Oxford Univ. Press, Oxford, (1976)
5. H. Haberland, H.G. Schindler, D.R. Worsnop: Ber. Bunsenges. Phys. Chem. **88**, 270 (1984)
6. H. Haberland, C. Ludewigt, H.G. Schindler, D.R. Worsnop: Surf. Sci. **156**, 157 (1985)
7. G.H. Lee, S.T. Arnold, J.G. Eaton, H.W. Sarkas, K.H. Bowen, C. Ludewigt, H. Haberland: Z. Phys. D **20**, 9 (1991)
8. C.P. Schulz, R. Haugstätter, H.U. Tittes, I.V. Hertel: Phys. Rev. Lett. **57**, 1703 (1986)
9. C.P. Schulz, R. Haugstätter, H.U. Tittes, I.V. Hertel: Z. Phys. D **10**, 279 (1988)
10. C.P. Schulz, A. Gerber, C. Nitsch, I.V. Hertel: Z Phys. D **20**, 65 (1991)
11. I.V. Hertel, C. Hüglin, C. Nitsch, C.P. Schulz: Phys. Rev. Lett. **67**, 1767 (1991)
12. C. Nitsch, C.P. Schulz, A. Gerber, W. Zimmermann–Edling, I.V. Hertel: Z. Phys. D **22**, 651 (1992)
13. F. Misaizu, K. Tsukamoto, M. Sanekata, K. Fuke: Chem. Phys. Lett. **188**, 241 (1992)
14. R.N. Barnett, U. Landmann, C.L. Cleveland, J. Jortner: Phys. Rev. Lett. **59**, 811 (1987)
15. R.N. Barnett, U. Landmann, C.L. Cleveland, J. Jortner: Chem. Phys. Lett. **145**, 382 (1988); **148**, 249 (1988)
16. M. Tranary, H.F. Schaefer III, P. Kollman: J. Am. Chem. Soc. **99**, 3885 (1977)
17. J. Bentley: J. Am. Chem. Soc. **104**, 2754 (1982)
18. L.A. Curtiss, E. Kraka, J. Gauss, D. Cremer: J. Phys. Chem. **91**, 1080 (1987)
19. N.R. Kestner, S. Dhar: *In Large Finite Systems*, ed. by J. Jortner, A. Pullmann, B. Pullmann, D. Reidel Publishing Company, 1987, p. 209
20. W.E. Blumberg, T.P. Das: J. Chem. Phys. **30**, 251 (1959)
21. J. Jortner: J. Chem. Phys.: **34**, 678 (1961)
22. H. Kistenmacher, H. Popkie, E. Clementi: J. Chem. Phys. **59**, 5842 (1973)
23. V.A. Nicely, J.L. Dye: J. Chem. Phys. **52**, 4795 (1970)
24. J. Greer, Chr. Hüglin, I.V. Hertel, R. Ahlrichs: Z. Phys. D (1993) submitted
25. J. Jortner: private communication
26. C.P. Schulz, Chr. Hüglin, C. Nitsch, I.V. Hertel: (1993) in preparation
27. A. Migus, Y. Gauduel, J.L. Martin, A. Antonelli: Phys. Rev. Lett. **58**, 1559 (1987)
28. F.H. Long, H. Lu, K.B. Eisenthal: Phys. Rev. Lett. **64**, 1469 (1990)
29. C. Dedonder-Lardeux, C. Jouvet, M. Richard-Viard, D. Solgadi: Chem. Phys. Lett. **170**, 153 (1990)
30. R.N. Barnett, U. Landman: Phys. Rev. Lett. **70**, 1775 (1993)
31. K. Hashimoto, S. He, K. Morokuma: Chem. Phys. Lett. **206**, 297 (1993)

2.2 IR Spectroscopy of Solvated Molecules

F.G. Amar, S. Goyal, D.J. Levandier, L. Perera, and *G. Scoles*

2.2.1 Introduction

This chapter describes the infrared spectroscopy of small molecules seeded in, or adsorbed on, argon clusters that range in size up to 10^3 atoms and the molecular dynamics simulations which have been used to elucidate the different types of behaviour observed for various chromophores in the clusters. By studying the infrared spectrum of a molecule in clusters of varying size we are effectively monitoring the properties of the system as they evolve from those of an isolated species to those of a bulk phase.

In the limit of large cluster sizes this type of spectroscopy is analogous to normal matrix isolation spectroscopy (MIS). Originally developed as a means for trapping reactive molecules in an inert environment for the spectroscopic examination of their properties, MIS has more recently been applied to the study of guest–host interactions (for a review of this area see [1]). This has resulted in part because with the improvements afforded by modern instrumentation, it was realized that the relatively small perturbations induced by the matrix on the guest species can be studied in great detail. A series of experiments carried out in the eighties by *Jones* and *coworkers* [2–7] illustrates clearly the wealth of information regarding guest–host interactions which can be derived from the sometimes complicated infrared spectra of molecules isolated in cryogenic matrices. The cluster experiments described in this chapter can, therefore, be regarded as the study of guest–host interactions carried out by "gas phase" matrix spectroscopy.

The use of infrared spectroscopy in this work is important for a couple of reasons. The infrared spectrum is a good probe of the properties of the system since perturbations of the vibrational levels of the chromophore are sensitive to the packing and dynamical behaviour of the surrounding solvent particles. In addition, these experiments are amenable to theoretical simulations because such calculations involve ground state (neutral) potentials of relatively simple molecules. These potentials already exist or can be derived theoretically (or empirically) using reasonably straightforward methods.

The experimental technique, which was first implemented by *Gough* et al. [8] and is described in some detail in Chapter 3, is photodissociation spectroscopy of molecular beams and makes use of infrared lasers and bolometric detection.

Briefly, a beam of clusters is generated in a supersonic expansion in which the large number of collisions produces, early in the expansion, a high degree of condensation. The expansion produces a distribution of cluster sizes, the average of which can be controlled by adjusting the stagnation pressure before the nozzle. Mixed clusters are formed either by expanding a dilute mixture of the probe species and the rare gas, or by using a cross-beam (pick-up) technique [9] in which solute molecules are deposited on already-formed neat argon clusters by crossing the primary condensed beam with an effusive flux of the chromophore.

The beam of mixed clusters is admitted through a skimmer to the experimental chamber where the infrared spectrum is measured by crossing the molecular beam with the output of a line-tunable CO_2 laser and by detecting the fragments that evaporate when the excited chromophores relax by coupling to vibrational modes of the solvent clusters.

Since the temperature of large clusters is determined by evaporative cooling [10], and in contrast to what one would naïvely expect on the basis of purely statistical considerations, it is possible to apply the photoevaporation detection method to the examination of clusters much larger that those for which that technique was originally developed (e.g., dimers and oligomers). Indeed, as the absorbed photon energy is dissipated in the weakly bound system, its temperature rises above the critical value at which evaporation may again occur on a time scale comparable with the time-of-flight through the apparatus. Only when the cluster is large enough for the loss of one atom to produce a negative change in temperature smaller than the positive change in temperature caused by the photon will a loss of sensitivity be experienced.

Although the experimental technique employed is very sensitive, some limitations are still present. One of these, the low resolution of line-tunable CO_2 lasers, the use of which is mandated by the relatively high power level needed, is compensated by employing as many as three lasers with different isotopomers of CO_2 as the active media (namely, $^{12}C^{16}O_2$, $^{13}C^{16}O_2$ and $^{12}C^{18}O_2$) resulting in slightly different lasing frequencies due to the isotope effects.

A second limitation stems from the limited knowledge of the cluster size distribution which is a consequence of studying neutral clusters, as opposed to charged ones. Although mass spectrometers have been used in experiments involving laser photofragmentation of seeded rare gas clusters [11, 12], the lack of knowledge of the extent of fragmentation that occurs, even when soft ionization techniques are used, casts considerable doubt on the mass determination of the neutral species. Indeed, in experiments in which CO_2 laser photofragmentation spectra of CH_3F/Ar_n clusters were measured using an on-axis mass spectrometer, the data could not rule out contributions, to the particular ions monitored, from clusters with greater numbers of argon atoms; only contributions from clusters with smaller numbers of argon atoms than in the ion observed could be eliminated [12]. In another experiment, in which small Ar clusters containing a single SF_6 or SiF_4 molecule were studied by time-of-flight mass spectroscopy, ionization by electrons with energies as low as 20 eV resulted in "anomalous" fragmentation of the seed molecule, as well as the evaporation of all the Ar atoms [13]. For clusters of up to a half-dozen atoms scattering methods can be used [14, 15], however, these methods are not useful in the size range of interest here (see Chapter 3).

The average cluster size has been estimated, however, by using the $pd^{0.8}$ law (p = stagnation pressure, d = nozzle diameter) to scale the expansion conditions [16, 17] to those used in the electron diffraction experiments which included size determinations of neat Ar clusters [18, 19], as discussed in Chapter 2. All references to average cluster size in the ensuing discussion of spectra are derived by this scaling procedure. In the experiments described here cluster beams have been generated from a 30 μm nozzle with a range of stagnation pressures from 207 to 2756 kPa which corresponds to a range in average cluster size (average number of Ar atoms) of from < 10 to ~ 700, respectively.

2.2.2 Solvation Studies

2.2.2.1 SF_6/Ar_n Clusters

Figures 1a, b are typical spectra of SF_6/Ar_n clusters determined using the conventional coexpansion method when a 0.025% SF_6/Ar mixture is expanded at stagnation pressures of 861 and 2756 kPa, respectively. The data were measured in the range between 935 and 949 cm^{-1} and involve the v_3 vibration of SF_6, which occurs at 948 cm^{-1} in the gas phase.

The actual data, obtained in these spectra using the $^{12}C^{16}O_2$ and $^{13}C^{16}O_2$ lasers only, comprise the individual points which correspond to the intensity of the photofragment flux produced at the discrete laser frequencies. The spectra are composites produced by a simple scaling of the data measured with either laser. The curves in each of the spectra (and in the other figures, as noted) are the results of fitting Gaussian functions to the data.

The spectra are characterized by relatively broad bands (typically ~ 0.5–4 cm^{-1}) shifted to the low-frequency side of the gas phase absorption, due to the interaction between the chromophore and the surrounding medium. In fact, the extent of red-shifting increases with the number of nearest neighbours in the cluster, until the absorption frequency approaches the "bulk" value [20–22].

The coexpansion SF_6/Ar_n spectra in Figs 1a, b are dominated by the band located at ~ 938 cm^{-1}. This frequency is nearly coincident with the location of a matrix isolation doublet which disappears upon annealing the sample and which is red-shifted by ~ 1 cm^{-1} with respect to the main annealed matrix absorption doublet [1]. This cluster absorption has, therefore, been attributed to SF_6 residing in matrix-like sites in solid argon clusters [21]. Since the red-shift is proportional to the average polarizability of the solvent medium [20], which is proportional to the argon density, it can be concluded that the environment hosting the SF_6 molecule in the cluster is more closely packed than in a well annealed solid. It is worth noting at this point that the spectra in Figs. 1a, b, which involve average cluster sizes of ~ 60 and ~ 700, respectively, are not identical since the absorption of radiation between 941 and 942 cm^{-1} is relatively more important in the high pressure (large clusters) spectrum than in the low pressure (small clusters) one.

Figures 1c, d are pick-up spectra of SF_6/Ar_n clusters measured with expansion pressures of 861 and 2756 kPa, respectively, using all three CO_2 lasers to give greater resolution. Comparison of these two spectra indicate a dramatic difference with the change in cluster size (which is seen to occur rather smoothly in the spectra at intermediate pressures [23]). While the 861 kPa coexpansion and pick-up spectra are nearly identical, the 2756 kPa spectra differ because of the prominent band occurring around 941/942 cm^{-1} in the pick-up spectrum. The location of this feature, between the absorptions due to the "matrix-like" SF_6/Ar_n cluster and the gas phase molecule, indicates that the environment of the SF_6 molecules responsible for this band involves interactions with fewer Ar atoms than required to fully solvate the chromophore. In light of this and the nature of the pick-up experiment, in which the chromophore is attached to the cluster via its surface, the band at 941/942 cm^{-1} has been assigned to species in which the SF_6 is located at the cluster surface [9]. Indeed, it has been shown that the spectrum of a spherical molecule, half-buried in the surface of a cluster, will exhibit an absorption located halfway between the gas phase and fully-solvated species [20].

Comparison of the coexpansion and pick-up spectra of SF_6/Ar_n clusters in Fig. 1, and those measured at intermediate stagnation pressures [23], indicate that at the lowest pressure, or for smaller clusters ($n \le 60$), the SF_6 molecule prefers to be located inside the argon clusters. The "surface" absorption is seen to increase in relative intensity with cluster size, in the pick-up experiments, until it dominates. This suggests that, in the pick-up experiments, the SF_6 diffuses inside smaller clusters but remains at the surface of larger clusters.

A third absorption located at ~ 939 cm^{-1}, and evident in both coexpansion and pick-up spectra, has been assigned to a fully solvated species in which the density of the surrounding cluster is less than the density of the crystalline Ar seen in matrix isolation spectroscopy [23]. This band, whose assignment was aided by the coincidence of this absorption with the spectrum of SF_6 in liquid Ar [24] (as shown in Fig. 1), is evidence of the existence of either liquid or amorphous medium-to-large Ar clusters (where the amorphous structure has the same density as the liquid).

2.2.2.2 CF_3Cl/Ar_n Clusters

Figure 2 shows typical spectra of the CF_3Cl/Ar_n system obtained with the $^{12}C^{18}O_2$ laser in the region between 1094 and 1110 cm^{-1}, which relate to the ν_1 mode of the molecule. In each spectrum, the positions of the CF_3Cl absorptions

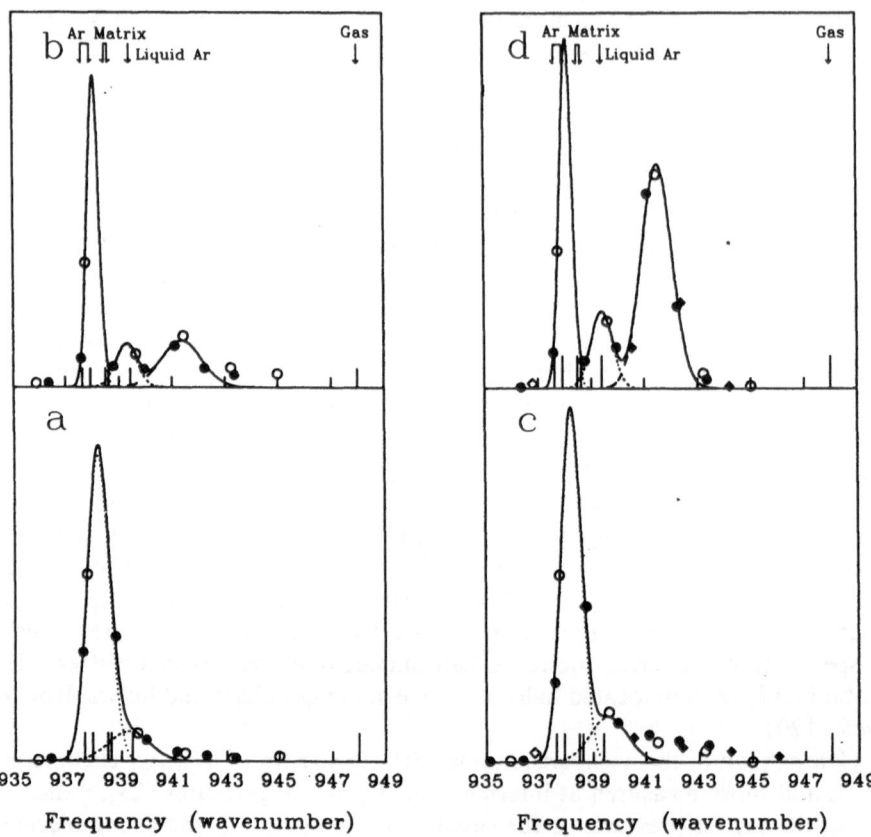

Fig. 1. Spectra of SF_6/Ar_n clusters. In each spectrum the locations of the absorptions of SF_6 in crystalline Ar, liquid Ar and in the gas phase are indicated. **a** and **b** are coexpansion spectra measured when 0.025% SF_6/Ar was expanded at 861 and 2756 kPa, while **c** and **d** are pick-up spectra measured with cluster beams generated at the same stagnation pressures, respectively. The data points measured with the different lasers are represented as follows: $^{12}C^{16}O_2$ – *diamonds*, $^{13}C^{16}O_2$ – *filled circles* and $^{12}C^{18}O_2$ – *open circles*. The curves are the results of fitting Gaussian functions to the spectra

in the gas phase and in liquid Ar are indicated and Gaussian curves are fitted to the bands. The coexpansion spectra, in Figs. 2a–c, involve expansions of a 0.1% CF_3Cl/Ar mixture at pressures of 689, 1378 and 2756 kPa, respectively. The pick-up spectra, in Figs. 2d–f, were collected using the same stagnation pressures (respectively).

The coexpansion spectrum at 689 kPa is dominated by the peak at ~ 1099.4 cm^{-1}. Because of the proximity of this band to the absorption of CF_3Cl in liquid Ar, it is attributed to a fully solvated species where the molecule resides inside the Ar clusters. That this absorption is just to the blue of the frequency measured for CF_3Cl in liquid Ar (1098.6 cm^{-1} [25]) suggests this species involves a site in which the molecule is only loosely solvated (i.e., the

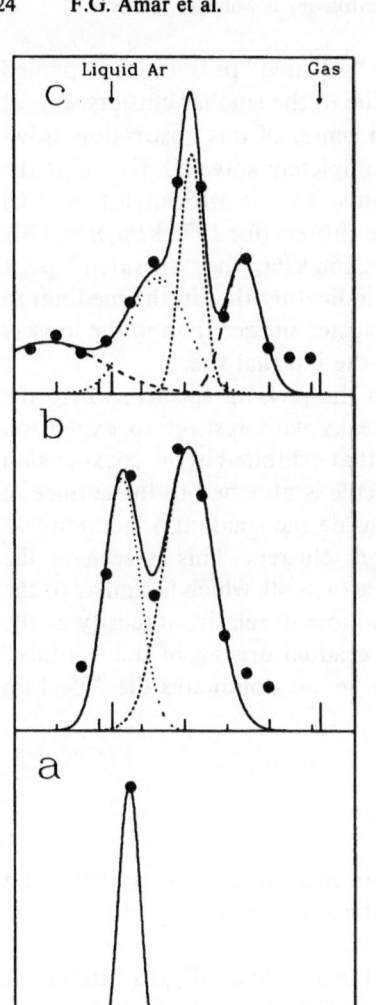

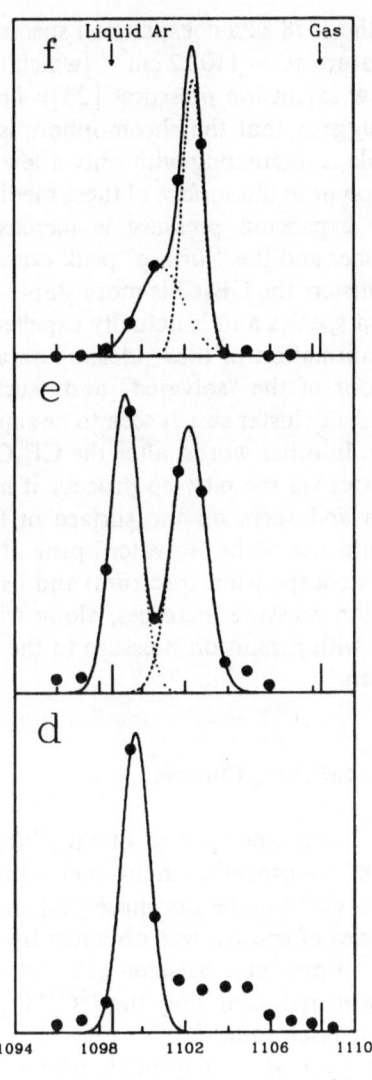

Fig. 2. Spectra of CF_3Cl/Ar_n clusters. In each spectrum the locations of the absorptions of CF_3Cl in liquid Ar and in the gas phase are indicated. **a–c** are coexpansion spectra measured when 0.1% CF_3Cl/Ar was expanded at 689, 1378 and 2756 kPa, while **d–f** are pick-up spectra measured with cluster beams generated at the same stagnation pressures, respectively. The data were measured with the $^{12}C^{18}O_2$ laser only. The curves are the results of fitting Gaussian functions to the spectra

average CF_3Cl-Ar distance is greater in the cluster than in liquid Ar). This could be due either to a large cavity site in a regular (icosahedral) lattice or to an amorphous structure which is less dense than the liquid, in which case the perturbation of the vibration should also be less. A high resolution matrix isolation spectrum of CF_3Cl in Ar is not available at this time.

In the 1378 kPa coexpansion spectrum the "solvated" peak is accompanied by a feature at ~ 1102.2 cm^{-1} (which is not due to the smaller clusters seen at very low expansion pressures [23]). The band center of this absorption, however, suggests that the chromophore is not completely solvated, (i.e., that the molecule is interacting with only a few Ar atoms), so it is attributed to CF_3Cl sitting on or in the surface of these medium-size clusters (for 1378 kPa, $n \approx 170$). As the expansion pressure is increased to 2756 kPa, the "solvated" peak diminishes and the "surface" peak dominates, indicating that in the medium to large clusters the CF_3Cl is more stable at the cluster surface than in the loosely solvated species and is actually expelled from the internal site.

Confirmation of these ideas is obtained in the pick-up spectra, where the behaviour of the "solvated" and "surface" peaks with respect to expansion pressure, or cluster size, is seen to be similar to that exhibited in the coexpansion spectra. In other words, after the CF_3Cl molecule is attached to the surface of the cluster via the pick-up process, it moves inside the small and medium-size clusters and stays on the surface of the larger clusters. This is seen in the predominance of the "solvated" peak at 689 kPa ($n \approx 40$; which is similar to the 689 kPa coexpansion spectrum) and its gradual loss in relative intensity as the expansion pressure increases, along with the gradual arising of the "surface" feature with expansion pressure to the point where it dominates the 2756 kPa spectrum.

2.2.2.3 SiF$_4$/Ar$_n$ Clusters

Figure 3 contains spectra of SiF$_4$/Ar$_n$ clusters measured between 1015 and 1040 cm^{-1}. Absorption in this region involves the v_3 mode of SiF$_4$, which occurs at 1031.4 cm^{-1} in the gas phase [26] and at ~ 1022.9 cm^{-1} in an Ar matrix [3]. This series of spectra was obtained by expanding a 0.15% SiF$_4$/Ar mixture at expansion pressures between 689 and 2756 kPa. The spectra in Figs. 3a and b were measured using only the $^{13}C^{16}O_2$ laser, while Figs. 3c and d also include $^{12}C^{16}O_2$ laser data.

The peak at 1026.9 cm^{-1}, which dominates all the spectra in Fig. 3, is assigned to a "surface" species in which a SiF$_4$ molecule resides at the surface of the medium-to-large Ar$_n$ clusters. That is, the position of this band is approximately half-way between the gas phase SiF$_4$ frequency and the center of the doublet observed in the matrix isolation spectrum of SiF$_4$ [3]. This is, again, the "half buried" chromophore [20].

Evidence for a smaller second peak at ~ 1024 cm^{-1} is also clear in these data. This feature grows with stagnation pressure until it reaches its greatest relative intensity in the 964 kPa spectrum (Fig. 3b). This less intense peak is just to the blue of the matrix isolation absorption and is, therefore, attributed to a species in which the SiF$_4$ molecule is completely solvated, but by an amorphous "matrix" that is less dense than the crystalline sample seen in matrix isolation spectroscopy.

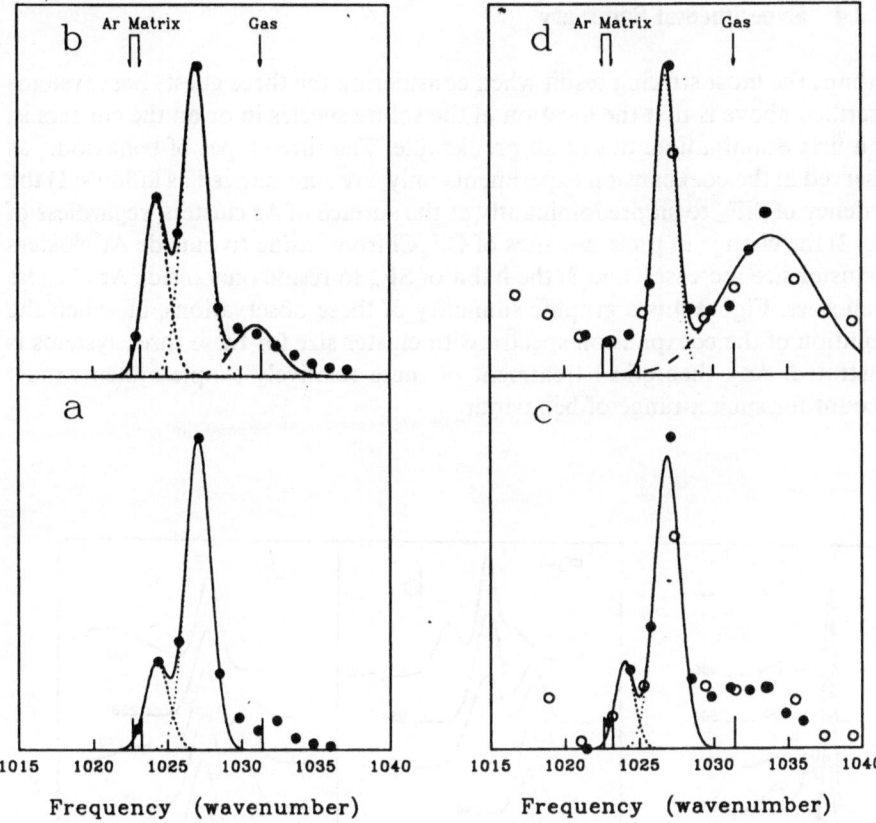

Fig. 3. Spectra of SiF_4/Ar_n clusters. In each spectrum the locations of the absorptions of SiF_4 in crystalline Ar and in the gas phase are indicated. **a–d** are coexpansion spectra measured when 0.15% SiF_4/Ar was expanded at 689, 964, 1378 and 2756 kPa, respectively. The data points measured with the different lasers are represented as follows: $^{12}C^{16}O_2$ – *open circles*, $^{13}C^{16}O_2$ – *filled circles*. The curves are the results of fitting Gaussian functions to the spectra

The broad feature at ~ 1033.9 cm^{-1} in the 400 psi spectrum (Fig. 3d, and also apparent at 1378 kPa) appears to be due to a dimer species, $(SiF_4)_2/Ar_n$, which should display two bands split by ~ 19 cm^{-1} (by comparison to the gas phase dimer spectrum [27]). Part of a second broad band at ≤ 1015 cm^{-1} is seen in Fig. 3d (neither laser operates just below 1015 cm^{-1}).

The assignment of the main spectral peak for SiF_4/Ar_n to a surface species is confirmed by pick-up spectra which are very similar, in terms of the bands observed and their relative intensities, to the coexpansion spectra involving similar stagnation pressures (the pick-up data are, therefore, not displayed here). The same small amount of "matrix" absorption (at 1024 cm^{-1}) observed in the pick-up spectra indicates that it is still possible to access this amorphous site when SiF_4 is attached to the surface of Ar clusters.

2.2.2.4 Experimental Summary

Perhaps the most striking result when considering the three guest–host systems described above is that the location of the solute species in or on the clusters is, on a first examination, not at all predictable. The three types of behaviour, as observed in the coexpansion experiments only, are summarized as follows: 1) the tendency of SiF_4 to sit predominantly at the surface of Ar clusters, regardless of size; 2) the change in preferred sites of CF_3Cl from inside to outside Ar clusters as cluster size increases; and 3) the habit of SF_6 to reside only inside Ar clusters of all sizes. Figure 4 is a graphic summary of these observations, in which the evolution of the coexpansion spectra with cluster size for these three systems is illustrated. Any theoretical treatment of these relatively simple systems must account for such a range of behaviour.

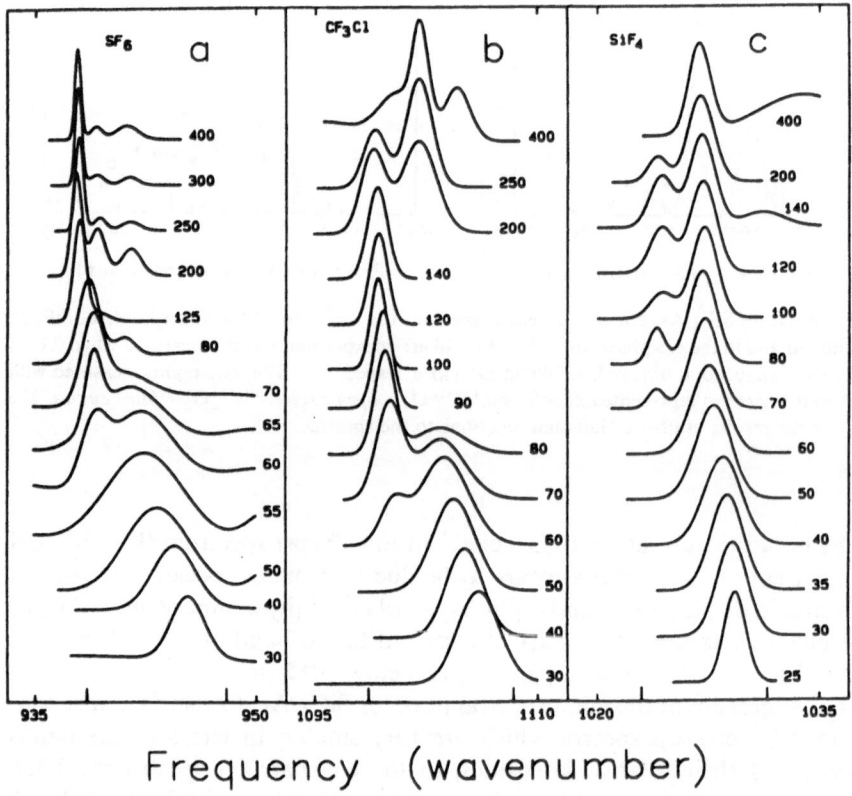

Fig. 4. Summary of **a** SF_6/Ar_n, **b** CF_3Cl/Ar_n, and **c** SiF_4/Ar_n data. For clarity only the fitted Gaussian curves have been shown. The figure includes spectra from more exhaustive studies of these systems [22, 23]. The stagnation pressures associated with each spectrum is shown in units of psi and the positions of the bulk phase and gas phase absorptions of each molecule are indicated

2.2.3   Computer Simulations

2.2.3.1   Interaction Potentials

Simulation methods offer the possibility of understanding the observed spectra from a microscopic perspective. Ideally, the goal of the simulation is to calculate the experimental observable as a function of the control parameter(s) of the experiment using, as input, a functional description of the interaction potentials between the constituent particles.

The first illustration of the potential usefulness, in terms of interacting with theory, of the infrared spectroscopy of a small probe molecule in clusters involved efforts by the LeRoy group to discern whether the evolution of experimentally observed spectral features with size, for small SF_6/Ar_n clusters [28], was due to solid-liquid phase bistability [20, 29, 30]. Molecular dynamics simulations, using realistic potentials, led to the conclusion that the broadening of the absorption profile for clusters with $n \approx 11-14$ (estimated) was due to the combination of the facts that the supersonic expansions, used to generate clusters, produce a distribution of cluster sizes, and that the red-shift of the absorption band is very sensitive to the number of Ar atoms in the cluster when $n < 18$. This work also resulted in the development of a method for simulating the band positions, as a function of the degree of immersion of the chromophore into the cluster, in the IR spectrum of a (spherical top) molecule in large clusters [20]. This model (the EL model) was useful in supporting experimental evidence that SF_6 can reside at the surface of medium-to-large argon clusters, as well as inside [9].

Extending the simulation techniques of LeRoy and coworkers to the other chromophores discussed in Sect. 2.2.2 is complicated by the lack of potentials for SiF_4–Ar and CF_3Cl–Ar which are of comparable quality to the accurate anisotropic potential for Ar–SF_6, which has been given by *Pack, Valentini*, and *Cross* [31] and used by *Eichenauer* and *LeRoy* (we shall abbreviate reference [20] by EL) in their work. It is important to ask the question, however, whether the success of the EL model in accounting for the experimental results is really all that dependent on the use of very accurate anisotropic potentials or rather, do the overall size and strength of the interaction suffice to explain the results?

Perera and *Amar* [32] have recently sought to answer this question by repeating the calculations of the LeRoy group but with the Ar–SF_6 interaction represented by the less accurate Lennard–Jones form:

$$V(r) = 4\varepsilon[(\sigma/r)^{12} - (\sigma/r)^6] \ .$$

This potential has the virtue that its two parameters, ε and σ represent the binding energy of the interacting pair and the size of the pair, respectively; in addition, these parameters have been tabulated for a great many molecular and atomic pairs. The parameters used here were chosen by fitting to the spherical part of the *Pack, Valentini, Cross* potential for Ar–SF$_6$ [31]. Table 1 gives these parameters as well as those for the Ar–Ar interaction, also represented by the Lennard–Jones form.

The simulated spectral shifts obtained for SF$_6$/Ar with the simpler potentials are shown in Fig. 5 for a range of cluster sizes corresponding to the experimental conditions of the lower half of Fig. 4a. Note that these spectral shifts reproduce, qualitatively at least, the trends observed in Fig. 4a. Detailed comparison with the EL results shows that the lineshapes obtained with the poorer potentials are virtually the same as those obtained with the accurate anisotropic potential. This gives us confidence that a knowledge of the variation in size and interaction strength parameters from one chromophore to another can explain the basic physics of cluster solvation for this class of molecules.

The behaviour of different chromophores in argon clusters can now be simulated by varying the parameters, ε and σ, which represent the interaction between the solvent argon and the chromophore. The parameters given in Table 1 are obtained from the literature (see [32]) by using the combining rules given by *Hirschfelder* et. al. [33].

These potential parameters have been used to investigate, in a systematic way, the changes in structural and spectral properties associated with changing the identity of the chromophore species from SF$_6$ to SiF$_4$ and CF$_3$Cl. Before presenting some results, it will be useful to review some computational tools which have been found useful in extracting structural information from MD simulations.

Table 1. Lennard–Jones parameters for the interaction of guest–Ar pairs of molecules. The third and fourth columns of the table are reduced values of the guest–Ar interaction parameters relative to the Ar–Ar interaction. The following combining rules have been used: $\sigma_{AB} = (\sigma_A + \sigma_B)/2$, $\varepsilon_{AB} = \sqrt{\varepsilon_A \varepsilon_B}$

Guest (– Ar)	σ(Å)	ε(K)	σ^*	ε^*
Ar[a]	3.405	119.8	1.000	1.000
SF$_6$[b]	3.960	236.8	1.162	1.977
CF$_3$Cl[c]	4.240	165.0	1.242	1.377
SiF$_4$[a]	4.178	133.2	1.227	1.112
CH$_3$F[c]	3.570	200.0	1.048	1.669
HCl[c]	3.440	180.0	1.010	1.503
Xe[d]	3.650	177.6	1.072	1.482

[a] L–J parameters from [33] or derived therefrom using combining rules
[b] Fit to spherical part of Pack, Valentini and Cross potential [20, 31]
[c] See [22, 23]
[d] See [48]

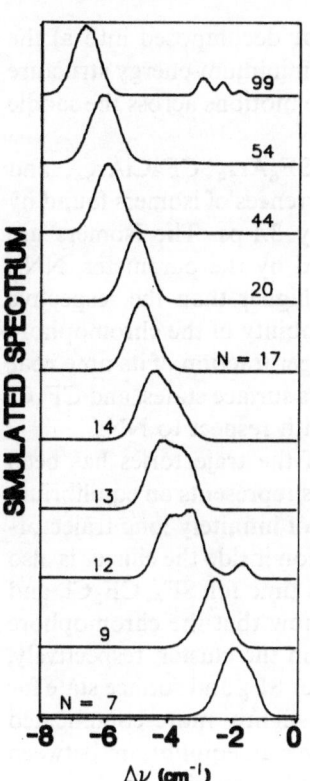

Fig. 5. Simulated spectral shifts (relative to the gas phase) for the v_3 mode of SF_6 in Ar_n clusters. Spectra are averaged over trajectories of 6.22 ns which are equilibrated at temperatures slightly above the melting temperature (about 35 K for $n > 12$)

2.2.3.2 Quenching and Structural Classes

It is natural for chemists to interpret spectra in terms of structure. Within a simulation protocol, we can quantify the terms "matrix" state and "surface" state discussed in Sect. 2.2.2 by considering the number of solvent atoms which are found distance to be $1.1 r_{eq}$ (where $r_{eq} = 2^{1/6} \sigma$). We will make use of two nearest-neighbour parameters, NN_t and NN_q.

The first parameter, NN_t, is the number of nearest neighbours to the chromophore for an arbitrary configuration of the cluster, chosen from an MD trajectory at a given total energy (subscript "t" for thermal or time-dependent). As the trajectory is propagated in time, NN_t fluctuates as the chromophore diffuses in or on the cluster. The second parameter, NN_q, is the number of nearest neighbours of a quenched configuration obtained by stopping the cluster trajectory and rapidly quenching the system to the nearest local minimum (or isomer) on the potential hypersurface of the system.

This type of quenching procedure has been used successfully to correlate isomerization rates with the melting properties of clusters [34, 35] and to understand the "hidden structure of liquids" [36, 37]. The basic notion is that the complicated thermal motions of a cluster can be decomposed into a) the vibrational or librational motions around or near a minimum-energy structure (isomer) and b) the longer time-scale, large-amplitude motions across the saddle points separating isomers.

Figure 6 shows three "skyscraper" plots for SF_6Ar_{54}, CF_3ClAr_{54}, and SiF_4Ar_{54}. The vertical axis gives the number of occurrences of isomers found by quenching trajectories of 7.78 ns total length every 3.1 ps. The isomers are characterized (horizontal axes) by their energies and by the parameter, NN_q. These trajectories have been run at total energies higher than the respective melting points of these clusters, allowing for high mobility of the chromophore in the cluster. It is clear that SF_6 tends to spend a larger fraction of its time near isomers with a larger NN_q (matrix states), SiF_4 prefers surface states, and CF_3Cl displays *ambistructural* (more uniform) behaviour with respect to NN_q.

In the skyscraper plots, the time dependence of the trajectories has been suppressed so that the relative number of occurrences represents an equilibrium probability of finding a given structure (in the limit of infinitely long trajectories). However, the *dynamics* of the chromophore motion inside the cluster is also of interest. In Fig. 7, we plot the parameter NN_t vs time for SF_6, CF_3Cl, and SiF_4 in Ar_{54}. For both SiF_4 and SF_6, these plots show that the chromophore particle, initially at an unfavourable location in or on the cluster, respectively, rapidly diffuses to the preferred state (solvated state for SF_6 and surface state for SiF_4) and stays there. For CF_3Cl, the situation is slightly more complicated since the chromophore, initially at the surface, achieves an equilibrium between an embedded surface state $\langle NN_t \rangle \sim 8$ and a solvated state with $\langle NN_t \rangle \sim 11$. This solvated state is one in which the CF_3Cl particle is found in the first shell of the cluster, not at the center.

These data show that the simple variation of interaction strength and size parameters serves to explain the different features and trends observed in the IR spectra of a number of chromophores in argon clusters. A detailed understanding of each of a number of chromophores in argon clusters. A detailed understanding of each spectrum will require a better understanding of the contribution of the anisotropic portion of the chromophore–solvent potential to the spectral shift. The close agreement of the simulated spectra obtained with the anisotropic potential for SF_6–Ar and the spherical potential for the same system, suggests that additional terms in the spectral line shift mechanism may be just as important as modifications of the potential.

2.2.3.3 A Systematization of the Structural Interpretation

By varying the parameters ε and σ of the chromophore–Ar interaction, it is possible to identify regions of the $\{\varepsilon, \sigma\}$ parameter space that correspond to matrix and surface states. NN_{max} is the largest number of solvent argons that can be packed uniformly around a chromophore of a given size. One hundred independent trajectories have been run for a set of different values of ε and σ

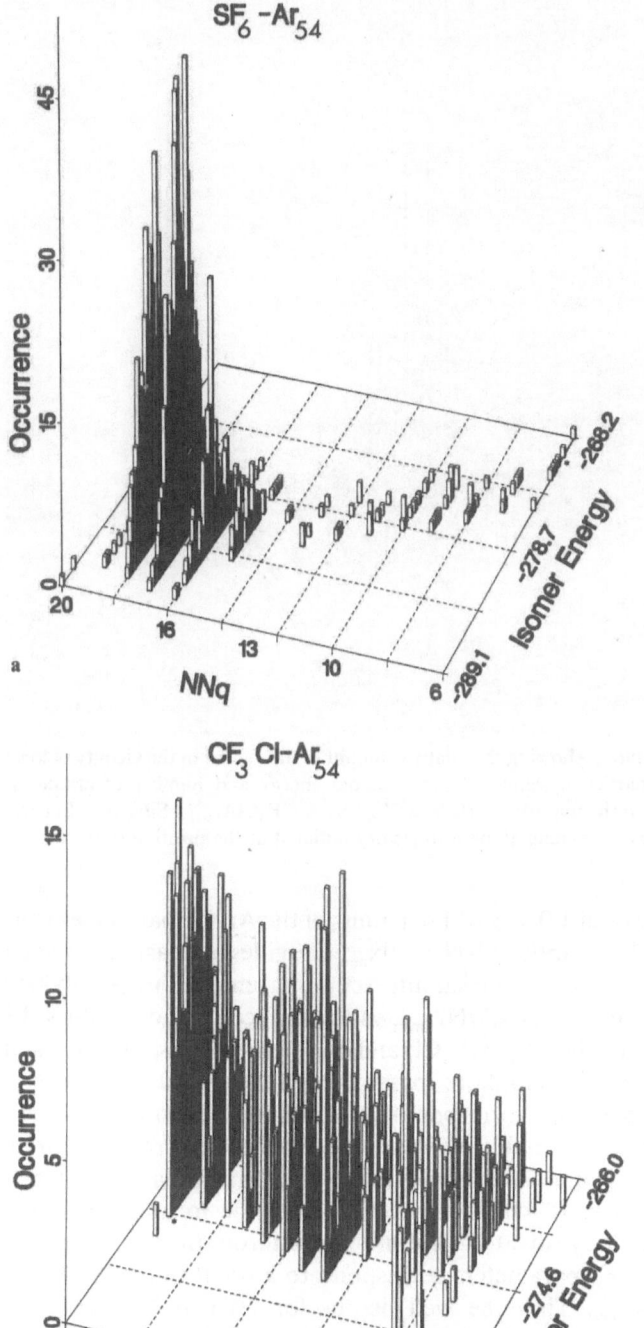

Fig. 6a, b.

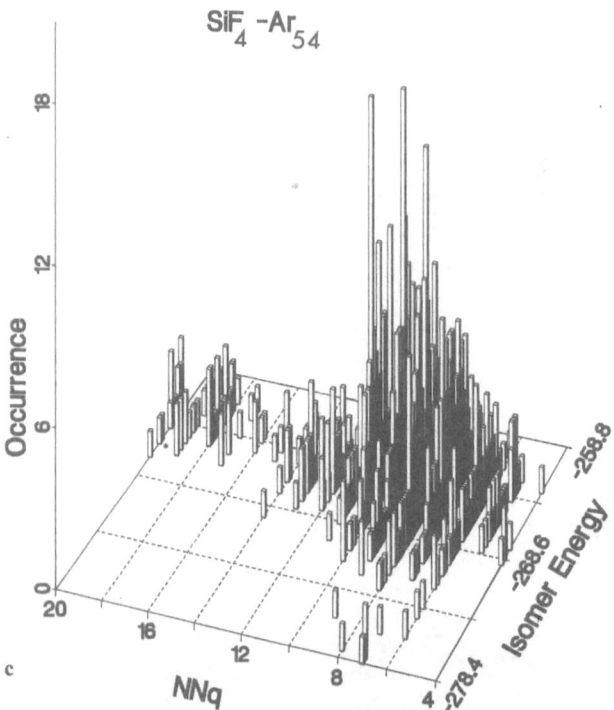

Fig. 6. "Skyscraper plots" showing the relative amount of time spent in the vicinity of local minima or isomers having particular values of the potential energy and number of argons in nearest neighbour positions to the impurity particle: **a** SF_6Ar_{54}; **b** CF_3ClAr_{54}; **c** SiF_4Ar_{54}. In each case the data are generated by quenching along a trajectory initiated at the position of the asterisk

with $1.0 < \varepsilon < 2.0$ and $1.0 < \sigma < 1.4$ in units of the Ar–Ar parameter values. For each trajectory, the quantity $\langle NN_t \rangle/NN_{max}$ provides a measure of the tendency of a chromophore with the chosen interaction parameters to be solvated in the cluster. Figure 8 shows $\langle NN_t \rangle/NN_{max}$ vs the reduced ε and σ values. Listed in Table 1, along with the SF_6, CF_3Cl, and SiF_4 parameters, are the "spherical" Lennard–Jones Ar-chromophore parameters for several other systems. As indicated by the positions corresponding to these L–J pairs (as marked) in Fig. 8, the theory agrees reasonably well with the experimental results. For example, SF_6, HCl and CH_3F are all molecules (see below) that, in large clusters, occupy stable "matrix" sites, which is corroborated by the location of the relevant points, i.e., in the "fully solvated" region removed from the "escarpment" in Fig. 8. Similarly, the SiF_4 parameters correspond to a point in the "surface" region (low $\langle NN_t \rangle/NN_{max}$). That the well depths for HCl and CH_3F ($-$ Ar) are intermediate between those for SiF_4 and SF_6 suggests that the small size of these first two contributes a great deal towards the stability of their "matrix" sites.

The location of the point corresponding to CF_3Cl is interesting in that it is actually in a region of the plot that indicates location *in* a surface site (i.e., with

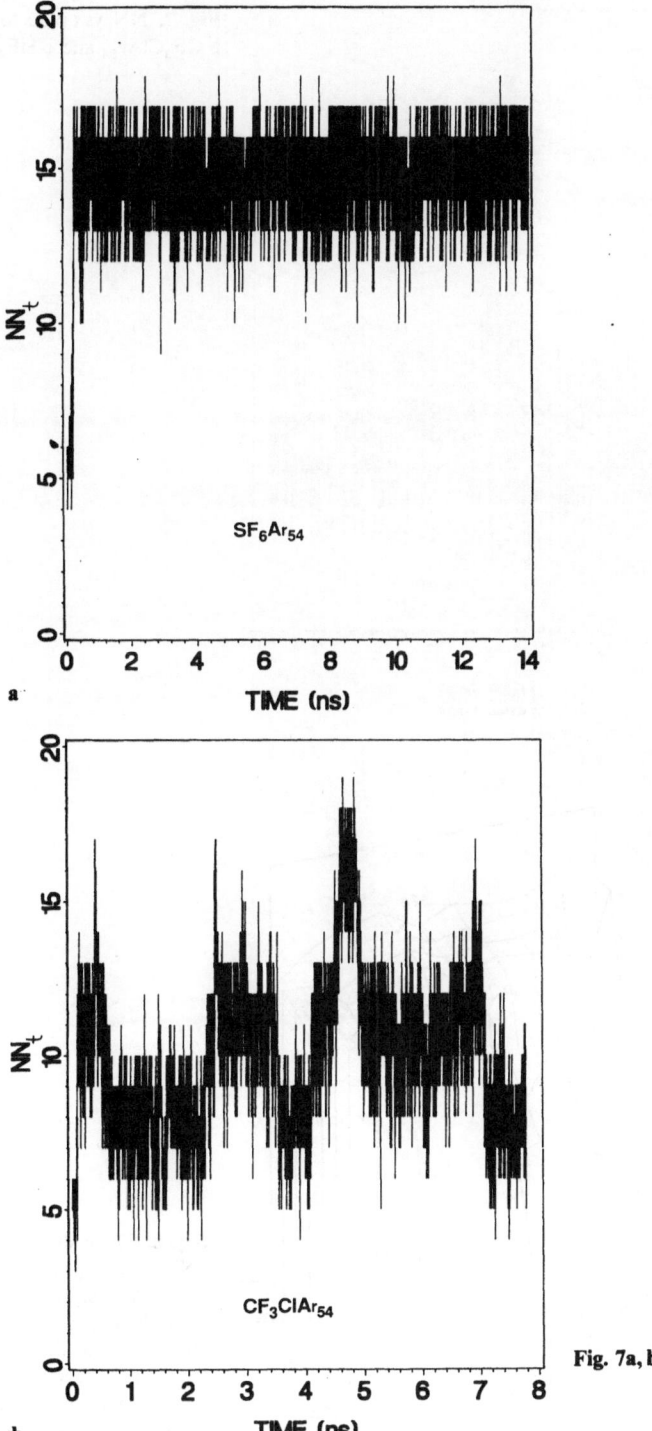

Fig. 7a, b.

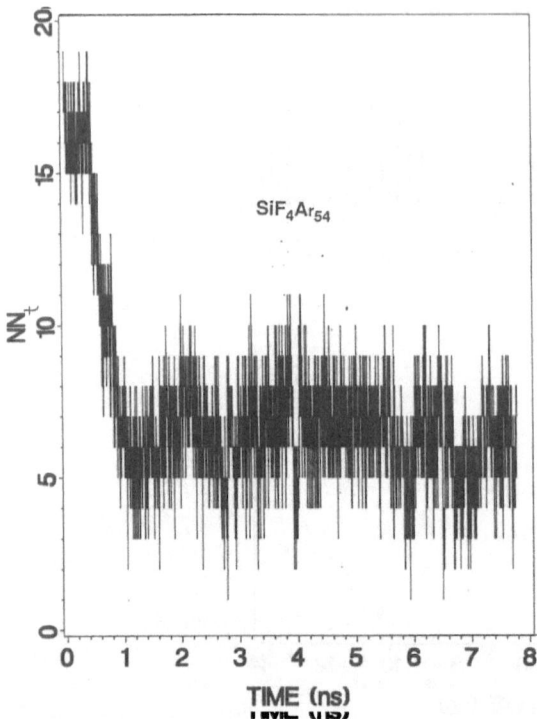

Fig. 7. NN_t vs t plots for **a** SF_6Ar_{54}, **b** CF_3ClAr_{54} and **c** SiF_4Ar_{54}

Fig. 7c

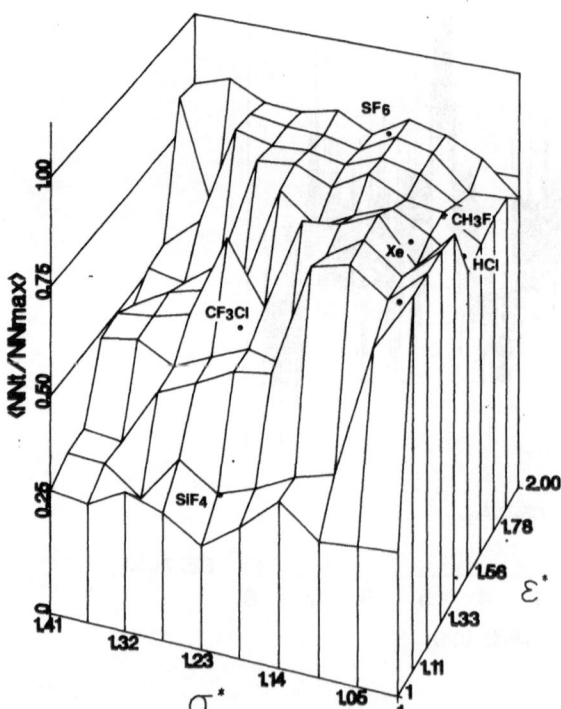

Fig. 8. Three dimensional plot of $\langle NN_t \rangle / NN_{max}$ vs reduced ε and σ values for Impurity·Ar_{54} clusters. Note the appearance of an escarpment marking the transition zone between surface behaviour (low $\langle NN_t \rangle$) and matrix behaviour (high $\langle NN_t \rangle$). Positions of ε^* and σ^* for several small molecules in argon clusters are marked

$\langle NN_t \rangle / NN_{max} \approx 0.5$, as opposed to *on* a surface or in a solvated site), correlating well with the structural interpretation of Figs. 6b and 7b. This interpretation is consistent with the experimentally observed behaviour, though it is not clear how the plot of Fig. 8 would be affected by cluster size.

2.2.3.4 Larger Clusters

So far simulations of mixed systems are available only for relatively small clusters. The behaviour of the impurity species in larger clusters, however, requires further consideration. For instance, it is not clear that the theory, as it stands, will be able to simulate the shift in the preference of CF_3Cl for being solvated in small clusters to occupying surface sites in large species.

The question arises of whether the change in location by CF_3Cl, for example, is in some way connected to the reasons for the change in cluster structure with size. In the largest clusters studied in our experiments, the average size of the clusters is approximately that in which the electron diffraction of Ar clusters indicates a change from multilayer icosahedra to fcc [38] or decahedral ordering [39]. Calculations aimed at discovering the cause of the shift in structure [40, 41] (which is inevitable since bulk Ar has an fcc lattice) have shown that the crossover is due to the fact that in the icosahedral structure, which is inherently strained in such a way that intralayer Ar–Ar distances are shorter than the equilibrium lengths and the intralayer distances are longer, as more layers are added the center of the cluster experiences greater compression. This occurs to the extent that the central atom (at least) of a neat cluster is compressed significantly into the repulsive region of the potential [40]. At the point where this strain overcomes the lower surface energy characteristic of the icosahedral arrangement, the fcc lattice becomes the preferred structure. In the clusters observed in molecular beams this does not lead to a phase transition but, rather, to clusters whose cores have icosahedral structures and whose outer layers grow with an fcc-type lattice [38]. The change from the purely icosahedral structures, at $n \approx 750$, also marks what is considered the boundary between medium-sized and large clusters, as determined by electron diffraction results [19].

To be complete, the answer to the above question must also take into account the mobility of the guest species in the clusters. That is, so far the discussion has ignored the role of kinetics in the observed behavior of the mixed clusters. Very recent, yet unpublished, results obtained in Princeton, with an apparatus that can produce even larger clusters than those studied here, indicate that at very large expansion pressures SF_6 coexpansion spectra also show a tendency for the SF_6 to be expelled from the cluster; a trend already hinted at in the results presented here, as noted in our comments to Fig. 1 (see above). A decisive step towards unraveling the kinetic behaviour will be made when the clusters' temperature will be determined independently. This will require the application of continuously tunable lasers to this type of experiment which, given the continual improvements in the power level of tunable lasers and the improved detection limits of the new Princeton apparatus, appears to be a very real possibility.

2.2.4   Complex Forming Reactions on the Surface and in the Interior of a Cluster

The potential for the extension of the techniques described above to "real chemistry" can be illustrated in the following examples which involve "reactions" that occur between two molecules in or on argon clusters. The study processes in clusters provides the opportunity to examine chemical dynamics in solutions of finite size in which only two solute molecules are present.

2.2.4.1   Formation of $(CF_3Cl)_2$ on the Surface of Ar Clusters

Further consideration of the CF_3Cl/Ar_n system reveals an interesting phenomenon. Figure 9a shows three superimposed spectra obtained when a beam of pure Ar clusters, generated by a stagnation pressure of 2756 kPa, is crossed with CF_3Cl pick-up beams of different densities. The most obvious effect on the 2756 kPa "pick-up" spectrum is the appearance and growth, with increasing pick-up flux, of the peak at ~ 1104.8 cm^{-1} (which is also seen in the 2756 kPa

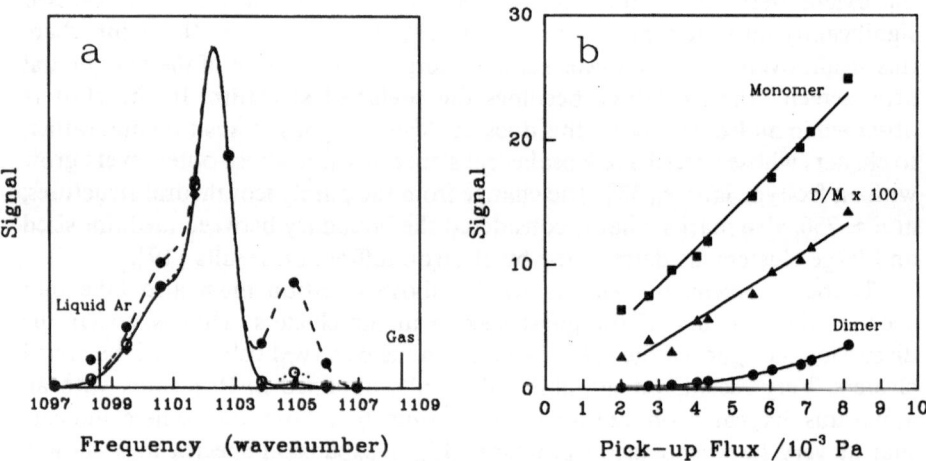

Fig. 9. a Superimposed pick-up spectra of CF_3Cl/Ar_n clusters measured using a stagnation pressure of 2756 kPa and pick-up "fluxes" producing partial pressures (in the source chamber) of 2.0 $\times 10^{-3}$ Pa (*filled diamonds*; normal flux), 6.0×10^{-3} Pa (*open circles*) and 1.3×10^{-2} Pa (*filled circles*). The three spectra are normalized on the "surface" peak at the point at 1102.79 cm^{-1}. The positions of the gas phase and liquid Ar absorptions of CF_3Cl are indicated **b** "Surface dimer" signal (*circles*), "surface monomer" signal (*squares*) and the ratio, dimer/monomer ($\times 100$; *triangles*), versus pick-up flux for CF_3Cl/Ar_n pick-up spectra measured using a 2239 kPa stagnation pressure

coexpansion spectrum; see Fig. 2). Note that the three spectra are normalized to the signal at 1102.79 cm^{-1} (9R(30)^{12}C^{18}O$_2$), i.e., to the "surface" peak.

Figure 9b shows a plot of data from spectra collected with a 2239 kPa cluster beam while varying the pick-up beam density. The plot shows that growth of the "surface" peak (as represented by the signal at 1102.79 cm^{-1} 9R (30)^{12}C^{18}O$_2$) is linear with pick-up flux, while for the peak at 1104.9 cm^{-1} (represented by the signal at 1104.91 cm^{-1} 9R(34)^{12}C^{18}O$_2$) the growth is quadratic. This concentration behaviour indicates that the absorption at 1104.9 cm^{-1} is due to a (CF$_3$Cl)$_2$/Ar$_n$ species, i.e., a dimer of CF$_3$Cl, the location of which can be shown [42] to be in all likelihood on the surface of the clusters.

It must be realized that the formation of the surface dimer, after the independent attachment of two chromophores, points to a high likelihood of mobility of the molecules on the surface of the clusters (the unlikely alternatives being the pick-up of dimers or the attachment of a second molecule at the site already occupied by the first). The hypothetical presence of CF$_3$Cl dimers inside the large clusters, these having the tendency to push the monomer outside, would require that the dimers form on the cluster surface (in a pick-up experiment) and then move inside. The possibility that the dimerization would alter the Ar–CF$_3$Cl interaction enough to result in (CF$_3$Cl)$_2$ species moving inside the large Ar clusters is quite remote. It is quite clear that this type of experiment has a high potential for the study of the sticking to and the mobility on the surface of molecular clusters.

2.2.4.2   Formation of Hydrogen Bonded Complexes in Ar Clusters

Figure 10 shows two superimposed spectra which illustrate the effect of crossing a beam of CH$_3$F/Ar$_n$ clusters, produced by coexpansion, with a "pick-up" beam of HCl (a situation in which both methods of seeding clusters is employed). The

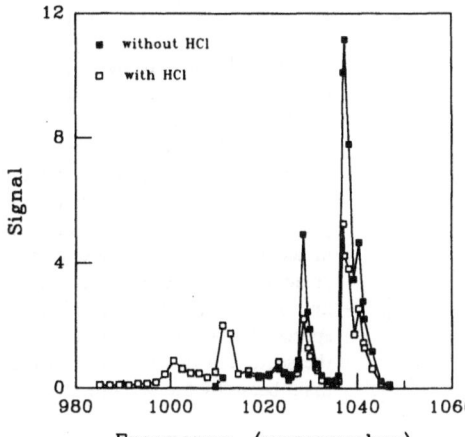

Fig. 10. Spectrum of 0.1% CH$_3$F/Ar mixture expanded at 689 kPa (*filled squares*). The spectrum measured with the same seeded expansion, but with HCl also being deposited onto the clusters by the pick-up source is shown by *open squares*

"a" spectrum was measured with no pick-up beam, i.e., is a simple "coexpansion" spectrum of the CH_3F/Ar_n clusters, obtained in the region of the v_3 absorption of CH_3F (1048 cm^{-1} in the gas phase). It is characterized by peaks at ~ 1039 and ~ 1030 cm^{-1}, which have been assigned to CH_3F/Ar_n and $(CH_3F)_2/Ar_n$, respectively [42]. Another characteristic of the CH_3F/Ar_n system is that no "surface" absorption is observed under any conditions, indicating that CH_3F is always found inside Ar clusters (see above). Spectrum "b" was recorded using the same CH_3F/Ar seeded expansion conditions but with the HCl pick-up flux on.

The most obvious differences when the HCl beam is on are the dramatic decrease in the intensity of the monomer and dimer CH_3F adsorptions and the appearance of new absorptions at 1012 and 1001 cm^{-1}. Comparison to studies of the spectra of hydrogen-bonded CH_3F–HCl complexes in Ar matrices [43, 44] indicates that the absorption at 1012 cm^{-1} is due to the formation of a CH_3F–HCl complex in the Ar clusters. The second, low frequency, band has been assigned to the reaction of HCl with $(CH_3F)_2$ [42].

The dramatic drop in intensity at 1039 and 1029 cm^{-1} can then be understood as caused by the reaction of HCl with the CH_3F species. Closer examination, however, shows that a substantial part, but not all, of the signal loss is due to scattering of clusters from the beam by collisions with the pick-up species. Figure 11 shows the results of a study [45] in which the drop in monomer signal when a pick-up beam of HCl is present is compared to the drop occurring when Ar (since it has a similar mass) is used as the pick-up species. This plot shows that the extent of scattering by the pick-up beam is related to the average cluster size (i.e., the scattering, or fragmentation of the small clusters, is more pronounced at lower stagnation pressures). Indeed, this is not surprising and is the reason why, in general, pick-up spectra are not measured using cluster beams produced with expansion pressures lower than ~ 689 kPa.

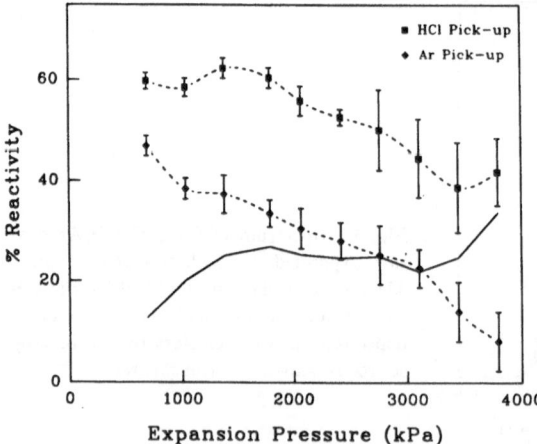

Fig. 11. Plot of percent reactivity, defined as unity minus the ratio of the reacted/unreacted peak heights for the CH_3F/Ar_n "monomer" absorption, as a function of expansion pressure. The solid curve represents the true reactivity as determined by the difference between the "reaction" with HCl (*open circles*) and Ar (*filled circles*) as pick-up species. This difference accounts for the scattering of clusters from the main beam by the pick-up flux

Figure 11 also provides evidence that the HCl molecule diffuses into the Ar clusters very quickly. That is, the extent of lowered monomer signal due to the reaction of CH_3F with HCl is independent of cluster size (stagnation pressure). Even in the largest clusters ($n \approx 10^3$), therefore, HCl reaches the CH_3F molecule in a time shorter than the travel time of the cluster beam between the pick-up region and the laser crossing (< 150 μs).

We close this chapter noting that the pick-up technique could also be applied to problems of practical importance. For example, the facility for placing individual reacting species in (or on) a water (ice) cluster would allow the detailed study of chemical processes occurring in the upper atmosphere, where ice microcrystals are found in considerable abundance [46, 47]. At present, information of this kind is obtained in flow reactors, the walls of which have been coated with a thin layer of ice.

2.2.5 Recent Experimental Results

A new apparatus has recently been built of improved sensitivity and available range of cluster sizes. With this apparatus, this type of measurements have been extended to larger cluster sizes and to several other systems. The results are described here briefly and further details may be found in the cited references.

2.2.5.1 SF_6 in Larger Clusters of Ar, Kr and Xe

The spectra of SF_6 was recorded in clusters in a larger range of sizes than those discussed in Sect. 2.2.2.1 [49, 50]. For argon (in agreement with the earlier work) and krypton, a spectral feature characteristic of the SF_6 solvated in a cluster with an amorphous structure is seen at lower sizes, when a mixture of the chromophore in Ar and Kr was coexpanded. This feature disappears on producing larger sizes, since the chromophore prefers to reside on the surface of the cluster. However, on producing still larger clusters, a different absorption appears which is accurately located at the same position as the main absorption in a well annealed matrix of Ar or Kr. This behavior may be related to the transition of clusters from a Mackay icosahedral structure, shown to be more stable for smaller clusters, to a face-centred cubic structure which is observed in the bulk phase. This structural transition occurs at a nozzle stagnation pressure, corresponding to an average cluster size of about 1800 atoms for both Ar and Kr. Further, some scattering studies performed on argon suggest that the fcc type clusters correspond to the largest sizes in the cluster size distribution at that pressure. The structural transition of xenon clusters was not established as the SF_6 appears not to be solvated in Xe at all sizes. In addition, the spectroscopic behavior of SF_6 at the surface of Ar and Kr clusters was clarified by experiments in which the chromophore is deposited on the cluster surface by the pick-up source. The so measured spectra were analyzed by a simple theoretical model [51]. These deposition experiments also helped in establishing the approximate cluster sizes at which the diffusion of the SF_6 drops, confirming and extending the earlier studies. Finally, it was established that the mobility of the SF_6 on the clusters' surface is high for all cluster sizes.

2.2.5.2   Spectra of SF_6 Seeded in Quantum Clusters

The infrared spectrum of $(SF_6)_{n=1,2}$ was also recorded in clusters which are expected to show a more quantum behavior, such as those of He, H_2 and D_2.

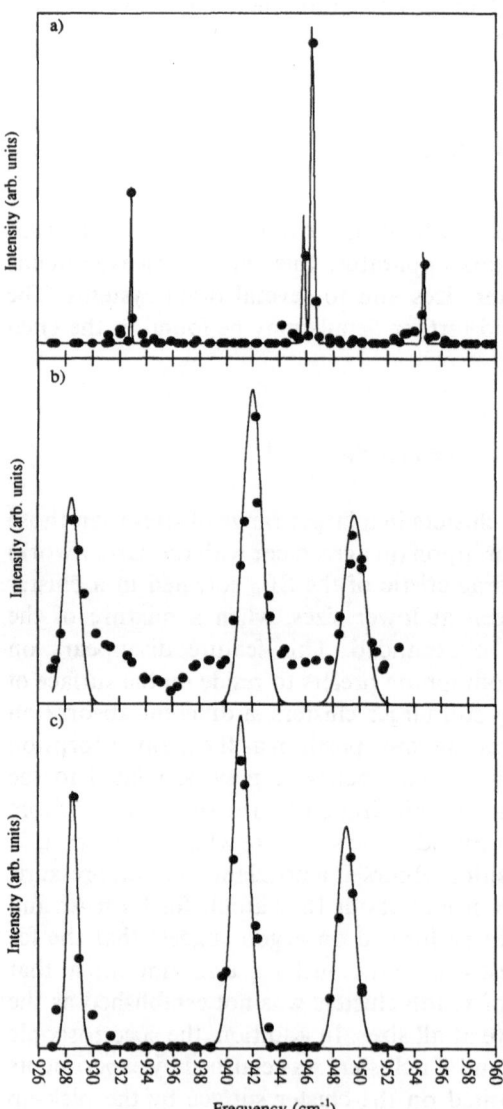

Fig. 12. The spectrum of SF_6 attached to clusters of: **a** He at 100 bar nozzle pressure and 15 K, **b** H_2 at 30 bar nozzle pressure and 50 K, **c** D_2 at 30 bar nozzle pressure and 50 K. The *darkened circles* are the measured intensities and the *solid line* is a trace of gaussian curves fitted to the data

These spectra were analyzed using an instantaneous dipole-induced dipole mechanism developed by Eichenauer and LeRoy [51]. The results showed that there is an increase in the local density of the solvent around the impurity over the density of the solvent in the bulk. This density increase is argued to arise from the lowering of zero point energy of the solvent atoms or molecules in the potential field of the impurity. Further, these studies explored the suitability of this method as a probe for investigating superfluidity in finite size systems. In He [52], the impurity was found to be located on the surface of the cluster (because of the presence of two peaks due to the absorption of the SF_6 monomer; see Fig. 12a) despite the solvation energy gain which is available to the molecule if it were to solvate inside the cluster. This expulsion of SF_6 was discussed as a possible consequence of the superfluidity of the He. This behavior is in contrast to that of the non-superfluid normal hydrogen (Fig. 12b) and deuterium (Fig. 12c) clusters [53], which are found to solvate the impurity.

References

1. B.I. Swanson, L.H. Jones: in *Vibrational Spectra and Structure*, ed. J.R. Durig, (Elsevier) 1993
2. B.I. Swanson, L.H. Jones: J. Chem. Phys. **74**, 3205 (1981)
3. L.H. Jones, B.I. Swanson, S.A. Ekberg: J. Chem. Phys. **81**, 5268 (1984)
4. L.H. Jones, B.I. Swanson, S.A. Ekberg: J. Phys. Chem. **88**, 1285 (1984)
5. L.H. Jones, B.I. Swanson: J. Chem. Phys. **80**, 3050 (1984)
6. L.H. Jones, B.I. Swanson: J. Chem. Phys. **80**, 2980 (1984)
7. B.I. Swanson, L.H. Jones, S.A. Ekberg, H.A. Fry: Chem. Phys. Lett. **126**, 455 (1986)
8. T.E. Gough, R.E. Miller, G. Scoles: J. Phys. Chem. **69**, 1558 (1978).
9. T.E. Gough, M. Mengel, P.A. Rowntree, G. Scoles: J. Chem. Phys. **83**, 4958 (1985)
10. C.E. Klots: Nature **327**, 222 (1987)
11. K. Janda: Adv. Chem. Phys. **60**, 201 (1985)
12. F.G. Celii, K. Janda: Chem. Rev. **86**, 507 (1986)
13. N.R. Isenor, J. Qi: Chem. Phys. Lett. **155**, 283 (1989)
14. U. Buck, H. Meyer: Phys. Rev. Lett. **52**, 109 (1984)
15. U. Buck, H. Meyer: Surface Science **156**, 275 (1985)
16. O.F. Hagena, W. Obert: J. Chem. Phys. **56**, 1793 (1973)
17. O.F. Hagena: Z. Phys. D **4**, 291 (1987)
18. J. Farges, M.F. deFeraudy, B. Raoult, G. Torchet: J. Chem. Phys. **78**, 5067 (1983)
19. J. Farges, M.F. deFeraudy, B. Raoult, G. Torchet: J. Chem. Phys. **84**, 3491 (1986)
20. D. Eichenauer, R.J. LeRoy: J. Chem. Phys. **88**, 2898 (1988)
21. T.E. Gough, D.G. Knight, P.A. Rowntree, G. Scoles: J. Phys. Chem. **90**, 4026 (1986)
22. X.J. Gu, D.J. Levandier, B. Zhang, G. Scoles: D. Zhuang. J. Chem. Phys. **93**, 4898 (1990)
23. D.J. Levandier, S. Goyal, J. McCombie, B. Pate, G. Scoles: J. Chem. Soc. Faraday Trans. **86**, 2361 (1990)
24. J.J. Turner, M. Poliakoff, S.M. Howdle, S.A. Jackson, J.G. McLaughlin: J. Chem. Soc. Faraday Disc. **86**, 271 (1988)
25. T.D. Kolomiitsova, V.A. Kondaurov, S.M. Melikova, D.N. Shchepkin: J. Appl. Spec. **43**, 999 (1985)
26. R.S. McDowell, M.J. Reisfeld, C.W. Patterson, B.J. Krohn, M.C. Vasquez, G.A. Laguna: J. Chem. Phys. **77**, 4337 (1982)
27. J. Geraedts, M.N.N. Snels, S. Stolte, J. Reuss: Chem. Phys. Lett. **106**, 377 (1984)

28. T.E. Gough, D.G. Knight, G. Scoles: Chem. Phys. Lett. **97**, 155 (1983)
29. R.J. LeRoy, J.C. Shelley, D. Eichenauer: in *Large Finite Systems*, ed. J. Jortner and B. Pullmann (D. Reidel, 1987)
30. D. Eichenauer, R.J. LeRoy: Phys. Rev. Lett. **57**, 2920 (1986)
31. R.T. Pack, J.J. Valentini, J.B. Cross: J. Chem. Phys. **77**, 5486 (1982)
32. L. Perera, F.G. Amar: J. Chem. Phys. **93**, 4884 (1990)
33. J.O. Hirschfelder, C.F. Curtiss, R.B. Bird: *Molecular Theory of Liquids and Gases* (Wiley, New York, 1964)
34. F.G. Amar, R.S. Berry: J. Chem. Phys. **88**, 2898 (1988)
35. J.D. Honeycutt, H.C. Andersen: J. Phys. Chem. **91**, 4950 (1987)
36. F.H. Stillinger, T.A. Weber: Phys. Rev. A **25**, 978 (1988)
37. F.H. Stillinger, T.A. Weber: J. Phys. Chem. **87**, 2833 (1983)
38. J. Farges, M.F. deFeraudy, B. Raoult, G. Torchet: in *Large Finite Systems*, ed. J. Jortner and B. Pullman (D. Reidel, 1987)
39. B. Raoult, J. Farges, M.F. deFeraudy, G. Torchet: Z. Phys. D **12**, 85 (1989)
40. J. Xie, J.A. Northby, J.D. Doll: J. Chem. Phys. **91**, 612 (1989)
41. J.A. Northby, J. Xie, D.L. Freeman, J.D. Doll: Z. Phys, D **12**, 69 (1989)
42. D.J. Levandier, M. Mengel, R. Pursel, J. McCombie, G. Scoles: Z. Phys. D **10**, 337 (1988)
43. T.D. Kolomiitsova, Z. Milke, K.G. Tokhadze, D.N. Shchepkin: Opt and Spect. (USSR) **46**, 391 (1979)
44. M.F. Barri, K.G. Tokhadze: Opt. Spectros. (USSR) **51**, 70 (1981)
45. D.J. Levandier, J. McCombie, R. Pursel, G. Scoles: J. Chem. Phys. **86**, 7239 (1987)
46. M.J. Molina, T.L.T. Tso, L.T. Molina, F.C.Y. Wang: Science **238**, 1253 (1987)
47. M.A. Tolbert, M.J. Rossi, R. Malhotra, D.M. Golden: Science **238**, 1258 (1987)
48. D. Scharf, J. Jortner, U. Landman: J. Chem. Phys. **88**, 4273 (1988)
49. S. Goyal, G.N. Robinson, D.L. Schutt, G. Scoles: J. Phys. Chem. **95**, 4186 (1991)
50. S. Goyal, D.L. Schutt, G. Scoles: (to be published)
51. D. Eichenauer, R.J. LeRoy: J. Chem. Phys. **88**, 2898 (1988)
52. S. Goyal, D.L. Schutt, G. Scoles: Phys. Rev. Lett. **69**, 933 (1992)
53. S. Goyal, G.N. Robinson, D.L. Schutt, G. Scoles: Chem. Phys. Lett. **196**, 123 (1992)

2.3 IR Spectroscopy of Hydrogen Bonded Charged Clusters

M.W. Crofton, J.M. Price, and *Y.T. Lee*

2.3.1 General Background and Motivation

"[N]o rest is given to the atoms in their course through the depths of space. Driven along in an incessant but variable movement, some of them bounce far apart after collision while others recoil only a short distance from the impact. From those that do not recoil far, being driven into a closer union and held there by the entanglement of their own interlocking shapes, are composed firmly rooted rock, the stubborn strength of steel and the like . . . "

– *Titus Lucretius Carus,* "On the Nature of the Universe", 55 B.C. [1].

The above reference demonstrates that our interest in the forces that hold groups of atoms together is nearly as old as the concept of atoms themselves. Progress in our understanding of the strong interactions that constitute chemical bonds has been considerable, but it has not been until relatively recently that it has been possible to probe the much weaker forces between molecules that give rise to many of the interesting properties of bulk matter. One important approach in the investigation of these forces has been to study cluster systems.

Charged clusters, the subject of this chapter, constitute an important class about which little structural or dynamical information has been measured so far in the gas phase. These species are represented by the generic formula: $A^{\pm q} \cdot L_n$, where $A^{\pm q}$ is most commonly a singly charged ion and L_n are the surrounding n neutral ligands. A large body of data is available concerning the thermodynamics and kinetics of these species however, (see Chapters 2.4–2.7, 2.9) and *Castleman* in addition to his own contributions has compiled the thermodynamic results of many investigators into an atlas, [2] listing the $\Delta H°$, $\Delta S°$, and $\Delta G°$ of individual clustering steps for a wide range of ions.

Structural information about charged clusters can sometimes be found by mapping $\Delta H°$ as a function of cluster size. Performing this operation can yield a curve with discontinuities at masses where the ion's first solvation shell can reasonably be expected to fill. This is the case for the ammoniated ammonium ion, $NH_4^+ (NH_3)_n$, where a strong discontinuity is observed between $n = 4$, the complete first solvation shell, and $n = 5$. A discontinuity is not obvious for the hydrated hydronium ions, $H_3O^+(H_2O)_m$, which have quite high binding energies even for second shell ligands and whose first shell is filled at $m = 3$. (See

Table 1.) Although these measurements can give qualitative information about the location of the solvent molecules around the ion, and geometries can be calculated from *ab initio* methods [3], the accurate equilibrium geometry of a given cluster can still only be obtained from spectroscopic measurements.

In nature, gas phase ionic clusters are abundant in cold environments with ionizing particles or photons present, including interstellar space and the earth's atmosphere [4, 5]. It is known from mass spectrometer studies, for example, that the hydrated hydronium ($H_3O^+(H_2O)_n$) species are the dominant ions in some regions of the atmosphere [6, 7]. Ionic clusters play essential roles in atmospheric chemistry, [8] nucleation [9] and biology, [10] are important components of many solutions and crystalline materials. Ions in solution can be thought of at some level of approximation as a collection of ionic clusters.

Charged clusters generally have a higher binding energy than their neutral counterparts. For the small species with $n = 1$, binding energies similar to those of weak covalent chemical bonds are not uncommon: [2] ΔH° is ≈ -36 kcal/mol for the reaction $H_3O^+ + H_2O \rightarrow H_5O_2^+$ [11, 12] and ≈ -27 kcal/mol for the reaction $OH^- + H_2O \rightarrow HOHOH^-$ [13, 14]. For Li^+ ions solvated by H_2O and a variety of basic ligands, the binding energy is extremely high, often 60 kcal/mol or even more [2, 15, 16]. Because of the small size of the Li^+, the ion dipole interaction energy is unusually large, and interactions arising from polarization effects can play a considerable role as well. Binding energies for a number of the ionic clusters discussed in this chapter appear in Table 1.

Among charged clusters, those species involving hydrogen bonds are of particular importance. Hydrogen bonded species are among the most strongly bound ionic clusters. The strongest hydrogen bond measured in nature, that of

Table 1. $- \Delta H^\circ$ for the stepwise solvation of a number of ions in this work (Units are in kcal/mol)

Ref	Ion	Neutral	$-\Delta H^\circ$:Ion(Neutral)$_{n-1}$ + Neutral $\rightarrow$ Ion(Neutral)$_n$						
			$n=1$	2	3	4	5	6	7
a	H_3^+	H_2	9.6	4.1	3.8	2.4			
b	H_3O^+	H_2O	36	22.3	17	15.3	13	11.7	10.3
c	NH_4^+	H_2O	19.9	14.8	12.2	10.8	10.6	9.1	8.4
d	$NH_4^+(NH_3)$	H_2O	12.9	12.7	12.2				
d	$NH_4^+(NH_3)_2$	H_2O	12.4	11.7					
d	$NH_4^+(NH_3)_3$	H_2O	11.7						
e	NH_4^+	NH_3	27	17	16.5	14.5	7.5		
f	NH_4^+	NH_3	21.5	16.2	13.5	11.7	7.0	6.5	

[a] K. Hiraoka and P. Kebarle: J. Chem. Phys. **62**, 2267 (1975)
[b] P. Kebarle, S.K. Searles, A. Zolla, J. Scarborough, and M. Arshadi: J. Am. Chem. Soc. **89**, 6393 (1967)
[c] M. Meot-Ner: J. Am. Chem. Soc. **106**, 1265 (1984)
[d] J.D. Payzant, A.J. Cunningham, and P. Kebarle: Cand. J. Chem. **51**, 3242 (1973)
[e] S.K. Searles, and P. Kebarle: J. Phys. Chem. **72**, 742 (1968)
[f] M.R. Ashardi and J.H. Futrell: J. Phys. Chem. **78**, 1482 (1974)

FHF$^-$, has a dissociation energy of 58 kcal/mol [17] or 2.5 eV, weaker than the average covalent bond but well within the range of covalent bond energies. As one of the most ubiquitous interactions in chemistry, the hydrogen bond is profoundly influential in determining the structural organization and processes that occur in a variety of environments, from hydrogen bonding solvents to entire living organisms. The proton transfer reaction, [18–23] perhaps the most general and important reaction in chemistry, is governed in hydrogen bonding solvents by hydrogen bonded "cluster" species.

A great many studies have been made of the solvated proton in the liquid and solid phases, using X-ray and neutron diffraction, infrared and nuclear magnetic resonance (NMR) spectroscopy and other techniques [18, 23]. While much has been learned from these studies, the presence of a distribution of species in the sample, the impossibility of studying most samples at very low temperatures and other problems, often result in serious ambiguities in data interpretation. Particularly in solution, the structures of $H_3O^+(H_2O)_m$ have not been well understood. In the solid phase, convincing evidence for $H_5O_2^+$ with a centralized bridging proton and $H_9O_4^+$ with C_{3v} symmetry have been found. Many ammonium salts such as the series $NH_4^+X^-$ and $NH_4^+(NH_3)_nX^-$, where $X = Cl^-, Br^-, I^-$ have well-characterized vibrational spectra [24]. The solid phase results, however, are difficult to extrapolate to the liquid or gas phase.

In spite of the substantial effort devoted to the study of the solvated proton in the bulk phases, predictions of the structures and dynamical processes relating to hydrogen bonded systems are often unreliable or computationally intractable [25, 26]. For larger systems, it is usually impractical to attempt any detailed predictions. Hydrogen bond lengths, the distance between the X atoms in X–H . . . X, and stretching frequencies are in general far less accurately determined for cluster systems that for molecules of similar size and type but containing only covalent bonds. This is due in part to the flatness of the potential well even near the bottom, and the greater anharmonicity of the potential surface associated with the hydrogen bond. There is a need for experimental data from which accurate potential surfaces can be derived, in order to guide theoretical efforts.

The understanding of neutral hydrogen bonded systems has been considerably enhanced recently by the detailed spectroscopic studies of species such as $(H_2O)_2$, [27–29] $(NH_3)_2$, [25] and a host of van der waals complexes [30, 31]. From these, accurate equilibrium geometries can be determined. While some low resolution data exists for a few of the larger clusters of $(NH_3)_n$ [32] and $(H_2O)_n$ [33–35] which are of direct interest to the question of the onset of the liquid state, these systems have resisted detailed analysis. The study of these neutral cluster species is complicated experimentally by difficulties in obtaining the spectrum for a given solvation number n without contributions from clusters of different sizes [32, 36]. This is due to the lack of an unambiguous technique for mass selection, which is not a problem for charged clusters. The spectroscopic study of hydrogen bonded ionic clusters can answer many of the same questions about hydrogen bonding as similar studies of neutral clusters and further sheds light on the topics of proton transfer and the solvation of ions in solution.

2.3.2   Some Previous Studies

While there are many attractive features to the study of ionic clusters, including the potential for mass selectivity and the abundance of data on the stepwise heats of formation for the clusters, there is one significant drawback for the spectroscopist: low number density of the sample for study. Due to space charge effects, the production of ions in abundance higher than about $10^{12}/cm^3$ becomes difficult to maintain. While this concentration is certainly sufficient for the spectroscopic study by direct absorption of many molecular ionic species, [37] the production of ionic clusters requires a relatively gentle environment. As in the case of neutral species, it is highly desirable to simplify the complicated spectra associated with all but the simplest species, by use of the supersonic expansion. For clusters, the expansion is also of great utility in simply forming the species for study. Due to the space charge effect, low ionization efficiency by electron impact and high rates of destruction of ions by recombination etc., the number of ionic clusters of a given mass that can be produced from conventional cluster sources is orders of magnitude lower than what can easily be obtained for neutral clusters. Any technique for spectroscopy of charged clusters must take the low density into account.

   Although the problem of low number density for ionic clusters makes direct absorption spectroscopy difficult, studies have been made for a few systems. The landmark studies of $H_3O^+(H_2O)_m$ ($m = 3$–5) in 1977 [38] and $NH_4^+(NH_3)_n$ ($n = 2$–4) in 1980 [39] by *Schwarz* represent the only gas phase direct absorption spectroscopic measurements of the n, $m \geq 1$ species. *Schwarz* obtained direct infrared absorption spectra by pulse radiolysis of a gas cell containing argon and small amounts of H_2O or NH_3. He made use of the known equilibrium constants for the clustering reactions to deconvolute the absorption signal obtained as a function of pressure into that of the individual cluster ions. The experiments suffered from limited spectral resolution (40 cm^{-1} full-width-half-maximum (FWHM)), relatively high vibrational and rotational temperatures (> 300 K), and some ambiguity regarding the distribution of cluster sizes in the sample. Nevertheless, *Schwarz* was able to observe the v_3 and $2v_4$ vibrational bands of the $n = 4$ ammonium ion core, and to show that the $n = 3$ spectrum was consistent with C_{3v} symmetry. Through isotope substitution measurements, he was also able to establish that the symmetry of the $n = 4$ ammoniated ammonium ion was tetrahedral. For the hydrated hydronium ions, he concluded that the first solvent shell fills at $m = 3$, and observed bands associated with both primary and secondary solvent shells, as well as with the H_3O^+ core.

Direct absorption methods have generated much more information regarding strongly bound molecular ions than for ionic clusters. The technique of velocity modulation detection, developed in 1983, is an extremely efficient means to detect infrared spectra of molecular ions formed in a glow discharge [37]. By means of a bipolar symmetric oscillating electric field between anode and cathode, ions acquire an oscillating drift velocity in the field which is approximately equal in magnitude to their thermal velocity. Due to the Doppler effect, infrared laser spectroscopy of these ions is then equivalent to frequency modulation of the laser in the ion frame of reference. Because the neutrals, to a very good approximation, are not modulated, the use of velocity modulation greatly simplifies the infrared absorption spectrum of the discharge. The resulting spectra are still very complex when many ionic species are present in abundance. Such is the case, for example, in discharges containing carbon and hydrogen. Large amounts of CH_3^+, $C_2H_2^+$ and $C_2H_3^+$ have been found to be present under various conditions, with the probable presence of other species such as CH_5^+ and heavier ions containing more than two carbon atoms [40]. The analysis of the resulting spectrum was difficult and time consuming, since the spectrum of $C_2H_3^+$ is unusual and the carrier of any of the hundreds of spectral transitions could be CH_3^+, $C_2H_2^+$, $C_2H_3^+$ or some other species. The carrier of each transition therefore had to be assigned on the basis of factors such as chemistry, linewidth, and spectral patterns involving the frequency of the transition in question relative to all others. While these methods eventually led to assignments of quantum numbers as well as carrier for many of the spectral transitions in this particular case, an experimental technique which readily discloses the carrier of any given spectroscopic transition is obviously of great utility.

One nontraditional method which nonetheless retains the traditional feature of direct absorption spectroscopy is intracavity laser absorption spectroscopy (ICLAS) of fast ion beams. Infrared spectra of species such as NH_4^+ and HN_2^+ have been obtained with very high resolution because of the kinematic compression of the infrared Doppler profile in the fast beam [41]. In order to obtain sufficient ion beam current for direct absorption studies, the technique uses a discharge followed by expansion with a low backing pressure (10 torr is typical) through a large orifice (1 or 2 mm diameter) with several kV/cm extraction voltage to generate μA of ion beam current. Under these conditions, the ions can be quite hot vibrationally, since they are accelerated before entering the collisionless regime. Due to the high internal temperature and large partition functions, it is still an open question whether the method can be successfully applied to ionic clusters.

The more successful attempts to gain spectroscopic information about ionic clusters have made use of "consequence" techniques rather than direct absorption methods. Consequence techniques use a direct consequence of photon absorption rather than attenuation of the incident radiation as the signal that an absorption event has taken place. Such methods include, among others, photodetachment, fluorescence and vibrational predissociation spectroscopies.

2.3.3 Consequence Spectroscopy of Charged Clusters

Consequence spectra of ionic clusters are relatively few in number. Negative cluster ion photodetachment studies have been used to probe the solvated electron [42–44]. Inert gas clusters of $C_6F_6^+$ have been studied with laser induced fluorescence and time-of-flight mass spectroscopy to determine solvation effects [45]. Photodissociation studies of various species with intense visible or UV lasers are quite numerous [46]. Picosecond experiments have recently been performed on mass-selected gas phase cluster species [47], giving the proton transfer rate between cluster subunits after the excitation of one subunit.

Vibrational predissociation techniques have generated the most detailed information so far on hydrogen bonded ionic cluster systems. The first spectroscopic data for the hydrogen cluster ions $H_3^+(H_2)_n$ in any wavelength region was obtained by such means as well [48]. Other studies of the smaller hydrated hydronium ions ($m = 1$–3) were made, that were able to resolve rotational structure. From these data, and hydrogen "messenger" studies, it was established that the structure of the smallest hydrated hydronium ion ($H_5O_2^+$) is symmetric with respect to the position of the proton between the two waters [49, 50]. This data is valuable in that it provides information relevant to the analysis of the proton transfer potential.

Recently, *Lisy* and coworkers have studied $Cs^+(CH_3OH)_n$ ($n = 4$–25) clusters with an apparatus similar to the apparatus to be discussed in detail in the next sections [51]. A line-tunable CO_2 laser produces enough vibrational predissociation to observe depletion of a mass-selected cluster ion beam. It has been concluded from the spectral data that the first solvation shell about the cesium ion consists of ten methanol molecules, and that larger species are composed of small hydrogen bonded clusters of methanol bound to the surface of the solvated ion. Monte Carlo simulations with pairwise potentials were found to be consistent with these interpretations. They have studied the Na^+ $(CH_3OH)_n$ system very recently.

2.3.4 Basic Principles of the Experiment

Our work makes use of a vibrational predissociation technique involving a mass selected ion beam and a radio frequency (RF) ion trap. The experimental apparatus is shown schematically in Fig. 1. We shall refer to it, at times, as a MAMICS or Mass Analyzed Molecular Ion Consequence Spectroscopy apparatus. The principle elements are an ion source which incorporates a supersonic expansion, two stages of mass selection (a magnetic mass spectrometer and a quadrupole mass filter), an octupole ion trap and an infrared laser source. Several of the important components of the device will be discussed in detail below, but a brief overview is in order.

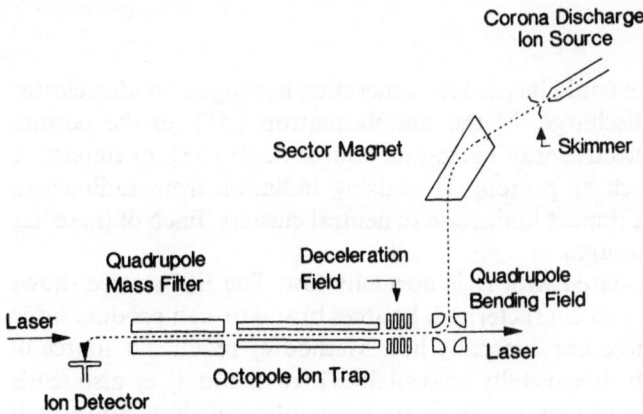

Fig. 1. Schematic diagram of the experimental apparatus

Cluster ions are generated by one of a number of sources. Of the distribution of cluster ions created by the source, one is selected for study by means of a sector-type mass analyzer. The mass selected beam is then directed into a radio frequency ion trap which holds the cluster ions of interest under collision free conditions while they interact with a tunable infrared laser. If sufficient energy is absorbed from the laser through the excitation of one of the high frequency stretching motions of the ion core or the solvent molecules, the weakly bound clusters may undergo vibrational predissociation along a channel involving the loss of one or more solvent molecules.

After excitation, the contents of the trap are then directed into a second mass spectrometer tuned to pass fragment ions of a particular mass. Since the appearance of the mass selected daughter ions is a direct result (or consequence) of the absorption of one of the tunable infrared photons by the larger parent cluster ions, these smaller mass product ions can be used as a signal that an absorption has taken place. Plotting the number of daughter ions produced as a function of the tunable infrared wavelength yields, to a first approximation, the absorption spectrum of the parent ion. More details of the excitation scheme and the approximations involved in this technique are given below.

In contrast to direct absorption methods, the sensitivity of this technique is extremely high. Ion counting permits the detection of single absorption events with essentially zero background at the mass of interest. For typical conditions in the experiments described below, we have an infrared absorption cross section $\sigma \approx 10^{-17}$ cm^2, an absorption pathlength $l \approx 100$ cm, and an ion beam density $n \approx 10^{-3}$/cm^3. The fractional absorption, given by $n\sigma l$, is then on the order of 10^{-12}, at least five orders of magnitude more sensitive than the detection limit of the best direct absorption experiments [52].

2.3.4.1   Ion Sources

Among the sources which could be used for generating hydrogen bonded cluster ions are a) the slow discharge, b) the duoplasmatron [53], c) the corona discharge [54], d) the electrospray ionization source (ESI) [55], e) impact of high energy particles such as protons, f) ionizing radiation from radioactive isotopes, and g) electron impact ionization of neutral clusters. Each of these has advantages and disadvantages.

The ion density associated with f) is normally low. The ESI source shows great promise but is not well characterized. Sources b) and d) can produce large numbers of ions, but these are internally hot. Method e) requires a source of high energy particles, which is usually unavailable. Like b) and d), e) also tends to produce large numbers of ions, but these are again internally hot. However, if the ions are made to drift through a region containing a buffer gas at high pressure and expanded through a nozzle, they can be made internally quite cold.

The corona discharge followed by a supersonic expansion is the source used in these experiments and appears in Fig. 2. It is able to produce internally cold ionic clusters at a medium flux level, and therefore represents a good compromise. Our typical conditions for the operation of the corona discharge involve tens of microamps of discharge current and very low extraction voltages (< 5 V) between the 70 μm source nozzle and the skimmer to the second differential region of the machine. The low voltage difference is used to prevent the ions from accelerating into the neutrals during the supersonic expansion between the nozzle and the skimmer, resulting in vibrational heating of the ions and possible dissociation of the clusters. To further improve cooling, the source may be kept in thermal contact with a liquified gas reservoir. Ion flux is low

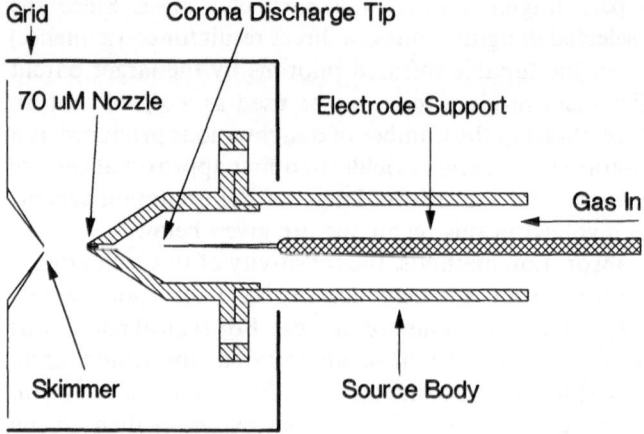

Fig. 2. Schematic diagram of the ion source, which consists of a corona discharge and supersonic expansion

under these conditions (on the order of 10^7 ions/s summed over all masses) but the internal temperature of the clusters is also low (less than 30 K rotational temperature and substantially below room temperature vibrationally).

The ion flux can be increased by enlarging the nozzle diameter and/or increasing the extraction voltage (the voltage difference between nozzle and skimmer). The price paid for this is higher internal temperature and a shift in ion abundance distribution toward the smaller clusters. With a nozzle diameter of 1 mm or larger, extraction voltage of several kV and a low pressure, high current discharge, ion currents of 10 µA can be obtained as mentioned in Sect. 2.3.2, instead of the pA levels of the current experiments. With such large currents, direct absorption spectroscopy of some of the species in an ion beam could be possible.

2.3.4.2 Mass Selection and Ion Storage

The distribution of clusters produced by the corona discharge source is directed onto the entrance slits of a 60-degree sector type mass spectrometer with a mass resolution of $m/\Delta m = 150$ (see Fig. 2). This level of resolution is routinely achieved by the experimental apparatus, but relies strongly upon the monochromaticity of the ion beam. Variations in the energy of the ions must be minimized, and to some extent, the resolution is dependent upon the conditions of the ion source. Large voltage differences between the nozzle of the source and the skimmer shown in Fig. 1 lead to degradation of the mass resolution due to fringing field effects.

Following the mass selection, ions of the desired mass are deflected 90° by a dc quadrupole bending field, decelerated and focused into the octupole ion guide whose axis is that of the infrared laser beam. The octupole consists of eight molybdenum rods, equally spaced on a 1.5 cm radius. The rods of the octupole carry approximately 350 V dc, with applied radio frequency voltages of typically 300 V peak to peak. Adjacent rods are maintained 180° out of phase. The capacitance of the rods together with the inductance of a copper coil atop the machine form an LC tank circuit which is made resonant in the range 5–20 MHz by the choice of the coil. Smaller ions such as H_5^+ are trapped most efficiently at frequencies toward the high end of the range.

Trapping is accomplished by means of entrance and exit repellers, each of which carries a positive ≈ 350 V dc. Upon these a negative 12 V pulse may be applied to allow the entrance and exit, respectively, of ions into and out of the octupole. The relative timing is adjustable to vary the storage time and the ions can be easily stored for $t < 1$ min with only about 50% losses. The upper limit to our trapping time is determined by the $\approx 10^{-9}$ torr background pressure in the UHV region of the octupole. The number of trapped ions is optimized by adjusting the dc voltage on the rods: the dependence is rather sharp. One thousand ions are routinely trapped at a time, with a repeat frequency of 40 Hz. The major limitation in the number of stored ions seems to be the low ion flux from the source. The trapped ion energy in the longitudinal direction is < 0.5 eV.

The choice of an octupole ion trap over the simpler quadrupole design stems from the superior shape of the effective potential that an octupole field generates. For the present case of a fast oscillating field, the effective potential governing the motion of a slow moving particle with charge q in an ideal electric multipole field with $2n$ poles is given by:

$$U_{\text{eff}}(r) = n^2 K \left(\frac{r}{r_0}\right)^{2n-2} ,$$ (1)

where K is described by:

$$K = \frac{q^2 V_0^2}{4m\omega^2 r_0^2}$$ (2)

and V_0 is the rms rf voltage, r_0 is the trap radius, ω is the rf angular frequency and m is the mass of the particle. For an octupole, the effective potential varies as $(r/r_0)^6$ [56]. This dependence is $(r/r_0)^2$ for a quadrupole, with a well depth only 25% of the octupole depth for a given voltage, mass, and frequency.

2.3.4.3 Laser Excitation and Consequence Detection

The trapped, mass-selected ionic clusters are vibrationally excited by a pulsed infrared laser derived from the difference frequency mixing, in a $LiNbO_3$ crystal (Quanta Ray IR WEX), of a Nd:YAG laser fundamental (1064 nm) and tunable dye laser radiation (typically 740–835 nm). With the aid of a feedback loop, this commercial unit automatically adjusts the tilt of the crystal to maintain the phase matched condition for difference frequency generation as the dye laser is scanned. Typical laser power is 3.5 mj/pulse at 4000 cm^{-1}.

For neutral clusters or weakly bound ionic clusters such as the hydrogen cluster ions (see Table 1), a single photon of tunable infrared light in the 4000 cm^{-1} region provides sufficient energy to promote the system over the dissociation threshold along the solvent molecule loss channel. Absorption of a photon into one of the high frequency stretching modes of the ion core in $H_3^+(H_2)_3$ results in the loss of up to two of the three available H_2 ligands [57]. For weakly bound systems, excitation of a single quantum of a high frequency vibrational stretch can couple to the continuum of levels of the dissociating products.

For weakly bound systems like these, the absorption spectrum of the parent cluster ion should be well represented by the number of daughter ions produced as a function of the tunable infrared light. The frequency dependence of the reaction cross section for the photodissociation reaction

$$H_3^+(H_2)_n + h\nu_1 \rightarrow H_3^+(H_2)_{n-1..2} + (1 .. 2)H_2$$ (3)

comes only from the frequency dependence of the absorption cross section of the parent cluster.

The situation becomes somewhat more complicated when the ionic cluster in question has a binding energy larger than the energy of a single tunable infrared photon. Under these circumstances, more energy must be deposited in the system to induce dissociation unless there are large amounts of internal excitation already present. Virtually all of the systems studied in this work fall under this category. A glance at Table 1 reveals that for the hydrated hydronium or ammoniated ammonium ion dimer species (either a hydronium or an ammonium ion solvated by a single ligand) the $\Delta H°$ of formation is roughly a factor of three larger than the energy of a 3000 cm^{-1} photon. Only for the largest clusters is the binding energy small enough for a single photon to be effective in removing a solvent molecule. To put the required extra energy into the system we have turned to a multiphoton excitation scheme.

For all but the largest clusters in this study, a second laser is also used. Once the cluster has absorbed a tunable infrared photon, it is in an energy regime characterized by a very high density of states referred to as the quasi-continuum. Stepwise excitation of the cluster in this region by a line tunable CO_2 laser (MBP Technologies, 8 W cw at 10.6 μm) is a reasonably facile process, and through this stepwise excitation the cluster can obtain sufficient energy to undergo loss of one or more solvent molecules. This reaction is represented in its most general form for positive ionic clusters by the following:

$$A^+(M)_n(hv_1) + m(hv_2) \rightarrow A^+(M)_{n-i} + iM . \tag{4}$$

In this case, (hv_1) represents a tunable infrared photon of frequency v_1 and $m(hv_2)$ represents m number of fixed frequency photons of frequency v_2. The parent cluster in this case has absorbed sufficient energy from the two lasers to result in the evaporation of i solvent molecules to produce a smaller mass daughter ion.

Since the spacing between energy levels is much larger in the low energy regimes, the absorption cross section for the tunable infrared photons is large only at a strong resonance with a fundamental vibrational transition. The total cross section for the photodissociation reaction above is assumed to have wavelength dependence only from the absorption cross section for the first tunable infrared photon which excites low lying vibrational states. Daughter ions are counted using a Daly ion detector, and spectra are usually normalized for the tunable infrared power using a simple linear power dependence.

2.3.5 Typical Results and Interpretations

Presented below are representative data obtained using the MAMICS technique for a number of ammoniated ammonium ions ($NH_4^+(NH_3)_n$, hydrated hydronium ions ($H_3O^+(H_2O)_m$), and an example of a cluster involving both ammonia and water around an ammonium core ($NH_4^+ (NH_3)_3(H_2O)$). While the MAM-ICS apparatus can be adapted to study high frequency stretching overtones or electronic spectra below 2.5 μm, the data to be discussed here pertain only to the 2.5–4.0 μm region of the spectrum. The spectral features which appear here are due primarily to vibrational fundamentals involving the high frequency NH or OH stretches of the solvent molecules and the ion cores of the clusters. Other, weaker features have been assigned to bending overtones or combination bands involving bends and NH or OH stretches.

When a molecule is involved in a hydrogen bond, its vibrational frequencies will be shifted from the isolated subunit values. Among the high frequency stretches, the shifts are largest for those modes involving the hydrogen atom directly associated with the hydrogen bond. For hydrogen atoms not directly involved, the shift in X–H (X = N, O, etc.) vibrational frequency is relatively small. The hydrogen bonded ionic clusters to be discussed here, are characterized by an ionic core species, NH_4^+ or H_3O^+, which form extremely strong

Table 2. Molecular constants for NH_4^+, and NH_3 (All units are in cm^{-1} except ζ_3 which is dimensionless)

Molecule or Ion	Symmetry	Molecular Constants	References
NH_4^+	T_d	v_1: 3270 ± 25	[a]
		v_2: 1669	[c]
		v_3: 3343.26	[e]
		v_4: 1447.22	[f]
		A_0: 5.9293 ± 0.0002	[e]
		B_0: "	[e]
		C_0: "	[e]
		ζ_3: 0.0604	[e]
NH_3	C_{3v}	v_1: 3336.2, 3337.2	[b, g]
		v_2: 932.5, 968.3	[d, g]
		v_3: 3443.6, 3443.9	[d, g]
		v_4: 1626.1, 1627.4	[d, g]
		A_0: 6.196	[g]
		B_0: 9.444	[g]
		C_0: "	[g]
		ζ_3: 0.06	[g]

[a] P. Botschwina: J. Chem. Phys. **87**, 1453 (1987), scaled *ab initio* value
[b] W.S. Benedict, E.K. Plyler and E.D. Tidwell: J. Chem. Phys. **32**, 32 (1960)
[c] D.J. DeFrees and A.D. McLean: J. Chem. Phys. **82**, 333 (1985), scaled *ab initio* value
[d] T. Shimanouchi, *Tables of Molecular Vibrational Frequencies*, U.S. Dept. of Commerce, Natl. Stand. Ref. Data Ser. Natl. Bur. Stand. **39** (U.S. GPO, Washington, D.C., 1972)
[e] M.W. Crofton and T. Oka, J. Chem. Phys. **79**, 3157 (1983); **86**, 5983 (1987)
[f] M. Polak, M. Gruebele, B.W. DeKock and R.J. Saykally: Mol. Phys. **66**, 1193 (1989)
[g] G. Herzberg: *Electronic Spectra of Polyatomic Molecules* (Van Nostrand, Princeton, N.J., 1967)

hydrogen bonds to the solvating NH_3 or H_2O species. As mentioned above, the dissociation energy for loss of the NH_3 or H_2O is about 30 kcal/mol for the protonated dimer cases (See Table 1.). The hydrogen bonded oscillators of the ionic core have associated red shifts of about 400 cm^{-1} and 1000 cm^{-1} for the largest $NH_4^+(NH_3)_n$ and $H_3O^+(H_2O)_m$ clusters discussed here, respectively, and shifts much larger than 1000 cm^{-1} for the $n = m = 1$ case. Hydrogen bonds to the ionic core are stronger than those to first or second shell ligands, because of the larger electrostatic contribution to the bonding interaction.

Due to the effects described above, the spectra of X–H vibrational stretching fundamentals of ammoniated ammonium $(n \geq 4)$ or hydrated hydronium $(m \geq 3)$ are naturally divisible into two well-separated regions. The strongly red-shifted region contains stretching bands of the ionic core, while the stretching fundamentals of the ligands appear at higher frequencies. Fundamentals in the latter region can similarly be divided into relatively unshifted X–H ligand stretches where H is unbound, and moderately strongly red-shifted stretches where H is hydrogen bonded to an outer ligand. These natural divisions have greatly simplified the process of spectral assignment of the vibrational bands.

Vibrational frequencies for the fundamental vibrations of NH_4^+, H_3O^+, NH_3 and H_2O appear in Tables 2 and 3. Where available, experimentally

Table 3. Molecular constants for H_3O^+ and H_2O
(All units are in cm^{-1} except ζ which is dimensionless. Rotational constants are for the ground vibrational state)

Molecule or Ion	Symmetry	Molecular Constants	References
H_3O^+	C_{3v}	ν_1: 3570	[a]
		ν_2: 954.40 $(1^- \leftarrow 0^+)$	[b]
		ν_2: 525.83 $(1^+ \leftarrow 0^-)$	[b]
		ν_3: 3530.17, 3513.84	[c]
		ν_4: 1564	[a]
		$A_0 = B_0$: 11.2540 (0^+)	[b]
		C_0: 6.1	[c]
H_2O	C_{2v}	ν_1: 3656.7	[d]
		ν_2: 1594.6	[d]
		ν_3: 3755.8	[d]
		A: 27.877	[e]
		B: 14.512	[e]
		C: 9.285	[e]

[a] P.R. Bunker, W.P. Kraemer and V. Şpirko: J. Mol. Spectrosc. **101**, 180 (1983); scaled *ab initio* value
[b] D.-J. Liu, T. Oka, and T.J. Sears: J. Chem. Phys. **84**, 1312 (1986); and references therein
[c] M.H. Begemann, G.S. Gudeman, J. Pfaff, and R.J. Saykally: Phys. Rev. Lett. **51**, 554 (1983); M.H. Begemann and R.J. Saykally: J. Chem. Phys. **82**, 3570 (1985)
[d] W.S. Benedict, N. Gailar, and E.K. Plyler: J. Chem. Phys. **24**, 1139 (1956)
[e] G. Herzberg, *Electronic spectra of polyatomic Molecules* (Van Nostrand, Princeton, N.J., 1967)

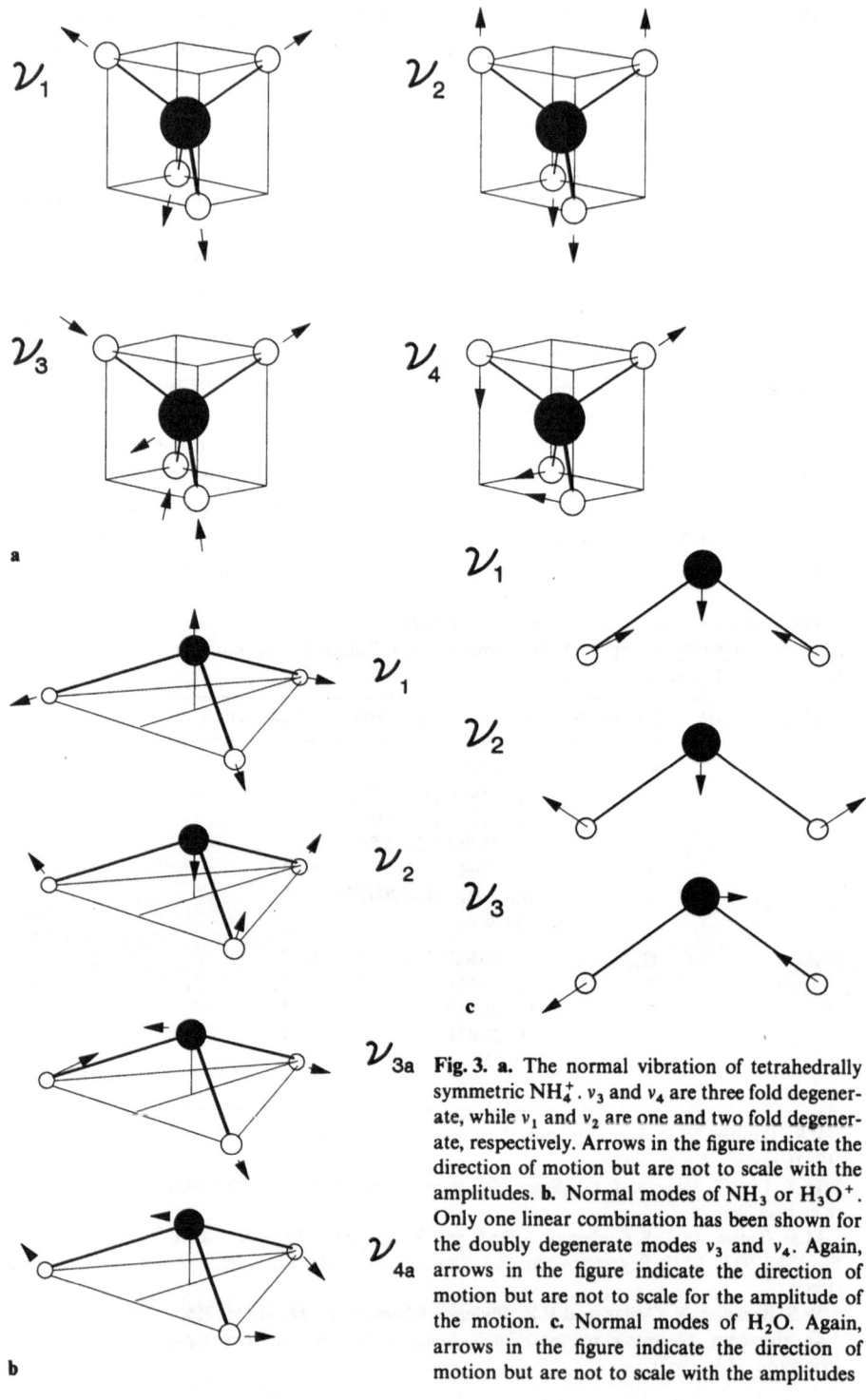

Fig. 3. a. The normal vibration of tetrahedrally symmetric NH_4^+. ν_3 and ν_4 are three fold degenerate, while ν_1 and ν_2 are one and two fold degenerate, respectively. Arrows in the figure indicate the direction of motion but are not to scale with the amplitudes. b. Normal modes of NH_3 or H_3O^+. Only one linear combination has been shown for the doubly degenerate modes ν_3 and ν_4. Again, arrows in the figure indicate the direction of motion but are not to scale for the amplitude of the motion. c. Normal modes of H_2O. Again, arrows in the figure indicate the direction of motion but are not to scale with the amplitudes

measured values have been used. In some instances *ab initio* estimates of the constants are presented. Pictorial representations of the normal vibrational modes appear in Fig. 3.

2.3.5.1 $NH_4^+(NH_3)_n$, $n=1$ to 10

The infrared vibrational predissociation spectra for the ammoniated ammonium ions appear in Figs. 4 and 5 and are discussed below. A more extensive discussion of these results may be found elsewhere [58]. Our preferred structures for some representative clusters may be found in Fig. 6.

The spectra of the ammoniated ammonium ion series show a number of common features. Both strong absorptions due to the ammonium core stretches and weaker ones due to the ammonia solvent molecules can be assigned in every case. Features of the NH_4^+ core assigned to vibrations involved in hydrogen bonds with the solvent NH_3's are strongly red-shifted from the gas phase values obtained from velocity modulation spectroscopy [59, 60] due to the strong interactions of the N–H bonds of the ion with the solvent molecules. Vibrational transitions involving the NH_3 molecules in the first solvation shell (1°) of the complex (see Fig. 6) are significantly less perturbed from their gas phase values,

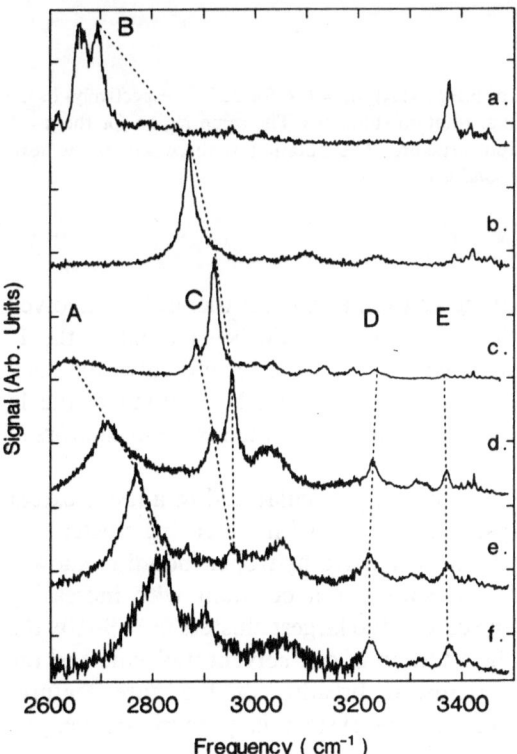

Fig. 4. Vibrational predissociation spectra of $NH_4^+(NH_3)_n$ ($n = 3-8$ for 4a–4f, respectively) in the 2600–3500 cm^{-1} region. Bands forming clear series in n are connected by *dashed lines*

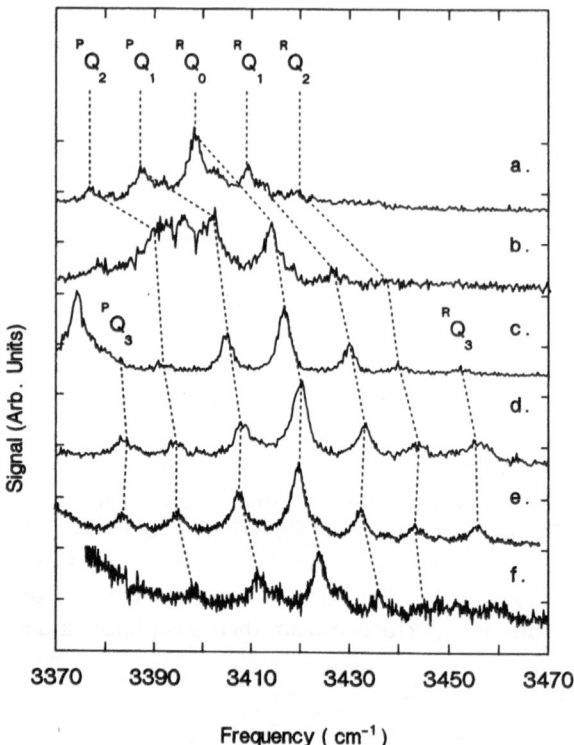

Fig. 5. Vibrational predissociation spectra of $NH_4^+(NH_3)_n$ ($n = 1$–6 for 5a–5f, respectively) in the 3370–3470 cm^{-1} region, showing the internal rotation subbands. The same region for the $n = 8$ cluster does not show the internal rotation structure. The notation is discussed in the text. Corresponding subbands are connected by *solid lines*

however, because the N–H bonds of the ammonias are not themselves involved in the hydrogen bond. This situation is changed for an N–H bond of the 1° ammonia which is involved in a hydrogen bond with a second solvation shell (2°) ammonia. A larger red-shift will take place for that N–H bond of the 1° ammonia, although not as large as that for hydrogen bonded N–H oscillators of the ion core.

The spectral features due to the ion core in particular and to a lesser extent the solvent molecules show some systematic trends with increasing cluster size. The ion core frequencies associated with extensive hydrogen bonding show a gradual blue shift. The amount of this shift is not constant with increasing solvation however, and in the extreme case of the largest clusters ($n = 9$–10), the spectra change very little between the addition of one solvent molecule and the next, indicating along with the increased breadth of the core features (> 100 cm^{-1} FWHM) that the core of the system is perhaps converging towards a liquid-like environment.

A.

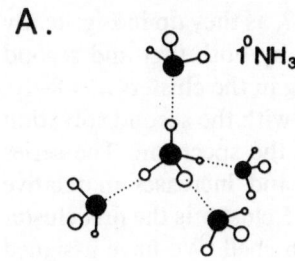

Fig. 6. Proposed structures for the $n = 4, 5$ and 8 complexes. The hydrogen bonds are denoted by *dashed lines*. It is not certain that the hydrogen bonding between $1°$ and $2°$ NH_3 is linear

B.

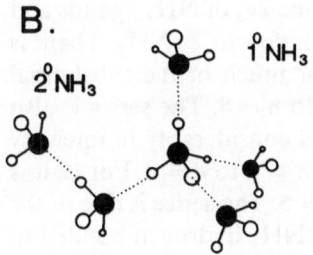

C.

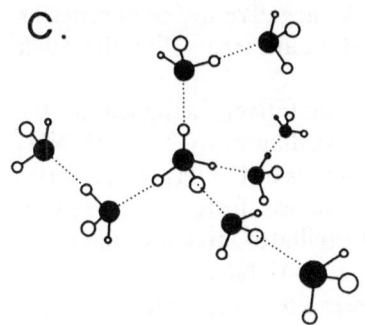

Figure 4 shows the spectrum for a series of ammoniated ammonium ions ($n = 3$–8) from 2600 to 3500 cm^{-1}. The series labeled B in this figure arises from an antisymmetric stretching mode involving two ($n = 2, 6$), three ($n = 3, 5$) or four ($n = 4$) equivalent N–H oscillators of the NH_4^+ core, where H is hydrogen bonded to a $1°$ NH_3. The series C arises from *symmetric* stretches of the same oscillators. The symmetric stretch carries no infrared absorption transition dipole for tetrahedrally symmetric $n = 4$, and is not observed in Fig. 4b.

If the structure of the $n = 7$ cluster is the same as the $n = 8$ cluster shown in Fig. 6c less one $2°$ NH_3 ligand, then the $n = 7$ NH_4^+ core contains only one N–H oscillator with *only* a $1°$ NH_3 attached by hydrogen bonding. The remaining N–H oscillators of the ion core are influenced by both $1°$ and $2°$ NH_3's. As a result, one expects that the antisymmetric and symmetric stretching modes of n

$= 6$ collapse to a single N–H stretching band for $n = 7$, as they do in Fig. 4e. By $n = 8$, all the N–H bonds of the core are influenced by both first and second shell ammonias and we find that this feature is missing in the clusters $n = 8$–10.

As the clusters increase in size, features associated with the second solvation shell should be expected to make a contribution to the spectrum. The series labeled 'A' in Fig. 4 begins at the $n = 5$ spectrum and increases in relative intensity to dominate the spectrum by $n = 8$. The $n = 5$ cluster is the first cluster thought to have an ammonia in the second solvation shell. We have assigned this feature to components of the N–H stretching modes of the core where the N–H bonds are involved in hydrogen bonding to both 1° and 2° NH_3's.

Between the B and E series lies the bending overtone $2v_4$ of NH_3 ligands and the N–H oscillator of 1° NH_3 which is hydrogen bonded to 2° NH_3. There is little doubt that the latter contribution accounts for much of the substantial increase in relative intensity of series D from $n = 5$ to $n = 8$. The series E also arises from N–H oscillators of 1° NH_3 and increases considerably in intensity relative to the most intense band for a given n, from $n = 5$ to $n = 8$. For E, it is much more obvious that the first appearance is at $n = 5$. The series is due to the two equivalent N–H oscillators of 1° NH_3 with a 2° NH_3 hydrogen bonded to its remaining N–H. Series D and E correspond in some measure then to symmetric and antisymmetric stretching modes of 1° NH_3 with attached 2° NH_3. As such, there is some possibility of Fermiresonance interaction between the modes. Such an interaction may account for the negative $\Delta v/\Delta n$ of series D. Even so, D and E have their smallest $|\Delta v/\Delta n|$ values at large n, like the other series.

Between D and E, weaker features appear, tentatively assigned to the symmetric stretch of the three equivalent N–H oscillators of 2° or 1° NH_3 and/or symmetric stretches of the two equivalent bonds of 1° NH_3 with 2° NH_3 attached to the third oscillator. The rather intense feature centered at ≈ 3370 cm^{-1} in Fig. 4a is due to the single N–H oscillator stretch of the $n = 3$ NH_4^+ core, where the H is *not* hydrogen bonded to a 1° NH_3.

The most striking feature of the spectra between 2600 and 3500 cm^{-1} is a weak band originating at roughly 3400 cm^{-1} composed of a number of resolved subbands having a separation between adjacent components of about 12 cm^{-1}. Shown in Fig. 5 for $n = 1$ to 6, this spacing is about twice the rotational constant for NH_3 about its C_3 symmetry axis. We have been able to show that this structure arises from a vibrational transition of the solvent molecules in the first solvation shell involved in a nearly free rotation about the hydrogen bond to the ion core. The vibration is analogous to the doubly degenerate antisymmetric stretch, v_3, of free ammonia, which occurs at 3444 cm^{-1} in the gas phase [61] (see Fig. 3).

Such a vibrational band carries the selection rule $\Delta K = \pm 1$, where K is the quantum number for angular momentum about the NH_3 C_3 axis. The assignments of the internal rotation subbands are given in Fig. 5 and the notation is that used for a perpendicular vibration-rotation transition of a strongly prolate symmetric top molecule. While this notation is not strictly correct for the case of internal rotation, we use it for convenience as it follows easily from the picture of an ammonia molecule attached to a "wall" perpendicular to its C_3 axis. Such structure cannot be expected for any vibrational band of the core or any solvent vibrational band, such as the symmetric stretch v_1 of NH_3, which carries a $\Delta K = 0$ selection rule. The fact that the structure is observed for $n = 5$ and 6 indicates that some first shell solvent molecules can still be quite unhindered even when the second solvation shell has begun to be filled. Further discussion of internal rotation within ionic clusters can be found in later sections.

2.3.5.2 $H_3O^+(H_2O)_m$, $m = 3$ to 8

Recently, we have extended our early studies of the hydrated hydronium ions to the larger cluster ions where m is greater than 3 and the second solvation shell is formed [62]. Figure 7 contains the spectra of $m = 3–8$ in the 2600–4000 cm^{-1}

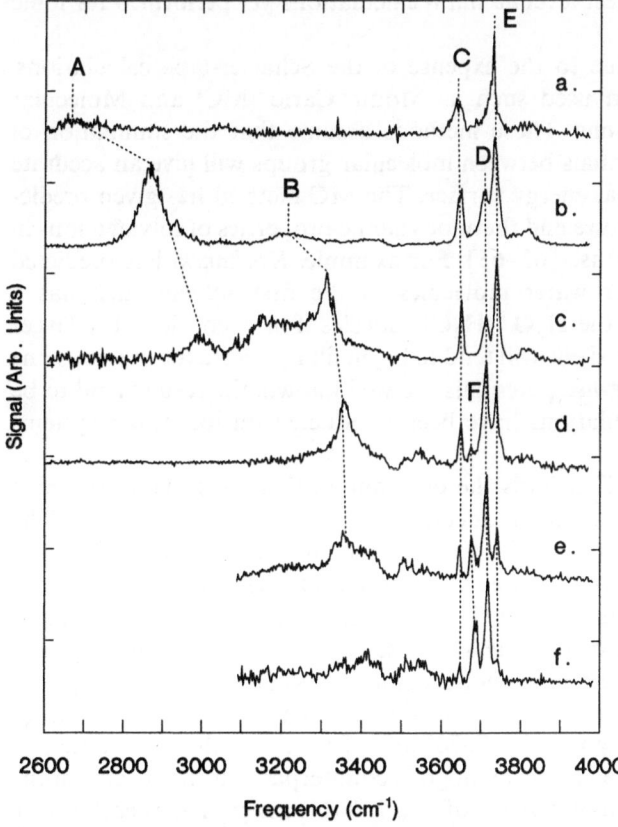

Fig. 7. Vibrational predissociation spectra of $H_3O^+(H_2O)_m$ ($m = 3–8$ for 7a–7f, respectively) in the 2600–4000 cm^{-1} region. Bands forming a clear series in m are connected by *dashed lines*. Structures are observed, supporting the notion that the first solvation shell is filled at $m = 3$, and that the third shell may start to fill before the second shell is full

region (with the exception of $m = 7$ and 8 where only the region 3200–4000 cm^{-1} was obtained). It is instructive to compare some of the general features of these data, and the data for the ammoniated ammonium ions discussed above. The spectra of $H_3O^+(H_2O)_m$ are similar to those of $NH_4^+(NH_3)_n$ in several respects, including a) the presence of band series decreasing in $\Delta v/\Delta n$ as n increases, b) a clear separation between those bands associated with non-bonded oscillators, 1° 2° ligand bonded oscillators, and hydrogen bonded oscillators of the ion core, and c) agreement with the rough empirical correlation of width and frequency of the bands with the strength of the hydrogen bond [18].

Previously, theorists and experimentalists collaborated heavily to determine that $H_5O_2^+$ ($n = 1$) has a single minimum potential with respect to the proton involved in the hydrogen bond. The minimum energy position was found to be midway between the oxygen atoms. The Schaefer ground has calculated geometries, vibrational frequencies and intensities for $H_3O^+(H_2O)_m$ ($m \leq 3$) using full configuration-interaction with single and double excitations in the simulation. These are the highest level *ab initio* calculations yet performed for ionic clusters.

For larger systems, due to the expense of the Schaefer-type calculations, other methods have been used such as Monte Carlo (MC) and Molecular Dynamics (MD) simulations. These methods assume that the summation of pairwise interaction potentials between molecular groups will give an accurate description of the potential energy surface. The MC method has given predictions concerning the structure and thermodynamic properties of solvated ions in solution and in the gas phase [63–65]. For example, *Kochanski* has predicted that the presence of four water molecules in the first solvent shell has a significant probability in the $H_3O^+(H_2O)_m$ species if m is considerably larger than 4, and that successive shells will tend to begin filling before the previous one is full [63]. When the pairwise potentials are well known, the results tend to be reliable. Recent MD simulations have been conducted on ion–water systems [66, 67].

The $m = 3$ spectrum (Fig. 7a) is the only one in this set of data for which vibrational frequencies have been calculated at a high level of theory. The vibrational transition of the H_3O^+ core with the largest predicted intensity is the antisymmetric stretch of the three O–H oscillators. The *ab initio* frequency for this mode was 2997 cm^{-1} for a C_{3v} structure, with the symmetric stretch 23 cm^{-1} higher and a factor of 40 less intense [68]. The experimental spectra from *Schwarz*'s previous direct absorption measurements and our own consequence technique show a rather broad band centered at 2670 cm^{-1} and a weaker one at 3050 cm^{-1}. *Schwartz* assigned the 2670 cm^{-1} feature to the core antisymmetric stretch. This feature and its analogous counterparts in the spectra of the larger clusters is designated by series A of Fig. 7. If this assignment is correct (the antisymmetric stretch band is expected to be the most intense feature of the spectrum below 3000 cm^{-1}, which this band is), it means that the theoretical calculation is wrong by more than 300 cm^{-1} for the frequency of this vibration, or more than 10%. The usual accuracy for small, polyatomic molecules is better than 2% or 3%. The poor accuracy for the cluster systems results from an inadequate treatment of the effect of hydrogen bonding. One obvious difficulty is that the most sophisticated calculations still do not consider the influence of anharmonicity directly, and instead apply a semi-empirical scaling factor to a calculated harmonic frequency. The scaling factor is often taken to be the same for all modes.

The symmetric stretch of the $m = 3$ core has been a source of some controversy, especially since *Schwarz* and *Newton* [69] identified it with the weak band at $\approx 3000 \text{ cm}^{-1}$ but the recent *ab initio* calculation placed it only 23 cm^{-1} above the antisymmetric stretch [68]. We believe the 3050 cm^{-1} band to be the same feature *Schwarz* observed, but assign it to either a bending overtone, most probably of the solvent molecules, or a combination band which would likely involve the core antisymmetric stretch and a hydrogen bond stretching vibration (the harmonic frequency of the antisymmetric stretch involving v_{O-H-O} in the $m = 3 \text{ C}_{3v}$ structure is of the order of 300 cm^{-1}, [68]). In Fig. 7b, the band of series A has a less intense companion about 70 cm^{-1} to the blue; this is probably the core symmetric stretch. For a structure of type similar to the ammoniated ammonium case but with 3 instead of 4 solvent molecules in the first and second shells, series A should consist of a single band for $m = 5$ (it does), probably with significantly reduced intensity since stretches of the core will only carry high intensity for antisymmetric stretches of two or more equivalent oscillators. Indeed, the intensity is low, as it was for the analogous band of $n = 7$ ammoniated ammonium.

Series B, which begins at $m = 4$, and seems to continue through $m = 8$, can be identified quite readily with the stretch of O–H oscillators of 1° H_2O which are hydrogen bonded to 2° H_2O. The width of the $m = 4$ peak seems a bit large compared to the rest of the series, and might arise from an overlapping band on the low frequency side; $m = 5$ also has a broad feature in this region. A comparison of $m = 5$ with $m = 4$ and 6, however, after mentally removing the presence of series B, suggests that the $m = 5$ spectrum might be anomalous. It has broad absorption in the 3150–3450 cm^{-1} region which bears little resemblance to the $m = 6$ or $m = 4$ spectrum. One possible explanation is the presence of a second isomer of $m = 5$.

According to Newton, a structure for the $m = 5$ ion involving an $H_5O_2^+$ ion core should exist [70]. Designated $H_5O_2^+(H_2O)_4$, this species should have a global minimum energy only 2.5 kcal/mol above that of the $H_3O^+(H_2O)_5$ structure. Newton proposed this structure to be an important intermediate in the proton transfer mechanism in aqueous acid solution, with the symmetry of the $H_5O_2^+(H_2O)_4$ complex being essential to the mechanism. The expected stretching frequencies of the four equivalent O–H oscillators in the $H_5O_2^+$ core are, of course, unknown, but should certainly fall somewhere between 3000 and 3400 cm^{-1} and carry a large intensity. The relative intensity of the series transitions in the $m = 5$ spectrum, compared to the trend of $m = 4$–8, may offer some support for the $H_5O_2^+(H_2O)_4$ hypothesis. If the hypothesis is correct, a very intense, very broad band should exist somewhere in the 500–2000 cm^{-1} region, arising from the vibration of the central $H_5O_2^+$ proton.

Series C and E arise, respectively, from the symmetric and antisymmetric stretching modes of the ligands which have none of their O–H bonds involved in a hydrogen bond. As solvent molecules with all O–H oscillators free, the O–H stretching frequencies are shifted very little from their isolated gas phase values. In $n = 5$, for example, the shifts are ≈ 4 and ≈ 15 cm^{-1} to the red for symmetric and antisymmetric stretch, respectively (see Table 3). Because of the small shifts, contributions from ligands in more than one shell tend to overlap. The decreasing relative intensity from $m = 3$–8 suggests, however, that the absorption cross section decreases in the order $1° > 2° > 3°$ for these oscillator types.

The assignment of a given spectral feature is made primarily on the basis of the lowest solvation number for which it is present, the frequency at which it appears, and the change in relative intensity with solvation number. Series D begins as the second solvation shell begins to form with the $m = 4$ cluster and increases in relative intensity as a function of cluster size to dominate the spectrum by $m = 6$. We have assigned this feature to the free O–H stretch of an H_2O that has one of its O–H oscillators involved in a hydrogen bond with another solvent molecule. Since molecular species tend to carry a transition dipole moment which increases with partial charge, a general trend with respect to absorption cross section of core $> 1°$ is expected when similar oscillators are compared. Theoretical calculations show that the partial charge decreases in the order core $> 1° > 2°$ so we would expect that most of the intensity associated with this peak is due to transitions involving the H_2O's in the $1°$ shell bound by $2°$ H_2O's.

A similar series begins very weakly at $m = 6$ and is very apparent at $m = 7$, where the third solvation shell is expected to begin filling. This feature is attributed to the free O–H stretch of species involved in hydrogen bonding to both $2°$ and $3°$ H_2O's. It is labeled F in the figure. If this assignment is correct it confirms the Monte Carlo prediction mentioned above, that the higher solvation shells may begin to fill before the lower shells have been completely occupied.

While the vibrational predissociation spectra of $NH_4^+(NH_3)_n$ become somewhat indifferent to an increase in n above $n = 7$, the spectra of $H_3O^+(H_2O)_m$ evolve considerably from $m = 6$ to 7 and 7 to 8, in spite of the fact that each shell is thought to contain only three ligands instead of four. This can be attributed to at least two factors: 1) the larger intrinsic hydrogen bonding interaction of H_2O as compared to NH_3, and 2) a larger absorption cross section for H_2O than for NH_3 in the secondary and tertiary solvation shells. Even the spectrum of the gas phase cluster $(H_2O)_{17}$ is known to have a sharp feature at about 3710 cm^{-1} [71], corresponding to the non-bonded O–H stretch of H_2O ligands where the other O–H is part of the hydrogen bonded network. The $(H_2O)_{17}$ spectrum is more reminiscent of liquid water in the 3100–3600 cm^{-1} region, however, which consists of a very broad, featureless absorption. It is not really difficult to imagine that the spectrum even of $H_3O^+(H_2O)_{17}$ is similar in this region.

The spectrum by $m = 8$ has several broad peaks, arising from various hydrogen bonded O–H oscillator stretching vibrations. These could include 1° O–H to 2° H_2O, 1° O–H to 2° H_2O to 3° H_2O, and 2° O–H to 3° H_2O. Bending overtones may appear in the 3100–3300 cm^{-1} region as well. As the cluster size grows, the number of possible types of hydrogen bonds increases further, with a resultant filling effect in the 3100–3500 cm^{-1} spectral region. Further, the ratio of hydrogen bonded to free O–H oscillators is certain to rise, with a resultant decrease in the relative intensity contribution of free O–H oscillators in the 3600–3800 cm^{-1} region. The difference between spectra of hydrated hydronium and neutral water clusters should become small when the size of the clusters is very large (hundreds or thousands of H_2O subunits). For the intermediate case, vibrations of the ion core and the inner shell(s) account for most of the difference.

2.3.5.3 $NH_4^+ (NH_3)_3 H_2O$

In addition to studying the proton solvated by just one type of solvent molecule, we have investigated the spectra of a number of "mixed" cluster species containing both NH_3 and H_2O [72]. These species are of particular interest, as they provide information about chemically heterogeneous systems and can shed light upon the sorts of solvent–solvent and solvent–solute interactions encountered in bulk mixtures. We present here the spectrum of $NH_4^+ (NH_3)_3 H_2O$ (see Fig. 8) over the frequency range of 2600–4000 cm^{-1}, which serves to illustrate many of the essential features observed.

When a hydrated hydronium species $H_3O^+ (H_2O)_m$ encounters NH_3, a proton transfer reaction is likely to take place resulting in the formation of the species $NH_4^+ (H_2O)_{m+1}$. Since the proton affinity of H_2O is 6.5 eV vs. 8.9 eV for NH_3 [73], the energy released by the process is in the neighborhood of 2 eV, or about 50 kcal/mol. In our experiment, this excess energy is removed through collisions in the supersonic expansion or through the evaporation of solvent molecules from the cluster. The strong favoring of the proton transfer reaction to the ammonium side leads us to assign all mixed cluster peaks in our mass spectra to the isomers containing an ammonium rather than a hydronium core.

Competition occurs between the heterogenous solvent molecules for sites around the ammonium ion during the cluster's formation. A number of experiments have been made studying the ΔH^0 of formation for the stepwise solvation of the ammonium ion by both water and ammonia. Some of these results appear in Table 1. It has been shown that the hydrogen bond of NH_4^+ (NH_3) is much stronger than that of $NH_4^+ (H_2O)$ [74]. The affinity of NH_3 for $NH_4^+ (NH_3)_n$ is higher than that of H_2O for $NH_4^+ (H_2O)_n$ in the first solvation shell ($n \leq 4$) because of the greater proton affinity of NH_3. The difference decreases with increasing n however, and is nearly zero for $n = 4$, which represents a crossing point. In the second solvation shell, additional H_2O's bind more strongly to the $NH_4^+ (H_2O)_n$ than NH_3's do to $NH_4^+ (NH_3)_n$. This is

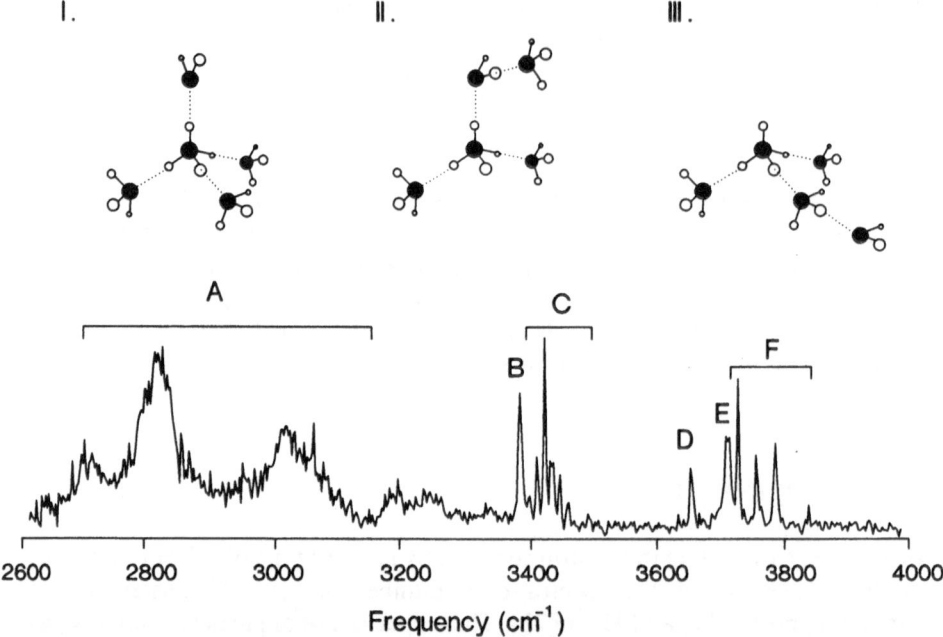

Fig. 8. Vibrational predissociation spectrum for the NH_4^+ $(NH_3)_3(H_2O)$ mixed cluster species in the 2600–4000 cm^{-1} region. Bands are observed due to the water, ammonia and ammonium subunits. At least three different isomers are possible, shown above the figure

because the hydrogens involved in second shell hydrogen bonding are more similar to the hydrogen atom than a proton since the charge has been delocalized over the entire cluster; the hydrogen bonding interaction of H_2O with neutral H_2O or NH_3 is stronger than that of NH_3. The $NH_4^+(NH_3)_3H_2O$ cluster is particularly interesting as one of the systems where ammonia and water are bound with nearly the same energy.

The spectrum of the $n = 3$, $m = 1$ cluster appears in Fig. 8 over the frequency range of 2600 to 4000 cm^{-1}. Above the spectrum are shown three isomers which we consider significant to our discussion. Structure I has the three ammonia molecules oriented about the ammonium ion with their C_3 axes along the axes of the N–H bonds of the core. In this case, the water molecule is oriented along its C_2 symmetry axis. This orientation is stabilized by a favorable ion–dipole interaction involving the N–H bond and the dipole moment of the H_2O. In structure II, the water molecule is still attached to the ion core, but one of the ammonia molecules has adopted a 2° solvation site, and has attached itself to the water. Structure III shows the opposite configuration, where the water has adopted the 2° solvation site and attaches itself to one of the 1° ammonia molecules.

A variation on isomer I merits some discussion. It could happen that the water molecule binds to the ammonium ion oriented along an orbital containing a lone pair of electrons. In this case the water molecule would be substantially tilted from the orientation shown in the figure. A tilted orientation of H_2O is observed, for example, in the hydrogen bonding of ice and in $HF-H_2O$ [75].

As in the case of the ions studied previously, features are observed that can be assigned to transitions arising from the vibrational motions of both the solvent molecules and the ion core of the cluster. It is evident from Fig. 8 that nearly free internal rotation of NH_3 occurs in the $NH_4^+(NH_3)_3H_2O$ complex, as the same structure (labeled C in the figure) observed for $NH_4^+(NH_3)_n$ ($n = 1-6$) is present in the $3400\ cm^{-1}$ region. The frequency of the "$^RQ_0$" subband of the $NH_3\ \nu_3$ fundamental is identical for $NH_4^+(NH_3)_3H_2O$ and $NH_4^+(NH_3)_4$. to within $2\ cm^{-1}$. In the higher frequency region, structures are present which obviously correlate to the symmetric (labeled D) and antisymmetric (labeled F) stretches of the solvent molecules in the $H_3O^+(H_2O)_m$ ($m = 1$ to 8) systems, labeled C and E in Fig. 7. The strong features between 2600 and $3200\ cm^{-1}$, (labeled A) have been assigned to modes of the ion core. While the spectrum is similar in areas to both the hydrated hydronium and the ammoniated ammonium spectra there are significant differences as well.

In the case of $NH_4^+(NH_3)_4$, the evidence is strong that the local C_3 axis of each NH_3 coincides with an $N-H$ bond axis of NH_4^+. The evidence includes *Schwarz*'s determination of a tetrahedral equilibrium structure, the very low barrier to internal rotation of NH_3, the relative intensities of the $NH_3\ \nu_3$ subbands, and the numerical value of the characteristic subband spacing. In the spectrum of Fig. 8 ($NH_4^+(NH_3)_3H_2O$ spectrum) structure I is clearly present as a constituent. We see three narrow peaks (F in the figure) with $\approx 29\ cm^{-1}$ separation, in the $3700-3800\ cm^{-1}$ region where the ν_3 antisymmetric stretch of H_2O is expected. If this structure is due to the same sort of internal rotation seen for NH_3 in the ammoniated ammonium ions, the separation observed should be roughly twice the constant for rotation about the internal rotation axis. Indeed, the constant for rotation about the local C_2 axis of H_2O is 14.5 or $29/2\ cm^{-1}$.

As was the case for internal rotation in the ammoniated ammonium ions, only a perpendicular band of the solvent H_2O can show such internal rotation structure. The spin weight with respect to the quantum number K (once again, we are adopting a much simplified picture of the complex, essentially H_2O attached to a wall) for the angular momentum about the $H_2O\ C_2$ axis, by analogy with formaldehyde-type molecules, is 1 for $K = $ even and 3 for $K = $ odd. As a result, $^RQ_0$ is less intense than $^RQ_1$ or $^PQ_1$, in spite of the Boltzmann factor which, for $kT = 25\ cm^{-1}$, is $I(K = 1)/I(K = 0) \approx \exp(-14.5/25) = 0.56$ (while $^PQ_1$ has a greater peak height than $^RQ_1$, the integrated intensities appear to be comparable). This internal rotation structure could not be observed from isomer II, as the $2°$ ammonia would quench the free internal rotation of the water. Isomer III is also unlikely to produce this structure, particularly in view of the observation that only first shell subunits in ammoniated ammonium complexes seem to give rise to a nearly free internal rotation band structure.

The sharp band labeled B is not present in the $n = 4$ ammoniated ammonium ion cluster, but is observed for the $n = 3$ ammoniated ammonium ion.

There, it was assigned to a core stretching band primarily associated with the single free N–H oscillators of the core. It appears in the spectrum of the $n = 3$, $m = 1$ cluster with nearly the same frequency and intensity relative to the ammonia internal rotation structure. This points to isomers of type II or III being present, where one of the sites on the ammonium ion is free, resulting in one high frequency core N–H stretching band. The presence of a sharp, strong band at about $3380 \, \text{cm}^{-1}$ is a signature of this vibrational mode, since the frequency will not depend strongly upon the solvation of the other 3 binding sites.

Recall that in the section on the hydrated hydronium ions, a high frequency feature was observed in the $m = 4$ spectrum correlated with the addition of a solvent molecule into the second shell. Labeled D in Fig. 7b, it was assigned to the free O–H stretch of a 1° water bound by a 2° water. This band also provides a signature, since it must appear quite close to the average of symmetric and antisymmetric stretching frequencies of an H_2O at that binding site. Isomer II of the $n = 3$, $m = 1$ mixed cluster should show a similar, free O–H stretch. The peak at $3710 \, \text{cm}^{-1}$ labeled E in Fig. 8 has been assigned to this. Isomer I is clearly more stable than isomer II, from the observation that if the gas pressure behind the expansion nozzle is increased to lower the cluster ion internal temperature, a significant intensity reduction of peak E relative to peaks F takes place. Given the greater competitiveness of NH_3 for the inner solvent shell and H_2O for the outer solvent shell from the thermodynamic measurements cited above, one might predict an isomer with 2° H_2O hydrogen bonded to a 1° NH_3 to be more stable than isomer II. There is no clear evidence in the spectrum for such an isomer, however, nor the variation on isomer I discussed above where the water is tilted, and no clear evidence for isomer III. Another possible isomer involves one free core oscillator, and the H_2O subunit binding equally to two of the core hydrogens via highly nonlinear hydrogen bonds involving the oxygen lone pairs.

2.3.6   Concluding Remarks

2.3.6.1   Aspects of Internal Rotation

It was not known whether nearly free internal rotation (barrier height $< 20 \, \text{cm}^{-1}$) would be observed in the hydrogen bonded ionic cluster species, which are relatively strongly bound cluster systems. The influence of internal rotation tunneling effects has been observed in many of the van der Waals molecules investigated to date, but relatively few species have been found to have barrier heights less than $100 \, \text{cm}^{-1}$, especially if we ignore clusters containing rare gas atoms or H_2 [76]. NH_3–CO_2 is one of the exceptions which falls into the category of nearly free internal rotators [77]. Internal rotation seems to have been observed in the spectrum of H_3^+; the band at $4250 \, \text{cm}^{-1}$ must be the v

$= 1, J = 2 \leftarrow v = 0, J = 0$ transition of the loosely bound H_2 subunit (see Ref. 48 for the spectrum). Due to the large rotational energy of $> 300 \text{ cm}^{-1}$ in the upper state, however, the barrier to internal rotation of H_2 in this case could conceivably be 100 cm^{-1} or more – the dependence of such transitions upon the barrier height is not rigorously known. It is likely that the majority of ionic clusters containing loosely bound H_2 will undergo a similar transition. It is interesting to note that the barrier to rotation of H_2 in crystalline (solid) H_2 is on the order of 1 cm^{-1} [78].

For the smallest clusters, internal rotation of the subgroups can be given the same sort of rigorous theoretical treatment previously applied to analogous neutral molecules. The spectrum of the $n = 1$ ammoniated ammonium ion has been modeled with the same formalism applied to molecules like dimethyl-acetylene, where the methyl groups rotate relative to each other through a three-fold potential [58, 79]. Fig. 9 shows the results of this modeling for $NH_4^+ (NH_3)$. It was found that in the case of $NH_4^+ (NH_3)$ the internal rotation is nearly free ($< 10 \text{ cm}^{-1}$ barrier height to the motion) and that the spectrum at 1 cm^{-1} resolution is consistent with a symmetric equilibrium structure, although it was not shown conclusively that the symmetric structure corresponds to a potential minimum.

The discussion of hydrated hydronium spectra in section 2.3.5.2 is conspic-uous for the absence of any mention of internal rotation. Indeed, the internal rotation barrier for H_2O in hydrated hydronium is clearly much higher than it is in $NH_4^+ (NH_3)_3 H_2O$ or for ammonia in the $NH_4^+ (NH_3)_n$ complexes ($n \leq 7$). On the other hand, the barrier is very likely to be less than several hundred

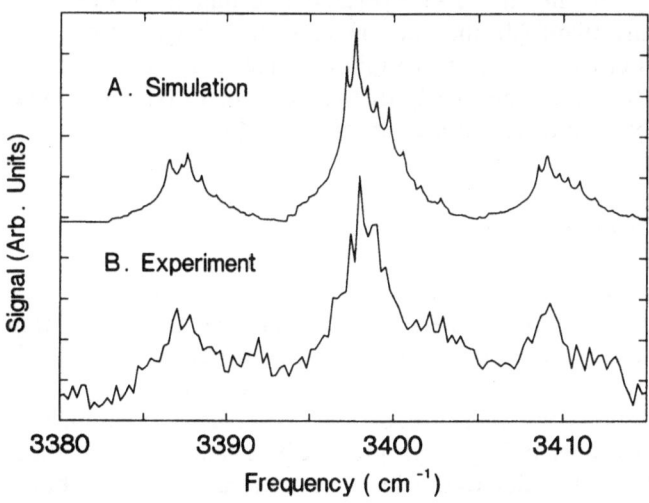

Fig. 9. Simulated and experimental spectrum of $N_2H_7^+$ for the D_{3h} equilibrium structure. The simulation includes only Q branch transitions, since these are expected to dominate the spectrum. Rotational constants used in the simulation: $A' = 3.20 \text{ cm}^{-1}$, $A'' = 3.16 \text{ cm}^{-1}$, $B' = B''$ $= 0.16 \text{ cm}^{-1}$, $T = 30 \text{ K}$, $\Delta V = -1.0$, $\zeta = 0.0$

wavenumbers, and there is even a hint in the spectrum that the excitation of a low frequency torsional vibration involving H_2O may result in a greatly increased tunneling rate [62]. The presence of a nonbonding electron pair in the p_x and p_y orbitals of water may be related to the fact that the 1° NH_3 solvent subunits in the ammoniated ammonium species rotate freely ($n \leq 7$) but the waters in hydrated hydronium do not. Since the H_2O in $NH_4^+(NH_3)_3H_2O$ clearly rotates with a very low barrier, the interaction of a single H_2O subunit with the rest of the ammoniated ammonium complex is not sufficient to create a substantial barrier. When additional NH_3 subunits are replaced by H_2O, however, the features of the H_2O antisymmetric stretching band are broadened considerably. The system with $n + m = 3$ ($NH_4^+(NH_3)_n(H_2O)_m$) exhibits a very similar phenomenon, although the symmetry of the ion is different. When all three or four subunits are H_2O the broadening is greatest. The broadening and complication of the structure is presumably associated with a more complex potential function and a higher barrier height. It seems to be the case, therefore, that the H_2O subunits interact with each other. This interaction could be primarily through their strong dipole moments, for example, making the hydrogen bonds slightly nonlinear and therefore complicating the internal rotation, or it might be more directly related to the electronic structure as already mentioned. In any case, the barrier even in $NH_4^+(H_2O)_4$ is clearly lower than in $H_3O^+(H_2O)_4$ or the other hydrated hydronium species, indicating that other factors also play a role. One very important factor is the symmetry with respect to the internal rotation. Six-fold barriers are inevitably lower than three-fold or two-fold barriers [80]. In the $n = 3, m = 1$ cluster the barrier with respect to H_2O rotation is six-fold, provided that we neglect to consider the hydrogens of the other subunits. For the $m = 1-3$ hydrated hydronium clusters, the analogous barrier is only two-fold, and the higher barrier height is not unexpected. Since the barrier can be either two-fold or three-fold in ammoniated ammonium complexes which exhibit nearly free rotation of 1° NH_3 subunits, both symmetry and subunit interaction are important factors.

2.3.6.2   The Transition from the Gas to the Liquid Phase

The properties of clusters evolve with size and become similar to those of bulk materials in the limit of very large clusters (see, for example, Sect. III of this volume). Trends in the measured vibrational frequencies are observed for both core and solvent vibrations in all of the systems investigated in this work. For the ammoniated ammonium ions, these trends are particularly apparent. For the most part, observed features undergo a progressive blue shift as the solvation number, n, increases. This shift is due to the decrease in average hydrogen bond energy for a given bond type as the solvation number is increased. Delocalization of the charge over the cluster is the primary factor which reduces the "per-ligand" interaction strength with the ionic core of the cluster. In the limit of complete dispersal of the charge (large cluster limit), hydrogen bonding energies of successive subunits approach a constant value, which is the intrinsic hydrogen bonding energy of that subunit with the subunit(s) to which it attaches. Since the hydrogen bonding interaction of H_2O is intrinsically much higher than NH_3, the convergence can be expected to be relatively slow for hydrated hydronium as compared to ammoniated ammonium.

The v_3 antisymmetric stretching band for the solvent molecules in $NH_4^+(NH_3)_n$ was first observed in the $n = 1$ cluster. Plotted in Fig. 10a are the frequencies of the central $^RQ_0$ bands in the internal rotation structure for this transition as a function of cluster size. Convergence of this plot is observed with increasing n, as is a discontinuity in the curve at the $n = 5$ cluster where the second solvation shell of the system begins to form.

Similar plots for the core vibrational features appear in Figs 10b and 10c. Figure 10b shows the position of the ammonium core N–H stretch with attached 1° NH_3 (The first point in this plot, labeled with a square, is from *Schwarz*'s measurement of the $n = 2$ cluster [38].) Fig. 10c shows the position of the bands of the same oscillators with associated 1° *and* 2° NH_3's. Both of these features converge at large n.

For the hydrogen bonded ionic clusters considered here, there is no precise analogy of the gas phase cluster to the bulk phase, which always contains counter ions. For weak counter ions and/or high molar ratios of solvent to ion core in the bulk, the possibility remains that properties can be somewhat similar

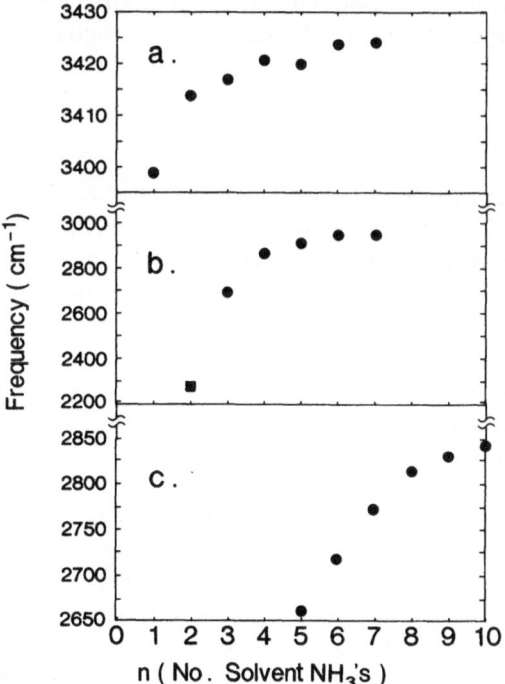

Fig. 10. a Plot of the $^RQ_0$ bands for the internal rotation structure associated with the 1° NH_3's antisymmetric stretching transition as a function of cluster size. **b** Plot of the peak frequency of the antisymmetric stretching band for the ammonium ion core bound only to 1° NH_3 molecules as a function of cluster size. The series shows eventual convergence as n increases. See text for discussion. **c** Plot of the peak frequency of the antisymmetric stretching band for the ammonium ion core bound to both 1° and 2° NH_3 molecules as a function of cluster size. Converged values are similar to those observed for some solutions of ammonium salts dissolved in liquid ammonia. See text for discussion

at least for portions of the gas phase cluster. The environments of the ion core and innermost solvent shells can be expected to approach their bulk counterparts more closely than outermost solvent shells, especially when the counter ion is excluded from close proximity to the core and inner shell solvent species. The spectrum of $n = 8$, which is quite similar to the spectra of $n = 9$ and 10 in the regions where they overlap, shows definite similarity to the spectra of $NH_4^+ClO_4^-$ in NH_3 with a $1:3$ mole ratio [81]. When the ratio of NH_3 is higher, even the salt with the strongly interacting Cl^- counter ion has spectra similar to the gas phase ionic cluster. Most of the vibrational bands show a correspondence which is close enough to assign the solution phase spectra. The most intense band in the solution phase spectrum is centered at about 2800 cm^{-1}, and can be assigned by this means to a vibration of an extensively NH_3 solvated NH_4^+ ion with a mode closely analogous to the intense v_3 mode of the $n = 8$ ion core. For $n \geq 7$ in the gas phase cluster, $\Delta v/\Delta n$ is rather small (see Fig. 10c), indicating convergence. For the $NH_4^+ClO_4^-$ in NH_3 with a $1:3$ mole ratio, this may mean that NH_4^+ ions are connected by a bridge of NH_3 solvent molecules. It seems that for the large ammoniated ammonium ions, from the standpoint of the ion core and to some extent the first solvent shell, a liquid-like environment is reached by about the $n = 8$ cluster. Of course, the gas phase environment is still much less fluxional – for example, hydrogen bonds are not being continuously broken and reformed.

Hydrated hydronium spectra, even at $m = 8$, show relatively little resemblance to aqueous acid solutions. The larger hydrogen bonding interaction and greater degree of ion-induced structuring evidently dictate comparison with much larger gas phase clusters if one is to see good spectral correspondence.

2.3.6.3 Future Developments

We have described the spectroscopic study of only a small subset of the ionic clusters that may be probed in the 2.5–3.9 μm region of the spectrum, at lower resolution than what is routinely possible now. Enhanced laser resolution and improved power from newer pulsed dye laser systems makes possible studies of the smaller cluster ions $n, m = 1$–4 at high resolution (< 0.03 cm^{-1}). Studies of this sort should provide fully resolved vibration-rotation spectra for some of these species. Analysis of these spectra will yield improved geometries and provide essential information regarding the position of the proton between solvent molecules, and the degree of distortion that the substituents undergo due to the various interactions.

MAMICS is, of course, not limited to the study of hydrogen bonded species; certainly the majority of the hundreds of cluster ions listed in compilations such as Castleman's [2] are amenable to the technique. Spectroscopy of species such as $Na^+(H_2O)_n$ is an obvious development which requires minimal change to the MAMICS apparatus. This species presumably becomes hydrogen bonding after the first solvent shell is filled. While it is somewhat less sensitive, the depletion detection technique of *Lisy* using a line tunable CO_2 laser has already demonstrated that $Cs^+(CH_3OH)_n$ has a first solvent shell size of $n = 10$ [51]. The $(H_2O)_n e^-$ species or hydrated electron is another example of a charged cluster system of considerable chemical interest.

MAMICS is not at all limited to the infrared, and we can envision it being performed from vacuum ultraviolet to 20 μm with resolution and accuracy better than 0.05 cm^{-1}. Nor is MAMICS necessarily limited to the study of the $m < 500$ amu species described in this article. With the use of a more suitable ion source and improved mass analysis, we envision the MAMICS study of large systems, possibly even molecules of biological interest such as solvated peptides of 10000 amu molecular weight!

Our understanding of the way atoms and molecules are held "by the entanglement of their own interlocking shapes" has been greatly enhanced in recent years by the study of cluster systems. Studying the infrared spectroscopy of ionic clusters has been shown to be a valuable way of learning about the organization of molecules around an ion, and the onset of solution phase properties. The MAMICS technique is a useful method for obtaining this information.

Acknowledgements. The authors would like to thank Dr. Gereon Niedner-Schatteburg for a critical reading of the manuscript. His helpful comments were greatly appreciated.

This work was supported by the Director, Office of Energy Research, Office of Basic Energy Sciences, Chemical Sciences Division of the U.S. Department of Energy, under Contract No. DE-AC03-76SF00098.

References

1. T.L. Carus: *On the Nature of the Universe*, trans. R.E. Latham, Ed. B. Radice (New York: Penguin Books, 1956)
2. R.G. Keesee, A.W. Castleman, Jr.: J. Phys. Chem. Ref. Data, Vol. 15, No. 8 (1986)
3. See Refs. 17, 19, and 23
4. J.V. Coe, R.J. Saykally: Infrared Laser Spectroscopy of Molecular Ions in *Ion and Cluster Ion Spectroscopy and Structure*, ed. J.P. Maier (Elsevier, Amsterdam, 1989)
5. H. Heitmann, F. Arnold: Nature **306**, 747 (1983)
6. R.S. Narcisi, A.D. Bailey: J. Geophys. Res. **70**, 3687 (1965)
7. E.E. Ferguson, F.C. Fehsenfeld, and D.L. Albritton, in *Gas Phase Ion Chemistry*, Vol. 1, M.T. Bowers, ed. (Academic Press, New York, 1979)
8. D.S. Smith, N.G. Adams: Top. Curr. Chem. **89**, 1 (1980); R.G. Keesee, A.W. Castleman, Jr.: J. Geophys. Res. **90**, 5885 (1985)
9. A.W. Castleman, Jr., P.M. Holland, R.G. Keesee: J. Chem. Phys. **68**, 1760 (1978); A.W. Castleman, Jr.: Adv. Colloid and Interface Science **10**, 73 (1979)
10. T. Heinis, S. Chowdhury, S.L. Scott, P. Kebarle: J. Am. Chem. Soc. **110**, 400 (1988); R.P. Grese, R.L. Cerny, M.L. Gross, **111**, 2835 (1989)
11. Y.K. Lau, S. Ikuta, P. Kebarle: J. Am. Chem. Soc. **104**, 1462 (1982); A.J. Cunningham, J.D.

Payzant, P. Kebarle, **94**, 7627 (1972); P. Kebarle, S.K. Searles, A. Zolla, J. Scarborough, M. Arshadi, **89**, 6393 (1967)

12. M. Meot-Ner, F.H. Field: J. Am. Chem. Soc. **99**, 998 (1977)
13. J.C. Gary, P. Kebarle: J. Phys. Chem. **94**, 5184 (1990)
14. M. Meot-Ner, L.W. Sieck: J. Phys. Chem. **90**, 6687 (1986); L.W. Sieck: J. Phys. Chem. **89**, 5552 (1985)
15. I. Dzidic, P. Kebarle: J. Phys. Chem. **74**, 1466 (1970)
16. M. Alcami, O. Mo, M. Yanez, F. Anvia, R.W. Taft: J. Phys. Chem. **94**, 4796 (1990)
17. "Calculated Vibrational Spectra of Hydrogen Bonded Systems" by P. Janoschek in *The Hydrogen Bond*, Vol. I, ed. by P. Schuster, G. Zundel, C. Sandorfy (North-Holland, New York), 1976, p. 183
18. *The Hydrogen Bond*, Vols I, II, and III, ed. by P. Schuster G. Zundel, C. Sandorfy (North-Holland Publishing, Amsterdam 1976)
19. S. Scheiner: Acc. Chem. Res. **18**, 174 (1985), and references therein
20. R.A. Copeland, S.I. Chan: Annu. Rev. Phys. Chem. **40**, 671 (1989); F.M. Koswer, D. Huppert: Annu. Rev. Phys. Chem. **37**, 127 (1986)
21. G.C. Pimentel, M.O. Bulanin, M. Van Theil: J. Chem. Phys. **36**, 500 (1962)
22. E.F. Caldin, V. Gold, (eds.): *Proton Transfer Reactions* (Chapman and Hall, London), 1975
23. C.I. Ratcliffe, D.E. Irish, in *Water Science Reviews* 2, ed. by F. Franks (Cambridge University Press, Cambridge 1986)
24. J. Corset, P.V. Huong, J. Lascombe: Spectrochim. Acta A **24**, 2045 (1968)
25. D.D. Nelson, Jr., G.T. Fraser, W. Klemperer: J. Chem. Phys. **83**, 6201 (1985)
26. *Water Science Reviews* 3, F. Franks, ed. (Cambridge University Press, Cambridge 1988)
27. T.R. Dyke, J.S. Muenter: J. Chem. Phys. **57**, 5011 (1972); Z.S. Hyang, R.E. Miller: J. Chem. Phys. **88**, 8008 (1988)
28. K.L. Busarow, R.C. Cohen, G.A. Blake, K.B. Laughlin, Y.T. Lee, R.J. Saykally: J. Chem. Phys. **90**, 3937 (1989)
29. L.H. Coudert, J.T. Hougen: J. Mol. Spectr. **130**, 86 (1988); L.H. Coudert, F.J. Lovas, R.D. Suenram, J.T. Hougen: J. Chem. Phys. **87**, 6290 (1987); J.T. Hougen: J. Mol. Spectr. **114**, 395 (1985)
30. S.E. Novick, K.R. Leopold, W. Klemperer: *The Structures of Weakly Bound Complexes as Elucidated by Microwave and Infrared Spectroscopy*, ed. by E.R. Bernstein (Elsevier, New York, 1989)
31. R.E. Miller: J. Phys. Chem. **90**, 3301 (1986); *Structure and Dynamics of Weakly Bound Molecular Clusters*, ed. A. Weber, NATO ASI Series (Reidel, Dordrecht, 1987); D.J. Nesbitt: Chem. Rev. **88**, 843 (1988); M.A. Duncan, D.H. Rouvray: Sci. Am. **261**, 110 (1989)
32. M.F. Vernon: Ph.D. Thesis, University of California, Berkeley, 1982
33. M.F. Vernon, D.J. Krajnovigh, H.S. Kwok, L.M. Lisy, Y.R. Shen, Y.T. Lee: J. Chem. Phys. **77**, 47 (1982)
34. R.H. Page, M.F. Vernon, Y.R. Shen, Y.T. Lee: Chem. Phys. Lett., Vol. 141, No. 1, 2, 1 (1987)
35. D.F. Coker, R.E. Miller, R.O. Watts: J. Chem. Phys. **82**, 3554 (1984)
36. U. Buck, C. Lauenstein: J. Chem. Phys. **92**, 4250 (1990)
37. C.S. Gudeman, R.J. Saykally: Ann. Rev. Phys. Chem. **35**, 387 (1984)
38. H.A. Schwarz: J. Chem. Phys. **67**, 5525 (1977)
39. H.A. Schwarz: J. Chem. Phys. **72**, 284 (1980)
40. M.W. Crofton, M.-F. Jagod, B.D. Rehfuss, T. Oka: J. Chem. Phys. **91**, 5139 (1989)
41. W.H. Wing, G.A. Ruff, W.E. Lamb, Jr., J.J. Spezeski: Phys. Rev. Lett. **36**, 1488 (1976); and ongoing experiments in the R.J. Saykally laboratory at University of California, Berkeley
42. N.E. Levinger, D. Ray, M.L. Alexander, W.C. Lineberger: J. Chem. Phys. **89**, 5654 (1988)
43. L.A. Posey, M.A. Johnson: J. Chem. Phys. **89**, 4807 (1988); M.J. DeLuca, B. Niu, M.A. Johnson: J. Chem. Phys. **88**, 5857 (1988)
44. J.T. Snodgrass, J.V. Coe, C.B. Freidhoff, K.M. McHugh, K.H. Bowen: J. Chem. Phys. **88**, 8014 (1988)
45. C.Y. Kung, T.A. Miller: J. Chem. Phys. **92**, 3297 (1990)

46. J.P. Maier, ed.: *Ion and Cluster Ion Spectroscopy and Structure* (Elsevier, Amsterdam, 1989)
47. J. Steadman, J. Syage: J. Chem. Phys. **92**, 4630 (1990); J.J. Breen, L.W. Peng, D.M. Willberg, A. Heikal, P. Cong, A.H. Zewail: J. Chem. Phys. **92** 805 (1990)
48. M. Okumura, L.I. Yeh, Y.T. Lee: J. Chem. Phys. **88**, 79 (1988)
49. L.I. Yeh, M. Okumura, J.D. Myers, J.M. Price, Y.T. Lee: J. Chem. Phys. **91**, 7319 (1989)
50. M. Okumura, L.I. Yeh, J.D. Meyers, Y.T. Lee: J. Phys. Chem. **94**, 3416 (1990)
51. W.-L. Liu, J.M. Lisy: J. Chem. Phys. **89**, 605 (1988); J.A. Draves, Z. Luthey-Schulten, W.-L. Liu, J.M. Lisy: J. Chem. Phys. **93**, 4589 (1990).
52. G.C. Bjorklund: Opt. Lett. **5**, 15 (1980); D.J. Nesbitt, H. Petek, C.S. Gudeman, C.B. Moore, R.J. Saykally: J. Chem. Phys. **81**, 5281 (1984)
53. M. Okumura, Ph.D. Thesis, University of California, Berkeley (1986)
54. L.I. Yeh, Ph.D. Thesis, University of California, Berkeley (1988); Engleking, P.C.: Rev. Sci. Instrum. **57**, 2274 (1986)
55. M. Yamashita, J.B. Fenn: J. Phys. Chem. **88**, 4451 (1984); **88**, 4671 (1984)
56. E. Teloy, D. Gerlich: Chem. Phys. **4**, 417 (1974); D. Gerlich, *Electronic and Atomic Collisions*, ed. D.C. Lorents, W.E. Meyerhof, J.R. Peterson (North-Holland press, Amsterdam 1985)
57. M. Okumura, L.I. Yeh, Y.T. Lee: J. Chem. Phys. **83**, 3705 (1985); **88**, 79 (1988)
58. J.M. Price, M.W. Crofton, Y.T. Lee: J. Chem. Phys. **91**, 2749 (1989); J.M. Price, M.W. Crofton, Y.T. Lee: J. Phys. Chem. **95**, 2182 (1990)
59. M.W. Crofton, T. Oka: J. Chem. Phys. **79**, 3157 (1983); **86**, 5983 (1987)
60. E. Schafer, R.J. Saykally: J. Chem. Phys. **79**, 3159 (1983); E. Schafer, R.J. Saykally, A.G. Robiette: J. Chem. Phys. **80**, 3969 (1984)
61. W.S. Benedict, E.K. Phyler, E.D. Tidwell: J. Chem. Phys. **32**, 32 (1960)
62. M.W. Crofton, J.M. Price, G. Niedner-Schatteburg, D.W. Boo, Y.T. Lee: manuscript in preparation
63. W.L. Jorgensen: J. Am. Chem. Soc. **103**, 341 (1981); J. Chandrasekhar, W.L. Jorgensen: J. Chem. Phys. **77**, 5080 (1982)
64. E. Kochanski: J. Am. Chem. Soc. **107**, 7869 (1985)
65. G.G. Malenkov: The Chemical Physics of Solvation. in *Studies in Physical and Theoretical Chemistry*, ed. by R.R. Dogonadze, E. Kalman, A.A. Kornyshev, J. Ulstrup (Elsevier, New York, 1984)
66. B. DeRaedt, M. Sprik, M.L. Klein: J. Chem. Phys. **80**, 5719 (1984); R.W. Impey, M. Sprik, M.L. Klein: J. Am. Chem. Soc. **109**, 5900 (1987)
67. M. Mezei, D.L. Beveridge: J. Chem. Phys. **74**, 6902 (1981); M. Rao, B.J. Berne: J. Phys. Chem. **85**, 1498 (1981)
68. R. Remington, H.F. Schaefer III., unpublished results
69. M.D. Newton: J. Chem. Phys. **67**, 5535 (1977)
70. M.D. Newton, S. Ehrenson: J. Am. Chem. Soc. **93**, 4971 (1971)
71. See refs 32–35
72. J.M. Price, M.W. Crofton, Y.T. Lee: unpublished results
73. S.G. Lias, J.F. Liebman, R.D. Levin: J. Phys. Chem. Ref. Data **13**, 695 (1984)
74. A. Pullman, A.M. Armbruster, Chem. Phys. Lett. **36**, 558 (1975); Intern. J. Quantum Chem. **8S**, 169 (1974)
75. G. Cazzoli, P.G. Favero, D.G. Lister, A.C. Legon, D.J. Millen, Z. Kisiel: Chem. Phys. Lett. **117**, 543 (1985), and references contained therein
76. A.R.W. McKellar: J. Chem. Phys. **93**, 18 (1990), and references contained therein
77. G.T. Fraser, K.R. Leopold, W. Klemperer: J. Chem. Phys. **81**, 2577 (1984), see also: K.I. Peterson, W. Klemperer, J. Chem. Phys. **80**, 2439 (1983); **85**, 725 (1986)
78. M. Okumura, M.-C. Chan, T. Oka: Phys. Rev. Lett. **62**, 32 (1989); J. van Kranendonk: *Solid Hydrogen* (Elsevier, Amsterdam, 1983)
79. W.B. Olson, D. Papousek: J. Mol. Spectr. **37**, 527 (1971), and references contained therein
80. C.H. Townes, A.L. Schawlow: *Microwave Spectroscopy* (McGraw-Hill, New York, 1955); N.L. Owen: Studies of Internal Rotation by Microwave Spectroscopy, in *Internal Rotation in Molecules*, ed. W. J. Orville-Thomas (Wiley, New York, 1974)
81. J. Corset, P.V. Huond, J. Lascombe, Spectrochim. Acta A **24**, 2045 (1968)

2.4 Solvated Cluster Ions

A.W. Castleman, Jr.

2.4.1 Introduction

One of the most challenging and scientifically exciting problems in the field of chemical physics is to elucidate the influence which solvation, degree of aggregation, and the proximity of solute complexes have on the reactions and properties of reactive species, to contrast differences in their behavior between the gaseous and condensed phase, and to attempt to relate the large body of detailed information known about the gas phase to understanding phenomena in the condensed state [1–10]. The results of studies of the interaction of ions and molecules provide information on the forces involved, and gas-phase studies of interactions within a cluster can also yield insight into the structure and bonding of complexes having analogies to those existing in solutions of electrolytes, for example.

In addition to their application in elucidating the molecular details of condensed matter, studies on ionic clusters also serve to reveal the microscopic aspects of nucleation phenomena [11–14], including the formation of highly dispersed media having a large surface-to-volume ratio, e.g., aerosol particles [13, 15]. Investigation of both the thermochemical properties as well as the kinetics of association reactions has been particularly important in the subject of phase transitions, where progress has been impeded by lack of fundamental data for comparison with molecular theories [11, 12, 14]. Related both to studies of condensed phases as well as surfaces are investigations involving the scattering of high-energy neutrals and ionic particles from surfaces, where cluster ions are often the observed reaction products [16]. Determining factors which influence their size, stability, and mechanisms of formation provide a basis for explaining the results of such experiments.

Ionic clusters are observed in flames and related combustion processes [17], in both the lower and upper atmosphere which may be viewed as a weakly ionized plasma surrounding the earth [18, 19], in certain regions of outer space [20], and in fact in most situations where matter is exposed to ionization and there are collisions of sufficient number and duration to enable stable complexes to be formed either by three-body association reactions or through radiative transitions. Several other areas where consideration of clustering about ions is important include the effluents of nuclear reactors, fusion and magnetohyd-rodynamic devices [21], and even the heat exchangers of high-temperature energy sources where cluster ions are suggested to originate as part of a gas-phase corrosion phenomenon [22, 23].

In addition to providing details on the energetics of interactions, studies of cluster ions and of the "ensuing solvation" which takes place upon ionizing a moiety within a cluster yield information on basic mechanisms of ion reactions in the cluster and also give considerable insight into those which can occur in the bulk liquid state. The entire course of a chemical reaction following either a photophysical or ionizing event, depends on the mechanisms of energy transfer and dissipation away from the primary site of absorption. Neighboring solvent or solute molecules can influence this by collisional deactivation (removal of energy), through effects in which dissociating molecules are kept in relatively close proximity for comparatively long periods of time due to the presence of the solvent, and in other ways where the solvent influences the energetics of the reaction coordinate. Through the use of supersonic molecular beams, it is now possible to produce and tailor the composition of virtually any system of interest. Utilizing these methods, coupled with laser spectroscopy, one can selectively solvate a given chromophore (site of photon absorption/site of ionization) and investigate changes between the gas and the condensed phase by selectively shifting the degree of solvent aggregation, i.e., the number and location of solvent molecules attached to or bound about the site of absorption of the electromagnetic radiation [1].

As discussed in what follows, through cluster work it is becoming possible to unravel several important phenomena which demonstrate the ability to eluci-date basic processes in condensed matter in terms of their gas-phase counter-parts. Spectroscopic results have shown the influence which solvation has on the electronic properties of molecules, and such methods have been used to investi-gate the transition from the gas to the condensed state. In particular, recent results have shown that it is possible to relate the bonding of gas-phase complexes to those of analogous complexes present in the condensed phase and to determine the energetics involved in their solvation [7]. Importantly, a wide range of studies have demonstrated that reactions proceed in complexes follow-ing an initial ionization event, which in many cases very closely parallel those of gas-phase ion–molecule reactions [1, 6, 9]. In several important examples, it has been possible to identify very specific reactions which are solvation driven and that have direct analogies to those taking place in the liquid phase [1]. In addition to the relationship to liquid reactions, cluster ion research also pro-vides detailed insight into some molecular aspects of surface chemistry and even the electronic properties of metals. These aspects of the subject will be briefly discussed herein as well.

2.4.2   Experimental Techniques

As with many other areas of chemistry, the advent of new experimental and theoretical techniques have paved the way for the advances which have been made in understanding the dynamics and properties of ionic clusters. For instance, the flowing afterglow, stationary afterglow, and ion cyclotron resonance methods have been extensively utilized to investigate association reactions responsible for cluster ion formation, as well as the reactivities of ionic clusters through such processes as ligand exchange reactions. The aforementioned techniques also have been employed to derive thermochemical properties, although the most extensive sets of thermodynamic data have been obtained via the high-pressure mass spectrometer technique [7, 24, 25]. Through the use of lasers, studies of cluster ion dissociation spectroscopy and processes involved in the dynamics of photodissociation are now being unraveled at the molecular level [26–29]. Concomitant investigations of cluster ion unimolecular and collision-induced dissociation are further contributing to an understanding of dynamical processes involved in energy transfer and reactivity [30–32]. Finally, ab initio calculations are contributing to an understanding of the structure and bonding of both strongly and weakly bound cluster ions.

2.4.2.1   Molecular Beam-Photoionization-Time-of-Flight Mass Spectrometry

The time-of-flight (TOF) mass spectrometer technique is experiencing a resurgence in popularity due to the advent of pulsed lasers which supply efficient and short durations of ionization in a small volume, and fast-timing electronic circuitry. In a typical time-of-flight mass spectrometer, either a two element or alternately a single element accelerating field may be used in the region of ionization. This is followed by a field-free drift region, whereafter the ions are detected. Using the conventional TOF method, dissociation which occurs with rates in the neighborhood of 10^5–10^8 s^{-1} can be investigated by either of two methods. One involves analyzing the peak shape (arrival spectrum) of ions created in a dual field accelerating *Wiley–McLaren* [33], in which cases a knee is observed [34] since the ions spend far more time in the first low-field region where ionization is initiated, than in the second high-field region where the bulk of the acceleration occurs. An alternate method is to operate under single field conditions and deduce rates from the shape of the late arriving tail of the peak [35].

One of the most useful methods of studying dissociation employs a reflecting electrical field (reflectron). Although originally designed to enhance the resolution of the TOF method [36], a reflectron can also be employed to investigate dissociation in the field-free drift region, so that slower dissociation processes may be observed. Such experiments are performed by subjecting the cluster beam to multiphoton ionization, oftentimes using a tunable dye laser with various optical components that provide further frequency selection capabilities. The ions are accelerated in the accelerating field to several keV, whereafter they enter a field-free region and then are electrically reflected and detected in a manner such as the one depicted in Fig. 1. With appropriate potentials applied to the reflectron grids, non-dissociating parent ions can be separated from those that dissociate while within the field-free region. A unique identification of these daughter ions can be accomplished by the time separation and by an energy analysis with the reflectron. The separation of the parent and daughter ions is possible as a result of the loss in kinetic energy with essentially no change in velocity of the cluster ion packet upon dissociation, whereby the parent species with greater kinetic energy have a longer path to the detector than do the daughter (dissociation) products.

Supersonic expansion techniques including both continuous sources as well as pulsed jets are commonly used to produce beams of neutral clusters. In both cases cooling of the beam is accomplished through the conversion of the random thermal energy of a high pressure source gas into a directed beam velocity [37]. Since the latent heat of condensation which is released during the clustering process leads to internal vibrational and rotational heating of the aggregate, clusters do not generally attain temperatures as low as unclustered species. But, cooling collisions with an inert gas serve to reduce the internal temperature of the cluster and enable ones to be produced that have sufficiently long lifetimes to be interrogated in an experiment.

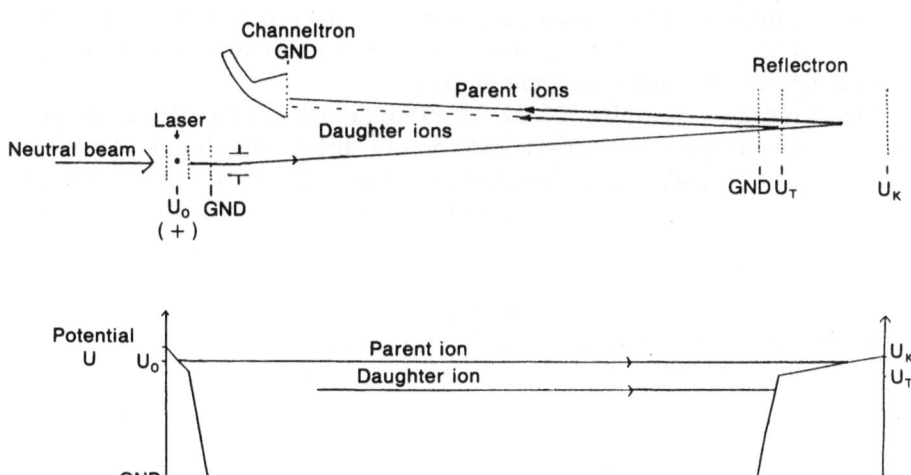

Fig. 1. Schematic of time-of-flight mass spectrometer with reflectron. Lower figure depicts the electrostatic potentials. With a judicious selection of potentials the daughter ions arising from metastable decay arrive at the detector prior to the parent ions which have higher kinetic energy; adapted from [72]

2.4.2.2 Technique for Deducing Cluster Bond Energies from Unimolecular Dissociation Studies

The general versatility of the evaporative ensemble approach for deducing bond energies was first demonstrated [31, 32] for ammonia cluster ions, though it is a technique of general applicability. Following ionization of neutral clusters, ammonia for example, internal ion–molecule reactions lead to specific ion clusters. These, which are comprised of protonated ammonia in the example under discussion, undergo rapid fragmentation and dissociation in the TOF region (a time window of typically no more than 1 μs) and, thereafter undergo subsequent evaporative cooling through dissociation which extends to longer times. A general cluster ion evaporative dissociation process can be expressed as

$$IL_n \rightarrow IL_{n-x} + xL \; , \tag{1}$$

Here, I designates the ion core (NH_4^+ in the case of ammonia) and L the clustering ligand (e.g., NH_3). The intensity and width of the metastable ion peaks carry information on the internal energy of the parent cluster ions.

In the measurement of decay fractions of dissociating cluster ions, the parent and daughter ions are decelerated in the first region and reflected in the second field of the reflectron. Since the daughter ions have an energy of $U_d = (M_d/M_p)U_0$ as a result of metastable decomposition [U_0 is the birth potential and M_d and M_p are the masses of the daughter and parent, respectively], they do not penetrate into the reflective field as deeply as the corresponding parent ions, although a critical aspect of deducing accurate kinetic energy release and rate measurements is to vary potential settings on the second and last grids of the reflectron to cause parent and daughter ions to follow the same flight paths. The integrated intensities of the peaks are then used to compute the decay fraction of the original parent cluster.

The evaporative ensemble [31, 32, 38, 39] approach assumes that each cluster ion has undergone at least one evaporation before entering the field-free region of the time-of-flight mass spectrometer. For each cluster ion, t_0 is defined as the flight time that the parent ion spends from the point of ionization to the last TOF lens, whereas t is the flight time that the parent ion spends from the last TOF lens to the first grid of the reflectron unit. At time t, the remaining population of dissociating cluster ions is given by $P = P_0 - D$, where, P_0 is the population of parent ions at time t_0 and D is the population of daughter ions at time t. The evaporative ensemble predicts that the normalized population of daughter ions at time t is given by

$$D = (C_n/\gamma^2)\ln\{t/[t_0 + (t - t_0)\exp(-\gamma^2/C_n)]\} \; , \tag{2}$$

where C_n is the heat capacity of the cluster ion (in units of Boltzmann constant k_B), and γ is the Gspann parameter. Electron diffraction experiments of clusters containing many thousands of atoms suggest that γ is ~ 25, usually independent of cluster size [32, 38, 39]. However, for smaller clusters, the Gspann parameter requires modifications:

$$\gamma'^2 = \gamma^2/[1 - (\gamma/2C_n)^2] \ . \tag{3}$$

Replacing γ with the modified Gspann parameter γ' in Eq. (2), leads to

$$D = (C_n/\gamma'^2) \ln \{t/[t_0 + (t - t_0)\exp(-\gamma'^2/C_n)]\} \ . \tag{4}$$

For systems comprised of non-linear molecules, the heat capacity of the cluster ion of size n is chosen to be $6(n-1)k_B$ by considering (only) the cluster modes. The binding energy of a molecule in a cluster ion of size n can be calculated using the equation:

$$\Delta E_n = \gamma \langle E_r \rangle / [1 - (\gamma/2C_n)] \ . \tag{5}$$

2.4.2.3 Thermodynamics of Cluster Reactions via High Pressure Mass Spectrometry

Where both forward and reverse reactions are possible, cluster formation can be represented by the general reactions:

$$IL_{n-1} + L \rightleftharpoons IL_n \ . \tag{6}$$

Collisions of energetically activated intermediates in the clustering sequence with a third-body are necessary for stabilization of the complexes. Taking the standard state to be 1 atm, and making the usual assumptions [40] concerning ideal gas behavior and the proportionality of the chemical activity of an ion cluster to its measured intensity, the equilibrium constant $K_{n-1,n}$ for the nth clustering step is given by

$$\ln K_{n-1,n} = \ln \frac{[IL_n]}{[IL_{n-1}][P_L]} = -\frac{\Delta G^o_{n-1,n}}{RT} = -\frac{\Delta H^o_{n-1,n}}{RT} + \frac{\Delta S^o_{n-1,n}}{R} \ . \tag{7}$$

Here, $[IL_{n-1}]$ and $[IL_n]$ represent the respective measured ion intensities, P_L the pressure (in atm) of the clustering species L, $\Delta G^o_{n-1,n}$, $\Delta H^o_{n-1,n}$, and $\Delta S^o_{n-1,n}$ the standard Gibbs free energy, enthalpy, and entropy changes, respectively, R the gas-law constant, and T absolute temperature. By measuring the equilibrium constant $K_{n-1,n}$ as a function of temperature, the enthalpy and entropy change for each sequential association reaction can be obtained from the slope and intercept of the van't Hoff plot ($\ln K_{n-1,n}$ versus $1/T$). Alternatively, under some circumstances, thermodynamic information can be obtained by studying switching or exchange reactions [7, 25]. Experimental techniques that employ van't Hoff plots lead to enthalpy changes derived from slopes which are often representable as straight lines over moderate temperature ranges. In actual fact, enthalpy change is a weak function of temperature due to the difference in heat capacity, ΔC_p, between products and reactants.

$$\Delta H_{T_2} = \Delta H_{T_1} + \int_{T_1}^{T_2} \Delta C_p(T) \, \mathrm{d}T \ . \tag{8}$$

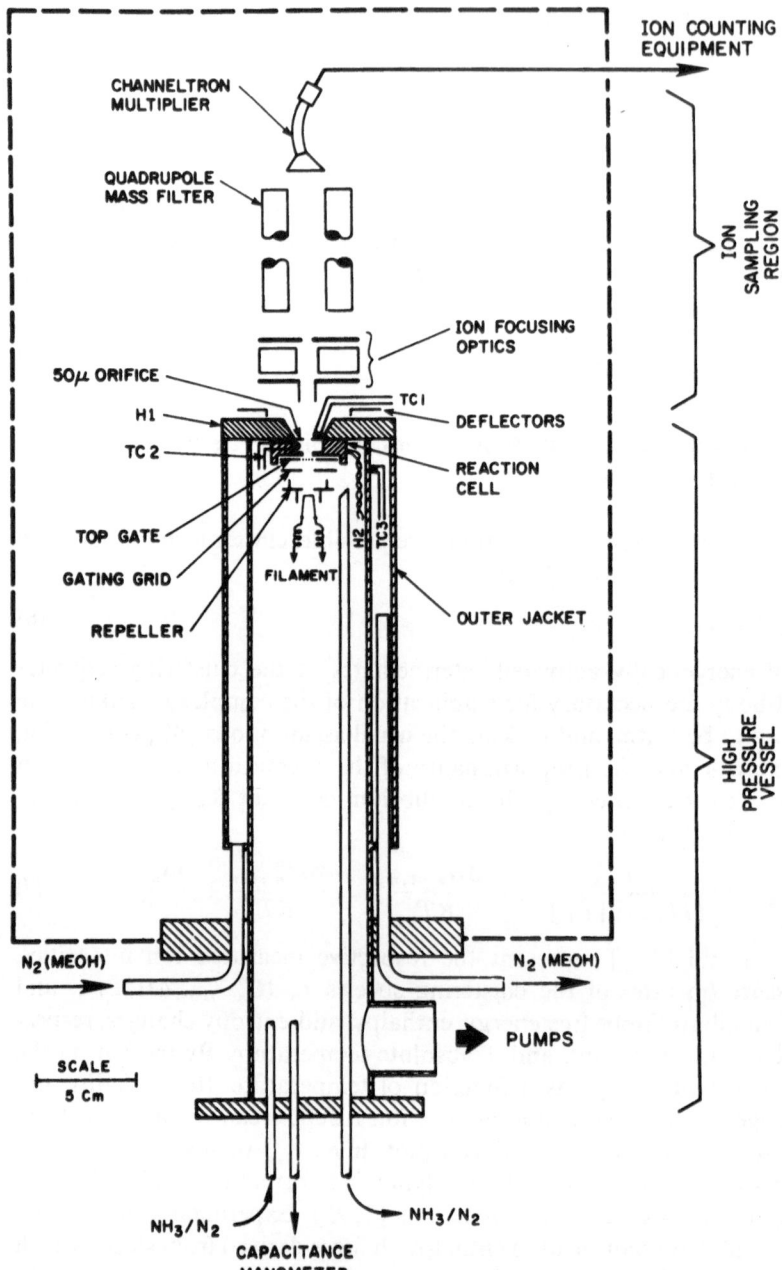

Fig. 2. Schematic of the ion clustering apparatus. The figure is drawn approximately to scale, with internal supports omitted for clarity, ---, a region of high vacuum. Positive ions are emitted from a thermionic emission source by application of a positive voltage. The ions are focused by means of a repeller assembly whose potential is set a few volts below that of the filament. Ion energies are further controlled by means of two electrodes, placed before and just inside the reaction cell. The potential of the "top gate" is adjusted to ensure field-free conditions in the reaction cell, where clustering reactions and the final equilibration processes take place. (Taken from [40])

Correction to a zero of temperature is necessary in order to obtain bond energies, but numerically these values are generally only slightly different from the measured enthalpies.

High pressure mass spectrometry (HPMS) [7, 24, 40] has been a particularly valuable tool in quantitatively determining the thermochemical properties of ion clusters. In this technique, ions effuse from a high pressure source such as the one shown in Fig. 2 (typically a few torr through a small aperture into a mass filter where the equilibrium distribution of ion clusters is determined). Ionization may be initiated by various methods including radioactive sources, heated filaments, electric discharges, or laser ionization. The pressure of the ion source is maintained sufficiently high such that ions reside in a region of well-defined temperature for a time adequate to ensure the attainment of equilibria among the various ion cluster species of interest, but the pressure and/or concentration of the clustering component must be low enough to avoid additional clustering via adiabatic expansion as the gas exits the sampling orifice.

In the case of positive ions, a thermionic emission source is often employed with the application of a positive voltage to a heated filament onto which the selected source material is dispersed. In the case of negative ions, the source is usually comprised of a platinum filament on which barium zirconate is dispersed. Upon heating, this filament becomes a copious source of electrons, whereupon desired anions are formed by associative or dissociative attachment. The ions are directed by means of a repeller assembly whose potential is set a few volts attractive relative to that of the filament, and their energies are further controlled by means of electrodes placed before and just after the reaction cell. The potential of the "top gate" is adjusted to establish field free conditions in the reaction cell where clustering reactions and the final equilibration processes take place. The temperature of the high thermal conductivity reaction cell (usually a gold plated copper block) is established by a combination of electrical heating wires and a jacket with circulating cooling or heating fluid surrounding the high pressure vessel. Thermocouples monitor the temperature of the reaction cell, the value of which must be well known. The high pressure vessel is mounted inside a vacuum chamber delineated by the broken line in the figure. Ions and cluster ions diffuse through a 50–75 μm hole in a metal disc into the vacuum chamber, pass through suitable ion optics for focusing the ionic species into the mass spectrometer, and quantified using pulse-counting techniques. Extensive tests such as measurements at different dilutions, pressures, carrier gas composition and electrical lens potentials are made to ensure that equilibrium distributions in the cell are properly determined.

2.4.2.4   Fast Flow Reactors for Studying Reaction Kinetics

The flowing afterglow technique (FA) developed by Ferguson, Fehsenfeld, and Schmeltekopf [41] and other related flow reactors such as the selected ion flow tube (SIFT) [42] have provided a wealth of data on general ion–molecule reactions [43, 44], with some attention to ion clusters. A typical fast flow apparatus is shown in Fig. 3; the flow tube is generally about 1 m long and 8 cm in diameter. Flow velocities are on the order of 10^2 ms^{-1} and pressures in the reactor region are typically around 1/2 torr. While most of the gas is pumped away, a small fraction is sampled through an orifice where the ions are mass identified and counted. Reactant gases are added uniformly into the flow, so kinetic data (or the approach to equilibrium) can be determined by varying the position of the rate of reactant into the tube, or the bulk flow velocity. In comparison to HPMS, the flow tube technique affords more versatility in making kinetic measurements and identifying mechanisms, whereas high pressure mass spectrometry is more amenable to temperature control and enables measurements at higher pressures. Generally the former is more useful for rate measurements while the HPMS technique is the preferred one for thermochemical determinations.

The ions or cluster ions are thermalized by collisions with an inert carrier gas (usually helium), although oftentimes argon or even nitrogen is employed. Neutral reactant gas is added through a reactant gas inlet (RGI) at an appropriate location downstream in the flow tube; and allowed to react with the injected ions. Ions on the flow tube axis are sampled through a small orifice where they are mass analyzed by a (second in the case where a first is used in parent selection) quadrupole mass spectrometer and detected. A large volume Roots

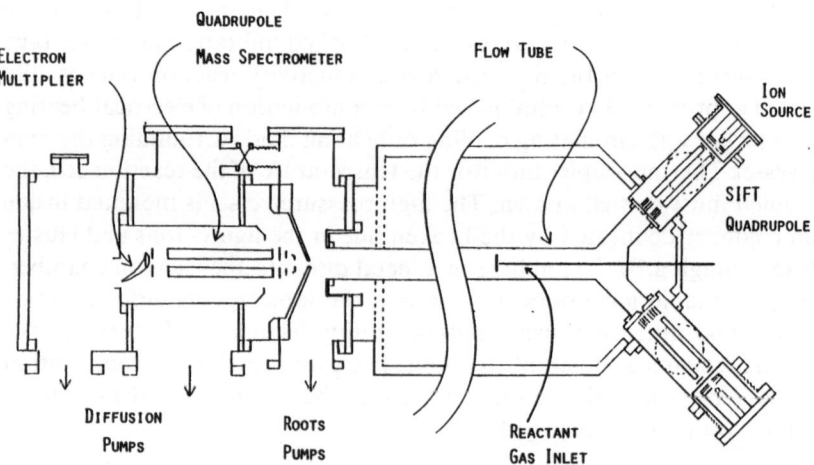

Fig. 3. SIFT-DRIFT fast-flow reaction kinetics apparatus. Ions or ion clusters are introduced into the flow tube from various sources and reactions proceed after they encounter the reactants added through a ring injector located at a selected position in the flow tube. The disappearance of the reactant ions and formation of products are monitored with the quadrupole mass spectrometer shown. An electric drift field can be positioned in the flow tube to investigate the influence of translational energy on the reaction. (From R.J. Shul, B.L. Upschulte, R. Passarella, R.G. Keesee and A.W. Castleman, Jr., J. Phys. Chem. **91**, 2556 (1987))

blower is used to maintain the flow velocity and pump away the bulk of the gas in the flow tube.

Rate coefficients, k, are determined by pseudo-first order kinetics in which the reactant ion concentration is small compared to the reactant neutral concentration. Bimolecular rate coefficients, k, are obtained from the slope of the natural logarithm of the measured signal intensity, I, of the reactant ion versus the flow rate Q_B of reactant gas [41, 45, 46]

$$\ln\left(\frac{I}{I_0}\right) = -\frac{kz\,Q_B}{V_i Q_c}\frac{P}{k_B T}, \tag{9}$$

where I_0 is the reactant ion intensity at Q_B equal to zero (no reactant gas flow) and is usually constant over the course of an experiment, z is the reaction distance (from reactant gas inlet to sampling orifice), Q_c the carrier gas flow rate, P the average pressure (or number density) in the flow tube, V_i the measured ion velocity, k_B the Boltzmann constant, and T the absolute temperature. The latter can be determined by applying a pulsed potential on the reactant gas inlet and measuring the arrival time of the resulting disturbance in the ion intensity. Where appropriate, termolecular rate coefficients [46] are determined from the slope of the apparent bimolecular rate coefficient plotted versus the pressure P.

In addition to the determination of rate coefficients, product identification and measurement of branching ratios can also be accomplished [41, 42, 47, 48]. Where studies of the influence of translational (center-of-mass) energy on cluster reactions are desired, a homogeneous electric field is established along the flow tube axis by a series of drift rings that are inserted in the flow tube. The reactant ions injected into the flow tube are transported into the drift region by the carrier gas and travel approximately 1/3 of the way down the flow tube before entering the drift field. Thereby they undergo numerous collisions with the carrier gas, assuring a thermal distribution of the ions as they enter the drift field. The ions reach a mean drift velocity in addition to the bulk flow velocity as they move under the influence of the electric field. A basic parameter of importance is the electric field energy (E/N), where E is the electric field intensity and N is the number density of the carrier gas particles. Varying either E or N changes the drift velocity, which is directly related to the center-of-mass kinetic energy of the reactants. Rate coefficients evaluated as a function of this energy can often be used to bridge the range studied individually in thermal and beam experiments. The center-of-mass kinetic energy is determined from the kinetic energy of the ion which, in turn, is evaluated from the Wannier equation [49]. The reliability of the Wannier equation for the evaluation of the ion kinetic energy has been verified by theoretical and experimental studies [50–52]. *Viehland* [53] has discussed how drift experiments can be employed to separate effects of internal and translational energy on ion–molecule reactions.

2.4.3   Reactions Influenced by Solvation

2.4.3.1   Reactions of Clusters Following Ionization

It has been recognized for a long time that, following the ionization of one moiety within either a single component or mixed cluster system, internal reactions proceed whose study can often facilitate unraveling similar processes in the condensed phase. The observed mechanisms also frequently bear direct analogy to those observed for isolated gas phase ion–molecule reactions; indeed, this analogy has provided a useful starting point for predicting the stable product ions resulting during the course of the dynamical events. In beam experiments, evaporative dissociation typically dominates at long times and current interest centers on unraveling the role, if any, of the ionization process and determining whether the rates of dissociation and the kinetic energy release are in accord with statistical theories of unimolecular dissociation. Klots [38, 39] has proposed an evaporative ensemble approach as a basis for understanding the commonality observed for systems of greatly disparate properties, and data on evaporative dissociation provide a test of his model.

A related subject of considerable interest in the field of cluster research concerns the origin of magic numbers, a term which has been used to describe the anomalous abundance of certain sizes of clusters in an otherwise smoothly varying trend of intensity versus size. Mass spectrometry is most often employed in cluster detection, sometimes of clusters formed from pre-ionized species, but more frequently of ones produced upon the ionization of preformed neutral aggregates. Whether the anomalous abundances can be attributed to especially stable structures in neutrals or to the inherent stabilities of the resulting cluster ions has been a controversial subject, but there is strong evidence that in most systems comprised of weakly bound neutrals the magic numbers arise due to especially stable structures of the (detected) cluster ions.

Among the first cluster magic numbers to be extensively discussed in the literature was the protonated 21-mer of water which had been observed in a number of different experiments including ones involving molecular beam expansions of protons in water vapor [54, 55], secondary ion emission from ice following bombardment with argon ions [56], and electron impact ionization of neutral water clusters [57–61]. The fact that this one anomalously abundant species is observed in both neutral and pre-ionized experiments is strongly suggestive that it arises because of the stability of the cluster ion, a fact which is borne out by findings [62, 63] under well defined thermal conditions that support the idea of thermodynamic stability [64]. Likewise, in studies of systems comprised of rare gas atoms, early experimenters ascribed the findings to particular stabilities of neutral clusters [65], but more recent work has further substantiated the importance of dissociation [66–70]. Despite mounting evidence about the importance of the cluster fragmentation upon ionization, the literature remains replete with references [71] to the importance of neutral structure stability in effecting the magic numbers observed in mass spectrometric studies of clusters.

Clusters of ammonia have become another test case for unraveling the factors governing the dynamics of dissociation and elucidating the origin of magic numbers. During the course of a systematic study of the dissociation of ammonia clusters following their ionization by laser multiphoton ionization techniques, direct proof was obtained of the importance of the metastability of the cluster ions and of the extensive dissociation processes which influence the resulting cluster distributions [72]. In particular, a combination of collision-induced and unimolecular (evaporative) loss processes was observed for a range of cluster sizes up to more than 50 molecules. Direct evidence was obtained for the loss of as many as six monomer units from the protonated 9-mer following its production, at least two monomer units being lost directly by evaporation with the others due to collisional processes. Although all ammonia clusters undergo fragmentation following ionization, only the protonated pentamer displays an anomalously large abundance in the locality of its size range [31, 32, 73].

An example of the usefulness of the reflectron technique discussed earlier in this chapter is seen for the case of ammonia clusters in Figs. 4a, b. Upon ionization, ammonia undergoes an internal ion-molecule reaction leading to protonated cluster ions, and concomitant evaporative unimolecular dissociation.

$$(NH_3)_n + h\nu \rightarrow NH_3^+(NH_3)_{n-1} + e , \tag{10}$$

$$NH_3^+(NH_3)_{n-1} \rightarrow [NH_4^+(NH_3)_{n-2}]^* + NH_2 , \tag{11}$$

$$[NH_4^+(NH_3)_{n-2}]^* \rightarrow NH_4^+(NH_3)_{n-2-m} + mNH_3 . \tag{12}$$

Figure 4a shows a conventional time-of-flight spectrum, obtained under total hard reflection conditions, while Fig. 4b shows a spectrum obtained with the reflectron where a daughter ion (from the loss of one ammonia molecule during the flight in the field-free region prior to the reflectron) precedes each of the corresponding parents. By reducing the applied potential at the end of the reflectron (i.e., $U_K < U_T$, see Fig. 1), only the lower kinetic energy products are reflected. The non-dissociating ions are eliminated from the spectrum as shown in the lower part of Fig. 4c. Further reduction of the reflecting potential U_T (with $U_K < U_T$) improves the ability to discern small contributions arising from more extensive dissociation.

Most systems comprised of (at least partially) hydrogen bonded constituents undergo similar processes, although some display additional rather interesting and revealing solvation driven competitive reaction channels. Consider, for example, the case of clusters comprised of methanol. Following multiphoton ionization (MPI), neutral methanol clusters are also found [74–76] to undergo a well known ion–molecule reaction which leads to the production of protonated clusters and the evolution of CH_3O. In accord with observations for most other systems, there is a general trend for the evaporation rates to decrease with

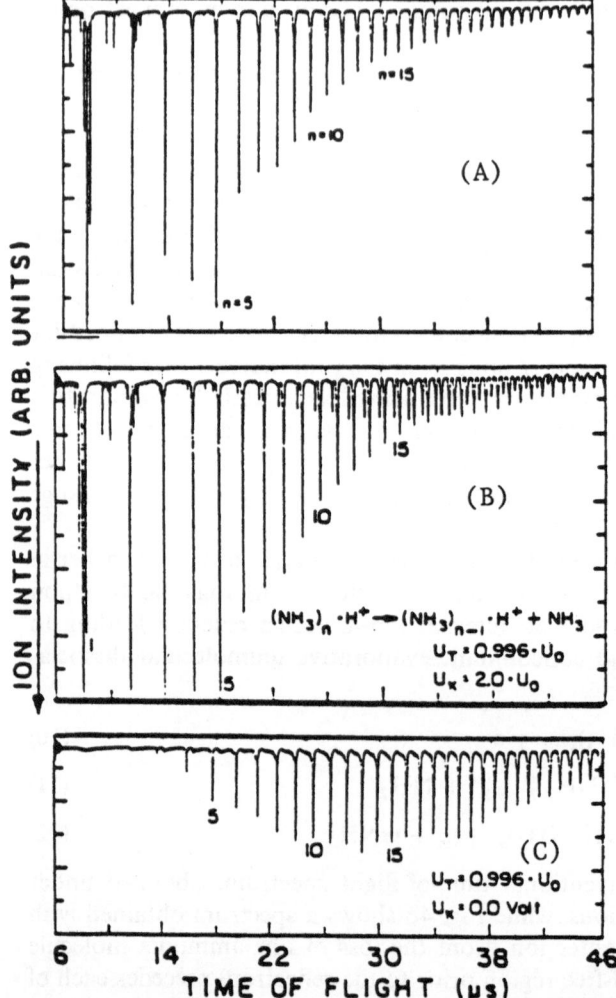

Fig. 4. **A** A conventional TOF mass spectrum of $H^+(NH_3)_n$ clusters from the multiphoton ionization of ammonia clusters at 266 nm. **B** Spectrum taken using the reflectron shown in Fig. 1. The potentials U_T and U_K are chosen such that all ions are reflected but with the daughter ions arriving at the particle detector earlier than their corresponding parent ions. **C** U_K is lowered to eliminate the parent ions which pass through the reflectron. Only the daughter ions are reflected to the detector. Adapted from [72]

time after the initial ionization event and, for a given observational time window, to display an increase in rate with cluster size.

The cluster ions, $H^+(CH_3OH)_n$, are also found to undergo several other intracluster reaction pathways which show a dependence on the degree of aggregation. A conventional time-of-flight mass spectrum is displayed in Fig. 5a. The figure shows typical results from the expansion of a 1:7 mixture of methanol vapor and argon at a source pressure of 450 torr. Two particularly interesting sequences of cluster ions include those labeled a which consist of protonated methanol clusters of form $H^+(CH_3OH)_n$ and those labeled b which become clearly evident after the $n = 7$ peak in the a series, and are of the form $H^+(H_2O)(CH_3OH)_n$.

The origin of the sequence a, corresponding to protonated methanol peaks, is a rapid intracluster proton transfer reaction following ionization of the neutral clusters. This reaction has a well known bimolecular counterpart that proceeds at near collision rate [77].

$$(CH_3OH)_n^+ \rightarrow H^+(CH_3OH)_{n-1} + CH_3O . \tag{13}$$

The formation of sequence b, $H^+(H_2O)(CH_3OH)_n$, is envisioned to occur via:

$$[H^+(CH_3OH)_n]^* \rightarrow H^+(H_2O)(CH_3OH)_{n-3} + (CH_3)_2O + CH_3OH . \tag{14}$$

The value of a reflectron in unraveling the reaction mechanisms is seen by referring to the daughter-only ion spectrum in Fig. 5b. Five types of peaks (labeled d–h) are seen clearly in the spectrum where those labeled d correspond to a mass loss of 32 amu from a parent ion cluster that enters the drift region as $H^+(CH_3OH)_n$ (these are analogous to the a peaks in Fig. 5a) as shown below,

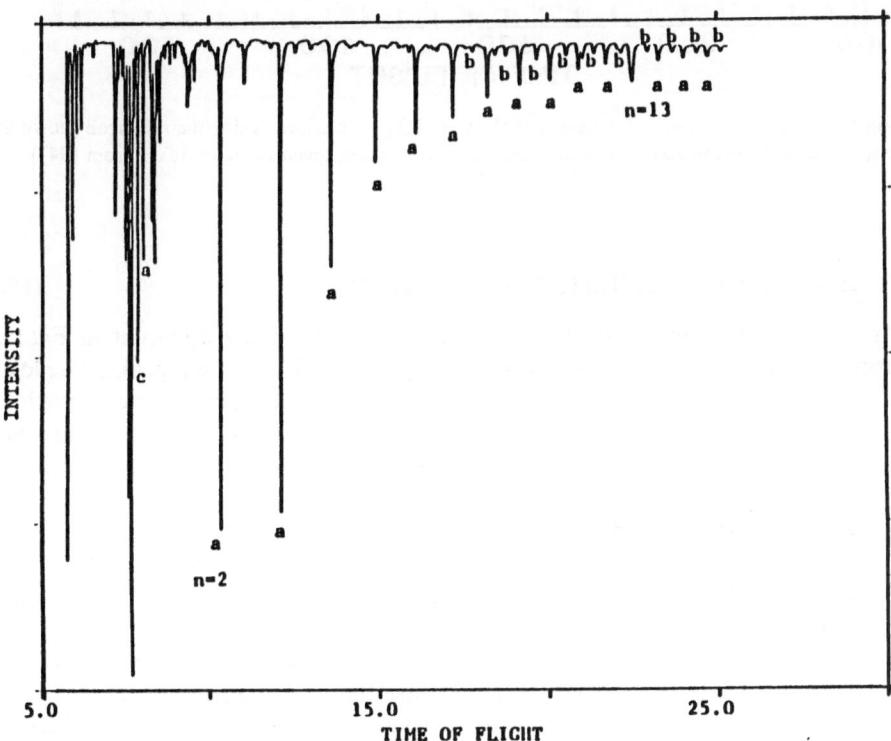

Fig. 5a. A conventional TOF spectrum of $H^+(CH_3OH)_n$. Neutral alcohol clusters are ionized with a pulsed Nd:YAG laser operated at 266 nm

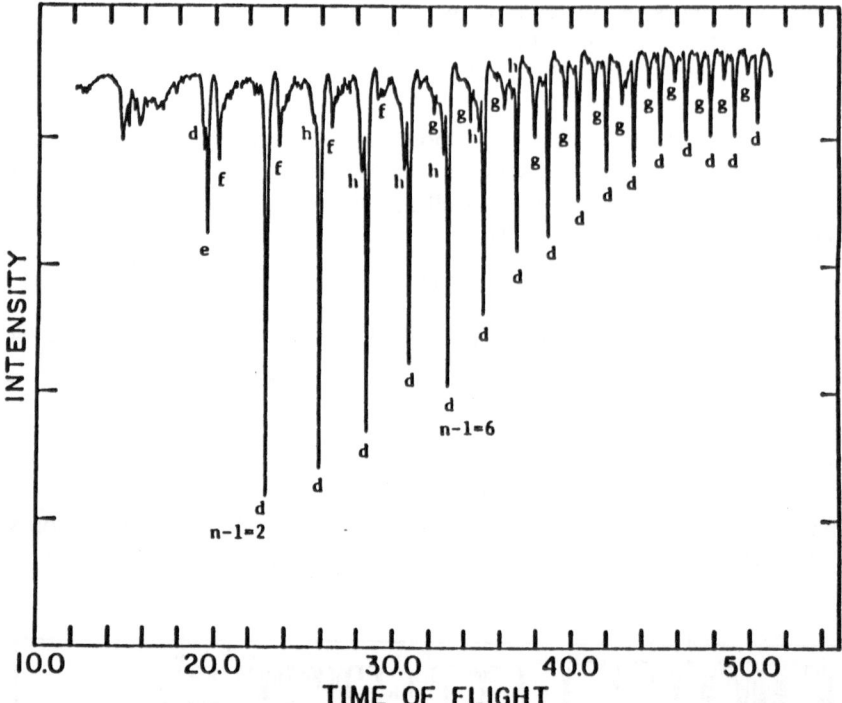

Fig. 5b. A daughter-ion TOF spectrum of $H^+(CH_3OH)_{n-1}$ measured using the reflectron shown in Fig. 1. See text for discussion of peaks identified in the mass spectra (Figure taken from [94])

$$H^+(CH_3OH)_n \rightarrow H^+(CH_3OH)_{n-1} + CH_3OH \ . \tag{15}$$

Mass losses of more than one monomer unit, not readily apparent in Fig. 5, appear as unresolved shoulders on the early arrival side of the d peaks. The loss of up to five methanol monomers from the protonated octamer is observed.

The peak designated e in Fig. 5b, corresponds to loss of water from the protonated methanol dimer ion via the dehydration reaction

$$H^+(CH_3OH)_2 \rightarrow (CH_3)_2OH^+ + H_2O \ , \tag{16}$$

a process requiring an induction time of at least several tenths of a microsecond [75]; also see [94].

The reflected ions which do not dissociate in the field-free region account for the peaks labeled f. Of particular interest are the peaks labeled g and h. Peaks labeled h correspond to loss of 78 amu from $H^+(CH_3OH)_n$, indicating loss in the drift region of both a methanol monomer and C_2H_6O (dimethyl ether) for parent ions $n = 4-9$ [via Eq. (14)]. Peaks designated as g in Fig. 5b represent the loss of one methanol moiety from the mixed cluster species

$H^+(H_2O)(CH_3OH)_n$. If these clusters are also formed by Reaction (14), they initially contained 9 or more methanol molecules. Interestingly, Reaction (16) which corresponds to the protonated methanol dimer ion eliminating H_2O, while retaining dimethyl ether, recently has been also found in thermal reaction experiments and is evidently also due to solvation effects. Analogous water elimination reactions are not observed for the parent cluster ions larger than $n = 2$.

In the case of these large clusters, H_3O^+ is solvated more strongly by methanol than is protonated dimethyl ether, $(CH_3)_2OH^+(25)$. Hence for these clusters, water retention and dimethyl ether elimination leads to production of mixed clusters of form $H^+(H_2O)(CH_3OH)_n$. Prompt (immediately following ionization) loss of dimethyl ether in the TOF ion lens, results in the mixed cluster sequence labeled b in Fig. 5a, a conventional time-of-flight mass spectrum. In contrast, a "slower" elimination of dimethyl ether in the drift region (along with CH_3OH) from cluster sizes $H^+(CH_3OH)_n$, $n = 4–9$ results in the peak sequence labeled h in Fig. 5b. Thus, a significant difference in reactivity with cluster size is observed; the smaller clusters lose dimethyl ether on a longer (field-free region) timescale while the larger clusters undergo comparatively rapid loss of C_2H_6O in the ion lens.

Some direct evidence of how the presence of a solvent can influence a reaction is seen from data for the acetone system, where the solvent H_2O can dramatically affect the course of one of the (dehydration) reaction channels [78]. Observed major cluster ions resulting from prompt fragmentation following multiphoton ionization include $[(CH_3)_2CO]_m \cdot H^+$, $m = 1–15$, $[(CH_3)_2CO]_m \cdot C_2H_3O^+$, $m = 1–17$, and $[(CH_3)_2CO]_m \cdot CH_3^+$, $m = 1–10$. In a time window of a few tens of microseconds, all these three classes of cluster ions unimolecularly decompose, losing only one acetone monomer (designated T), similar to the ammonia and alcohol systems discussed above. The results of the present study show that most of the internally excited $(T)_m^{+*}$ ions fragment to form $(T)_{m-1-x} \cdot H^+$, $(T)_{m-1-x} \cdot C_2H_3O^+$, $(T)_{m-1-x} \cdot CH_3^+$. The three major prompt fragmentation processes are expressed as follows:

$$(T)_m^{+*} \rightarrow (T)_{m-1-x} \cdot H^+ + C_3H_5O + x(T) , \tag{17}$$

$$(T)_m^{+*} \rightarrow (T)_{m-1-x} \cdot C_2H_3O^+ + CH_3 + x(T) , \tag{18}$$

$$(T)_m^{+*} \rightarrow (T)_{m-1-x} \cdot CH_3^+ + C_2H_3O + x(T) . \tag{19}$$

The loss of neutral fragments by evaporation proceeds in clusters following ionization in a manner analogous to the processes discussed above; also see [94].

Interestingly, a reaction corresponding to the dehydration of $[(CH_3)_2CO]_m \cdot H^+$ and leading to the production of $[(CH_3)_2CO]_{m-2} \cdot C_6H_{11}O^+$ is observed for $m = 2–6$. The most striking finding is that the presence of water molecules in a cluster suppresses this dehydration reaction. This can be seen for example in the case of ion peaks at masses 157 and 215 (shown in Fig. 6), 273, and 331, which correspond to $(T)_{m-2} \cdot C_6H_{11}O^+$, $m = 3$ and 4 and are the

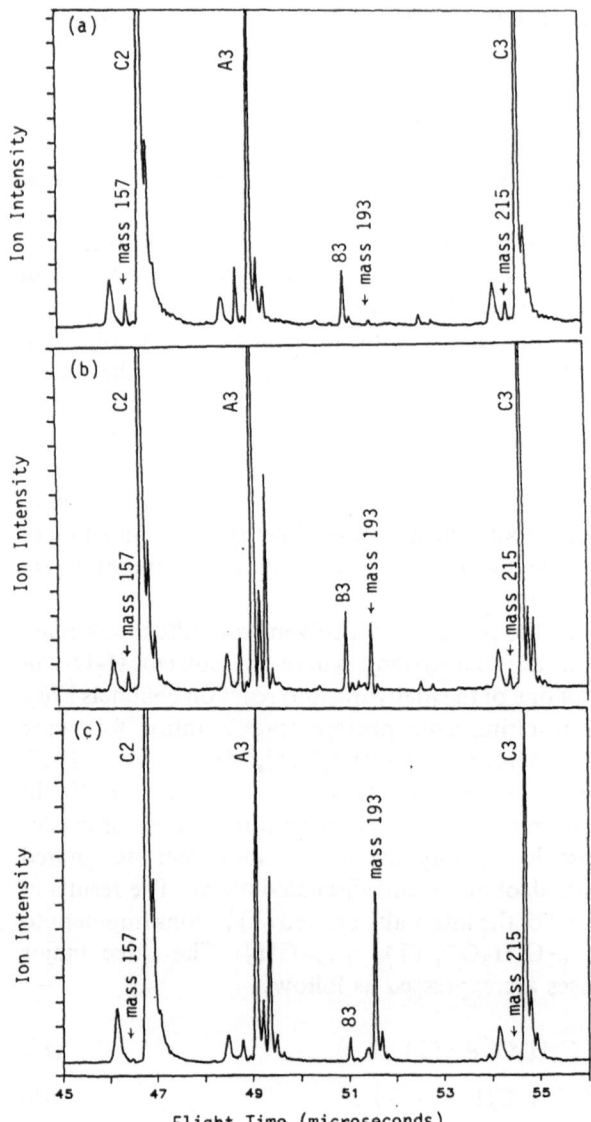

Fig. 6. Reflectron-TOF spectra of acetone cluster ions. $AM \equiv (T)_m \cdot H^+$, $B_M \equiv (T)_m \cdot CH_3^+$, $C_M \equiv (T)_m \cdot C_2H_3O^+$. (a) 0.4% water in the acetone sample, $(T)_1 \cdot C_6H_{11}O^+$ (mass 157) and $(T)_2 \cdot C_6H_{11}O^+$ (mass 215) are seen, whereas ion signal corresponds to $(T)_3 \cdot H_3O^+$ (mass 193 is not found). (b) 0.7% water in the acetone sample, all peaks corresponding to masses 157, 215 and 193 are observed. (c) 1.0% water in the acetone sample, ion peaks corresponding to $(T)_1 \cdot C_6H_{11}O^+$ (mass 157) and $(T)_2 \cdot C_6H_{11}O^+$ (mass 215) are not seen; however, $(T)_3 \cdot H_3O^+$ (mass 193) is clearly identified. Neutral clusters are ionized at 355 nm using a pulsed Nd:YAG laser. (Figure taken from [78])

dehydration products of the protonated species, $(T)_m \cdot H^+$. These $(T)_{m-2} \cdot C_6H_{11}O^+$ ions have been observed in previous studies of ion–molecule reactions [79] made on this system. However, *Luczynski* and *Wincel* [80] have demonstrated that protonated acetone can associate with $(CH_3)_2CO$ and H_2O molecules to form the proton-bound species $(T)_m \cdot H^+$ through sequential clustering reactions, and $(T)_{m-1} \cdot (H_2O) \cdot H^+$ (with m up to 6) through exchange processes. In addition, the $(T)_m \cdot H^+$, $m \geq 2$, decomposes by eliminating a water molecule to form $(T)_{m-2} \cdot C_6H_{11}O^+$.

Figure 6 shows a portion of the TOF spectrum of the acetone cluster ions in the case where the water content in the acetone sample is 0.4% or less. There are no peaks corresponding to $(T)_3 \cdot (H_2O)^+$ (mass 192) or $\{(T)_3 \cdot (H_2O)\}H^+$ (mass 193), and one can readily identify the ion peaks at masses 157 and 215 corresponding to the $(T)_{m-2} \cdot C_6H_{11}O^+$, $m \geq 2$ ions as clearly originating from the direct elimination of water molecules from the $(T)_m \cdot H^+$ via the reaction:

$$(T)_m \cdot H^+ \rightarrow (T)_{m-2} \cdot C_6H_{11}O^+ + H_2O \ . \tag{20}$$

Experiments conducted to study the influence of the presence of water in the cluster on the dehydration reactions were very revealing. Fig. 6b shows the same portion of the TOF spectrum of the acetone cluster ions as that in Fig. 6a, but in the case where 0.7% water is present in the acetone sample, the ion peaks corresponding to the $(T)_{m-2} \cdot C_6H_{11}O^+$, $m = 3$ and 4 (masses 157 and 215) and $\{(T)_3 \cdot (H_2O)\}H^+$ ion (mass 193) are evident. Additionally, Fig. 6c also displays the same portion of the TOF spectrum of the acetone cluster ions as that in Fig. 6a, but for the case where water content in the acetone sample is 1.0%. Interestingly, the peak corresponding to the $\{(T)_3 \cdot (H_2O)\}H^+$ ion (mass 193) is evident, whereas those corresponding to the $(T)_{m-2} \cdot C_6H_{11}O^+$ ions are not seen (no peaks appear at either mass 157 or mass 215). The findings strongly suggest that the presence of water inhibits the dehydration mechanism of $(T)_m \cdot H^+$ cluster ions.

Although Reaction (20) has been reported by several researchers [79–81] in the studies of ion–molecule reactions of the gas phase acetone system, others [82] did not observe this reaction in studies where the neutral clusters of acetone were ionized by electron impact. They suggested as possible explanation that the structures of the precursor ion $(T)_m \cdot H^+$ are different in these two different experiments. However, the results of the new studies show that the presence of water molecules in a cluster can significantly suppress the dehydration reaction, and although the earlier authors did not state the extent of water impurity in their system, if any, the new findings provide a plausible explanation for the discrepancy between the two studies mentioned above. The presence of water is believed to block a reaction site that otherwise enables the formation of the protonated mesityl oxide, $C_6H_{11}O^+$, to occur [78]. This finding not only clarifies the probable reason for the discrepancy between several earlier studies, but most importantly, provides evidence of another example for the influence of a solvent on ion reactions in clusters.

2.4.3.2   Intracluster Reactions Initiated through Resonant Enhanced Ionization

Resonance enhanced ionization is a valuable tool in the field of clusters, both for spectroscopic studies of solvation shifts, and as a way to initiate ion reactions of a specific nature. A typical ionization scheme is shown in Fig. 7. Applying this to the case of ammonia bound to the chromophore phenyl acetylene, protonated ammonia clusters of specific size have been observed [83] to form following the resonant enhanced ionization of the clusters $PA \cdot (NH_3)_n$ in the O_0^0 region of the S_1-S_0 transition of the (unclustered) PA. A detailed systematic investigation of the spectroscopic shifts upon the resonant enhanced ionization of $PA-NH_3$ clusters has revealed rather different trends depending on whether the clusters were produced by coexpansion or by attaching PA to the preformed ammonia clusters were produced by coexpansion or by attaching PA to the preformed ammonia clusters [83, 84]. Of particular interest was the finding that ionization also leads to the formation of protonated ammonia clusters of specific sizes, namely $H^+(NH_3)_n$, with $n \geq 2$.

The observed $H^+(NH_3)_n$ and $H^+(NH_3)_n(PA)$ clusters are thought to be formed in a two-step reaction sequence taking place after ionization of the $PA(NH_3)_n$ cluster. The first step is a charge transfer (CT) reaction between the resonantly ionized PA^+ and the NH_3 molecules in the cluster. The second step is an intracluster ion–molecule reaction (ICIMR) of the charged ammonia cluster leading to the formation of an $(n-1)$ protonated cluster ion. Step two has previously been established for NH_3 clusters [72] and is sufficiently exothermic for fragmentation of the cluster.

The proposed reaction sequence is depicted as follows:

$$(PA)(NH_3)_n \xrightarrow{h\nu} (PA^+)(NH_3)_n + e^- , \tag{21}$$

$$(PA^+)(NH_3)_n \xrightarrow{CT} (PA)(NH_3)_n^+ , \tag{22}$$

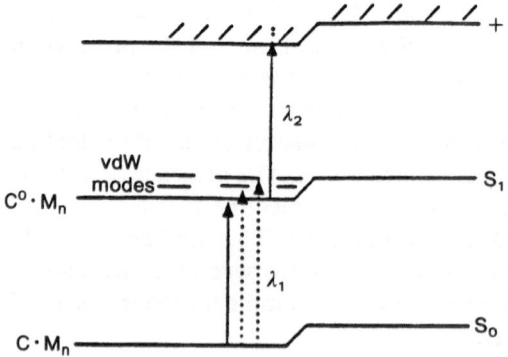

Fig. 7. Schematic of a typical ionization scheme for a solvated chromophore. S_0 and S_1 designate the ground and first electronically excited states of the chromophore, which are lowered by solvation due to clustering as shown to the left. Detection of the resulting cluster ions is made possible through resonantly ionizing from the S_1 state to the ionization continuum with a second photon, usually of a different color than the one leading to the first excitation step

$$(PA)(NH_3)_n^+ \xrightarrow{\text{ICIMR}} PA + NH_2 + H^+(NH_3)_{n-1-x} + xNH_3 , \qquad (23)$$

$$(PA)(NH_3)_n^+ \xrightarrow{\text{ICMIR}} NH_2 + H^+(NH_3)_{n-1-x}PA + xNH_3 . \qquad (24)$$

In this reaction scheme, the extent of x is related to the excess energy of the ionization process.

The observed overlap between the spectral features which appear in the one-color TOF mass channels corresponding to $PA(NH_3)_n$ clusters, where $n \geq 5$, with those which appear in masses corresponding to the ammonium cluster ions obtained in the same cluster attachment experiments, provides direct evidence for which cluster size the reaction first becomes exoergic. This definitive assignment is made possible by the different spectral shifts in the case of ammonia-PA clusters produced by coexpansion versus attachment, and the shift which this induces in the corresponding spectrum of the resulting protonated ammonia cluster ion products. The formation of these protonated cluster

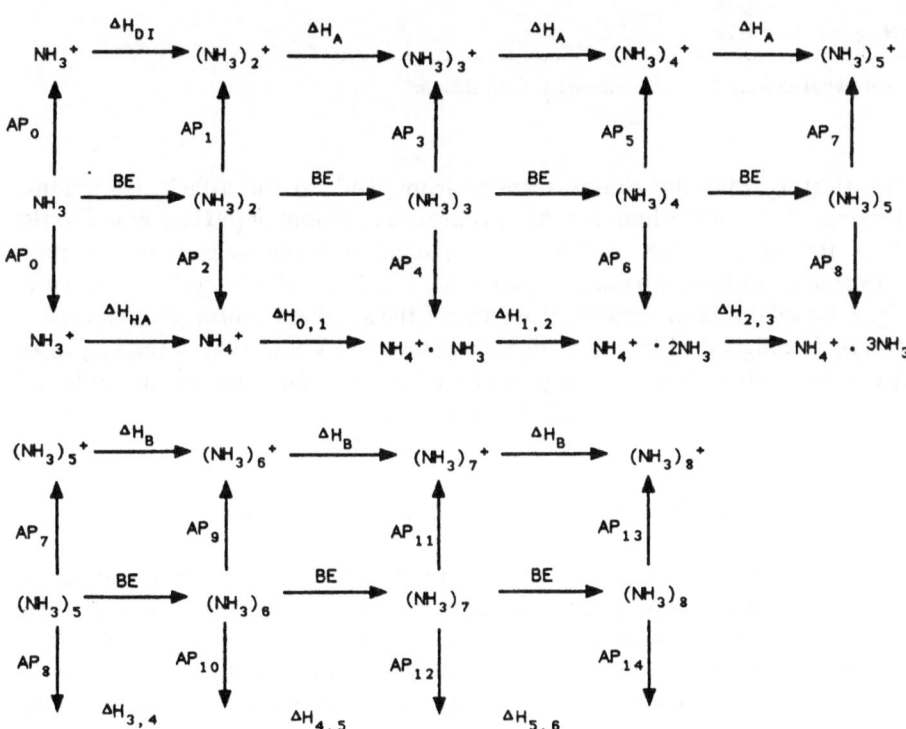

Fig. 8. Thermochemical Born–Haber cycle related to the ammonia cluster system discussed in the text. The thermochemical values corresponding to the various steps shown are listed in Table 1

Table 1. Thermodynamic values pertinent to the Born–Haber cycle for Fig. 8[a]

Reaction	Reference value/eV	Reaction	Reference value/eV
$E_x(P_0)$	10.166	$E_A(P_1)$	9.54
$E_{1,2}$	0.13	$E_A(P_2)$	9.59
$E_{2,3}$	0.24	$E_A(P_3)$	9.28
$E_{3,4}$	0.20	$E_A(P_4)$	8.73
$E_{4,5}$	0.17	$E_A(P_5)$	8.98
$E_{5,6}$	0.25	$E_A(P_6)$	8.17
$E_{6,7}$	0.25	$E_A(P_7)$	8.65
$E_{7,8}$	0.25	$E_A(P_8)$	7.72
ΔH_{DI}	-0.79 ± 0.5	$E_A(P_9)$	8.60
ΔH_A	-0.5 ± 0.2	$E_A(P_{10})$	7.46
ΔH_B	-0.3	$E_A(P_{11})$	8.55
ΔH_{HA}	-0.7 ± 0.4	$E_A(P_{12})$	7.37
$\Delta H_{0,1}$	-1.10	$E_A(P_{13})$	8.50
$\Delta H_{1,2}$	-0.76	$E_A(P_{14})$	7.34
$\Delta H_{2,3}$	-0.62		
$\Delta H_{3,4}$	-0.51		
$\Delta H_{4,5}$	-0.34		
$\Delta H_{5,6}$	-0.28		

[a] Values taken from [94] and references given therein

ions starting with the charge transfer from PA^+ to the attached ammonia clusters, sets an upper limit for the appearance potential of $(NH_3)_5^+$ equal to the IP of PA which is 8.8 eV. This is consistent with the ionization potentials estimated from thermodynamic data using the Born–Haber cycle as shown in Fig. 8. Based on the arguments given above, the adiabatic ionization potential of $(NH_3)_5$ is assigned to be (only slightly) below 8.8 eV and that of $(NH_3)_4$ to be above this value. These findings lead to the cycle for bonding energies and ionization potentials shown in Table 1.

2.4.3.3 Ionization Through Intracluster Penning Processes

Another interesting situation which demonstrates the role of solvation on cluster reactions, with particular relevance to identifying mechanisms of possible importance in the condensed phase, concerns the initiation of electron transfer reactions. Interesting examples include those initiated through Penning ionization [85], a process which is still awaiting specific identification in the liquid state.

Clusters provide interesting systems for investigating the influence of solvation on ionization and concomitant electron transfer processes, including Penning ionization, in terms of their gas phase counterparts. Analogous processes in paraxylene (PX) bound to $N(CH_3)_3$ were studied [86] following the absorption of photons through the perturbed S_1 state of paraxylene into high Rydberg states. An interesting comparison is provided by results of studies involving adducts of paraxylene bound to NH_3 with those involving trimethylamine, where the ionization potential of paraxylene is less than that of ammonia but greater than that of trimethylamine. In the case of $PX \cdot NH_3$, ionization by adsorption of a second photon into the perturbed S_1 state of paraxylene was found to begin near the ionization threshold of paraxylene, and to lead to the expected cluster ion $PX \cdot NH_3^+$; NH_3^+ is also observed at 1.8 eV above this threshold and NH_4^+ is seen in two-color experiments made at high fluence of the ionizing laser.

By contrast, absorption into high Rydberg states of paraxylene below its ionization potential in $PX \cdot N(CH_3)_3$ was found to produce primarily $N(CH_3)_3^+$, with $H^+N(CH_3)_3$ as a minor product. No $PX \cdot N(CH_3)_3^+$ ion was detectable. One conclusion is that photoexcitation of paraxylene leads to an intercluster ionization process bearing analogy to Penning ionization, where the perturbed high Rydberg states of paraxylene interact with the partner molecule $N(CH_3)_3$. A second, and more startling observation, was the finding of a slow ionization process as evident in the time-of-flight peak shapes shown in Fig. 9. Since the laser interacts with the molecules in the first of a two-field acceleration region, a long tail is only possible when the ionization process is slow. [It should be noted that fragmentation leads to a knee in the peak shape, and not to a long tail as observed in the figure.] Interestingly, the process is substantially slower when the energy of the ionizing photon is decreased. Questions arise whether the slow step is associated with the proton transfer channel (i.e., the $(CH_3)_3NH^+$ product) or an electron transfer process (i.e., the $(CH_3)_3N^+$ product), but careful measurements with deuterated species revealed that the tail is largely associated with the electron and not the proton transfer process. A plausible explanation is that a large geometry change is involved in the formation of the trimethylamine ion during the Penning-like ionization process. Interestingly, Hatano [87] has

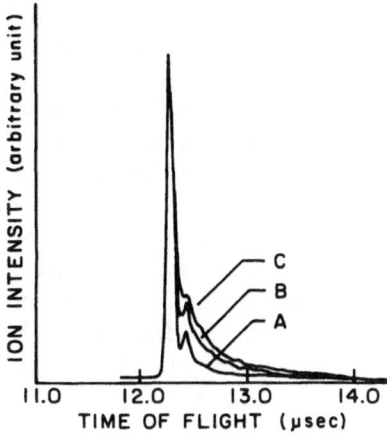

Fig. 9. TOF spectrum obtained by the photoexcitation of a paraxylene-trimethylamine (TMA) complex at selected photon energies. The tail is seen to increase with a lowering of the photon energy corresponding to lower excitation into the Rydberg manifold. Broadenings of TMA$^+$ ion peaks: B and C correspond to time constants of 160 ± 20 and 200 ± 20 ns, respectively. The peaks corresponding to TMA $\cdot$ H$^+$ are also observable. (Taken from [86])

found that orientational effects in the liquid phase, where motion is restricted, can lead to a significant reduction in the rate of a process which also apparently proceeds via Penning ionization.

2.4.3.4   Energy Disposal and Structure

2.4.3.4a)   Studies of Kinetic Energy Release (KER) of
a Cluster Ion Evaporative Ensemble

Recently, it has been shown by *Castleman* and coworkers [31, 32] that by measuring the average kinetic energy release and unimolecular dissociation rate constant during ion decomposition, and by using the peak shape analysis method described in the experimental section, bond energies for cluster ions can be derived. This procedure is based on the fact that during the decomposition of a metastable parent ion in the field-free drift region, the internal energy of a parent ion is converted to the translation energy of the daughter ion. As a result, the translation energy distribution of the daughter ion is broader than that of the parent ion. The importance of using a reflectron time-of-flight mass spectrometer in the studies of metastable unimolecular dissociation dynamics is that it enables simultaneous determination of the kinetic energy release and the dissociation rate. The method is found to lead to values of high precision, which in combination with various theoretical approaches, enables a determination of cluster bond energies.

An important example of the application of this method is seen for the case of ammonia. Referring to Fig. 10, the measured average kinetic energy release of metastable $(NH_3)_n H^+$, $n = 4-17$, is seen to display a maximum value of 9.5 meV at $n = 5$ and a gradual decrease to a value of 5.0 meV for $n = 17$. This finding of maximum kinetic energy release of $(NH_3)_5 H^+$ is consistent with recent KER measurements of $(NH_3)_n H^+$, $n = 2-8$, obtained with a double focusing mass spectrometry method [88]. The largest amount of kinetic energy release is found to be associated with the parent cluster ion $(NH_3)_5 H^+$ undergoing unimolecular decomposition. A small local maximum in the average kinetic energy release is also evident at $n = 12$.

Employing the KER data, and using a modified QET/RRK statistical analysis [88, 89], binding energies for large cluster ions can be readily determined [31, 90]. In this method, the deduced binding energy values must be scaled to match some determined value [the best available [25] for $(NH_3)_5 H^+$ is used here] whereupon the binding energies of all other clusters are found to be in good agreement with the reported literature values for cluster ions $(NH_3)_n H^+$, $n = 4, 6$ and 7 (see Fig. 11). The trend of the determined binding energies of metastable cluster ions $(NH_3)_n H^+$ shows a progressive decrease from $n = 4-5$, displays a precipitous drop from 5 to 6 and then slowly decreases thereafter. There are hints of slightly more stable structures, than neighboring ones, at $n = 12$ and 14.

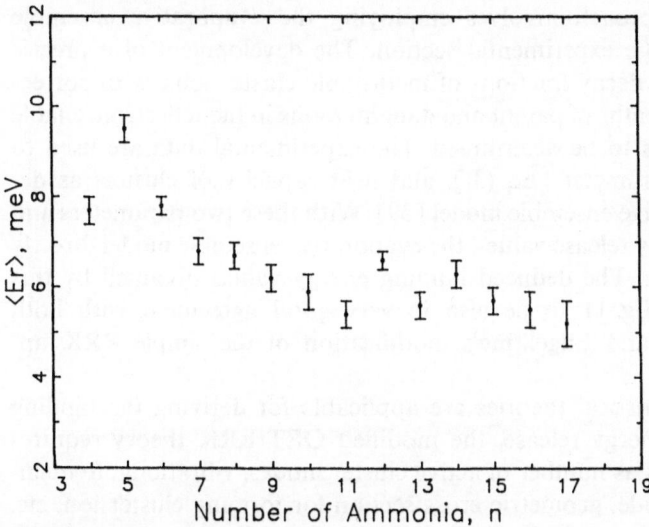

Fig. 10. A plot of the measured average kinetic energy release during the metastable unimolecular decomposition of $(NH_3)_nH^+$, $n = 4–17$, as a function of cluster size. The technique involves use of the reflectron shown in Fig. 1 (Figure taken from [31])

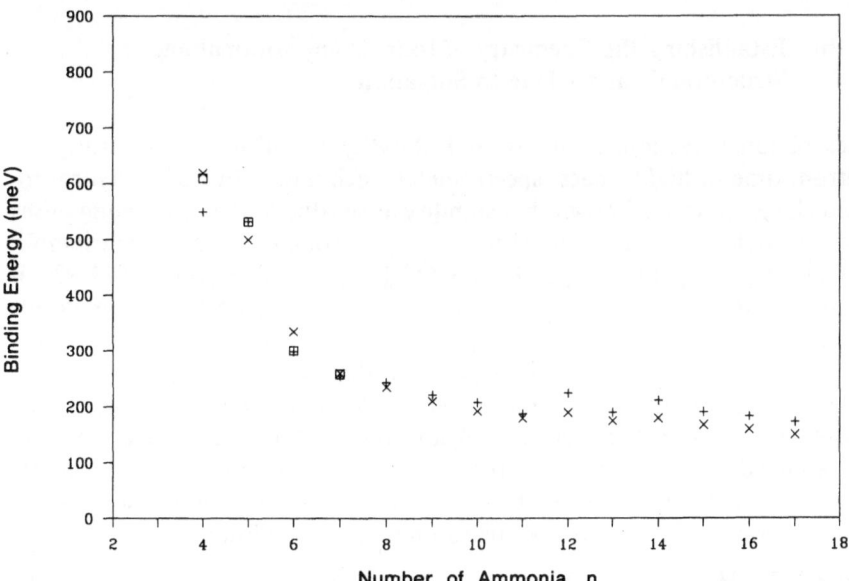

Fig. 11. A plot of calculated binding energies of $(NH_3)_nH^+$, $n = 4–17$, as a function of cluster size. $\square \equiv$ literature values based on high pressure mass spectrometry (see [25]); $\times \equiv$ deduced values using Klots' evaporative ensemble model; $+ \equiv$ deduced values using Engelking's modified QET/RRK model and the kinetic energy measurements shown in Fig. 10. (Figure taken from [32])

An alternative approach involves employing the evaporative ensemble method discussed in the experimental section. The development of a precise method for measuring decay fractions of metastable cluster ions, with corrections ensuring similar paths of parent and daughter ions in the reflectron, enable accurate bond energies to be determined. The experimental data are used to derive the Gspann parameter [Eq. (3)], and heat capacity of clusters as described in the evaporative ensemble model [39]. With these two parameters and measured kinetic energy release values, the evaporative ensemble model directly yields binding energies. The deduced binding energy values obtained by this method are seen in Fig. 11 to be also in very good agreement with both thermochemical data and Engelking's modification of the simple RRK approach.

Although both statistical theories are applicable for deriving the binding energy from kinetic energy release, the modified QET/RRK theory requires many parameters, such as number of active cluster modes, vibrational frequencies of each cluster mode, geometric cross section for forming cluster ion, etc. A greater limitation in its use is that it also requires some known thermochemical values in order to adjust the scaling parameters. By contrast, the evaporative ensemble model employs only two parameters which can be deduced from a precise measurement of decay fractions of cluster ions. Importantly, the deduced binding energies of cluster ions using the evaporative ensemble model agree very well with the reported literature values for smaller clusters.

2.4.3.4b) Establishing the Chemistry of Ions: Compositional and Structural Changes Due to Solvation

Studies of ion solvation, structure and stability are also possible using the reflectron time-of-flight mass spectrometer technique. In small hydrogen-bonded cluster ions it is known that stability arises due to the positioning of the ligand molecules around a central proton or protonated species, for example, $(CH_3COCH_3)_2H^+$ [91], $(CH_3OCH_3)_3H_3O^+$ [92], and $(NH_3)_4NH_4^+$ [73, 93]. It is interesting to consider two cases, one where the proton affinity of molecule X in the mixed cluster ions $(NH_3)_n(X)_mH^+$ is less than that of ammonia, and the other where that of molecule X is larger than that of ammonia.

The proton affinity of ammonia is 204.0 cal/mol and those of CH_3COCH_3, CH_3CN, and CH_3CHO are 196.7, 188.4, and 186.6 cal/mol, respectively [95]. In all three mixed cluster ion systems, the intensity distributions show that there is a maximum at $n + m = 5$ [96]. Results of metastable decomposition studies [97] of the mixed cluster ions can be summarized as follows:

$$(NH_3)_n(X)_mH^+ \rightarrow (NH_3)_n(X)_{m-1}H^+ + X, \quad \text{for } n = 1 , \tag{25}$$

$$(NH_3)_n(X)_mH^+ \rightarrow (NH_3)_{n-1}(X)_mH^+ + NH_3, \quad \text{for } n \geq 2 . \tag{26}$$

The results clearly indicate that NH_4^+ is the core ion with four available hydrogen-bonding sites to complete the first solvation shell. The four ligands can include any combination of ammonia and molecule X. The loss of an ammonia molecule resulting from the metastable decomposition for $n \geq 2$ indicates that X is more strongly bonded to the NH_4^+ than is another NH_3. This is not surprising since all molecules considered here have higher dipole moments and polarizabilities than those of ammonia, which leads to them having a greater ion-dipole and ion-induced dipole interaction.

The proton affinities [95] of pyridine (C_5H_5N) and trimethylamine (TMA, $(CH_3)_3N$) are 220.8 and 225.1 cal/mol which are larger than that of ammonia. The metastable decomposition studies of $NH_3(C_5H_5N)_mH^+$ ($m = 1-5$) show the following results:

$$NH_3(C_5H_5N)_mH^+ \rightarrow (C_5H_5N)_mH^+ + NH_3, \quad \text{for } m < 4 , \tag{27}$$

$$NH_3(C_5H_5N)_mH^+ \rightarrow NH_3(C_5H_5N)_{m-1}H^+ + C_5H_5N, \quad \text{for } m \geq 4 . \tag{28}$$

The higher proton affinity of pyridine leads to its stronger bonding to the proton, whereby the ammonia molecule is lost in the metastable dissociation for $m < 4$. The loss of pyridine in the case of $m \geq 4$ can be accounted for by the fact that a central NH_4^+ core ion is formed which provides four hydrogen bonding sites for ligands, a structure dictated by the net energies of bonding. This finding is consistent with the previous proposed stable structure for the cluster ion $(NH_3)_n(X)_mH^+$.

In the studies of mixed protonated clusters of ammonia and trimethylamine, the ion intensity distributions of $(NH_3)_n(TMA)_mH^+$ [98] display local maxima at $(n, m) = (1, 4), (2, 3), (2, 6), (3, 2)$, and $(3, 8)$. The fact that the maximum ion intensity occurs at $(n, m) = (1, 4), (2, 3)$, and $(3, 2)$ indicates that a solvation shell is formed around NH_4^+ ion with four ligands of any combination of ammonia and TMA molecules. In the case of the maximum ion intensity occurring at $(n, m) = (2, 6)$ and $(3, 8)$, the experimental results suggest that another solvation shell results which contains the core ions $[H_3N-H-NH_3]^+$ (6 available hydrogen bonding sites) and $[H_3N-H(NH_2)H-NH_3]^+$ (8 available hydrogen bonding sites). The observed metastable unimolecular decomposition processes are:

$$NH_3(TMA)_mH^+ \rightarrow (TMA)_mH^+ + NH_3, \quad m < 4 , \tag{29}$$

$$NH_3(TMA)_mH^+ \rightarrow NH_3(TMA)_{m-1}H^+ + TMA, \quad m \geq 4 , \tag{30}$$

$$(NH_3)_2(TMA)_mH^+ \rightarrow NH_3(TMA)_mH^+ + NH_3, \quad m < 6 , \tag{31}$$

$$(NH_3)_2(TMA)_mH^+ \rightarrow (NH_3)_2(TMA)_{m-1}H^+ + TMA, \quad m \geq 6 , \tag{32}$$

$$(NH_3)_3(TMA)_mH^+ \rightarrow (NH_3)_2(TMA)_mH^+ + NH_3, \quad m < 8 , \tag{33}$$

$$(NH_3)_3(TMA)_mH^+ \rightarrow (NH_3)_3(TMA)_{m-1}H^+ + TMA, \quad m \geq 8 . \tag{34}$$

The observed decomposition processes also support the above ion solvation shell model for these mixed cluster ions containing ammonia.

Strong evidence has also been obtained [99] for ring-like structures in studies on a series of mixed neutral clusters $(A)_n \cdot (M)_m$ (A is both the proton donor and acceptor such as ammonia and water; M is only a proton acceptor such as acetone, pyridine, and trimethylamine). The observed mixed cluster ions display a maximum intensity at $m = 2(n + 1)$ when $n \leq 5$ for $(NH_3)_n \cdot (M)_m H^+$, $m = n + 2$ when $n \leq 4$ for $(H_2O)_n \cdot (M)_m H^+$ under various experimental conditions indicating that the cluster ions with these combinations have the stable closed shell structures as discussed above. However, breakdown of the pattern occurs at $n > 5$ for ammonia system and $n > 4$ for water system, with the most intense peaks occurring for species with one molecule less than the expected pattern, i.e. $m = 2(n + 1) - 1$ when $n = 6$ for $(NH_3)_n \cdot (M)_m H^+$ and $m = (n + 2) - 1$ when $n = 5$ for $(H_2O)_n \cdot (M)_m H^+$. This can be explained based on the formation of the hydrogen-bonded ring structures (e.g. $(H_2O)_5(M)_6 H^+$ drawn below) which are believed to exist in the condensed phase. One possible ring structure consistent with the data is shown below.

2.4.3.4c)   Evidence for Isomeric Structure of Cluster Ions

The question of the existence of isomeric structures in small clusters is one of long-standing interest and isomeric forms have been proposed to explain the observed differences in the spectral properties for several neutral van der Waals systems [100–103]. Although isomeric forms of some ions and ionic dimers, such as HCO^+/HOC^+ [104, 105], HCN^+/HNC^+ [106], and $c\text{-}C_3H_6^+\text{-}OH_2$ [107], are well known, currently there is a paucity of evidence for isomeric cluster ion systems [108, 109]. Measurements of cluster ion dissociation and concomitant kinetic energy release measurements can be used to derive information on the existence of such structures and recently definitive proof of their existence has been obtained for the solvated cluster ion $NH_3((C_2H_5)_3N)_3H^+$.

Investigation of the dissociation dynamics of the cluster ion revealed four peaks which correspond to the following dissociation processes:

$$((C_2H_5)_3N)_3H^+ \rightarrow ((C_2H_5)_3N)_2H^+ + (C_2H_5)_3N \; , \tag{35}$$

$$NH_3((C_2H_5)_3N)_3H^+ \rightarrow NH_3((C_2H_5)_3N)_2H^+ + (C_2H_5)_3N \; , \tag{36}$$

$$NH_3((C_2H_5)_3N)_3H^+ \rightarrow ((C_2H_5)_3N)_3H^+ + NH_3 \; , \tag{37}$$

$$(NH_3)_2((C_2H_5)_3N)_3H^+ \rightarrow NH_3((C_2H_5)_3N)_3H^+ + NH_3 \; . \tag{38}$$

As discussed earlier, dissociation of a metastable cluster ion is usually accompanied with some kinetic energy release during the process [31, 32, 38, 39, 110, 111], a fact which was readily discernible for the present case.

Although observation that two metastable decomposition channels open simultaneously for the mixed cluster ions does not necessarily prove the isomer existence, in conjunction with the finding of a significant difference in the observed KER of these two channels, these data provide direct evidence for the existence of isomeric structures of $NH_3((C_2H_5)_3N)_3H^+$ and their influence on the dynamics of dissociation. If there were only one stable structure for $NH_3((C_2H_5)_3N)_3H^+$, the observation of two channels corresponding to the loss of $(C_2H_5)_3N$ and NH_3 proceeding at the comparable decomposition rates in the same metastable decomposition time window implies that the dissociation energies of these two channels would be about the same. Otherwise, only one channel with the lower dissociation energy would be observed. However, in terms of both the modified RRK and the evaporative ensemble model, the KER is directly proportional to the cluster ion dissociation (bond) energy. Therefore, on the premise of a single structure, the KER values of these parallel channels should be about the same since the dissociation energies of these two channels would have to be about equal (due to the fact that the two channels proceed at the same decomposition rate within the same metastable time window [about 10 μs]).

However, the measured KER for the TEA loss channel (2.70 ± 0.04 me V) is much larger than the ammonia loss channel (5.3 ± 0.7 me V) which definitely rules out the assumption of only one stable structure for cluster ion $(NH_3((C_2H_5)_3H^+$. Consequently, dynamical considerations prove the existence of more than one structure [98].

2.4.3.5 Ion Solvation Kinetics: Bimolecular and Association Reactions

The subject of cluster ion kinetics may be divided into several types of reactions. Association reactions are responsible for cluster formation and growth, whereas exchange reactions involve the replacement of one ligand by another. Additionally, catalytic reactions between a molecule and a ligand of a cluster ion, where the core ion may act as the catalyst, may also occur. Other reactions include those in which the cluster ion loses its charge via either recombination with an oppositely charged species or charge exchange. Dissociation of the cluster ions is also possible through either photodissociation, collision-induced dissociation or unimolecular processes.

Two other processes include internal ion–molecule reactions (half-reactions) which sometimes follow the ionization of a neutral cluster, discussed in detail above, and other reactions which take place at surfaces during ion bombardment and/or related sputtering processes. Collections of data available for rate constants of all classes of ion–molecule reactions are available [43, 44]. The techniques for measuring cluster ion reactions are quite varied, but the vast majority of the data have been determined using the flowing afterglow method or SIFT modification thereof, discussed in an earlier subsection.

Exchange reactions can be divided into several categories, depending on the nature of the process. Conversion reactions are defined as those in which the nature of the central ion core is changed during the clustering process; often a neutral product which is different from any of the initial molecular entities is formed. Ligand exchange reactions simply lead to a displacement of one ligand by another, a process which would imply stronger bonding for the replacing ligand in reactions where concentrations of the two are equal; but, such displacements can be driven by increasing the concentration of the replacing species.

An example of a conversion reaction driven by clustering involves the interaction of NO^+ with H_2O, a classic reaction [43, 44, 112, 113]. The reaction proceeds as follows:

$$NO^+ + H_2O \xrightarrow{M} NO^+ \cdot H_2O \ ,$$

$$NO^+ \cdot H_2O + H_2O \xrightarrow{M} NO^+ \cdot (H_2O)_2 \ ,$$

$$NO^+ \cdot (H_2O)_2 + H_2O \xrightarrow{M} NO^+ \cdot (H_2O)_3 \ ,$$

$$NO^+ \cdot (H_2O)_3 + H_2O \xrightarrow{M} H_3O^+ \cdot (H_2O)_2 + HNO_2 \ . \tag{39}$$

Energetic considerations suggest that chemical conversion occurs at the stage of hydration where the reaction becomes exothermic for the changing of the ion to a core of H_3O^+ [114].

Other cluster reactions have been found to undergo similar changes such as the reaction of $O_2^+ H_2O$ with H_2O to yield H_3O^+ [115]. In terms of simple atomic core ion conversion reactions, few are known but some examples are found [116, 117] for Ag^+ and Pb^+ interacting with various organic molecules. The influence of the extent of clustering on these reaction rates are particularly important to understanding the effects of solvation.

Study of the reactivity of cluster ions with gas phase molecules offers an approach for understanding the role of solvation in cluster reactions, which in turn bear on those of the condensed phase. For example, Bohme and coworkers [118–121] have studied the reactions

$$B^- \cdot S_n + AH = A^- \cdot S_m + (n - m)S + BH \ ,$$

$$B^- \cdot S_n + CH_3Br = Br^- \cdot S_m(n - m)S + CH_3B \ . \tag{40}$$

In these S_N2 reactions, it was found that stepwise increases in n led to a decrease in reactivity of the parent anion. In related studies [119] of cations, it was found that increases in the degree of hydration of H_3O^+ led to a decrease in the rate constant for proton transfer to H_2S. Further examples of the influence on proton transfer by the degree of aggregation (solvation) can be found elsewhere

[118, 120]. Other results [122] have further demonstrated the role of hydration in decreasing the reactivity of cluster ions and related studies of nucleophilic displacement reactions have been reported [121, 123]. An interesting example of the influence of hydration on reactivity is shown in Fig. 12.

Studies of ligand exchange reactions are more common. Such reactions provide interesting information on the energetics of the clustering process when allowed to proceed to completion so that the relative rates of forward and reverse steps can be assessed. These measurements are often used to derive important thermochemical properties for the clusters as considered in Sect. 2.4.4. As seen from the reaction rate constants reported [43, 44] ligand switching reactions are generally rather rapid when the processes are exothermic, i.e., $k \geq 10^{-10}$ cc/s. Depending on the energetics of solvation, the successive exchange may terminate at some intermediate stage of mixed solvation. On the other hand, for ligand exchange, some charge transfer processes of $O_2^-(H_2O)_n$, and reactions of hydrated hydronium ions, increased hydration of the reactant ion does not significantly decrease the reaction rate constant.

An enormous reactivity enhancement has been found by *Rowe* et al. [124] for a new class of ion-catalyzed reactions between neutrals occurring in cluster ions in which the central ion does not form chemical bonds, eg., the rate constant for the homogeneous gas phase reaction of N_2O_5 with NO is smaller than

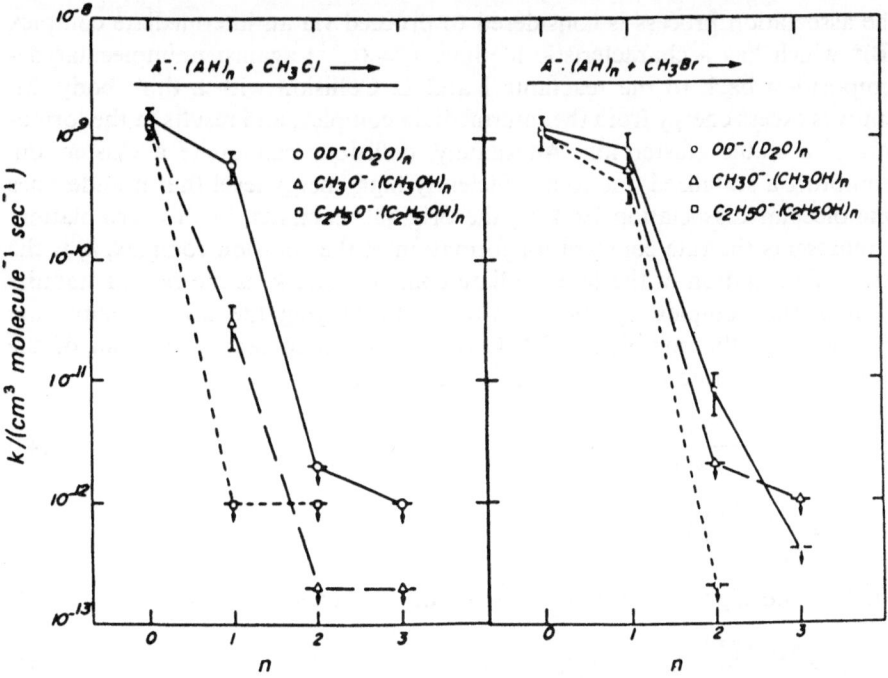

Fig. 12. Reaction rate constants for gas phase nucleophilic displacement reactions of solvated anions with methyl chloride and methyl bromide as a function of the extent of solvation. The dramatic effects due to solvation are evident. (Figure taken from [9]; results from work of Böhme.)

10^{-20} cm²/s, whereas the rate for this reaction when bound to an alkali ion is increased at least seven orders of magnitude in the case of Na^+ as central ion, and in excess of nine orders of magnitude in the case of Li^+ [125].

Except in those instances where cluster ions are generated by the ionization of neutral clusters or via surface bombardment/sputtering techniques, clusters are generally produced via a series of association reactions. These commence by way of an orbiting collision of a molecule with a cluster ion or bare ion precursor, requiring a subsequent collisional process with a third-body to remove energy from vibrational and/or rotationally excited intermediate cluster complexes [7, 9, 126–129].

Cluster formation between an ion I (or reactant cluster ion) and ligand B can be visualized to form through a sequence of reaction steps as follows:

$$I + L \underset{k_r}{\overset{k_c}{\rightleftharpoons}} (IL)^* \ , \tag{41}$$

$$(IL)^* + M \underset{k_a}{\overset{k_s}{\rightleftharpoons}} IL + M \ , \tag{42}$$

where the overall reaction is

$$I + L \underset{k_{ro}}{\overset{k_{fo}}{\rightleftharpoons}} IL \ . \tag{43}$$

The association process is considered to proceed via an intermediate complex $(IB)^*$ which has a characteristic lifetime, $\tau_r = (k_r^{-1})$ against unimolecular decomposition back to the reactants I and L. Collision with a third-body, M, removes excess energy from the intermediate complex, and results in the formation of a stable cluster ion. Alternately, collisions can excite a cluster ion, promoting a stabilized one to a sufficiently high energy level that it undergoes unimolecular dissociation back to the original reactants. In this formulation, k_c represents the rate constant for formation of the collision complex, k_r is the rate of dissociation of the intermediate complex, and k_a is the rate of stabilization of the complex by the third-body. Employing the usual steady-state treatment for the complex $(LB)^*$ leads to the Lindemann formalism of the overall forward rate constant, k_{fo}, and the overall reverse rate, k_{ro}.

$$k_{fo} = \frac{k_c k_s [M]}{k_r + k_s [M]} \ , \tag{44}$$

$$k_{ro} = \frac{k_r k_a [M]}{k_r + k_s [M]} \ . \tag{45}$$

The low and high pressure limits are readily obtained:

$$k_{fo} = \frac{k_c k_s [M]}{k_r} \quad k_{ro} = k_a [M] \ , \quad \text{(low)} \tag{46}$$

$$k_{fo} = k_c \quad k_{ro} = \frac{k_r k_a}{k_s} \ . \quad \text{(high)} \tag{47}$$

The rate constants for the association and stabilization steps can be computed by a number of different techniques. The collision rate for an ion with a polarizable atom, or a molecule having no permanent dipole moment, is given by the Langevin rate as detailed by Giomosis and Stevenson [130]. When a molecule with a permanent dipole moment serves as the clustering ligand, the formulation becomes more complex. There are several different formulations available at different levels of approximation including ADO and AADO theory [131]. *Hsieh* and *Castleman* [132] have considered the dynamics of the association step, leading to a slightly different formulation and in other studies, *Schelling* and *Castleman* [133] have developed an approach to account for the influence of the anisotropic polarizability on clustering. *Hase* and coworkers [134, 135] have analyzed the dynamics of formation of water clusters about alkali metal ions using a dynamical computer model, and other useful considerations on association complex formation are available in publications by *Herbst* [136, 137], *Bates* [138–142], and *Clary* [143]. Additionally, *Bowers* and coworkers [144] have developed a phase space approach for treating ion–molecule reactions, and *Su* and *Chesnavich* [145] developed a parametrization of the collision rate based on trajectory calculations which employed (angle) averaged polarizabilities. From a practical point of view the approach of *Su* and *Chesnavich* is especially useful.

Despite the extensive attention they have received in the past, association reactions are still a subject of considerable interest and study. For example, the effectiveness of collisions in removing energy, i.e., considerations of energy transfer between translation, and rotation and vibration (T–R and T–V) leading to the formation of stable clusters is far from a well-understood process. Particularly interesting fundamental questions derive from a consideration of both the pressure dependence and the temperature dependence of association reactions. One of the central issues pertains to the rate of dissociation of the intermediate excited complex, k_r. (Note, in the low pressure limit, the overall forward rate is inversely proportional to k_r; see Eq. (46).) Therefore, theoretical treatments are related to unimolecular dissociation involving the statistical redistribution of energy within the excited intermediate. In some cases simple RRK relationships are employed that do not consider the vibrational density of states in the cluster, while in others the more sophisticated RRKM formulations [146, 147] are used.

Although RRK theory is sometimes used because relatively little information is needed in its application, several important applications of RRKM theory which have given more insight into the fundamentals of association reactions have been published. For instance, *Olmstead* and coworkers [148] and *Jasinski* et al. [149] employed an RRKM treatment of ion–molecule cluster formation involving the proton bound dimers of H_2O, ammonia, CH_3NH_2, and $(CH_3)_2NH$ and for the formation of the benzene dimer cation $(C_6H_6)_2^+$. By making estimates of stretching and bending frquencies for the complexes based on values for the neutrals, the only adjustable parameter involved fitting the entropy of the complex which had been measured experimentally. Using this procedure, both the absolute value calculated for k_r, as well as the temperature dependence for the overall association reactions were found to be in good agreement with the experimental results.

An interesting example of the influence of aggregation (solvation in the most general sense of the word) on association reactions comes from recent results on the association of CO with copper cluster ions [150]. The interaction of carbon monoxide with metal surfaces is of interest with regard to elucidating the nature of bonding, identifying the nature of adsorption sites, and even the physical basis of catalytic processes. Laser vaporization was used to create an ensemble of copper cluster ions, which were thermalized at high pressures, and subsequently allowed to react with carbon monoxide in a flow tube apparatus. The CO molecule sticks to the copper cluster surface and shows a rapid increase in the association rate as a function of cluster size, clearly displaying transition from atomic to bulk behavior (see Fig. 13). Indeed, the kinetics change from low-pressure termolecular behavior for the small clusters ($n \leq 4$) to a surface or bulk-like regime for the larger clusters ($n \geq 7$). This transition is characterized by a change in the pressure dependence of the termolecular rate constant as the collision limited rate is approached. The overall trend is explainable, at least qualitatively, in terms of simple unimolecular decay theory and approach to bulk behavior was accounted for by incorporating into the theoretical collision limit an accommodation coefficient to account for the intrinsic surface site

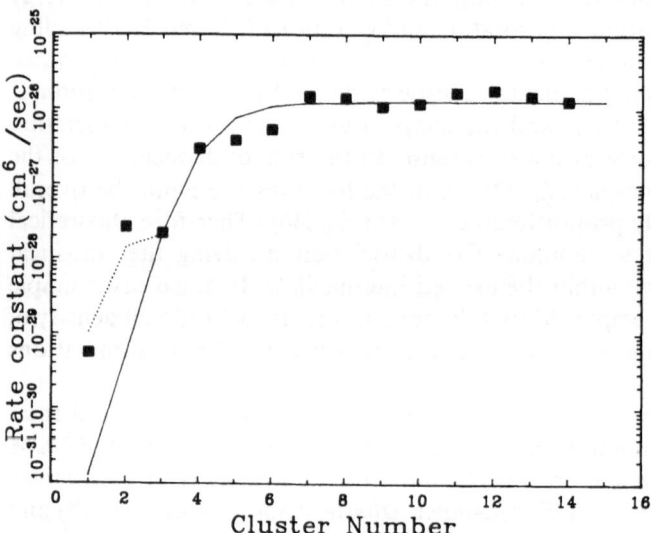

Fig. 13. Plot of experimental (■) and theoretical third-order rate constants for the attachment of CO to copper cluster cations as a function of cluster size. Note the dramatic increase in reactivity (almost four orders of magnitude) within the first seven clusters. The overall trend represents a transition from termolecular to effective bimolecular behavior. The *solid line* (theory) was obtained assuming a loose transition state while the dotted line shows the results for a tight transition state for monomer and dimer only (upper limit). (Figure taken from [150])

reactivity in the bulk. However, for certain clusters, namely, Cu^+, Cu_3^+, and Cu_9^+, a much smaller change in reactivity relative to their neighbors was evident which could be linked to structural and/or electronic effects.

There are a number of examples where the electronic character of clusters are found to play an important role in their behavior and reactivity. With respect to the thermodynamic stability of metal clusters, there are a large number of findings which offer support for the spherical jellium model (shell model of electronic structure) for the alkalies as well as for other metals, like copper and aluminum. An important example of the role of aggregation ("solvation") and electronic character on reactivity comes from studies [151] of aluminum cluster amions where the fast-flow-reactor technique has been used to investigate their reactions with O_2. Fig. 14a shows the cluster distributions following the introduction of a small amount of O_2 reactant, while Fig. 14b gives the distribution obtained following extensive reaction with O_2. Similar to other work [152] on cation clusters, the aluminum reactions proceed to etch atoms or dimers selectively away from the clusters, but with the charge being retained by

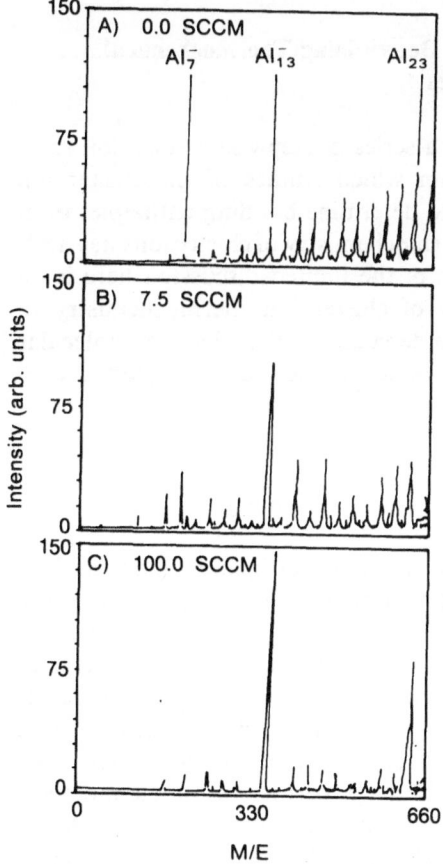

Fig. 14. Reaction of aluminum cluster anions with oxygen under thermal conditions in a flow-tube reactor (see Fig. 3). **A** typical cluster distribution; **B** oxygen added (7.5 standard cubic centimeters per minute) etches aluminum atoms from the anion; an odd–even alternation in the reactivity is evident; **C** at high oxygen additions, stable clusters corresponding to jellium calculations appear as the final products: Al_{13}^-, Al_{23}^-

the aluminum cluster anion. From the data it is readily discernible that the amounts of the charged species Al_{13}^- and Al_{23}^- are increasing dramatically as a result of selective etching of those clusters having a higher degree of aggregation than the ones being produced, i.e., the 13- and 23-mer are produced at the expense of larger cluster anions. These findings strongly point toward the jellium model [153, 154], although it is not certain whether the geometric nature of the clusters also partly contributes to these startling findings. More recent experiments [155] in which one niobium atom was inserted into the aluminum-anion clusters revealed an additional "magic number" comprised of four aluminum anions and one niobium anion bound with a single cluster. This is an 18-electron system which again, in terms of the jellium model, would be expected to be very unreactive. This continues to be a very active area of investigation, and further demonstrates the intriguing aspects of reactions influenced by aggregation and composition, namely solvation in the most general sense of the word.

2.4.4   Solvation Phenomena

2.4.4.1   Thermochemical Considerations: Determining Thermochemical Properties of Solvation Complexes

Cluster formation can be represented by a series of stepwise association reactions of the form shown in Eq. (6), from which studies of the cluster ion distributions at equilibrium can be used to determine bonding enthalpies from Eq. (7). Although a large portion of the thermochemistry of cluster ions has dealt with organic ions and solvents, the focus of the limited discussion here is on inorganic systems. For the applicability of cluster ion thermochemistry to problems of organic chemistry, such as understanding the effects of molecular structure, intrinsic acidity, and hydrogen bonding, the reader is referred elsewhere [156–158].

2.4.4.1a)   Consideration of Enthalpy Changes and Bonding

Comparing the relative bond strengths for a variety of ligands about a given positive ion is very instructive in elucidating the role of the ligand. For purposes of discussion, the enthalpy change, $-\Delta H_{0,1}^\circ$, is assumed to approximate the ion–molecule bond dissociation energy. This assumption is reasonable since ΔH° is expected to be only weakly temperature dependent for ion–molecule association complex formation [159, 160], a point which was discussed in Sect. 2.4.2.3. Data for a wide range of molecules of differing properties clustered to sodium are available, including ones with large permanent dipole moments and polarizabilities, as well as those having low permanent dipole moments but relatively large quadrupole moments. Trends for this simple ion with molecules having well known electric moments enable one to elucidate the role of the ligand's properties on bonding.

The ordering of the ligand molecules with respect to the magnitude [161] of the bonding of the first cluster with Na^+ is as follows: DME (dimethoxyethane) > NH_3 > H_2O > SO_2 > CO_2 > CO ≥ HCl > N_2 ≥ CH_4 (see Table 2). Additional considerations [162] suggest that DME forms a bidentate bond with Na^+; i.e., two functional groups on the ligand interact with the ions charge. Comparing the stronger bond for ammonia compared to water, in light of the electrostatic properties of the molecules, the importance of the ion-induced dipole interaction in governing bond strength is evident. Although ammonia has the smaller dipole moment, its much larger polarizability enhances its bonding at small cluster sizes. The role of the quadrupole moment is also seen by comparing the relative bonding of SO_2 with NH_3. Both have approximately equivalent dipole moments and polarizabilities, but the quadrupole moment of SO_2 leads to a repulsive interaction compared to the attractive one of NH_3, consistent with the comparatively low bonding strength of SO_2. Molecules without permanent dipoles typically have relatively small bond energies as seen for N_2 and CH_4, although polarizability can be important as evidenced by the relatively stronger bonding of CO_2 compared to N_2 and CH_4.

Another interesting result is found by comparing the association of molecules to ions of opposite polarity. Both Na^+ and Cl^- have closed electronic configurations and spherical symmetry (i.e., they are equivalent to noble gas-like atoms with a central charge), although the ionic radius of Na^+ is considerably smaller. Consequently, with other factors being equal, a neutral molecule would be expected to bind more strongly to the smaller Na^+. However, covalent bonding, charge transfer, and higher electrostatic moments can in some cases

Table 2. Stepwise heats of association, $-\Delta H_{n-1,n}$ in kcal/mole, for Na^+ with various neutral species[a]

L	$(n-1, n)$					
	(0, 1)	(1, 2)	(2, 3)	(3, 4)	(4, 5)	(5, 6)
$(CH_3OCH_2)_2$	47.2	35.1	23.2			
CH_3CN	–	24.4	20.6	14.9	12.7	
NH_3	29.1	22.9	17.1	14.7	10.7	9.7
H_2O	24.0	19.8	15.8	13.8	12.3	10.7
HNO_3	20.6					
SO_2	18.9	16.6	14.3	12.3		
CO_2	15.9	11.0	9.7	8.4		
CO	12.6	7.5				
O_3	12.5					
HCl	12.2					
N_2	8.0	5.3				
CH_4	7.2					

[a] Taken from [25] and references contained therein

also be important [163]. Considering experimentally determined stepwise heats of association, for the first complex it is evident [164] that only sulfur dioxide and hydrogen chloride bind more strongly to Cl^- than to Na^+. Ammonia exhibits the largest difference between Na^+ and Cl^-, with its binding to the negative ion being much weaker. The relative stability of the ion–molecule complexes for ammonia, water, and SO_2 are in reverse order for the two ions. In terms of magnitude, after the addition of the first ligand, the enthalpy change for the clustering of SO_2 onto Cl^- is much larger than for Na^+. Many trends are in accord with expectations from electrostatic considerations. For instance water, sulfur dioxide, and ammonia have similar dipole moments. However, when the dipoles are aligned in the electrostatic field of the ion, the small quadrupole moments of water and the considerably larger one of ammonia are repulsive to a negative charge and attractive to a positive one. In the case of sulfur dioxide, the situation is reversed. Thus, the ion–quadrupole interaction is consistent with the different ordering of the first bond energies for water, ammonia, and sulfur dioxide between the positive Na^+ and the negative Cl^-.

Another interesting effect is seen by comparing the relative bond energies for ions of the same sign, but with differing sizes, to a variety of ligands such as SO_2, H_2O, and CO_2 [169]. The hydrogen, sulfur, and carbon atoms are the centers which are attracted to a negative charge. Water has a slightly larger dipole moment than sulfur dioxide, but the quadrupole moment of water is repulsive when the dipole is attractive to a negative ion. Alternatively the quadrupole moment of SO_2 is attractive to a negative ion and carbon dioxide has no dipole but does have a significant quadrupole. Considering only charge–dipole or charge–multipole interactions, the bond strength for a ligand attached to a given ion would be expected to be in the order of H_2O about equal to SO_2, but both being greater than CO_2.

For a weakly basic or large ion like I^-, the order $SO_2 > H_2O > CO_2$ is observed. However, as the ions become smaller or more basic, SO_2 bonds relatively more strongly than water. With O_2^- the enthalpy change for addition of CO_2 becomes comparable to that of H_2O. Finally, for a small ion like O^-, the order $SO_2 > CO_2 > H_2O$ actually occurs. Interestingly, the mean polarizabilities for SO_2, CO_2, and H_2O follow the same order as their relative bond energies to small ions. Polarization energies are known to be relatively more important for smaller ions due to the ability of the neutral to closely approach the ion, and thereby become more influenced by the ionic electric field with the attendant result of a larger induced dipole. Qualitatively, consideration of the polarizabilities partially explains the increased bonding strength of SO_2 and CO_2 over that of water in clustering to smaller ions.

Evidence that the bonding of molecules to certain ions does involve some degree of covalent bonding comes from a number of experimental measurements which show that the bonding strengths exceed those expected on the basis of simple electrostatic considerations. Consider, for example, the bonding of H_2O to Sr^+ [166], and for H_2O and NH_3 to Pb^+ and Bi^+ [163(a), 167]. The stabilities of the cluster ions N_4^+, O_4^+, and $(CO)_2^+$ are all much greater than expected for ion–induced dipole interactions in the case of the first two, and for ion–dipole interaction in the case of the last species. It is suggested that bonding must arise due to the (partial) sharing of an electron by the two molecules [24, 168], although there are other possible contributing factors involving a closer ion-ligand approach such as in the case of Pb^+ [169].

Also, in the case of some anions, the bonds are sometimes strong enough to be considered as actual chemical ones instead of being due to merely weak electrostatic effects [7]; i.e., significant transfer of charge may occur between the original ion and the clustering neutral. An example is $OH^- \cdot CO_2$ which is more properly considered as the bicarbonate ion, and $Cl^- \cdot SO_2$ which has a rather strong bond energy for the first attachment, but a very weak one for the second addition. The first cluster addition may be thought of as forming a "molecule" over which the negative charge becomes relatively widely dispersed, thereby resulting in the significant energy difference between attachment of the first SO_2 molecule compared to the addition of the second [170].

For polyatomic ions, structural effects are evident in the stepwise clustering enthalpy changes [171, 172]. These ions have specific sites to which the addition of solvent molecules are expected to be most favorable. For instance, the successive enthalpy changes $-\Delta H_{n-1,n}$ for NO_3^- and CO_3^- decrease very slowly for the first three hydration steps, whereas for O_2^-, NO_2^-, and HCO_3^-, a sharper decrease is found, especially for the third hydration step. The difference in behavior reflects the fact that both NO_3^- and CO_3^- have three available oxygen atoms as clustering sites and the others only two (one of the oxygens of HCO_3^- is bound to the hydrogen). Other examples include the interaction of the hydronium ion, H_3O^+, with H_2O [173], where a break in the smoothly varying stepwise enthalpy changes has been observed between additions of the third and fourth ligand, and for NH_4^+ with both H_2O [174] and NH_3 [175, 176] where the break occurs between the fourth and fifth addition steps. The 'magic numbers" frequently observed upon ionizing neutrals of these systems are attributable to these particularly stable cluster ion structures.

The attachment of neutrals to some atomic ions have also shown irregularities which suggest that well-defined coordination shells exist in the gas phase. For example, as seen in Fig. 15a, both Li^+ and Na^+ show an irregularity in the stepwise enthalpy change for clustering of ammonia [40]. These data indicate a coordination number of four, in agreement with Raman spectra studies of Li^+ in liquid ammonia. In the larger alkali ions, more molecules apparently can be accommodated around the ion and no irregularities are observed for the addition of up to six molecules. The transition metal ions Ag^+ and Cu^+ have a strong preference for coordination by two water or two ammonia molecules [116].

Taft and coworkers [177–179] considered the problem of ion solvation, including concomitant effects on the binding to various ligands in the presence of solvent molecules analogous to interactions in the liquid phase. Related work by *Arnett* et al. [180] has also stressed the difference in relative bonding strengths in the presence of a solvent compared to the gas-phase binding of ions

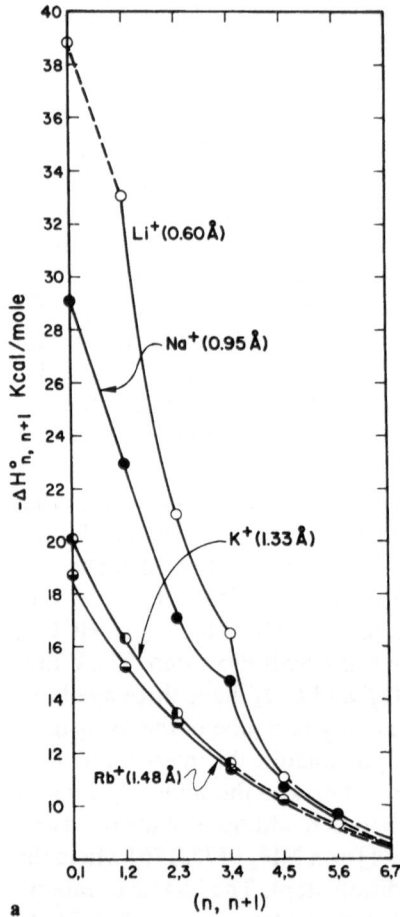

a

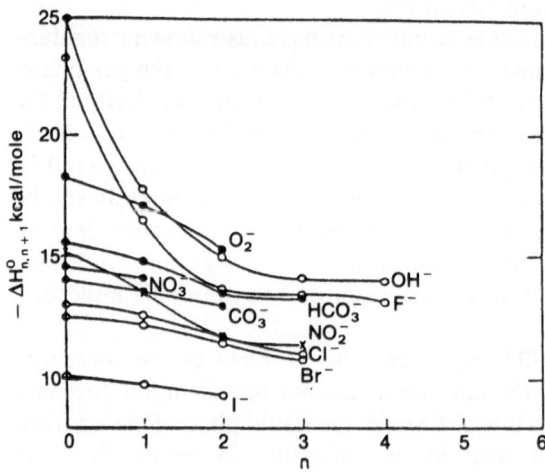

b

Fig. 15. a Plot of $\Delta H^{\circ}_{n,n+1}$ versus cluster size for the reaction $M^{+}(NH_3)_n + NH_3 \rightleftarrows M^{+}(NH_3)_{n+1}$ for several alkali metal ions (Figure taken from [40]). b Enthalpy changes for the stepwise gas-phase hydration of several negative ions (Data taken from [171, 172])

with ligands. Effects related to relative binding of mixed ligands in various solvation shells are by comparing data on relative stabilities. For instance, it has been found [181, 182] that the cluster ions formed by ionizing mixed neutral clusters of alcohols and water dissociate with preferred loss channels of either alcohol or water depending on the degree of aggregation, and similar trends have been observed [183] for mixed ammonia-water clusters. The data indicate that water is lost preferentially compared to ammonia for inner solvation shells where the proton affinity and the relative bonding of the ammonia molecules is stronger, while water is preferentially held compared to the ammonia for outer solvation shells where the latter binds less strongly. These are all examples of the manifestations of the thermodynamic stabilities of the mixed clusters.

A general tendency of the $\Delta H°$ values to approach the enthalpy of condensation at large cluster sizes is seen for ions of both signs clustered with water, as well as with other ligands [7]. (Fig. 15b.) Interestingly, however, data for water onto I^- show that it is possible for the absolute values of the heat of association to fall below that of heat of condensation at intermediate cluster sizes [7, 184]. This is understandable in the case of systems where the initial orientation of ligands about the ion hinders the development of further solvation by disruption of the preferred structure of the solvent molecules among themselves.

2.4.4.1b)   Entropy

The stability of small gas-phase clusters is influenced by entropy as well as bonding. There are three major contributions to the entropy change for a clustering reaction, namely translational, rotational, and vibrational effects; electronically excited states are rarely present. If distinguishable cluster structures exist that have similar energies [7, 185], a configurational contribution may also need to be considered. However, this contribution is essentially related to the number of isomers of comparable energy and these should be small for small ion clusters. The rotational, and particularly the vibrational contributions, are significant in that they reflect the details about the structure of the cluster ion.

In the case of an ion-neutral association reaction, the translational contribution is dominant since the combining of two particles into one results in the overall negative sign in the entropy change. The entropy changes for the various reactions differ primarily due to the rotational and vibrational contributions. Except in the case of an atomic ion, the rotational contribution is of the same sign as the translational one [12, 40]. These two contributions are partially offset by the vibrational frequencies. As an example, based on geometries and frequencies derived from a molecular orbital calculation on $Cl^- \cdot H_2O$ [186], the contributions to $\Delta S°$ at 470 K for the association of water onto Cl^- are ΔS (trans) $= -35.6$, ΔS(rot) $= +10.2$, and ΔS(vib) $= +6.3$ cal/K mol; the total entropy change is thereby deduced to be -19.1 compared to the experimentally determined value of -19.7 cal/K mol [187]. An accurate calculation of entropy changes, employing standard statistical mechanical procedures, requires information on the structure and vibrational frequencies of the ion clusters; unfortunately, these properties are not reliably known in most cases, especially for larger clusters. Although it is in principle possible to do so, rarely are the calculational methods extended to obtaining frequencies and entropy changes. As super computers become more readily available and the computational results become more reliable, calculation of entropy changes, in addition to bond energies, are expected to become more common.

As clustering proceeds, the motions of the ligands are constricted due to crowding by neighboring ones. As a result, internal rotational and bending frequencies become higher; but this is countered by a decrease in the stretching frequencies due to weaker bonding. Experimentally, the $\Delta S^{\circ}_{n-1,n}$ values are often found to at first become more negative upon successive clustering, an indication that crowding is overcompensating for the effect of weaker bonding. Such trends are valuable in gaining insight into cluster structure. In some cases a shift to smaller absolute entropy as well as enthalpy values happens upon reaching some degree of clustering. Sometimes this can be attributed to the onset of formation of a second shell of solvation around the ion, where the newly bound molecule is attached to the solvent shell and not the ion itself. Obviously, the first such molecule to occupy the periphery for the first solvation shell will have no restricted motions due to neighboring molecules.

2.4.4.2 Nucleation

Nucleation of supersaturated gases proceeds via the formation of gas-phase clusters, and information on their changing properties and behavior as a function of their degree of aggregation finds application in elucidating physical aspects of nucleation phenomena. A complete description of nucleation requires information on the kinetics of each individual forward and reverse step in the clustering reaction sequence:

$$IL + L \underset{}{\overset{M}{\rightleftharpoons}} IL_2 ,$$

$$IL_2 + L \underset{}{\overset{M}{\rightleftharpoons}} IL_3 ,$$

$$\vdots$$

$$IL_n^* + L \rightleftharpoons \text{nucleation step} . \tag{48}$$

Clustering is expected to evolve from kinetic processes dominated by third-order reactions at small sizes (in which a third-body M is required to remove the excess energy arising in the collision complex) to bimolecular kinetic mechanisms, dominated by the high pressure limit of the forward rate constant expressed by Eq. (47). This will occur because of the increased degrees of freedom and hence increased lifetime of the collision complex for large clusters. At present, there are no general theories of association reactions that can be used to quantitatively predict the rates and describe the transition from third-order to bimolecular dominated forward kinetics as cluster growth proceeds.

Classical nucleation theory [188] circumvents the difficulties of making a complete kinetic description of the phenomena, by assuming that a quasi-steady-state population of clusters exists, and that the rate of nucleation is determined by the rate of collision of gas molecules with clusters of a critical size, i.e., the last step in Eq. (48). At the critical size the clusters tend to grow spontaneously and smaller ones tend to evaporate. Theoretical formulations [189] need to be developed to quantify the free energy of formation of the cluster of critical size and thus determine the energy barrier to nucleation.

Experimental studies of the clustering of molecules onto ions in the gas phase have provided thermodynamic data for small clusters [3, 7]. A simple and often successful model used in nucleation theory is the liquid-drop model, in which the energetics of the critical nucleus are described in terms of macroscopic quantities such as surface tension and vapor pressure (194). However, comparison of the experimental results with theoretical treatments of ion-induced nucleation has suggested that the major failing of the classical liquid-drop theory lies in the neglect of ordered structure in small cluster ions [12]. Nevertheless, the classical liquid drop model does remarkably well in accounting for the enthalpy changes [14] and their trends for clustering onto ions, although the experimental entropy changes are considerably more negative than those deduced from the liquid-drop model [190].

The classical liquid drop formulation of Thompson [3, 191] for the case of ion-induced nucleation is based on the relationship

$$\Delta G_{0,n} = -nRT \ln S + 4\pi N r^2 \sigma + (q^2 N/2)(1 - 1/\varepsilon)(1/r - 1/r_i) \tag{49}$$

which is used to evaluate the free energy for formation of the nth cluster. Here ε is the dielectric constant of medium L, q is charge, and r and r_i are the effective radii of the cluster and ion, respectively. The cluster size at which $\Delta G_{0,n}$ obtains a local maximum defines the critical cluster. The first term on the right-hand-side accounts for the change in free energy due to the condensation of n molecules at a saturation ratio S (where S is the ratio of the partial pressure of the ligand to the vapor pressure of the bulk liquid at the same temperature). For the standard state, the partial pressure is conveniently taken to be 1 atm. The second term represents the work done in forming a droplet of radius r (where σ is the surface tension and N is Avogadro's number). The final rhs term gives the change in field energy due to the condensation of a dielectric about the ion. It should be noted that, in the Thomson equation, the nature of the ion is accounted for only by its radius and total charge; sign of the charge plays no specific in contrast to experimental findings [192, 193]. The stepwise change is given simply by $\Delta G_{0,n} - \Delta G_{0,n-1}$.

The Thomson equation can be rewritten in terms of the number of molecules by assuming the relation based on the volume of the droplet

$$n = 4\pi(r^3 - r_i^3)\rho N/3M_w \tag{50}$$

where M_w and ρ are the molecular weight and the bulk density of the liquid, respectively. Differentiation of Eq. (49) with respect to temperature enables an evaluation of entropy changes whereupon enthalpy changes are then deduced by substitution into well-known thermodynamic relationships. Studies of gas-phase clustering equilibria provide experimental data for the small site domain and also direct comparison to the theoretical treatment [14, 197].

2.4.4.3 Energetics of Ion Solvation

Using the above equations as well as the concepts discussed in the development of Eqs. (6) and (7), it is possible to deduce expressions which connect gas phase clustering and ion solvation in liquids. The overall energy associated with solvation is given by [1, 3, 7].

$$I(g) = nL(g) \rightarrow I \cdot nL \ , \tag{51}$$

where the standard free energy change is given by

$$G_{0,n}^{\circ} = \sum_{i=1}^{n} \Delta G_{i-1,i}^{\circ} \tag{52}$$

whereas in the latter case

$$L(g) + nL(l) \rightarrow I \cdot nL(l) \tag{53}$$

with a single free energy of solvation of $\Delta G_{(sol)}^{\circ}$ (at infinite dilution for large n). Here, L designates the ligand solvent and l the condensed phase. When n is sufficiently large, the product of Reaction (53) is a liquid droplet and so the free energies are related by

$$\Delta G_{0,n}^{\circ} = \Delta G_{sol}^{\circ} + \sum_{i=2}^{n} \Delta G_{i-1,i}^{\circ}(L) \tag{54}$$

Hence, the second term on the right-hand side is the free energy change for forming a droplet through the stepwise clustering of the solvent molecules. At sufficiently large n, $\Delta G_{i-1,i}^{\circ}(L)$ becomes the free energy of condensation for neutral L. Corresponding equations can be written for the enthalpy and entropy changes.

One of the first attempts to predict solvation energies in the condensed state was based on the Born relationship [194] which is the basis of the last term in Eq. (49). The Born relationship often leads to an overestimation of the heat of solvation when typical values of crystalline radii are used to account for the ion size. This failure has led to proposed modifications based on structural consider- ations, as well as other attempts to employ detailed ion-dipole, ion-quadrupole, and higher order interactions [195–197].

Although the experimental data for gas-phase clustering onto ions do not extend to sufficiently larger clusters where droplet formation can be expected to occur, an approach for relating the gas-phase data to solvation phenomena is to consider the differences between various ion types and compare these to the differences expected for the solution phase [198, 199]. By considering the differ- ences, the condensation contribution (the last term of Eq. (54)) to the gas-phase data is cancelled. Although the solvation term is still incomplete at this point, for very large clusters the gas-phase and solution data are expected to converge. Approximately 60% of the differences between single-ion heats of solvation for the halides are found to be accounted for by the first four gas-phase hydration steps [200]. Comparison of the difference between halide and alkali ions has been used as a consistency check on the results obtained with various methods from the solvation of salts [184].

Another approach has been to consider the ratio $\Delta H^\circ_{solv}/\Delta H^\circ_{0,n}$ as a function of cluster size [14]. Fig. 16 shows the ratio, as calculated by the Born (liquid

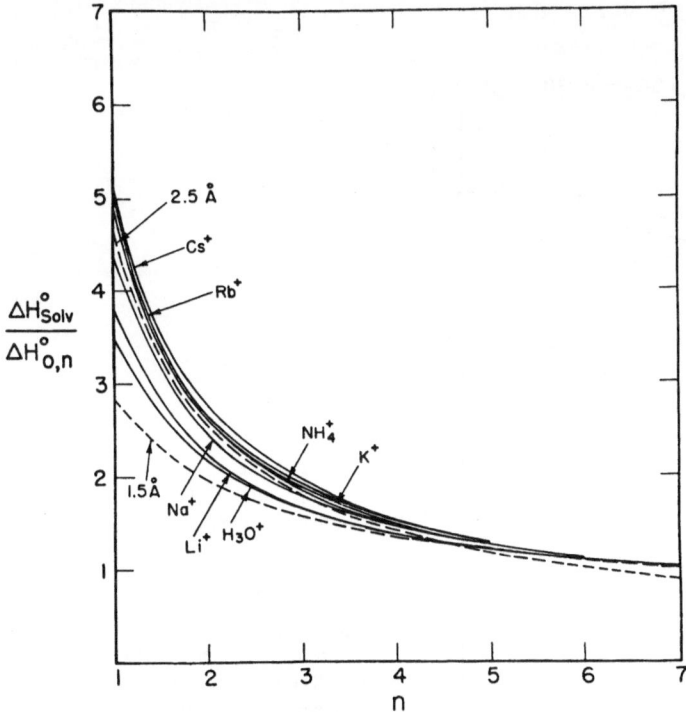

Fig. 16. Ratio of Randles' total enthalpy of solvation to the partial gas-phase enthalpy of hydration for positive ionic cluster size n. (Taken from [3])

phase) and Thomson (cluster) equations, compared to experimental data for the hydration of positive ions. The ratio is seen to converge for a range of ionic radii for clusters of about five solvent molecules. Similar results are also obtained for the ammoniation of positive ions and hydration of negative ions [14]. The findings suggest that valuable solvation data can be deduced from the gas-phase experiments.

Another interesting aspect deals with binary solvents. A question arises as to the influence of the composition of the solvent on the interactions in the neighborhood of the ion, and the extent to which it differs from the bulk solution. Insight can be gained by considering the total heat of clustering, $-\Delta H_{fc}^{\circ}$, as a function of the composition of the product cluster ion. As an example, Fig. 17 gives the heat for the mixed clustering of H_2O and SO_2 with Cl^-. Clearly, the values for the mixed systems are higher than the compositionally weighted average value of the pure systems (represented by the solid line) which suggests an enhancement in the overall clustering of the mixed system [201] due to interactions between the two different ligands. In contrast, for example, data [202] from the exchange reactions involving the mixed clusters of benzene and water bound to K^+ show a slight decrease or no deviation from the average value. Plots of this type are valuable in assessing compositional effects on the bonding of mixed cluster systems, and on trends

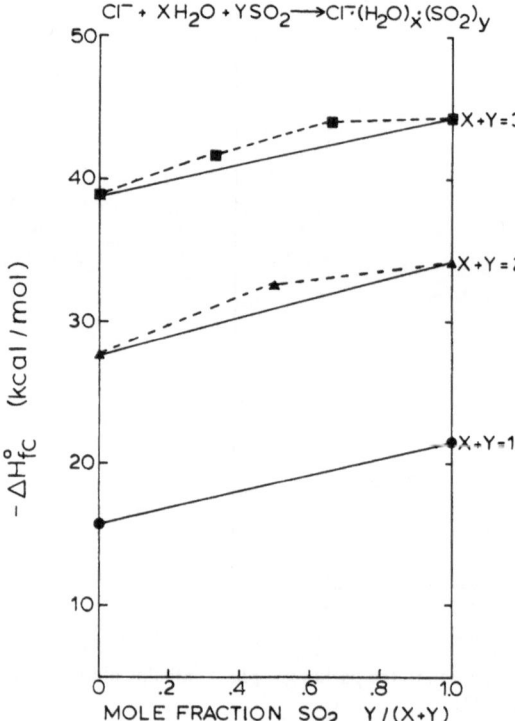

Fig. 17. The heats of clustering, $-\Delta H_{fc}^{\circ}$, for $Cl^-(H_2O)_x(SO_2)_y$. Note the enhancement, over a weighted-average, in the enthalpy of clustering for this system comprised of two different ligands. The enhancement is due to ligand–ligand attractive interaction; taken from Upschulte, B.L., Schelling, F.J., Keesee, R.G. and Castleman, A.W. Jr. Chem. Phys. Letters III, 389 (1984); also see [3]

expected for complexes of liquid phase interest and we can expect to see more use being made of them in the future.

2.4.5 Photodissociation Experiments of Solvated Cluster Ions

2.4.5.1 Spectroscopy

Photodissociation spectroscopy of solvated systems, which is briefly touched on in this chapter for completeness, is a valuable technique for probing the spectroscopic and dynamic structure of cluster ions [9, 26] at different degrees of aggregation. Study of these systems offers an opportunity to observe solvation effects, compare the behavior of members of homologous series, and investigate intramolecular energy transfer. A great variety of new questions dealing with the dynamics and the mechanisms of the photodissociation process itself are raised by such studies. The identity of the fragments into which the cluster dissociates following laser excitation, and the ways in which initial energy is partitioned among the various degrees of freedom of the fragments, are of particular interest. An advantage of studying cluster ions is the relative ease of incorporating methods to determine the energy release upon dissociation. This greatly enhances the power of the technique since analysis of the kinetic energy distributions of the fragment ions indicates where the excess energy of the photons is channeled by the dissociation process and, hence, helps identify the specific steps in the dissociation mechanism. Insight into the structure of clusters also can be gained from such measurements [26, 203, 204].

Vibrational predissociation can occur when one of the molecular subunits absorbs an amount of energy sufficient to dissociate a bond in the cluster. Coupling between the core vibrational modes and the cluster vibrations allows energy to be transferred from one mode to the other throughout the cluster. Photodissociation studies of cluster ions have been studied using a variety of instruments whose description is beyond the scope of the present chapter.

Some interesting examples [26] of the influence of solvation on the dissociation dynamics include studies of $SO_{2-} \cdot SO_2$, where dissociation results through a repulsive surface, and $CO_3^- \cdot (H_2O)_n$, where the presence of the water impedes the dissociation channel

$$CO_3^- + h\nu \rightarrow CO_2 + O^- \tag{55}$$

through solvation effects; the absorption of energy leads instead to the loss of water molecules. The SO_2 dimer anion has been observed [26, 205] to photodissociate to SO_2^- as the sole ionic photofragment; its photodissociation cross section has been reported [206] to be a smooth function displaying a maximum near 2.1 eV and having a total cross section of $1.9 \times 10^{-17} \text{ cm}^2$. Power studies at 600 nm (2.07 eV) indicate that the dissociation is a one-photon process. Based on measurements of the bond energy for $SO_2^- \cdot SO_2$ [1.04 eV] [207] and that for

$SO_2 \cdot SO_2$ [0.19 eV] [208], and taking electron affinity for SO_2 to be 1.097 eV [209], the photodissociation results lead to 1.95 eV as the electron affinity of the dimer ion.

Energy analysis experiments of $SO_2^-(SO_2)$ performed at different laser polarizations show that the measured laboratory frame energy distribution of the fragment ions is widest under conditions where the laser polarization is oriented along the ion beam axis. Experimental energy distributions of the SO_2^- photofragments are best fit by an anisotropy parameter which suggests that the transition is more parallel than isotropic, and that the dissociation is rapid with respect to rotation [26, 205]. Measurements of the average kinetic energy release upon dissociation show values far larger than those predicted from a statistical phase space theory, further pointing to dissociation on a repulsive surface.

Another example of a cluster system whose photodissociation dynamics have been studied in detail includes CO_3^- and its hydrates [26, 203, 204]. In the photon energy range 1.95 to 2.2 eV, photodissociation of CO_3^- leads to O^- and CO_2 fragments in accordance with Eq. (55). The photodissociation spectrum shows a well-defined and sharp structure which reproduces the spectra obtained in earlier experiments [210]. Excitation to a bound intermediate excited state ion has been confirmed. Dissociation of the core anion occurs by two mechanisms, one involving two-photon excitation to a repulsive surface and the other collisional dissociation of an excited anion which has undergone internal conversion from the excited intermediate state to high vibrational levels of the ground state. Photodissociation is initiated by a $^2A_1 \leftarrow {}^2B_1$ bound-bound transition that promotes the ion from the ground state to the intermediate weakly bound excited state ion. Sharp features observed in the photodissociation spectra were found to be identical to those which arise from the vibrational structure of the intermediate $^2A_1$ state. Analysis of the kinetic energy for the two-photon mechanism indicates that a large fraction (about 20%) of the available energy is deposited into relative translation of the photofragments.

The bond energies for the first three water molecules with CO_3^- have been measured to be approximately 0.6 eV each [200]. The only photodissociation channel observed for the CO_3^- hydrates is loss of all attached water molecules. It is an interesting observation that, in the larger clusters where partial dehydration is possible, the loss of all ligands is the only observed channel. The kinetic energy analysis of the CO_3^- photofragment ions from all three hydrates indicates that more energy is partitioned into translation than is predicted by phase space theory. The overall percentage of available energy that is channeled into relative translation is 27%, 18%, and 27% for the first, second, and third hydrates, respectively.

In the case of the hydrates, no O^- photofragment ion is detected. Conversion of energy from the core ion into the ion-water bond vibration leads to dissociation of the cluster. In the case of the CO_3^- hydrates, neither a $\Delta v = -1$ nor -2 transition would be sufficient to result in bond dissociation. Although vibrational predissociation would be expected to pertain in these systems, a surprisingly large amount of energy is partitioned into relative translation and hence the possibility that dissociation to some repulsive surface is operative in this cluster system cannot be ruled out [26].

Lineberger and coworkers [211] examined the photodissociation of $(CO_2)_n^+$ for $n = 2$–10. The relative cross-sections at 1064 nm were found to decrease sharply between the trimer and tetramer ion. However, at 644 nm, the relative cross-sections increased slowly and at 532 nm were essentially independent of cluster size. The chromophore in these clusters is believed to be the dimer ion $(CO_2)_2^+$. In subsequent studies, they [212] also observed that a constant average neutral mass loss per unit photon energy is approached for sufficiently large $(CO_2)_n^+$ clusters. The parent cluster size at which the constant value is reached increases with photon energy, but a constant mass loss per unit absorbed energy appeared to be reached when the parent cluster ion was larger than its average fragment cluster ion by about 13 molecules or more. From the number of CO_2 neutrals lost versus photon energy, an upper limit of 4.9 kcal/mol is found for the dissociation energy of a CO_2 molecule from these large clusters compared to 5.7 kcal/mol for the sublimation of bulk CO_2.

Gas-phase clusters also allow caging effects of solvents to be studied. For example, *Lineberger* and coworkers [213] have observed the cessation of Br_2^- photodissociation as the ion becomes clusters by CO_2 molecules; about ten molecules effectively trap the Br_2^- and prevent its photodissociation. A theoretical study [214] suggests that caging begins weakly at small sizes because of strong ion-induced dipole interactions, so that ionic clusters can have a caging effect due to attractive interactions. With larger clusters, the repulsive caging caused by collisions with surrounding solvent molecules dominates. Numerous other data on the structure of cluster ions are now becoming available from the spectroscopic studies of *Lee* and coworkers [215, 216] and *Saykally* and colleagues [217] and are discussed elsewhere in this book.

Electron solvation by nonmetallic species has been a problem of long-standing interest. For example, negative ions of CO_2, H_2O, and NH_3 are unstable, but electrons can be solvated in their condensed phases. The minimum size cluster which is required to bind the electron is a topic of considerable current interest [218, 219]. For CO_2, the negatively charged dimer is observed to be stable, and theoretical calculations [220, 221] indicate that the stable dimer ion is a bent CO_2^- solvated by a linear CO_2 molecule. Thus solvation of a molecular anion leads to its stability. Similarly, CO_2^- can be stabilized by a H_2O solvent molecule [222]. Negative ions [223] of NH_3 and H_2O clusters may involve diffuse electron states as opposed to localization on a given molecule, and the relative stability of surface states, trapped electrons, dipole-bound electrons are topics actively being pursued [219].

2.4.5.2 Investigating Basic Mechanisms

Study of the mechanisms of photodissociation of the rare-gas cluster ions has been the subject of several recent experimental and theoretical investigations [29, 224, 225]. Such studies have been stimulated by the important role these ions play in rare gas excimer lasers, as well as by the fact that their comparative simplicity allows comparison between experiments and theories.

There have been numerous experimental and theoretical studies of the argon trimer cation. Interest has been prompted by observations [29] that large cluster ions of argon readily photodissociate, and there are indications that the trimer may be the chromophore. From a practical viewpoint, Ar_3^+ is of importance in power loss in XeF excimer lasers. Models [224, 225] have suggested that Ar_3^+ is likely to be formed in the intense, high pressure discharges which occur in XeF lasers and that Ar_3^+ may be the transient absorber in these laser plasmas. Absorption by Ar_3^+ near 500 nm, with a cross section $> 10^{-17}\,cm^2$, would significantly reduced the output energy of the A ← C transition of a XeF excimer laser.

Based on the photodissociation spectrum of a mass-selected beam of Ar_3^+, the relative cross-section is known to rise with photon energy from 2.00 eV (620 nm) to 2.38 eV (520 nm) [27]. The value of the cross section is found to be about $1.8 \times 10^{-16}\,cm$ at 520 nm. Power studies at 590, 575, and 545 nm indicate that dissociation is a one-photon process.

There are two possible channels which could account for the formation of Ar^+ from Ar_3^+. One channel is a concerted loss of both Ar units, while the second involves the sequential loss of Ar. These two channels are indistinguishable on the time scale of the typical spectroscopic experiment, a few tens of microseconds from the laser interaction region to detection. However, the two can be distinguished by energy analysis of the photofragments.

In the photodissociation processes, the total energy is given by

$$E_{av} = h\nu - D_0 + E_{int}^P = E_{int}^F + Et_t \tag{56}$$

Here, $h\nu$ is the energy of the absorbed photon and D_0 is the dissociation energy of the ground state of the parent to the ground state of the fragments, and E_{int}^P is the internal energy of the Ar_3^+ ion. The D_0 for Ar_3^+ dissociation to Ar_2^+ ($^2\Sigma_u$) and Ar or Ar^+ ($^2P_{3/2}$) and 2 Ar have been determined experimentally [25] as 0.22 eV and 2.52 eV, respectively. The E_{int}^P is considered negligible because the internal temperature of the ion clusters is very low due to the method of the cluster ion production and the geometric cooling effect in the supersonic expansion beam experiments. How the available energy E_{av} becomes partitioned between the center-of-mass frame-translational energy of the photofragment, E_t, and the internal energy of the fragment, E_{int}^F, depends on the photodissociation mechanism.

The several energetically accessible dissociation mechanisms are

$$Ar_3^+ \xrightarrow{h\nu}$$

$$\begin{cases} [Ar \ldots Ar^+ \ldots Ar]^* \to Ar^+ + 2Ar & (57a) \\ [Ar^+ \ldots Ar \ldots Ar]^* \to Ar^+ + 2Ar \;(or\; Ar_2^*) & (57b) \\ [Ar \ldots (Ar-Ar)^+] \to [Ar-Ar]^{+*} + Ar \to 2Ar + Ar^+ & (57c) \end{cases}$$

If the lifetimes of the transition states in reactions (57a), (57b), and (57c) are shorter than their rotational periods, they should carry different translational energy which converts from excess electronic excited energy in the photodissociation reaction. It is found [27] that photodissociation involves all three types of mechanisms which result in different amounts of translational energy released from the electronic excitation. This translational energy is further confirmed by a study using different photon energies. In the photodissociation of Ar_3^+ via Mechanisms (a) and (b), the photodissociation may involve electronic direct dissociation. In Mechanism (c), photodissociation may involve electronic predissociation occurring at crossing electronic energy surfaces. The photodissociation mechanism will depend upon the energy of the photon used in the reaction as well as other thermodynamical conditions for the parent preparation.

Similar photodissociation studies have also been carried out for Kr_n^+ [226]. In these studies, the photodissociation of Kr_n^+ (up to $n = 11$) cluster ions have been studied over the range 565 to 630 nm. Kr_3^+ has a photodissociation cross-section of $(8.1 \pm 0.8) \times 10^{-17}$ cm^2 at 612 nm, while Kr_4^+ and Kr_5^+ has are found to have larger photodissociation cross-sections than Kr_3^+ with a slightly red-shifted spectrum. Only Kr^+ is detected as a photofragment of Kr_3^+, while in the Kr_n^+ ($4 \leq n \leq 7$) experiments, both Kr_2^+ and Kr^+ photofragments are seen. As size n increases from 7 to 11, Kr^+ disappears and Kr_3^+ appears. Laser power studies show that all are single photon photodissociation processes. For higher order Kr_n^+ clusters ($4 \leq n \leq 7$) the intensity ratio between the two photodissociation products Kr_2^+ and Kr^+ is dependent on the wavelength of the laser light used in the photodissociation, but independent of the polarization direction of the laser.

Translational energy analysis of the photofragments has been used to investigate the photodissociation mechanisms. In contrast to the photodissociation of Kr_3^+, where two types of Kr^+ photofragments with different values of translational energy release are observed, only one type of Kr_2^+ photofragment with zero kinetic energy release is found in the Kr_4^+ photodissociation. These findings are discussed in terms of the dynamics of photodissociation and possible structures of these cluster ions in [226].

Unimolecular decay of cluster ions has become a subject of extensive interest with metastability (with lifetimes up to several hundred microseconds) being observed in a wide variety of chemically different systems. The interest in such studies stems from the role of unimolecular decay in the origin of magic numbers [67] and in the testing of various theories and mechanisms for such processes discussed earlier. The role of rotational tunneling in effecting the very long metastable lifetime of Ar_3^+ has also been examined [227] and found to be a viable process of possible importance. Theoretical considerations indicate that there exists a few percent likelihood of observing tunneling lifetimes from 10^{-10} to 1 s. Thus, tunneling through rotational barriers may contribute to the long lifetimes observed in a wide variety of clusters. In an attempt to determine the likelihood that tunneling through rotational barriers is a general mechanism in clusters, the probability of tunneling in larger argon cluster ions was also investigated [228], and also found to be potentially significant.

Calculations performed using several different rotational distributions indicate that there exists a probability of at least several percent for having tunneling transmission coefficients ranging over ten orders of magnitude. Similar distributions are obtained for all argon cluster ions investigated, $Ar_{n=2-11}^+$. Tunneling can and does occur in all clusters investigated. From Ar_3^+ to larger clusters, the probabilities are fairly insensitive to size, although the form of the dependence changes with the form of the rotational distribution. This trend is consistent with predictions of size insensitivity [229]. Investigation of possible tunneling in concert with photodissociation through predissociating states will undoubtedly emerge as an important area for future studies.

2.4.6. Recent Developments

There have been a large number of recent contributions to this field. The interested reader can find a summary of recent developments in a Chapter entitled *Cluster Reactions* by *A.W. Castleman*, Jr. and *S. Wei*, in Annual Review of Physical Chemistry 1994.

Acknowledgments. Financial support by the U.S. Department of Energy, Grant No. DE-FGO2-88ER60648, and the National Science Foundation, Grant No. ATM-9015855, is gratefully acknowledged.

References

1. A.W. Castleman, Jr., R.G. Keesee: Science, **241**, 36 (1988)
2. A.W. Castleman, Jr., R.G. Keesee: Ann. Rev. Phys. Chem. **37**, 525 (1986)
3. A.W. Castleman, Jr., R.G. Keesee: Accts. Chem. Res. **19**, 413 (1986)
4. *Large Finite Systems*, Vol. 20, ed. by J. Jortner, A. Pullman, B. Pullman: D. Reidel Publishing Company (1987)
5. *Physics and Chemistry of Small Clusters*, ed. by P. Jena, B.K. Rao, S.N. Khanna, NATO ASI Series, Vol. 158 (1987)
6. A.W. Castleman, Jr., T.D. Märk: In *Gaseous Chemistry and Mass Spectrometry*, J.H. Futrell, ed. Wiley-Interscience, New York, p. 259 (1986)
7. A.W. Castleman, Jr., R.G. Keesee, Chem. Rev. **86**, 589 (1986)
8. (a) *Elemental and Molecular Clusters* (G. Benedek, T.P. Martin, and G. Pacchioni, Eds.) Springer-Verlag (1988). (b) *Surface Science* **156**, Part 1 and Part 2 (1985).
9. T.D. Märk, A.W. Castleman, Jr.: In *Adv. Atomic Molecular Phys.* **20**, 65 (1984)
10. P. Schuster, P. Wolschann, K. Tortschanoff: Dynamics of Proton Transfer in Solution, In *Chemical Relaxation in Molecular Biology* (I. Pecht and R. Rigler, Eds.) Springer-Verlag, 107–190 (1977)

11. A.W. Castleman, Jr.: Adv. Colloid Interface Sci. **10**, 73 (1979)
12. A.W. Castleman, Jr.: P.M. Holland, R.G. Keesee: J. Chem. Phys. **68**, 1760 (1978)
13. J. Heicklen: *Colloid Formation and Growth. A Chemical Kinetics Approach*, Academic, New York (1976)
14. N. Lee, R.G. Keesee, A.W. Castleman, Jr.: J. Coll. Interface Sci. **75**, 555 (1980)
15. A.W. Castleman, Jr., R.G. Keesee: Aerosol Sci. Technol. **2**, 145 (1983)
16. N.H. Turner, B.L. Dunlap, R.J. Colton: J. Anal. Chem. **56**, 373R (1984)
17. J.M. Goodings, S.D. Tanner, D.K. Bohme: Can. J. Chem. **60**, 2766 (1982)
18. D. Smith, N.G. Adams: Top Curr. Chem. **89**, 1 (1980)
19. E.E. Ferguson, F.C. Fehsenfeld, D.L. Albritton: In *Gas Phase Ion Chemistry*, ed. by M.T. Bowers, Academic Press, New York, Vol. 1, p. 45 (1974)
20. D. Smith, N.G. Adams: Int. Rev. Phys. Chem. **1**, 271 (1981)
21. W. Henkes: Phys. Lett. **12**, 322 (1964)
22. E.U. Franck, University of Karksruhe, personal communication (1978)
23. K.D. Pitzer: J. Phys. Chem. **87**, 1120 (1983)
24. P. Kebarle: Ann. Rev. Phys. Chem. **28**, 445 (1977)
25. R.G. Keesee, A.W. Castleman, Jr.: J. Phys. Chem. Ref. Data **15**, 1011 (1986)
26. A.W. Castleman, Jr., C.R. Albertoni, K.O. Marti, D.E. Hunton, R.G. Keesee: Fara. Discuss. Chem. Soc. **82**, pp. 261–273 (1986)
27. Z.Y. Chen, C.R. Albertoni, M. Hasegawa, R. Kuhn, A.W. Castleman, Jr.: J. Chem. Phys. **91**, 4019 (1989)
28. C.R. Albertoni, R. Kuhn, H.W. Sarkas, A.W. Castleman, Jr.: J. Chem. Phys. **87**, 5043 (1987)
29. (a) N.E. Levinger, D. Ray, M.L. Alexander, W.C. Lineberger: J. Chem. Phys. **89**, 5654 (1988);
    (b) N.E. Levinger, D. Ray, K.K. Murray, A.S. Mullin, C.P. Schulz, W.C. Lineberger: J. Chem. Phys. **89**, 71 (1988)
30. O. Echt, P.D. Dao, S. Morgan, A.W. Castleman, Jr.: J. Chem. Phys. **82** 4076 (1985)
31. S. Wei, W.B. Tzeng, A.W. Castleman, Jr.: J. Chem. Phys. **92**, 332 (1990)
32. S. Wei, W.B. Tzeng, A.W. Castleman, Jr.: J. Chem. Phys. **93**, 2506 (1990)
33. W.C. Wiley, I.H. McLaren: The Review of Scientific Instruments **26**, 1150 (1955)
34. J.L. Durant, D.M. Rider, S.L. Anderson, F.D. Proch, R.N. Zare: J. Chem. Phys. **80**, 1817 (1984)
35. H. Kuhlewind, U. Boesl, R. Weinkauf, H.J. Neusser, E.W. Schlag: Laser Chem. **3**, 3 (1983)
36. V.I. Karataev, B.A. Mamyrin, D.V. Shmikk: Sov. Phys. Tech. Phys. **16**, 1177 (1972); V.A. Mamyrin, V.I. Karataev, D.V. Shmikk, V.A. Zauglin: Sov. Phys. JETP **37**, 45 (1973)
37. J.B. Anderson, R.P. Andres, J.B. Fenn: Adv. Chem. Phys. **10**, 275 (1966); O.F. Hagena: Surf. Sci. **106**, 101 (1981)
38. C.E. Klots: J. Chem. Phys. **83**, 5854 (1985)
39. C.E. Klots: Z. Phys. D. **5**, 83 (1987)
40. A.W. Castleman, Jr., P.M. Holland, D.M. Lindsay, K.L. Peterson: J. Am. Chem. Soc. **100**, 6039 (1978)
41. E.E. Ferguson, F.C. Fehsenfeld, A.L. Schmeltekopf: Adv. At. Mol. Phys. **5**, 1 (1969)
42. D. Smith, N.G. Adams, Chapter 1 in: *Gas Phase Ion Chemistry*, Vol. 1, ed. by M.T. Bowers, Academic Press, p. 1044 (1979)
43. D.L. Albritton: At. Data Nucl. Data Tables **22**, (1978)
44. Y. Ikezoe, S. Matsuoka, M. Takebe, A. Viggiano: *Gas Phase Ion-Molecule Reaction Rate Constants Through* 1986, Maruzen Company, Ltd., Japan (1987)
45. B.L. Upschulte, R.J. Shul, R. Passarella, R.G. Keesee, A.W. Castleman, Jr.: Int. J. Mass Spectrom. Ion Proc. **75**, 27 (1987)
46. A.W. Castleman, Jr., S. Sigsworth, R.E. Leuchtner, K.G. Weil, R.G. Keesee: J. Chem. Phys. **86**, 3829 (1987)
47. R.J. Shul, R. Passarella, B.L. Upschulte, R.G. Keesee, A.W. Castleman, Jr.: J. Chem. Phys. **86**, 4446 (1987)
48. R.J. Shul, R. Passarella, X.L. Yang, R.G. Keesee, A.W. Castleman, Jr.: J. Chem. Phys. **87**, 1630 (1987)
49. G.H. Wannier: Bell Syst. Tech. J. **32**, 170 (1953)

50. H.R. Skullerud: J. Phys. **B6**, 728 (1973)
51. L.A. Vieland, E.A. Mason, J.H. Wheaton: J. Phys. **B7**, 2433 (1974)
52. I. Dotan, W. Lindinger, D.L. Albritton: J. Chem. Phys. **64**, 4544 (1976)
53. L.A. Viehland: Chem. Phys. **101**, 1 (1986)
54. J.Q. Searcy, J.B. Fenn: J. Chem. Phys. **61**, 5258 (1974)
55. R.R. Burke: Remarks at NATO Advanced Study Institute on Kinetics of Ion-Molecule Reactions held at La Baule, France, September 4–15, 1978
56. G.M. Lancaster, F. Honda, Y. Fukuda, J.W. Rabalais: J. Am. Chem. Soc. **101**, 1951 (1979)
57. S.-S. Lin: Rev. Sci. Instrum. **44**, 516 (1973)
58. V. Hermann, B.D. Kay, A.W. Castleman, Jr.: Chem. Phys. **72**, 2031 (1982)
59. A.J. Stace, C. Moore: Chem. Phys. Lett. **96**, 80 (1983)
60. O. Echt, D. Kreisle, M. Knapp, E. Recknagel: Chem. Phys. Lett. **108**, 401 (1984)
61. H. Shinohara, U. Nagashima, H. Tanaka, N. Nishi: J. Chem. Phys. **83**, 4183 (1985)
62. X. Yang, A.W. Castleman, Jr.: J. Am. Chem. Soc. **111**, 6845 (1989)
63. X. Yang, A.W. Castleman, Jr.: Production and Magic Numbers of Large Hydrated Anion Clusters $X^-(H_2O)_{N=0-59}$ X = OH, O, O_2 and O_3 Under Thermal Conditions, J. Phys. Chem. **94**, 8500 (1990); ibid. **94**, 8974 (1990)
64. P.M. Holland, A.W. Castleman, Jr.: J. Chem. Phys. **72**, 5984 (1980)
65. O. Echt, K. Sattler, E. Recknagel: Phys. Rev. Lett. **47**, 1121 (1982)
66. H. Haberland: In *Electronic and Atomic Collisions*, ed. by J. Eichler, I.V. Hertel, N. Stolterfoht, Elsevier Science Publishers, pp. 597–605 (1984)
67. O. Echt, M.C. Cook, A.W. Castleman, Jr.: Chem. Phys. Lett. **135**, 229 (1987)
68. D. Kreisle, O. Echt, M. Knapp, E. Recknagel: Phys. Rev. A. **33**, 768 (1986)
69. U. Buck, H. Meyer, Ber. Bunsenges: Phys. Chem. **88**, 254 (1984)
70. T.D. Märk: Int. J. Mass Spectrom: Ion Proc. **79**, 1–59 (1987)
71. For example, see Surface Science (H.C. Gatos, Ed.) Vol. **156**, Parts 1 and 2, North-Holland Publishing Co., Amsterdam (1985); Surface Science (H.C. Gatos, Ed.), Vol. **106**, North-Holland Publishing Co., Amsterdam (1981); Ber. Bunsenges. Phys. Chem. **88** (1984) and references therein
72. O. Echt, P.D. Dao, S. Morgan, A.W. Castleman, Jr.: J. Chem. Pys. **82**, 4076 (1985)
73. O. Echt, S. Morgan, P.D. Dao, R.J. Stanley, A.W. Castleman, Jr.: Ber. Bunsenges. Phys. Chem. **88**, 217 (1984)
74. S. Morgan, A.W. Castleman, Jr.: J. Phys. Chem. **93**, 4544 (1989)
75. S. Morgan, R.G. Keesee, A.W. Castleman, Jr.: J. Am. Chem. Soc. **111**, 3841 (1989)
76. S. Morgan, A.W. Castleman, Jr.: J. Am. Chem. Soc. **109**, 2867 (1987)
77. (a) M.T. Bowers, T. Su, V.G. Anicich: J. Chem. Phys. **58**, 5175 (1973). (b) L.M. Bass, R.D. Cates, M.F. Jarrold, N.J. Kirchner, M.T. Bowers: J. Am. Chem. Soc. **105**, 7024 (1983)
78. W.B. Tzeng, S. Wei, A.W. Castleman, Jr.: J. Am. Chem. Soc. **111**, 6035 (1989); J. Am. Chem. Soc. **111**, 8326 (1989)
79. L.W. Sieck, P. Ausloos: Radiat. Res. **52**, 47–58 (1972)
80. Z. Luczynski, H. Wincel: Int. J. Mass Spectrom. Ion Phys. **23**, 37 (1977)
81. K.A.G. MacNeil and J.H. Futrell: J. Phys. Chem. **76**, 409 (1972)
82. A.J. Stace, A.K. Shukla: J. Phys. Chem. **86**, 865 (1982)
83. J.J. Breen, W.-B. Tzeng, K. Kilgore, R.G. Keesee, A.W. Castleman, Jr.: J. Chem. Phys. **90**, 19 (1989)
84. J.J. Breen, W.B. Tzeng, R.G. Keesee, A.W. Castleman, Jr.: J. Chem. Phys. **90**, 11 (1989)
85. E.W. McDaniel, *Collision Phenomena in Ionized Gases*, p. 260, John Wiley & Sons (1964)
86. P.D. Dao, A.W. Castleman, Jr.: J. Chem. Phys. **84**, 1435 (1986)
87. Y. Hatano (personal communication); see also T. Wada, K. Shinsaka, H. Namba, Y. Hatano: Can. J. Chem. **55**, 2144 (1977)
88. P.C. Engelking: J. Chem. Phys. **85**, 3103 (1986)
89. P.C. Engelking: J. Chem. Phys. **87**, 936 (1987)
90. C. Lifshitz, F. Louage, J. Phys. Chem. **93**, 5633 (1989)
91. Y.K. Lau, P.P.S. Saluja, P. Kebarle: J. Am. Chem. Soc. **102**, 7429 (1980)

92. K. Hiraoka, E.P. Grimsrud, P. Kebarle: J. Am. Chem. Soc. **96**, 3359 (1974)
93. A.W. Castleman, Jr., I.N. Tang: J. Chem. Phys. **62**, 4576 (1975)
94. A.W. Castleman, Jr., W.B. Tzeng, S. Wei, S. Morgan: J. Chem. Soc. Faraday Trans. **86**, 2417 (1990)
95. S.G. Lias, J.F. Liebman, R.D. Levin: J. Phys. Chem. Ref. Data, **13**, 695 (1984)
96. W.B. Tzeng, S. Wei, D.W. Neyer, R.G. Keesee, A.W. Castleman, Jr.: J. Am. Chem. Soc. **112**, 4097 (1990)
97. W.B. Tzeng, S. Wei, A.W. Castleman, Jr.: Chem. Phys. Lett. **166**, 343 (1990)
98. S. Wei, W.B. Tzeng, A.W. Castleman, Jr.: The Structure of Protonated Solvation Complexes: Ammonia-Trimethylamine Cluster Ions, and Their Metastable Decompositions, J. Phys. Chem. **95**, 585 (1991)
99. S. Wei, W.B. Tzeng, A.W. Castleman, Jr.: Evidence of Cyclic Structures in Hydrogen-Bonded Complex, Chem. Phys. Lett. **178**, 411 (1991)
100. N. Halberstadt, B. Soep: J. Chem. Phys. **80**, 2340 (1984)
101. M. Castella, A. Tramer, F. Piuzzi: Chem. Phys. Lett. **129**, 105 (1984); **129**, 112 (1986)
102. P.D. Dao, S. Morgan, A.W. Castleman, Jr.: Chem. Phys. Lett. **111**, 38 (1984)
103. D.H. Levy, C.A. Haynam, D.V. Brumbaugh: Faraday Discuss. Chem. Soc. **73**, 137 (1982)
104. R.C. Woods, C.S. Gudeman, R.L. Dickman, P.F. Goldsmith, G.R. Huguenim, W.M. Irvine, A. Hjalmarson, L.A. Nyman, H. Olofsson: Astrophys. J. **270**, 583 (1983)
105. W. Wagner-Redeker, P.R. Kemper, M.F. Jarrold, M.T. Bowers: J. Chem. Phys. **83**, 1121 (1985)
106. F.W. McLafferty, D.C. McGilvery: J. Am. Chem. Soc. **102**, 4189 (1980)
107. J.D. Shao, T. Baer, J.C. Morrow, M.L. Fraser-Monteiro: J. Chem. Phys. **87**, 5242 (1987)
108. C.A. Deakyne, M. Meot-Ner, C.L. Campbell, M.G. Hughes, S.P. Murphy: J. Chem. Phys. **84**, 4958 (1986)
109. S.T. Graul, R.R. Squires: Int. J. Mass Spectrom. Ion Proc. **94**, 41 (1989)
110. T.L. Tai, M.A. El-Sayed: J. Phys. Chem. **90**, 4477 (1986)
111. H.J. Hwang, D.K. Sensharma, M.A. El-Sayed: Chem. Phys. Lett. **160**, 243 (1989)
112. F.C. Fehsenfeld, M. Mosesman, E.E. Ferguson: J. Chem. Phys. **55**, 2120 (1971)
113. C.J. Howard, H.W. Rundle, F. Kaufman: J. Chem. Phys. **55**, 4472 (1971)
114. A.M. Sapse, D.C. Jain: Int. J. Quant. Chem. **27**, 281 (1985)
115. F.C. Fehsenfeld, M. Mosesman, E.E. Ferguson: J. Chem. Phys. **55**, 2115 (1971)
116. P.M. Holland, A.W. Castleman, Jr.: J. Chem. Phys. **76**, 4195 (1982)
117. (a) S.W. Sigsworth, R.G. Keesee, A.W. Castleman, Jr.: J. Am. Chem. Soc. **110**, 6682 (1988); (b) S.W. Sigsworth, A.W. Castleman, Jr.: J. Am. Chem. Soc. **111**, 3566 (1989)
118. D.K. Bohme, *NATO Adv. Study Inst. Ser., Ser. C.* **118**, 111 (1984)
119. D.K. Bohme, G.I. Mackay, S.D. Tanner: J. Am. Chem. Soc. **101**, 3724 (1979)
120. D.K. Bohme, A.B. Rakshit, G.I. Mackay: J. Am. Chem. Soc. **104**, 1100 (1982)
121. D.K. Bohme, G.I. Mackay: J. Am. Chem. Soc. **103**, 978 (1981)
122. J.F. Paulson, M.J. Henchman, In *NATO Advanced Study Institute Series, Series C*, M.A. Almoster-Ferreira, Ed. Reidel, Boston, pp. 331–334 (1984)
123. M. Henchman, J.F. Paulson, P.M. Hierl: J. Am. Chem. Soc. **105**, 5509 (1983)
124. B.R. Rowe, A.A. Viggiano, F.C. Fehsenfeld, D.W. Fahey, E.E. Ferguson: J. Chem. Phys. **76**, 742 (1982)
125. M.M. Kappes, R.H. Staley: J. Phys. Chem. **86**, 1332 (1982)
126. M. Meot-Ner: In *Gas Phase Ion Chemistry*, ed. by M.T. Bowers, Academic Press, New York, Vol. 1 (1979)
127. N.G. Adams, D. Smith: In *Reactions of Small Transient Species, Kinetics and Energetics*, ed. by A. Fontijin, M.A. Clyne, Academic, New York, pp. 311–385 (1983)
128. E.E. Ferguson: *Ion Molecule Reactions*, ed. by J.L. Franklin, Butterworths, London, Vol. 2 (1972)
129. E.E. Ferguson, Ann. Rev. Phys. Chem. **26**, 17 (1975)
130. E.W. McDaniel, V. Cermak, A. Delgarno, E.E. Ferguson, L. Friedman, *Ion Molecule Reactions*, Wiley-Interscience, New York (1970)

131. W.J. Chesnavich, T. Su, M.T. Bowers: In *Kinetics of Ion-Molecule Reactions*, ed. by P. Ausloos, Plenum, New York, pp. 31–53 (1979)
132. E.T.-Y. Hsieh, A.W. Castleman, Jr.: J. Mass Spectrom. Ion Phys. **40**, 295 (1981)
133. F.J. Schelling, A.W. Castleman, Jr.: Chem. Phys. Lett. **111**, 47 (1984)
134. K.N. Swamy, W.L. Hase: J. Chem. Phys. **77**, 3011 (1982)
135. W.L. Hase, D.-F. Feng: J. Chem. Phys. **75**, 738 (1981)
136. E. Herbst: J. Chem. Phys. **70**, 2201 (1979)
137. E. Herbst: J. Chem. Phys. **72**, 5284 (1980)
138. D.R. Bates: J. Chem. Phys. **71**, 2318 (1979)
139. D.R. Bates: J. Phys. B. **12**, 4135 (1979)
140. D.R. Bates: J. Chem. Phys. **73**, 1000 (1980)
141. D.R. Bates: J. Chem. Phys. **83**, 572 (1985)
142. D.R. Bates: Chem. Phys. Lett. **112**, 41 (1984)
143. D.C. Clary: Mol. Phys. **54**, 605 (1985)
144. L.M. Bass, P.R. Kemper, V.G. Anicich, M.T. Bowers: J. Am. Chem. Soc. **103**, 5283 (1981)
145. T. Su, W.J. Chesnavich: J. Chem. Phys. **76**, 5183 (1982)
146. W. Forst: *Theory of Unimolecular Reactions*, Academic Press, New York (1983)
147. T.J. Robinson, K.H. Holbrook: *Unimolecular Reactions*, Wiley-Interscience, New York, 1575 (1972)
148. W.N. Olmstead, M. Lev-On, D.M. Golden, J.I. Brauman: J. Am. Chem. Soc. **99**, 992 (1977)
149. J.M. Jainski, R.N. Rosenfeld, D.M. Golden, J.I. Brauman: J. Am. Chem. Soc. **101**, 2259 (1979)
150. R.E. Leuchtner, A.C. Harms, A.W. Castleman, Jr.: J. Chem. Phys. **92**, 6527 (1990)
151. R.E. Leuchtner, A.C. Harms, A.W. Castleman, Jr.: J. Chem. Phys. **91**, 2753 (1989)
152. M.F. Jarrold, J.E. Bower: J. Chem. Phys. **87**, 5728 (1987)
153. (a) W.A. de Heer, W.D. Knight: In *Elemantal and Molecular Clusters*, Springer-Verlag, pp. 45–63 (1988); (b) M.Y. Chou, M.L. Cohen: *Phys. Lett.* **113A**, 420 (1986)
154. W. Ekardt: Phys. Rev. B **29**, 1558 (1984)
155. A.C. Harms, R.E. Leuchtner, S.W. Sigsworth, A.W. Castleman, Jr.: J. Am. Chem. Soc. **112**, 5672 (1990)
156. P. Kebarle: Mod. Aspects Electrochem. **9**, 1 (1974)
157. R.S. Mason: *NATO Adv. Study Inst. Ser., Ser. C.* **119**, 627 (1984)
158. M. Meot-Ner: Acc. Chem. Res. **17**, 186 (1984)
159. D.C. Conway, G.S. Janik: J. Chem. Phys. **53**, 1859 (1970)
160. A.W. Castleman, Jr., P.M. Holland, R.G. Keesee: Radiat: Phys. Chem. **20**, 57 (1982)
161. A.W. Castleman, Jr., R.G. Keesee: In *Swarms of Ions and Electrons in Gases*, ed. by W. Lindinger, T.D. Märk, F. Howorka, Springer-Verlag Wien New York, pp. 167–193 (1984)
162. A.W. Castleman, Jr., K.L. Peterson, B.L. Upschulte, J.F. Schelling: Int. J. Mass Spectrom. Ion Phys. **47**, 203 (1983)
163. (a) K.L. Gleim, B.C. Guo, R.G. Keesee, A.W. Castleman, Jr.: J. Phys. Chem. **93**, 6805 (1989); (b) B.C. Guo, A.W. Castleman, Jr.: The Clustering Reactions of Na^+ and Pb^+ with Several Important Ligands, Z. Phys. D. **19**, 397 (1990)
164. R.G. Keesee, A.W. Castleman, Jr.: In *Ionic Processes in the Gas Phase*, ed. by M.A. Almoster-Ferreiro, NATO Advanced Study Institute Series, Series C **118**, Reidel, Boston, p. 340 (1984)
165. I.N. Tang, M.S. Lian, A.W. Castleman, Jr.: J. Chem. Phys. **65**, 4022 (1976)
166. (a) I.N. Tang, A.W. Castleman, Jr.: J. Chem. Phys. **57**, 3638 (1972); (b) I.N. Tang, A.W. Castleman, Jr.: J. Chem. Phys. **60**, 3981 (1974)
167. A.W. Castleman, Jr.: Chem. Phys. Lett. **53**, 560 (1978)
168. H.H. Tang, D.C. Conway: J. Chem. Phys. **59**, 2316 (1973)
169. B.C. Guo, A.W. Castleman, Jr.: The Association Reaction of Pb^+ Ion with CH_3OH and CH_3NH_2 in the Gas Phase, Int. J. Mass Spectrom. Ion Proc. **100**, 665 (1990)
170. R.G. Keesee, N. Lee, A.W. Castleman, Jr.: J. Chem. Phys. **73**, 742 (1982)
171. J.D. Payzant, R. Yamdagni, P. Kebarle: Can. J. Chem. **49**, 3308 (1971)
172. N. Lee, R.G. Keesee, A.W. Castleman, Jr.: J. Chem. Phys. **72**, 1089 (1980)

173. Y.K. Lau, S. Ikuta, P. Kebarle: J. Am. Chem. Soc. **104**, 1462 (1982)
174. J.D. Payzant, A.J. Cunningham, P. Kebarle: Can. J. Chem. **51**, 3242 (1973)
175. S.K. Searles, P. Kebarle: J. Phys. Chem. **72**, 742 (1968)
176. M.R. Arshadi, J.H. Futrell: J. Phys. Chem. **78**, 1482 (1974)
177. R.W. Taft: *Kinet. Ion Mol. React.*, 271 (1975); R.W. Taft: in *Proton Transfer Reactions*, ed. by E.F. Cauldon, V. Gold, Chapman & Hall, London (1975)
178. R.W. Taft, J.F. Wolf, J.L. Beauchamp, G. Scorrano, E.M. Arnett: J. Am. Chem. Soc. **100**, 1240 (1978)
179. J. Bromilow, J.L.M. Abboud, I.C.B. Lebrilla, R.W. Taft, G. Scorrano, V. Lucchini: J. Am. Chem. Soc. **103**, 5448 (1980)
180. E.M. Arrnett, F.M. Jones, III, M. Taagepera, W.G. Henderson, J.L. Beauchamp, D. Holtz, R.W. Taft: J. Am. Chem. Soc. **94**, 4724 (1972)
181. A.J. Stace, A.K. Shukla: J. Am. Chem. Soc. **104**, 5314 (1982)
182. A.J. Stace: J. Am. Chem. Soc. **106**, 2306 (1984)
183. H. Shinohara, N. Nishi: Chem. Phys. Lett. **87**, 561 (1982)
184. M. Arshadi, R.Yamdagni, P. Kebarle: J. Phys. Chem. **74**, 1475 (1970)
185. S.H. Bauer, D.J. Frurip: J. Phys. Chem. **81**, 1015 (1977)
186. H. Kistenmacher, H. Popkie, E. Clementi: J. Chem. Phys. **59**, 5842 (1973)
187. R.G. Keesee, A.W. Castleman, Jr.: Chem. Phys. Lett. **74**, 139 (1980)
188. H.R. Pruppacher, J.D. Klett: *Microphysics of Clouds and Precipitation*, Reidel, Dordrecht (1978)
189. H. Rabeonv, P. Mirabel: J. Chim. Phys. Phys.-Chim. Biol. **83**, 219 (1986)
190. P.M. Holland, A.W. Castleman, Jr.: J. Phys. Chem. **86**, 4181 (1982)
191. J.J. Thompson: *Application of Dynamics to Physics and Chemistry*, 1st ed., Cambridge Press, Cambridge (1988)
192. B.J. Mason: *The Physics of Clouds*, Clarendon Press, Oxford (1971)
193. G.M. Pound: J. Phys. Chem. Ref. Data **1**, 119 (1972); J. Phys. Chem. Ref. Data **1**, 135 (1972)
194. M. Born: Z. Phys. **1**, 45 (1920)
195. J.E. Desnoyers, C. Jolicoeur: Mod. Aspects Electrochem. **5**, 1 (1969)
196. H.L. Friedman, W.T. Dale: in *Modern Theoretical Chemistry*, Vol. 5, ed. by B.J. Berne, Plenum, New York (1977)
197. M.H. Abraham, J. Lisze, L. Meszaros: J. Chem. Phys. **70**, 2491 (1979)
198. I. Dzidic, P. Kebarle: J. Phys. Chem. **74**, 1466 (1970)
199. C.E. Klots: J. Phys. Chem. **85**, 3585 (1981)
200. R.G. Keesee, N. Lee, A.W. Castleman, Jr.: J. Am. Chem. Soc. **101**, 2599 (1979)
201. B.L. Upschulte, F.J. Schelling, R.G. Keesee, A.W. Castleman, Jr.: Chem. Phys. Lett. **111**, 389 (1984)
202. J. Sunner, K. Nishizawa, P. Kebarle: J. Phys. Chem. **85**, 1814 (1981)
203. D.E. Hunton, M. Hofmann, T.G. Lindeman, A.W. Castleman, Jr.: J. Chem. Phys. **82**, 134 (1985)
204. D.E. Hunton, M. Hofmann, T.G. Lindeman, C.R. Albertoni, A.W. Castleman, Jr.: J. Chem. Phys. **82**, 2884 (1985)
205. H.-S. Kim, M.T. Bowers: J. Chem. Phys. **85**, 2718 (1986)
206. R.V. Hodges, J.A. Vanderhoff: J. Chem. Phys. **72**, 3517 (1980)
207. R.G. Keesee, N. Lee, A.W. Castleman, Jr.: J. Chem. Phys. **73**, 2195 (1983)
208. J.J. Breen, K. Kilgore, K. Stephan, R. Sievert, B.D. Kay, R.G. Keesee, T.D. Mark, J. van Doren, A.W. Castleman, Jr.: Chem. Phys. **91**, 305 (1984)
209. R.J. Celotta, R.A. Bennett, J.L. Hall: J. Chem. Phys. **60**, 1740 (1974)
210. J.T. Moseley, P.C. Cosby, J.R. Peterson: J. Chem. Phys. **65**, 2512 (1976); G.P. Smith, L.C. Lee, J.T. Moseley, J. Chem. Phys. **71**, 4034 (1979); J.F. Hiller, M.L. Vestal: J. Chem. Phys. **72**, 4713 (1980)
211. M.A. Johnson, M.L. Alexander, W.C. Lineberger: Chem. Phys. Lett. **72**, 285 (1984)
212. S.C. Ostrander, L. Sanders, J.C. Weissharr: J. Chem. Phys. **84**, 529 (1986)
213. W.C. Lineberger, M.L. Alexander, N.E. Levinger, M.A. Johnson, paper Phys. 0016 presented at the 191st Annual Meeting of the American Chemical Society, New York, 13 to 18 April 1986

214. F.G. Amar: In *Physics and Chemistry of Small Clusters*, Vol. 158 of NATO Asi Series B, Plenum, New York, p. 207 (1987)
215. M. Okamura, L.I. Yeh, Y.T. Lee: J. Chem. Phys. **83**, 3705 (1985)
216. (a) J.M. Price, M.W. Crofton, Y.T. Lee: J. Chem. Phys. **91**, 2748 (1989); (b) L.I. Yeh, M. Okumura, J.D. Myers, J.M. Price, Y.T. Lee: J. Chem. Phys. **91**, 7319 (1989); M. Okumura, L.I. Yeh, J.D. Myers, Y.T. Lee, J. Chem. Phys., submitted.
217. J.V. Coe and R.J. Saykally: In *Ion and Cluster Ion Spectroscopy and Structure*, ed. by J.P. Maier, Elsevier, pp. 131–154 (1989)
218. S.T. Arnold, J.G. Eaton, D. Patel-Misra, H.W. Sarkas, K.H. Bowen, In *Ion and Cluster Ion Spectroscopy and Structure*, ed. by J.P. Maier, Elsevier, pp. 417–472 (1989)
219. U. Landman, R.N. Barnett, C.L. Cleveland, D. Scharf, J. Jortner: in *Physics and Chemistry of Small Clusters*, Vol. 158 of NATO Asi Series B, Plenum, New York, p. 207 (1987)
220. A.R. Rissi, K.D. Jordan: J. Chem. Phys. **70**, 4422 (1979)
221. T. Kondow: J. Phys. Chem. **91**, 1307 (1987)
222. C.E. Klots: J. Chem. Phys. **71**, 4172 (1979)
223. H. Haberland, H.G. Schindler, D.R. Worsnop: Ber. Bunsenges: Phys. Chem. **88**, 270 (1984); H. Haberland, C. Ludewigt, H.G. Schindler, D.R. Worsnop: J. Chem. Phys. **81**, 3742 (1984)
224. Y. Nachson, F.K. Tittel, W.L. Wilson, Jr., W.L. Nighan: J. Appl. Phys. **56**, 36 (1984)
225. W.L. Nighan, R.A. Sayerbergy, Y. Zho, F.K. Tittel, W.L. Wilson, Jr., IEEE J. Quant. Elect. **QE-23**, 253 (1987)
226. Z.Y. Chen, C.D. Cogley, J.H. Hendricks, B.D. May, A.W. Castleman, Jr.: J. Chem. Phys. **93**, 3215 (1990)
227. E.E. Ferguson, C.R. Albertoni, R. Kuhn, Z.Y. Chen, R.G. Keesee, A.W. Castleman, Jr.: J. Chem. Phys. **88**, 6335 (1988)
228. C.R. Albertoni, A.W. Castleman, Jr., E.E. Ferguson: Chem. Phys. Lett. **157**, 159 (1989)
229. C.E. Klots: Chem. Phys. Lett. **10**, 422 (1971)

2.5 Solvated Electron Clusters

H. Haberland and *K.H. Bowen*

2.5.1 Introduction

An electron gains energy if it is brought from the vacuum into a macroscopic dielectric liquid or solid, as shown schematically in Fig. 1. The only exceptions being helium and neon, where after an initial attraction, about 1 eV is necessary to push the electron into the bulk medium. The concepts of a valence- and conduction-band might be better known for crystalline materials, but they can be extended also to amorphous and liquid matter. For semi-conductors and dielectrics the valence band is completely full, and the conduction band empty. Simple solid state theory tells us for this case that no net electric current can flow on application of an electric field. The field gives rise to a flow of electrons which

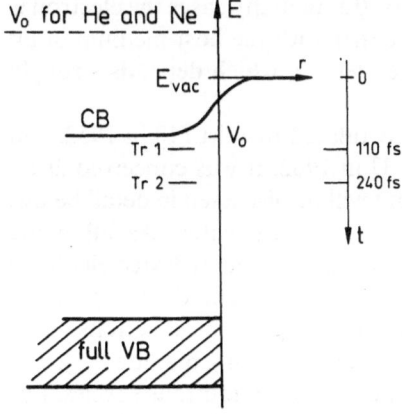

Fig. 1. The energy E of an electron is plotted as a function of its distance r from a bulk dielectric surface. A bulk dielectric has a full valence band (VB) and an empty conduction band (CB). The two bands are separated by so large a band gap, that no electron can be thermally excited across it. The energy of the "vacuum level" E_{vac} is the energy of an electron at rest at infinity. At large distances the interaction between an electron and a dielectric surface is always attractive, as the electron is attracted by the polarisation forces. The difference between the energy of the bottom of the conduction band and E_{vac} is called V_0. Only for He and Ne does the conduction band lie above E_{vac} as indicated by the dashed line. The localised trap states lie below the conduction band. The energies of a shallow (Tr1) and a deep (Tr2) trap state are indicated. The time it takes for an electron in condensed water to localise on the trap states is indicated on the right

is exactly cancelled by the flow of "defect-electrons" or "holes" in the opposite direction. Thus a substance where the highest occupied band is full, is an isolator. For dielectrics the bandgap which separates valence and conduction bands is so large, that no electrons can be thermally excited into the conduction band. But one can of course introduce electrons into the empty conduction band, by injecting them from the outside. Two different things can happen to the electron once it is inside the dielectric:

1. The electron can stay delocalised. It moves as a nearly free electron in the empty conduction band of the dielectric. This rare case will be discussed for Xe clusters below.
2. In the majority of cases the electron will become "trapped" or "localised", i.e. the atoms or molecules of its surrounding react to the changed force field, and move to new positions. The electron becomes bound to this "vacancy", or in chemical language it becomes solvated. This self trapping process has been described pictorially as "the electron digging its own trap". Examples which will be discussed below include He, H_2O, NH_3, and NaCl clusters. The energy of the trap states lies below the conduction band, as indicated in Fig. 1

If no chemical reaction occurs after localisation, solvated electrons can be stable for long times. This is the case of electrons in liquid ammonia, where they give the famous "blue solution" discovered in 1864, when Weyl dissolved sodium in liquid ammonia (NH_3). In modern notation $Na^+(NH_3)_n$ and $(NH_3)_n^-$ clusters, with n tending to infinity, are formed in the liquid. Under very clean conditions these clusters can be stable for years. But in many cases the electron is first solvated and then undergoes a reaction, either with the host medium or an impurity, giving the solvated electron a finite lifetime, which depends strongly on the impurity content of the solvent.

The celebrated "hydrated electron" was postulated to exist [1] in 1952, and was experimentally identified in the bulk [2, 3] in 1962. It was conceived as an electron self-trapped in bulk water. Its structure will be discussed in detail below. It is produced when ionising radiation interacts with water. As all living material, including us humans, consist mainly of water, some hydrated electrons are formed in our bodies, when e.g. an X-ray is taken, or the body is exposed to some other form of ionising radiation. Due to the strong chemical actions of the hydrated electron, it is important in radiation damage of living material [4]. The hydrated electron might be difficult to study, but it is an ubiquitous species. For example, the surface water of the oceans can contain up 10^9 hydrated electrons per liter at full sun light [5]. The history of the bulk chemical and biological studies of hydrated electrons has recently been summarised [4]. A collection of newer references can be found in [6, 7].

There exist two different classes of atoms and molecules, which form dielectrics in the bulk:

1. Those which have a negative ion state as a monomer like O_2, SF_6, N_2O, etc. The negative ions of these molecules are stable and well studied. The electron enters into a well defined molecular orbital of the molecule proper. The

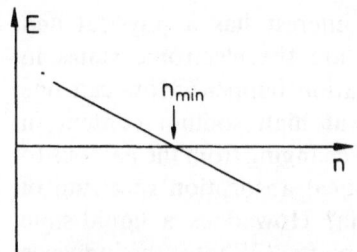

Fig. 2. The binding energy E of an extra electron is plotted against the cluster size n. For n below n_{min}, E is positive and the electron is not bound. One has a resonance in electron scattering. For $n > n_{min}$ the extra electron is bound and a stable cluster can be formed

electron affinity (EA) is positive, where EA is the minimal energy to eject an electron from the negative ion. (For a definition of EA see Fig. 9).

2. Some atoms and molecules on the other hand do not have a stable negative ion state as a monomer. Among these are: the rare gases with the possible exception of xenon, H_2O, NH_3 etc. The electron can only be bound by the *cooperative effect* of many atoms or molecules. It is for this latter class of substances only, that one speaks of a solvated electron proper. Some examples are collected in Table 1. All atoms and molecules in this second category have closed electronic shells.

If n is varied in, say, $(H_2O)_n^-$ from 1 to a very large number, the development of bulk hydrated electron can be studied. Several interesting questions and phenomena arise in this study:

1. Solvated electron clusters can only be formed in nonconducting, dielectric fluids and solids. But insulating materials are formed by condensing closed shell atoms and molecules, which most often do not have a bound negative ion state as a monomer. (Exceptions are the alkali-halides and possibly xenon.) The interaction of a slow electron with a closed shell atom or molecule leads to a resonance in the electron scattering continuum. A transition from a resonance in electron scattering to a bound electronic state must therefore occur as a function of cluster size, as indicated in Fig. 2.
2. What is the minimal number n_{min} of atoms and molecules necessary to support a bound state, and how does n_{min} vary from one substance to the other? (Compare Fig. 2 and Table 1.)
3. How is the macroscopic limit attained? Does the structure of the cluster between n_{min} and $n \to \infty$ evolve more or less continuously or are there interesting transitions?
4. What happens if the cluster size becomes equal to the diameter of the electronic charge distribution?
5. Is the electron inside or on the outside of the cluster?
6. Does the chemical reactivity change with cluster size?

2.5.1.1 Solvated Electrons in the Condensed Phase

A wealth of information is available from experiment and theory on solvated electrons in the condensed phase. Dipolar and nonpolar vapours, liquids and solids have all been studied [6, 7]. The high interest has a physical and a chemical origin. Questions asked are: What are the electronic states in disordered materials? How does electron localisation happen? How can one understand the metal to non-metal transition at high sodium content in Na/NH_3 solutions? How is the electron's mobility changing from the gaseous to the liquid state? Are there differences in the optical absorption spectrum of electrons solvated in liquid water and ammonia? How does a liquid-state environment influence the quantum behaviour of matter? What is the influence on electron transfer reactions in solutions, etc.? Because of its radiological relevance, a short summary is given here on how solvated electrons are formed in liquid water. Other applications are treated in the chapter by *Belloni* et al. in this book.

If a fast electron is injected into liquid water four different processes can be distinguished:

Thermalisation. The kinetic energy of the fast electron is quickly reduced by ionisation and electronic excitation until its energy is in the 5 eV range. Then excitation of vibrations (phonons) and hindered rotations become important. It takes about 10 to 20 fs to thermalise an electron in liquid water. (1 fs = 1 femtosecond = 10^{-15} s).

Localisation. The electron now moves for 110 to 180 fs in the empty conduction band of water (compare Fig. 1), until it finds a shallow trap state (Tr 1 in Fig. 1), on which it can localise. A strong absorption in the near infra red appears.

Solvation. The molecules around the newly trapped electron experience a new force field. The H_2O molecules rotate until they have found the new position of lowest energy. This takes about 240 fs. A strong absorption in the red part of the visible spectrum appears. The fast times cited above have been measured by generating the electron and probing its absorption by fs laser pulses [8, 9].

Recombination or Reaction. Not only the electron, but also the ion is solvated. They attract each other by the Coulomb force, which is screened by the water molecules lying between the two charge centers. If the attraction becomes strong enough, the electron and the ion can recombine. Alternatively the solvated charges can be destructed by a chemical reaction.

2.5.2   Case Studies

The different possibilities for electron localisation in and on small clusters of closed shell atoms and molecules will be discussed for several prototypes. Stress will be laid on the experimental observations.

2.5.2.1 Helium

Calculations indicate that excess electrons can be bound to the surface of very large He clusters [10]. These states are similar to the well studied electronic surface states on liquid helium. Due to the very low polarisability of He and Ne the interaction of an extra electron with bulk He or Ne is repulsive. For clusters containing more than 10^5 He atoms, the potential well between the attractive polarisation force and the repulsive bulk potential is just strong enough to support a bound state for an extra electron. There exist many interesting theoretical questions concerning these states. But definite experimental verification will become very difficult [11]. Large helium clusters are known to soak up any molecule they encounter on their way through a vacuum chamber. This makes for a terrible background problem. For example some O_2 is always present in a vacuum system. The O_2 molecule has a high positive electron affinity, thus forming easily negative ions. As O_2 has nearly the same mass as eight He atoms, one would need a mass spectrometer with a resolving power of nearly 10^9 in order to distinguish between He_{n+32}^- and $He_nO_2^-$, with $n \geq 10^5$.

2.5.2.2 Xenon

The bottom of the (normally empty) conduction band of solid xenon lies 0.4–0.5 eV below the vacuum level. Or stated differently, the electron affinity of solid xenon is 0.4 to 0.5 eV, a value obtained in two independent experiments [12, 13]. The extra electron is not self trapped, it can move freely in the otherwise empty conduction band.

For the atom it is not yet finally clear, if the recently observed [14] negatively charged xenon atom is in the electronic ground state or in an electronically excited metastable state. In any case the electron affinity of the Xe atom is either very small or negative.

Several theoretical treatments have attacked the problem of electron attachment to xenon clusters. The interaction of the extra electron with the Xe atoms has to be treated quantum mechanically. The interaction of the atoms among themselves is either treated classically [15], or a continuum model is used [16]. The calculated minimal cluster size is in the 6–10 atom range.

For the experimental test [14] of these predictions a mixture of Xe with Ar and N_2 was expanded from a pulsed supersonic beam and 100 eV electrons were injected at position A of Fig. 14 of "Experimental Methods". The fast electrons eject secondary electrons from the beam. These are quickly decelerated by collisions. The nitrogen gas is necessary, as it serves as a "moderator" for the electrons, extracting their kinetic energy via the short lived negative ion resonance: $N_2(v = 0) + e^- \rightarrow N_2^- \rightarrow N_2(v \geq 0) + e^-$. After the collision the electron has lost kinetic energy and the N_2 molecule is vibrationally excited. It seems, that only very low energy electrons can attach to the xenon clusters.

A Xe mass spectrum shown in Fig. 3. Some intensity is observed on all cluster masses down to the monomer. According to theory, the extra electron is repelled from the interior of an Xe atom. The space available to the extra electron are the "channels" between the atoms and the surface of the cluster. This is obviously a delocalised electron state. If there are more or deeper or wider channels for the electron to move in, its binding energy increases. For Xe_{13}^- a compact icosahedral structure has been calculated. The central atom with its two pentagonal caps of six atoms each can be seen in Fig. 4. For $N = 19$ one has an icosahedron with an additional pentagonal cap. These icosahedral shapes are more compact than their neighbours, with narrower channels between the xenon atoms and consequently lower electron binding energy. If the electron affinity is lower it is plausible that an electron is less often attached and more often detached in the expanding jet. This results in a low intensity on masses $N = 13$ and 19. Similarly the open structure of $N = 12$ and $N = 18$ leads to a higher electron affinity and to the higher experimental intensity. On this qualitative level experiment and theory are in agreement.

Rare gas atoms are known to have very long lived, metastable excited states which sometimes can attach an electron. For example the metastable triplet state of helium, $He(1s2s, {}^3S)$, can form a long lived negative ion state. If similar states would exist in the xenon atom and cluster, they could give rise to the mass spectrum in Fig. 3. An energy of about 10 eV is necessary to excite the metastable state. The kinetic energy of the electron beam is a factor of ten higher. So the excitation of a metastable Xe state, which could in principle attach an electron, cannot be ruled out.

An experiment with photo-electrons which utilised electron energies below 1 eV gave a smallest cluster of $n = 6$. Smaller Xe_n^- clusters could have been there,

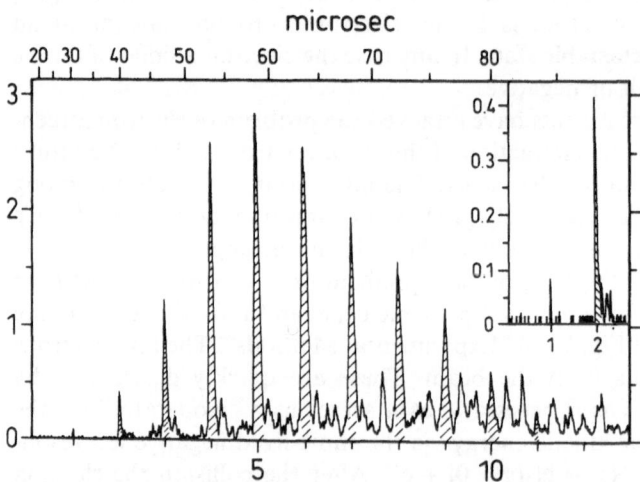

Fig. 3. Mass spectrum of negatively charged Xe clusters. The marked peaks are Xe_n^- clusters, the other ones are impurity peaks. The insert shows the $n = 1$ and 2 peaks on a large scale

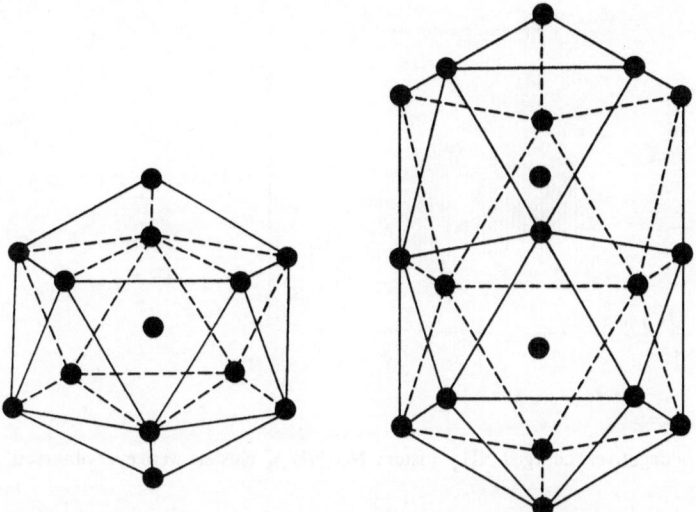

Fig. 4. Calculated structures for Xe_{13}^- and Xe_{19}^-. The $n = 13$ cluster has the shape of an icosahedron, an inner atom having two pentagonal caps. For $n = 19$ another pentagonal cap is added. As explained in the text the compact structure of the icosahedra can explain the high intensity on mass $n = 12$ and 18, and the low intensity on mass 13 and 19 in Fig. 3

but could not be seen due to a large interfering background. It is not impossible, that even the atom could support a negatively bound stable state. Some n_{min} values are collected in Table 1 below.

2.5.2.3   Water and Ammonia

Cluster models for electrons solvated in water and ammonia clusters had been popular with theorists long before experimental studies became feasible. As the electron localises on a few molecules it is theoretically often easier to treat a finite number of molecules, rather than the infinite bulk. By the early 1980's the first successful mass spectrometric studies had been performed, in a time when one had to learn how to synthesize these clusters. Later photo-detachment and photo-fragmentation studies proved a good meeting ground of theory and experiment.

2.5.2.3a)   Mass Spectra

Figure 5 shows a mass spectrum of $(NH_3)_n^-$ clusters. Above some $n = n_{min}$ the intensity increases rapidly by nearly 3 orders of magnitude, and tails off slowly for larger n. The decrease at high n is at least partially due to the detection

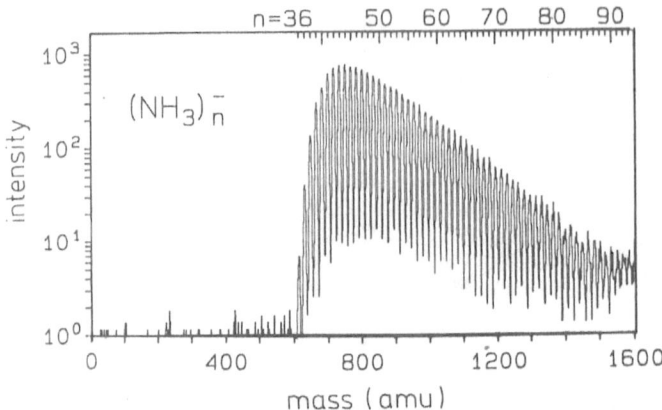

Fig. 5. Mass spectrum of negatively charged NH_3 clusters. No $(NH_3)_3^-$ clusters were ever observed with $n < 34$

problems discussed in Chapter 3.2. Larger clusters are always less efficiently detected.

The NH_3 clusters were produced by injecting very slow electrons into the very first part of a supersonic expansion (point A of Fig. 3.15). The clusters can grow around the electron. By playing with the diameter and temperature of the nozzle, and the percentage and nature of the seeding gas it was determined that below $n_{min} = 35$ no negatively charged NH_3 clusters could be synthesized. The n_{min} values of several molecules and atoms are collected in Table 1.

Alternatively one can attach electrons at point B of Fig. 3.15. The smallest cluster observed is always larger than n_{min}, even when the extremely low energy electrons available in Rydberg atoms are used [17]. The physical reason will be discussed below for the similar case of water clusters.

The case of water is more complex. First the results for a pure, unseeded water expansion will be treated. Attaching low energy electrons either at point

Table 1. The table gives the minimal number of atoms and molecules needed to support a bound negative ion state. Both the very large difference between water and ammonia, and the ammonia isotope effect are presently not understood

Molecule	n_{min}	Atom	n_{min}
NaF	1	Mg	3
H_2O	2	Hg	3
D_2O	2	Xe {	6
HCl	2		Possibly 1
$H_4C_2(OH)_2$	2		
NH_3	34		
ND_3	41		

A or B, the minimal cluster size is 11 to 12, as shown in the middle part of Fig. 6. If the electron's kinetic energy is increased, dissociative electron attachment

$$D_2O + e^- \rightarrow OD^- + D \tag{2.5.1}$$

can take place. The stable negative ions OD^- and OH^- can attach any number of water molecules, leading to the upper spectrum of Fig. 6. This proves that indeed all cluster sizes are present in the beam. It will be argued below, that smaller $(D_2O)_n^-$ are actually formed, but have too short a lifetime, to be visible in this experiment.

The authors [18] estimate with arguments similar to those leading to Eq. (3.21) that their cluster temperature is about 180 K. Once the electron is attached the water dipole moments start to rotate, seeking the new position of lowest energy. This will increase the clusters temperature further. What will be ejected by this high internal vibrational energy, an electron or a molecule? The two following processes are possible:

$$(D_2O)_n^- \rightarrow (D_2O)_{n-1}^- + D_2O \tag{2.5.2}$$

or

$$(D_2O)_n^- \rightarrow (D_2O)_n + e^- \tag{2.5.3}$$

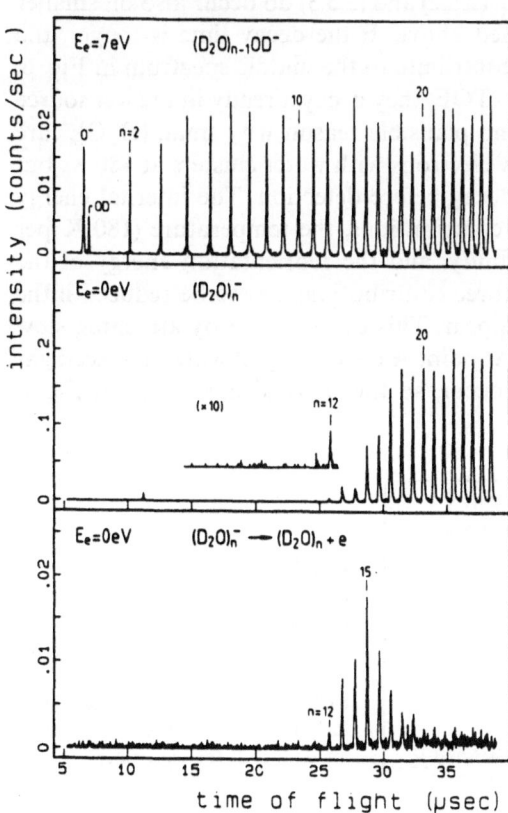

Fig. 6. The middle spectrum shows heavy water cluster anions obtained by attaching low energy electrons to cold water clusters. The lower spectrum shows the metastable clusters, those which have survived the acceleration into the mass spectrometer, but have subsequently ejected an electron. The upper spectrum is obtained if the energy of the electrons is raised

In the first case a neutral molecule, in the second one an electron is ejected. In the metastable time window $0.1 \cdot TOF \leq t \leq TOF$, where TOF is the flight time of the $(D_2O)_n^-$, the channel according to Eq. (2.5.3) is dominant. The mass spectrum of the remaining neutrals is shown in the lower part of Fig. 6. The clusters have lost their electron in the field free region of the TOF mass spectrometer. They have the same velocity as the charged ones, but as neutral particles no longer react to electric fields, and can easily be discriminated from the ions in the beam. Neutral particles with sufficient kinetic energy can also eject an electron from a metal surface, so that a mass spectrum can be recorded.

Figure 7 shows the electron attachment probability as a function of the kinetic energy of the electron. A resonance near zero kinetic energy is measured for the solvated electron cluster production, proving that the attachment is indeed a resonant process. This experiment proves that slow electrons can be attached to cold clusters, where they can be trapped, in what has been termed "preexisting traps", traps for the electron, which have been present in the neutral cluster. This question had controversially been discussed in the literature on solvated electrons in bulk media. For electron energies between 2 and 5 eV the electron attachment cross section is very low. Afterwards it increases due to the dissociative attachment processes (see Eq. (2.5.1)).

Decay processes according to Eq. (2.5.2) and (2.5.3) do occur also on smaller and longer time-scales than discussed above. If the decay time is longer, the cluster ions appear as stable. They contribute to the middle spectrum in Fig. 6. But if the lifetime is shorter than $0.1 \cdot TOF$, they decay already in the ion source or acceleration region. There exist no plausible reason why small $(D_2O)_n^-$ are not formed in the interaction of slow electrons with water clusters at 180 K, but their lifetime is so short that they decay before detection. The internal energy leading to the fast decay has three contributions, the temperature (180 K per degree of freedom), the electron affinity, and the reorientation energy of the water molecules. The sum of these three contributions has to be reduced if the smaller, less stable clusters are to appear. This can be done by attaching slow electrons at position A of Fig. 3.15. At point A the density of water and seed gas is so high, that multiple collisions occur, so that a yield curve like in Fig. 7

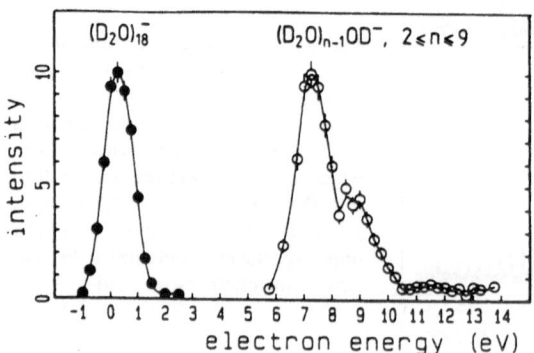

Fig. 7. Yield of specific cluster ions as a function of the kinetic energy of the attaching electrons. For electron energies below 1 eV solvated electron clusters are formed. Energies above 6 eV lead to dissociative electron attachment. The intensity below 0 eV is an artifact of the experiment

cannot be measured. The electron acts as a condensation point for the clusters, which grow around the charge in an already relaxed configuration. The clusters are cooled by collisions with the seed gas. The smaller the water content, the cooler the expansion will be, and the smaller clusters can be synthesized. This

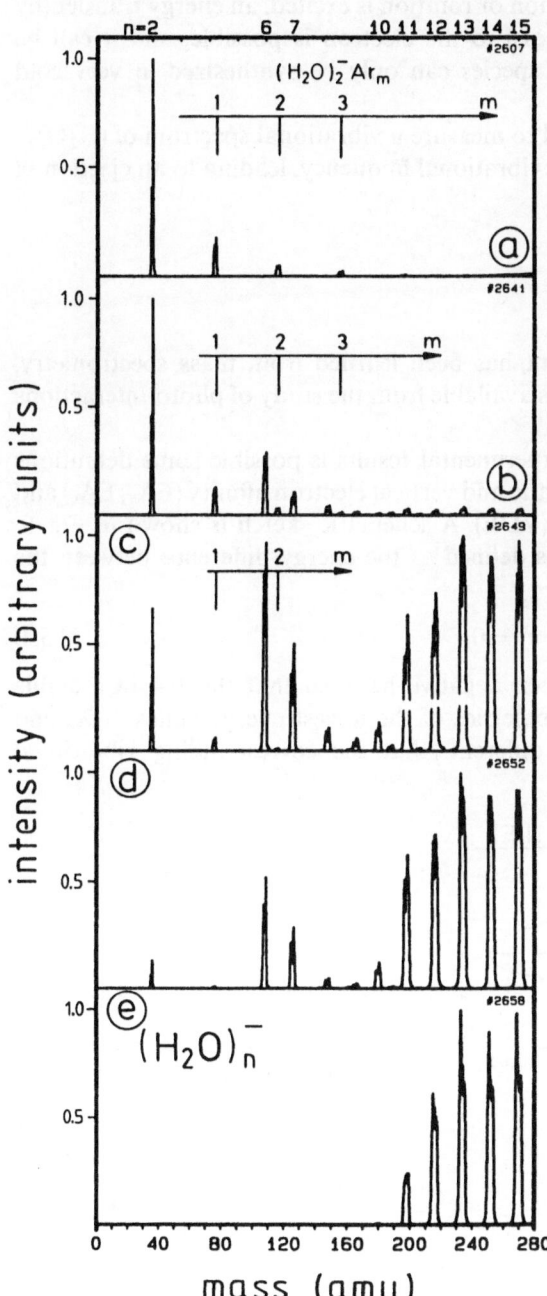

Fig. 8. For the production of small negatively charged water clusters a large surplus of rare gas is needed. The smaller the water content, the lower is the temperature in the expansion, and the smaller $(H_2O)_n^-$ clusters can be synthesized. From top to bottom the partial water pressure was only 0.0008, 0.008, 0.016, 0.022, 0.04 parts of the 7.5 bars argon pressure. If the water content is very low, mixed clusters like $(H_2O)_2^- Ar_n$ can be formed. Only if one sees these mixed clusters one can sure that the clusters are cold

expectation is confirmed by the experimental result shown in Fig. 8. At 4% water content $(H_2O)_n^-$, $n = 10$ is the smallest cluster visible. Reducing the water content all clusters with the exception of $n = 4$ can be observed. For the dimer, $(H_2O)_2^-$, the measured vertical detachment energy is only 47 meV [19]. If some vibration or rotation is excited, an energy transfer (by a non-Born–Oppenheimer process) to the electron is possible, and it can be ejected. Therefore, these fragile species can only be synthesized in very cold beams.

This feature has been utilised to measure a vibrational spectrum of $(H_2O)_2^-$. An IR-laser was used to excite a vibrational frequency, leading to an ejection of the electron [20].

2.5.2.3b) Photodetachment

Although a considerable amount has been learned from mass spectrometry, a more quantitative knowledge is available from the study of photo-interactions with mass selected clusters.

Before a discussion of the experimental results is possible some definitions have to be introduced: the adiabatic and vertical electron affinity (EA_a, EA_v) and the vertical detachment energy (VDE). A schematic sketch is shown in Fig. 9. The adiabatic electron affinity is defined as the energy difference between the ground states of X_n and X_n^-:

$$EA_a = E(X_n, v = 0) - E(X_n^-, v' = 0) \tag{2.5.4}$$

The binding energies E are taken negative here, so that the EA of a stable negative ion is positive. The geometries of the lowest energy states of X_n and X_n^- might occasionally be so different, that the corresponding vibrational

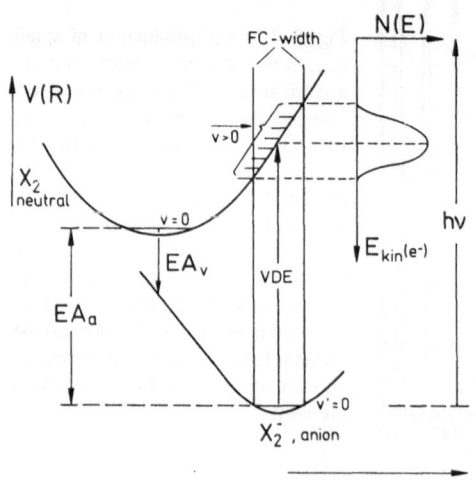

Fig. 9. Definition of the adiabatic and vertical electron affinity (EA_a, EA_v) and of the vertical detachment energy (VDE) for a diatomic molecule X_2. Electrons with different kinetic energies are emitted if a photon $h\nu$ induces an electronic transition. The minimal and maximal kinetic electron energies compatible with the Franck–Condon principle are indicated

wavefunctions $|v\rangle$ and $|v'\rangle$ do not have an overlap, i.e., $\langle v|v'\rangle = 0$. It is this overlap which controls the strength of an optical transition; its square is called the Franck–Condon factor [21]. If the Franck–Condon factor is zero between the two states on the right hand side of Eq. (2.5.4), the electron affinity cannot be measured optically, as the transition probability connecting the two states vanishes. But for higher vibrational states $v = n > 0$ the Franck–Condon (FC) factor can become finite (compare Fig. 9). The FC-factor increases at first, reaches a maximum and decreases afterwards. This is directly reflected in the kinetic energy spectrum of the emitted electrons. The vertical detachment energy (VDE) corresponds to the most probable part of this transition. The difference, $\delta E = \text{VDE} - \text{EA}_a$, is often called the rearrangement energy. It is the energy needed to rearrange the atoms in X_n so that the optical transition from X_n^- attains its largest value.

The word "vertical" in EA_v and VDE has the following origin. The Born–Oppenheimer approximation of molecular and solid state physics tells us, that the nuclei do not move during an electronic transition, i.e. $\delta R = 0$. In a potential energy diagram, where the potential energy $V(R)$ is plotted against the internuclear position R, the VDE corresponds to a *vertical* transition from the potential curve of X_n^- to the higher one for X_n, without any change in R. After these preliminaries the photo processes in

$$h\nu + (H_2O)_n^- \rightarrow (H_2O)_n + e^-$$
$$h\nu + (NH_3)_n^- \rightarrow (NH_3)_n + e^-$$

can be discussed. The geometry of the neutral or negatively charged clusters may be very different. If the $v' = v = 0$ vibrational wavefunctions would not overlap, only the VDE and not the EA_a could be measured optically.

Figures 10 and 11 show the measured VDE for water and ammonia clusters [19, 22, 23]. A continuous, mass selected negative cluster ion beam was crossed with a high intensity laser beam. The energy spectrum of the emitted electrons was measured. The peak maximum of the spectrum gives the VDE. In Figs. 10 and 11 the data are plotted against $n^{-1/3}$ which is proportional to the inverse of the cluster radius [24]. For water above $n = 11$ and for all ammonia data points, the experimental results are well represented by straight lines, which extrapolate to the correct bulk values for $n \rightarrow \infty$.

From a simple electrostatic calculation for the energy to eject an electron from the center of a homogeneous dielectric sphere one obtains [25]:

$$\text{VDE}(n) = \text{VDE}(n = \infty) - \frac{e^2}{2R}\left(1 + \frac{1}{D_{\text{opt}}} - \frac{2}{D_{\text{stat}}}\right)n^{-1/3}, \tag{2.5.5}$$

where D_{opt} and D_{stat} are the optical and static dielectric constants and R the effective radius of water or ammonia. Equation (2.5.5) tells us, that a straight line is expected for the VDE on a $n^{-1/3}$ plot if electrostatic effects prevail and if the electron is in an interior state inside the cluster. Save for the small water cluster sizes, the experimental results can be very well fit by straight line. Inserting the

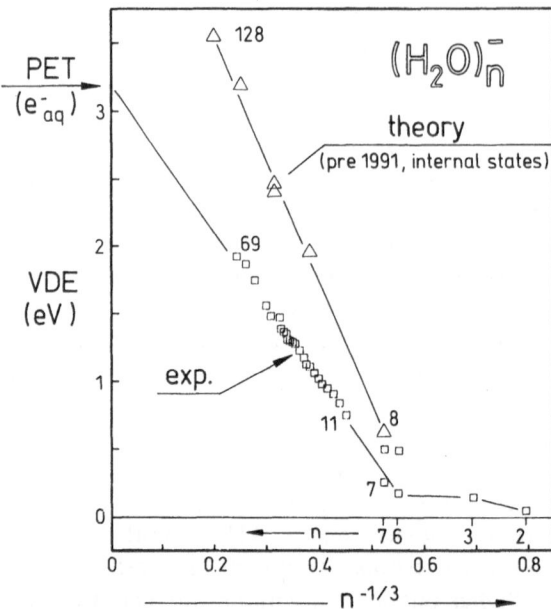

Fig. 10. The measured and calculated vertical detachment energies for $(H_2O)_n^-$ clusters are plotted against the inverse of the reduced cluster radius. A fit by Eq. (2.5.5) using bulk data gives a surprisingly accurate fit to the data. The calculated volume states show the same linear behaviour, but the slope and the asymptotic value are too high

known values into Eq. (2.5.5) one finds a surprisingly good agreement with the experiment for water. For ammonia the agreement is poorer. For example, the R-value for water is within 5% of that calculated for a density of 1 g/cm^3. The data for the small water cluster sizes do not fall on this line. But all other data can be well fit by a straight line, which has a slope calculated from Eq. (2.5.5) and the correct bulk limit. So this seems to be consistent with the picture of an electron solvated in ever increasing cluster sizes. A difficulty with this interpretation is discussed below.

The VDE ($n = \infty$) of Eq. (2.5.5) is not measured directly, but it should be equal to the photoelectric threshold (PET) energy of the bulk solvated electron. For water the PET is about 3.2 eV, and the photodetachment data extrapolate to a VDE ($n = \infty$) of 3.3 eV. Similarly one finds for ammonia, that an upper limit of PET(NH$_3$) is about 1.4 eV, while the detachment data extrapolate to about 1.25 eV. Above $n = 11$ for water and $n = 41$ for ammonia, the negatively charged clusters seem to evolve directly to the bulk solvated electron. No experimental evidence for a transition to another structure is seen.

One must be cautious in applying Eq. (2.5.5). It is derived by assuming that the solvated electron radius is much smaller than the cluster radius. This condition does certainly not hold for the smaller $(H_2O)_n^-$ clusters, where a straight line with the correct (bulk) parameters still gives a good fit. A recent

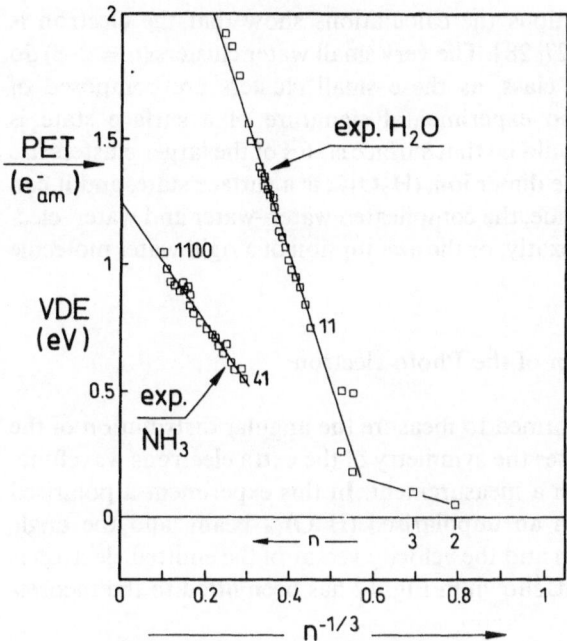

Fig. 11. Comparison of experimental vertical detachment energies for negatively charged water and ammonia clusters. While $(H_2O)_n^-$ is stable down to $n = 2$, for $(NH_3)_n^-$, $n = 34$ is the smallest one observable. Because of the low signal intensity no data exist for n between 34 and 41. The straight lines result from a least square fit to the data. The water data extrapolate to the correct bulk value, as shown in Fig. 10

model calculation using a mean spherical approximation for $I^-(H_2O)_n$ revealed that Eq. (2.5.5) might be obeyed only for $n \geq 125$, or so [26]. For the smallest clusters the VDE versus $n^{-1/3}$ curve oscillates. A close inspection of the data do in fact show some very weak oscillations. Whether they correspond to the calculated ones is not known. One is left with the dilemma, that Eq. (2.5.5) gives a beautiful fit to the data in a size region, where it should not be applicable.

Quantum path integral calculations have been performed on negatively charged water [27] and ammonia [27, 28] clusters. In essence the interaction of the assumed rigid water or ammonia molecules among themselves is treated classically, while quantum mechanics is used to describe the electron–molecule interaction. For large cluster sizes (e.g.: n above 64 for water) the calculations find "internal states" for the extra electron in water cluster anions [27]. The electron is primary localised in the center of the cluster for these volume states. The early calculated VDE values for the volume states of $(H_2O)_n^-$ are included in Fig. 10. They agree qualitatively with the experimental results, and show the same straight line behaviour on a VDE versus $n^{-1/3}$ plot. But their slope and extrapolated bulk value are too high. Very new results, using improved potentials, give a better agreement, but the slope and intercept are still too high [29].

For small water cluster anions the calculations show that the electron is localised on a "surface state" [27, 28]. The very small water clusters ($n = 2$–8) do quite possibly belong to this class, as these small clusters are composed of surface only. Why there is no experimental signature of a surface state is unknown for the moment. It could be that surface states of the larger clusters are not formed experimentally. The dimer ion, $(H_2O)_2^-$, is a surface state, and it can be formed. On the theoretical side, the complicated water–water and water–electron interaction is not known exatly, or the assumption of a rigid water molecule might not be tenable.

2.5.2.3c)   Angular Distribution of the Photo-electron

One experiment has been performed to measure the angular distribution of the emitted electrons. In simple cases the symmetry of the extra electrons wavefunction can be deduced from such a measurement. In this experiment a polarised laser beam is overlapped with an unpolarised $(H_2O)_{18}^-$ beam, and the angle θ between the laser polarisation and the velocity vector of the emitted electron is varied. The experimental result, shown in Fig. 12 has been fitted to the theoretical prediction:

$$I(\theta) = \sigma_{total}[1 + \beta P_2(\cos \theta)] .   \tag{2.5.6}$$

where $P_2(u) = (3u^2 - 1)/2$ is the second Legendre Polynomial and the asymmetry parameter β can take values between $+ 2$ and $- 1$. A fit to the data gives $\beta = 0.92$. For a pure s-type orbital (zero orbital angular momentum of the extra electron) one expects $\beta = 2$, while $\beta = - 1$ is obtained for a p-orbital. No calculation exists to which this experiment could be compared.

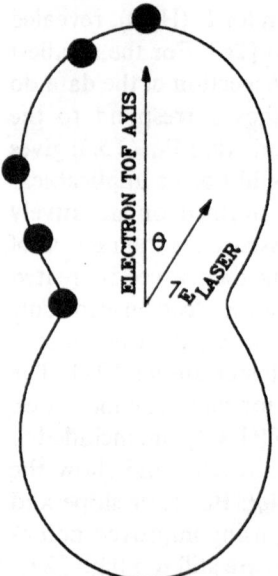

Fig. 12. Angular distribution of ejected photo-electrons. The angle Θ is spanned between the polarization vector of the photon and the velocity vector of the electron

2.5.2.3d) Photofragmentation

In the context of Eqs. (2.5.2) and (2.5.3) the question was asked, what does a highly excited negatively charged cluster emit, an electron or a molecule? This problem was studied in more detail for $(H_2O)_{25}^-$. The cluster was photoexcited and the two decay channels studied [30]. The energetic threshold for photodissociation (e.g. Eq. (2.5.2)), is ≈ 0.48 eV, while that for photodetachment is 0.8 eV. Photodetachment dominates for photon energies above 2 eV. Photofragmentation is seen only within about 1 eV above the detachment threshold. Probably the water molecules in the cluster have to move before the electron can leave at low photon energies. Otherwise the concurrence of the two time scales is difficult to understand. The influence of the clusters' internal excitation on the fragmentation pattern could be demonstrated [31].

2.5.2.3e)   Summary and Remaining Problems

Although enormous progress has been made in the understanding of the negatively charged water and ammonia clusters, several questions remain open, and are the focus of intense research:

1. What is the origin of the very different n_{min} values in Table 1?
2. What is the experimental signature of the calculated surface states? Why is there no experimental evidence for their existence at larger sizes?
3. What kind of physical information does the β value of the angular distribution contain?
4. Above which n do doubly negatively charged solvated electron clusters become stable? Are both the singlet and the triplet state bound?

2.5.2.4   Alkali-Halide Clusters

The excess electron states in alkali halides are somewhat different from those discussed above. The clusters are neutral and the monomer anion exists. But these systems, in both the cluster and the bulk, are similar in some ways to H_2O and NH_3, so that they are discussed here.

2.5.2.4a)   Excess Electron States for the Molecule and the Solid

The chemical bonding in alkali halide molecules, say NaF, is well described by treating the molecules as ion pairs, Na^+F^-. The valence electron of the metal fills the hole in the halogen's valence shell, giving two closed shell ions bound by the strong Coulomb attraction. In the monomer anions the extra electron goes into the alkali valence orbital, which is strongly polarized by the halogen ion [32]. The charge density of the extra electron is mainly located "behind" the positive metal ion core, well shielded from the halogen ion. Thus, in contrast to the examples discussed above, the monomer itself does have a positive electron affinity.

The excess electron states in the bulk are the well known F-centers. The electron occupies the position of a missing halogen molecule. Its charge density is concentrated mainly near the eight neighbouring positively charged metal ions. The F-centers are well studied and characterised, as explained in many textbooks on solid state physics [33].

2.5.2.4b) Calculations of Excess Electron States in Molecules and Clusters

The interaction potential used in calculations on alkali-halide molecules, clusters and crystals is the so called Rittner potential and its variants [34–36], which describes the Coulomb attraction of the ions counterbalanced by an exponential repulsion, which arises when the two closed electronic shells overlap. The calculated most stable structures often have cube or chair-like structures. Theoretically and experimentally it is observed, that cluster ions have especially stable cubic structures, if they contain an odd number $2n - 1 = j \cdot k \cdot l$ of atoms, with j, k, and l all odd integers. Examples are $Na_{14}F_{13}^+$ or $Na_{23}F_{22}^+$, with $2n - 1 = 3 \cdot 3 \cdot 3$, or $3 \cdot 3 \cdot 5$. The clusters form a filled cubic microlattice with 3 or 5 atoms on one side. Adding 1 extra electron to this cluster, makes a neutral cluster. The extra electron is so weakly bound that is has to be treated quantum mechanically [35, 36]. The electron distributes its charge density uniformly around the cluster, if the position of the nuclei is clamped fixed during electron attachment. If the atoms are allowed to move the excess electron localizes, forming a weakly bound surface state (binding energy ≈ 1.8 eV) . These states have been termed Class I in [36]. The cluster analog of the bulk F-center belongs to Class II. They again have the stable near cubic structures. Their detachment energies are in the 3–4 eV range. Excess electrons in noncubic clusters (Class III) have binding energies of ≈ 3.8 eV. The excess electron is probably bound to a single Na atom, similar to the alkali halide molecule. Most of these predictions have indeed been verified experimentally.

2.5.2.4c) Observation of Excess Electron States in Alkali-Halide Clusters

For the production of alkali-halide clusters one may use laser vaporisation of Na metal into a helium flow containing 1% SF_6 [36]. The molecule is dissociated in the hot, laser produced plasma and NaF molecules and clusters are formed. The electronic structure of the excess electrons and the cluster geometry is probed by one and two photon laser spectroscopy, giving a satisfactory agreement between theory and experiment [35, 36].

2.5.3 Outlook

Only a few of the questions asked in the introduction have been answered satisfactorily to date. The field is one of active research, which has profited much from the interaction of theory and experiment.

2.5.4 Recent Developments

(1) Recently, it has been pointed out [37] that the $(NH_3)_n$-VDE data fit Eq. (2.5.5) rather well when the dielectric constants for solid ammonia are used. Under the conditions utilized to form these cluster anions, they are very likely to be solid [38]. Also, more recent measurements [39, 40] of the bulk PET for ammoniated electrons in dilute solutions find it to range between 1.27–1.45 eV, in even closer agreement with the 1.25 eV extrapolated from experiment [19]. (2) The species $(H_2O)_{n = 2,6,7''}$, were observed via collisions of water clusters with laser-excited Rydberg atoms [41]. (3) Ethylene glycol cluster anions were studied by photodetachment [42]. (4) Mixed cluster anions of ammonia and water were studied by photodetachment [42]. Also, the dipole-bound dimer anion, $(H_2O)(NH_3)$, was predicted and then observed [43]. (5) Several alkali halide cluster anions were studied by photodetachment [44, 45] and evidence for both F and F' centers was seen.

Acknowledgments. The authors acknowledge the support of a NATO grant (#861307) and KB acknowledges the support of the US NSF under grant CHE-9007445.

References

1. G. Stein: Disc. Faraday Soc. **12**, 227, 1952
2. G. Czapski, H. Schwarz: J. Phys. Chem. 1962
3. E.J. Hart, J.W. Boag: J. Am. Chem. Soc. **84**, 4090, 1962
4. C. von Sonntag: *The Chemical Basis of Radiation Biology*, ed. by Taylor and Francis, London, 1988
5. A.J. Swallow: Nature **222**, 369, 1969
6. See many articles in J. Phys. Chem. **88**, 3699 and 3913 (1984)
7. See the review articles: *The Solvated Electron* by E.M. Itskovitch, A.M. Kusnetsov, J. Ulstrup; *Excess Electrons in Nonpolar Liquids* by L. Nyikos, R. Schiller: in, The Chemical Physics of Solvation, part C, ed. by R.R. Doganadze et al., Elsevier Amsterdam 1988
8. A. Migus, Y. Gauduel, J.L. Martin, A. Antonetti: Phys. Rev. Lett. **58**, 1559, 1987
9. F.H. Long, H. Lu, K.B. Eisenthal: Phys. Rev. Lett. 1990, 1990
10. M.V. Rama Krishna, K.B. Whaley: Phys. Rev. **B 16**, 1988, 1988
11. Northby et al. (T. Jiang, S. Sun and J.A. Northby: Physics and Chemistry of Small Clusters, eds.: P. Jena, B.K. Rao, and S.N. Khanna, NATO ASI-Series, Plenum Press 1992, New York 1992),

have recently bombarded a He beam with electrons. They observed a negative ion signal, which was not mass analyzed. It is interpreted as being due to an electron in a metastable "bubble state" inside a He-cluster. It cannot be the very weakly bound surface state, which would field detach in the electric fields of the mass spectrometer.

12. N. Schwentner, E.-E. Koch, J. Jortner: Electronic excitations in condensed rare gases. Springer Tracts in Modern Physics 107, 1985
13. L. Sanche: J. Phys. B **23**, 000, 1990
14. H. Haberland, T. Kolar, T. Reiners: Phys. Rev. Lett. **63**, 1219, 1989
15. G.J. Martyna, B.J. Berne: J. Chem. Phys. **90**, 3744, 1989
16. P. Stampfli, K. Bennemann: Phys. Rev. A **38**, 4431, 1988
17. T. Kondow, T. Nagata, K. Kudritse: Z. Phys. D **12**, 291, 1989
18. M. Knapp, O. Echt, D. Kreisle, E. Recknagel: J. Phys. Chem. **91**, 2601, 1987
19. G.H. Lee, S.T. Arnold, J.G. Eaton, H.W. Sarkas, K.H. Bowen, C. Ludewigt, H. Haberland: Z. Phys. D **20**, 9, 1991
20. A. Bar-on, R. Naaman: J. Chem. Phys. **90**, 5198, 1989
21. For a more precise treatment see any book on molecular electronic transtions
22. S.T. Arnold, J.G. Eaton, D. Patel-Misra, H.W. Sarkas, K.H. Bowen: *Ion and Cluster Ion Spectroscopy* ed. by J.P. Maier (Elsevier), 487, 1990
23. J.V. Coe, G.H. Lee, J.G. Eaton, S.T. Arnold, H.W. Sarkas, K.H. Bowen, C. Ludewigt, H. Haberland, D.R. Worsnop: J. Chem. Phys. **92**, 3980, 1990
24. If r is the (van der Waals) radius of an atom or molecule and R the radius of a spherical cluster containing n atoms, the cluster's volume is given by $V = 4\pi R^3 = n4\pi r^3$. Thus $n = (R/r)^3$, or $n^{-1/3} = 1/(R/r)$. The value R/r is often called the reduced cluster radius
25. R.N. Barnett, U. Landman, C.L. Cleveland, J. Jortner: Chem. Phys. Lett. **145**, 382, 1988
26. I. Rips, J. Jortner: *Ion solvation in clusters.* preprint, 1992
27. R.N. Barnet, U. Landman, C.L. Cleveland, J. Jortner: J. Chem. Phys. **88**, 4429, 1988
28. M. Marchi, M. Sprik, M.L. Klein: J. Chem. Phys. **89**, 4918, 1988
29. U. Landman, priv. communication
30. L.A. Posey, P. Campagnola, M.A. Johnson, G.H. Lee, J.G. Eaton, K.H. Bowen: J. Phys. Chem. **91**, 6536, 1989
31. P.J. Campagnola, L.A. Posey, M.A. Johnson: J. Chem. Phys. **95**, 7998, 1991
32. T.M. Miller, D.G. Leopold, J.K. Murray, W.C. Lineberger: J. Chem. Phys. **85**, 3268, 1986
33. For example: Ashcroft-Mermin: Solid State Physics, Chapter 30. The colour- or F-centers have broad, strong absorption bands in the visible, which led to their use in the F-center lasers. The letter F stands for Farbe, the German word for colour
34. T.P. Martin: Phys. Rep. **95**, 167, 1983
35. D. Scharf, U. Landman, J. Jortner. J. Chem. Phys. **87**, 2716, 1987
36. G. Rajagopal, R.N. Barnett, A. Nitzan, Landman, E.C. Honea, P. Labastie, M.L. Homer, R.L. Whetten: Phys. Rev. Lett. **64**, 2933, 1990
37. G. Makov, A. Nitzan: J. Phys. Chem. (in press)
38. L.S. Bartell (unpublished data)
39. F.A. Uribe, T. Sawada, A.J. Bard: Chem. Phys. Lett. **97**, 243 (1983)
40. A.J. Bard, K. Itaya, R.E. Malpas, T. Tehrani: J. Phys. Chem. **84**, 1262 (1980)
41. C. Desfrancois, N. Khelifa, A. Lisfi, J.P. Schermann, J.G. Eaton, K.H. Bowen: J. Chem. Phys. **95**, 7760 (1991)
42. S.T. Arnold: PhD Thesis (Johns Hopkins University, Baltimore, 1993)
43. C. Desfrancois, B. Baillon, J.P. Schermann, S.T. Arnold, J.H. Hendricks, K.H. Bowen: Phys. Rev. Lett. **72**, 48 (1994)
44. Y.A. Yang, L.A. Bloomfield, C. Jin, L.S. Wang, R.E. Smalley: J. Chem. Phys. **96**, 2453 (1992)
45. H.W. Sarkas: PhD Thesis (Johns Hopkins University, Baltimore 1993)

2.6 Internal Reactions and Metastable Dissociations After Ionization of van der Waals Clusters

T.D. Märk and *O. Echt*

2.6.1 Introduction

Electron impact or photon ionization in conjunction with mass spectrometry is the most common method to detect neutral clusters. If this method is used to analyze neutral van der Waals cluster abundance distributions (as produced by supersonic expansion techniques) quantitative information on the ionization process is necessary in order to unambiguously interpret and relate measured mass spectral distributions to the original neutral cluster distributions. Such ambiguities may arise from size dependent ionization cross sections, ion stabilities and detection efficiencies; a fact which is well known from the mass spectrometry of polyatomic molecules [1, 2]. Moreover, as the colliding target of an electron (or photon) advances from an atom or molecule to a cluster, a greater diversity of possible mechanisms for energy transfer, energy deposition and energy disposal are available. Whereas inelastic interaction of electrons with atoms and molecules results in changes of the electronic configuration and/or nuclear motion, respectively, interaction of electrons with van der Waals clusters involves – besides these intramolecular excitations – multiple electron collisions and intermolecular reactions within the cluster.

It is clear that these additional reactions will lead to even stronger fragmentation of the original clusters during the ionization event than in case of ordinary molecules. Curiously enough, despite this fact a number of earlier studies related on a one to one basis cluster ion mass spectra to neutral cluster distributions. It was only in 1982 that *Stephan* et al. [3] noted that "one has to account for the unimolecular dissociation when using impact ionization plus mass spectrometry to probe (quantitatively) neutral cluster beams". Today, it is widely accepted that abundance fluctuations present in mass spectra of van der Waals clusters are due to variations in the ionization efficiency and the properties of the ensuing ions rather than due to the distribution of their neutral precursors.

Here, we will summarize (i) electron impact ionization of van der Waals clusters, including the ionization mechanism and efficiency, and (ii) processes (and their properties) occurring after the primary ionization event in the cluster ion, i.e. internal ion molecule reactions and the various metastable dissociation processes (see also appropriate recent reviews on these subjects [4–10]).

2.6.2   Ionization Mechanisms and Processes

If the energy of an electron (or photon) colliding with a gas phase cluster is greater than a critical value (termed appearance energy) some of the neutral clusters will be ionized. The abundance and variety of the ions produced from a specific neutral precursor depends on the corresponding ionization cross sections (see Sect. 2.6.3) and will increase with increasing collision energy. The ionization event (leading to changes in the electron shell) and concomitant changes in vibrational and rotational excitation follows the Franck Condon principle and other selection rules. The outcome of the ionization process is depending on geometric and energetic properties of the neutral and ionized cluster. Some of the possible reaction channels are summarized in Table 1 for the simplest case of a dimer cluster.

Some of these reactions may be considered as single step ionization reactions (reaction (1)–(5) in Table 1), whereas others (reaction (6)–(9)) have to be viewed as two step processes where the initial interaction leads to an intermediate product which reacts subsequently via a unimolecular process. Fig. 1 illustrates these various possibilities. It is clear from this picture that measured cluster ion mass spectra are strongly depending on the time of detection after the ionization due to the occurrence of these two step processes with lifetimes in the metastable

Table 1. Various possible ionization reaction channels for a dimer cluster A_2 (After [5])

Reaction	Denotation
(1) $A_2 + e \rightarrow A_2^+ + 2e$	Single ionization
(2) $A_2 + e \rightarrow A_2^{z+} + (z + 1)e$	z-fold ionization
(3) $A_2 + e \rightarrow A^+ + A + 2e$	Dissociative ionization
(4) $A_2 + e \rightarrow A^+ + A^- + e$	Ion pair formation
(5) $A_2 + e \rightarrow A^- + A$	Dissociative electron capture
(6) $A_2 + e \rightarrow A_2^* + e$ $\rightarrow A_2^+ + e$	Autoionization
(7) $A_2 + e \rightarrow A_2^{+*} + 2e$ $\rightarrow A^+ + A$ $\rightarrow A_2^{2+} + e$	Metastable dissociation Autoionization
(8) $A_2 + e \rightarrow A_2^{2+*} + 3e$ $\rightarrow A^+ + A^+$	Coulomb explosion
(9) $A_2 + e \rightarrow A_2^{-*}$ $\rightarrow A_2^- + h\nu$ $\rightarrow A^- + A$	Resonance capture Metastable dissociation

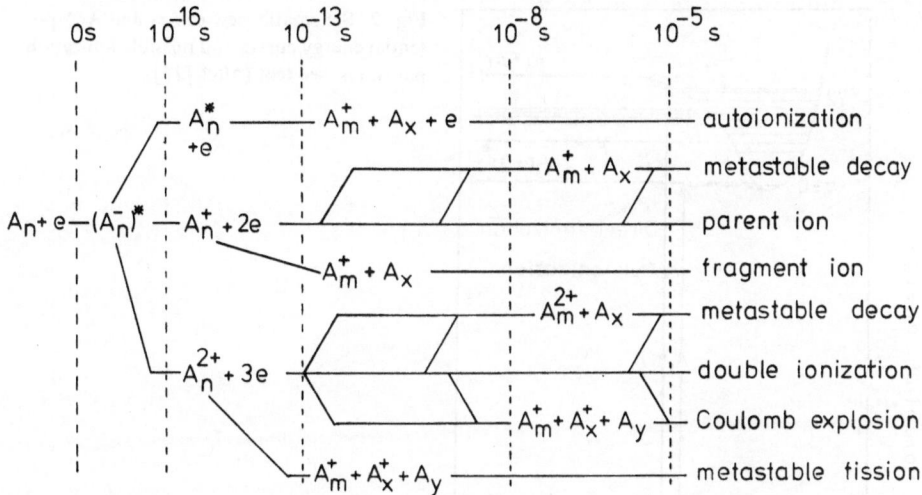

Fig. 1. Schematic genealogical time chart of electron impact ionization of van der Waals clusters A_n (After [12])

time regime (10^{-11} to 10^{-3} s [11]). For example, Ar^+ fragment ions will be produced (i) by a direct ionization process (reaction (3)) in the very first instance of an ionizing interaction via a repulsive state of the molecular ion (sometimes termed prompt fragmentation) and (ii) by the much slower two step mechanism of metastable dissociation via a quasi-bound (metastable) state of the molecular ion. In this simple case the measured ion ratio Ar^+/Ar_2^+ will increase with time passed since the primary electron/dimer interaction.

As an example, we shall briefly consider the sequence of events which lead to the production of an argon dimer ion by ionization of its neutral precursor (reaction (1) and (6)). Close to the (adiabatic) ionization threshold of Ar_2 electron impact most probably will only lead to the production of a molecular Rydberg state [13] due to the poor Franck Condon overlap between Ar_2^+ and Ar_2 in their ground states (see Fig. 2). Autoionization can then occur, i.e. a coupling between the Rydberg electron and the nuclear motion leads to the production of Ar_2^+ in its stable ground state. If the density of Rydberg states close to the adiabatic ionization threshold is high, little vibrational energy will be deposited in the dimer ion. A major contribution to the decrease in (adiabatic) ionization energy in going from the monomer to the dimer (and larger clusters) comes from the availability of this ionization process and the binding energy in the dimer ion (see Chapter 4.6, Vol. I). At higher electron energies this threshold mechanism will be replaced by that of direct ionization (vertical transition). Moreover, as the stabilizing entity in an argon cluster ion appears to be Ar_2^+ ($^2\Sigma_u^+$), or Ar_3^+, similar ionization events will take place in larger clusters [14, 15]. Whereas in the autoionization case very little excess energy will be available in the cluster after ionization, direct ionization and

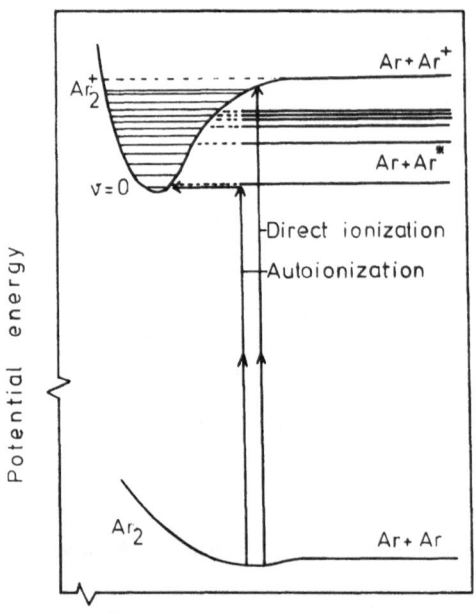

Fig. 2. Schematic view of Ar_2 and Ar_2^+ potential energy curves and possible ionization pathways, see text (After [14])

subsequent relaxation (trapping) will result in extensive heating of the cluster followed by an extended period of evaporative cooling (metastable dissociation).

Other variants of two step ionization mechanisms and their properties, including internal ion molecule reactions in the nascent cluster ion, will be discussed in Sects. 2.6.4 and 2.6.5. An especially exciting possibility arising from the interaction between an electron and a cluster is the occurrence of multiple collisions of the incoming electron leading to multiple ionization of the cluster (see Chapter 2.7).

2.6.3   Ionization Efficiency

Although detailed information about ionization cross section functions of van der Waals clusters would be necessary for their quantitative detection, little is known due to considerable experimental difficulties with respect to target density and composition. No absolute total, counting or partial ionization cross sections (for a definition see Refs. [1, 2]) for a specific cluster size (except for dimers, see Fig. 3) have been reported up to date.

Conversely, total, counting and partial ionization cross sections have been measured for cluster distributions of H_2 and CO_2 [16, 17]. Figure 3 shows the total ionization cross section functions divided by the averaged number of cluster constituents ("effective" cross section) for various H_2 cluster distributions. The position of the maximum of the cross section shifts to higher electron

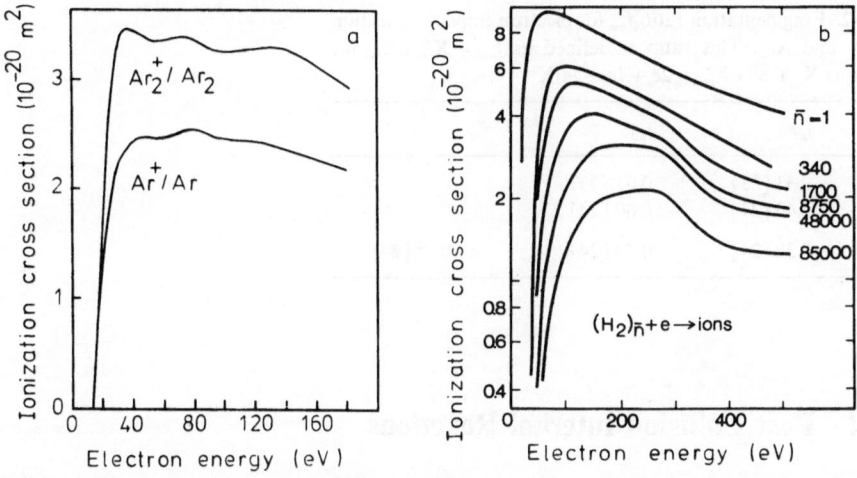

Fig. 3. a Partial ionization cross sections for the process $Ar + e \rightarrow Ar^+ + 2e$ and $Ar_2 + e \rightarrow Ar_2^+$ + 2e and as a function of electron energy after Märk [7]. **b** Effective total ionization cross sections for various H_2 cluster size distribution after *Henkes* and *Mikosch* [16]

energies with larger averaged cluster size n (due to the loss of energy as the electron passes through the cluster) and the magnitude of the effective cross section decreases for larger n (for a theoretical description of these phenomena see [6, 18]).

A few relative partial ionization cross section functions have been reported (the most extensive study concerns Ar clusters [19]). For the reasons given above no absolute values are available. Owing to this lack of data, the additivity rule [1, 2, 12] has been used occasionally to calibrate cluster ion signals detected by mass spectrometry, i.e. assuming for instance that the dimer ionization cross section is twice the monomer ionization cross section and so forth. This procedure, however, is at variance with the results presented above (Fig. 3). Furthermore it neglects possible fragmentation of the neutral under study and possible cascading from larger neutrals present in the distributions. Using a modified additivity rule taking into account dissociative channels it is possible to deduce at least for rare gas dimers absolute partial ionization cross sections [7] (e.g. see Fig. 3). Moreover, in a few cases accurate partial ionization cross section ratios have been determined for small clusters using either spectroscopic methods [20] or the method of producing size selected neutral clusters by momentum transfer via scattering [9]. As expected from the mass spectrometry of ordinary molecules [1, 2] appreciable fragmentation is occurring for some of the clusters studied [7, 9, 21, 22]. Fragmentation ratios deduced for Ar_2 and Ar_3 are given in Table 2: 40% of the ions produced (including prompt and metastable dissociation) from Ar_2 are Ar^+ fragment ions, and in case of Ar_3 almost 100% of the ions are ending up as Ar_2^+ or Ar^+ fragment ions. It is interesting to note that in the latter case (Ar_3) the ratio between the product ions Ar^+ and Ar_2^+ appears to be almost independent of the time (after ionization) [25].

Table 2. Fragmentation ratio f_{nm} for electron impact ionization of Ar_2 and Ar_3. This ratio is defined as $f_{nm} = X_m^+/\Sigma X_m^+$ for reactions $X_n + e \rightarrow X_m^+ + 2e + (n-m)X$

	f_{n1}	f_{n2}	f_{n3}
Ar_2	0.33 [23]	0.67 [23]	—
	0.40 [24]	0.60 [24]	—
Ar_3	0.30 [24]	0.70 [24]	$< 10^{-4}$ [24]

2.6.4   Post Collision Internal Reactions

As mentioned in Sect. 2.6.1 some of the ionization processes induced by electrons (or photons) proceed via a two step mechanism. In most of these cases (e.g. reaction (6), (7b), (9b)) the second step involves an electronic transition, i.e. autoionization (e.g. see Ref. [26, 27]) or radiative stabilization. There exists, however, a second class of two step processes involving in the second step the interaction between two or more cluster constituents. Several of these post collision internal reactions have been observed to occur:

Associative Ionization (or chemi-ionization), where bond formation is involved: Investigations of *Dehmer* and *Poliakoff* [28] of Ar_2 in the VUV using mass spectrometric detection showed that besides autoionization [26, 27] of Ar_2^*, Ar_2^+ is also formed in small amounts from the associative ionization reaction $Ar^* + Ar \rightarrow Ar_2^+ + e$ within the excited $(Ar^* \cdot Ar)$. *Ng* and coworkers [29] were able to identify and study the relative reaction probabilities for the formation of various product channels of the chemi-ionization process $CS_2^*(v, n) \cdot CS_2 \rightarrow$ yielding $(CS_2)_2^+$, and various fragment ions as a function of Rydberg level n.

Penning Ionization, where an electron is transferred: In this case initial energy deposition from the bombarding electron (photon) into an excited state of one cluster constituent (usually a rare gas atom) is followed by Penning energy transfer to a seed cluster constituent leading to its ionization. For instance, *Birkhofer* et al. [30] (see also [21]) have described Penning ionization of Ar clusters by metastable He (a process which has been also studied outside of the cluster [31]). *Kamke* et al. [32] have demonstrated the energy transfer from excited Ar atoms in clusters leading to the ionization of organic seed molecules, and *Dao* et al. [33] reported the possibility of Penning transfer of energy from an organic molecule to another molecule in a cluster involving delay times of up to 0.2 μs. Considering that this intermolecular Penning ionization may in principle occur as soon as the adiabatic ionization energy of the cluster as a whole is below the excitation energy of any of its (solvated) constituents *Hertel* and coworkers [34] recently suggested that this process is a rather common ionization channel, since the ionization energy of molecular clusters is usually much lower than for any of its constituents (see also Chapters 4.6 and 4.7 in Vol. I).

Ion Molecule Reactions. A very common phenomenon, especially in mixed and molecular clusters, are reactions within the energized ion–neutral complex after the primary ionization event; also referred to as ion–neutral half collision processes. These reactions (which usually lead to prompt fragmentation of the cluster ion) provide valuable energetic and dynamical information for specific ion molecule reactions which is difficult to measure by other means. There exists now a large body of molecular beam photoionization and electron impact ionization studies of these reactions (see the extensive reviews covering the literature up to 1986 [4, 8, 27]. Information about these reactions are usually deduced from the fragmentation mass spectra measured as a function of electron (photon) energy and stagnation conditions taking also into account known properties of the respective reactions in the gas phase. In these studies the fragmentation pattern is normally interrogated at times where not only prompt but also metastable dissociations (see 2.6.5) have already influenced the original pattern (see Fig. 1).

Recently, the photoion–photoelectron coincidence (PIPECO) techniques [35, 22], which utilize flight time correlation of an ion–electron pair, have been used extensively as an ideal approach to incorporate state and energy selection into these studies. For instance, the dissociation of $Ar^+(^2P_j) \cdot CO$ into $CO^+ + Ar$ (produced by photoionization of $Ar \cdot CO$ and studied with PIPECO) is rationalized by Ng and coworkers [36] by a stepwise mechanism involving the formation of a vibrationally excited $CO^+(\tilde{X}, v') \cdot Ar$ complex by near *resonance intermolecular charge transfer* and the subsequent unimolecular dissociation of the complex by vibrational predissociation (see Sect. 2.6.5). This is in line with the interpretation of the unimolecular (prompt) dissociation of single ethylene ions in association with large (up to Ar_{130}) argon clusters, where charge transfer on or close to the surface is made responsible for the observed fragment abundances [37].

The most common reactions are, however, *proton transfer reactions* and *association reactions* [4, 38]. The production of the protonated ammonia cluster series $NH_4^+ \cdot (NH_3)_n$ is an example for the occurrence of an exoergic proton exchange reaction, whereas $H_3O^+ \cdot (H_2O)_n$ for an endoergic reaction. The ammonia reaction has been studied extensively [39, 40] and *Castleman* and coworkers [41], using multiphoton ionization, found that the protonated pentamer is an unusually prominent species due to a presumed closed solvation shell. A similar prominent feature is the protonated 21-mer in H_2O cluster spectra, its structure has been ascribed to clathrate formation via an ion-induced mechanism [42]. It is interesting to note that in both cases ionization also leads to the production of stoichiometric cluster ion series, albeit with much smaller probability than in the protonated case. Moreover, evidence of multiple *Mc Lafferty rearrangements* in protonated *n*-butanoic acid clusters was reported [43]. Recently, the fragmentation pattern induced by such ion molecule reactions has been studied for several molecular clusters by *Buck* and coworkers [24, 44] using size selected neutral clusters as collision partners in the primary ionization process (see Chapter 4.7 in Vol. I). In case of ethylene clusters, it is assumed that one ethylene unit is ionized, which leads via an association reaction to a highly excited $C_4H_8^+$. This complex either decays to $C_3H_5^+$ or $C_4H_7^+$, or, in the case of clusters larger than the dimer, the $C_4H_8^+$ complex may be stabilized by a third body collision. The occurrence of $C_2H_5^+$ ions is rationalized by the endothermic *H atom transfer reaction* within the cluster $C_2H_4 \cdot C_2H_4 \rightarrow C_2H_5^+ + C_2H_3$. Another method to follow the ion chemistry has been presented by *Brutschy* [45] where the neutral precursor of product ions is assigned using the optical fingerprint in two color two photon ionization spectroscopy. Finally, it is quite noteworthy that similar ion molecule reactions are to be expected in negative molecular cluster ions [38, 46–48].

2.6.5   Metastable Dissociations

The remainder of this chapter is devoted to an interesting variation on the theme of unimolecular cluster ion reactivity, i.e., decomposition in the metastable time regime. Whereas prompt dissociation reactions in the cluster ion (as described above) cannot be observed directly and identified unambiguously from the measured fragmentation patterns, metastable dissociation reactions may be detected directly in the field free regions of certain mass spectrometer systems. This offers an opportunity to study in detail properties (energetics, kinetics and dynamics) of selected ion dissociation processes of uniquely selected cluster ion species. The existence of metastable cluster ions after electron (or photon) ionization of vdW clusters has been established only in the last decade [3, 46, 49, 50] despite the fact that (i) delayed unimolecular decay was to be expected [1, 2, 51] and that (ii) a series of cluster ions, differing only in size, constitute an ideal testing ground for statistical unimolecular decay theories (e.g. Quasi-Equilibrium-Theory in mass spectrometry [52]. Today, a large number of studies exists on the properties of metastable decay reactions of cluster ions (with lifetimes from 10^{-7} s up to several 100 μs [53–57]) produced by either electron or photon ionization of vdW clusters. It is interesting to note that there exist also several studies using other methods of metastable ion production (e.g. see Ref. 239 in [7]). Research carried out on this subject up to 1989 has been summarized in several reviews [4, 7, 8].

2.6.5.1   Experimental Techniques

Experimental techniques include for instance the use of a single [49] or double [25] focussing sector field mass spectrometer, time of flight mass spectrometer combined with energy analyzer (reflectron) [58], quadrupole mass spectrometer combined with energy analyzer [59], and quadrupole mass spectrometer combined with an ion trap (storage time up to 50 ms) [60].

The measuring principle of most experimental set-ups can be best described using, as an example, the technique used in *Innsbruck* [25]. Consider a monoenergetic ion beam issuing from a slit (e.g. D in Fig. 9 in Chapter 3.5 in Vol. I) and entering a field-free region before being analyzed in a magnetic sector field. If an ion m_1^+ decomposes in this field-free region, i.e.

$$m_1^+ \rightarrow m_2^+ + m_3 \tag{1}$$

and if this decomposition takes place without any conversion of internal into external energy, the daughter ion m_2^+ will continue to move along with the same original velocity. Thus, the initial kinetic energy $m_1 v^2/2$ will be equal to the total kinetic energy of the fragments, i.e. $m_1 v^2/2$ is shared between the fragments in the ratio of their masses. According to [51], these daughter ions, m_2^+, are transmitted through the magnetic sector with an apparent mass m^* where

$$m^* = m_2^2/m_1 . \tag{2}$$

The mass peak obtained due to the m_2^+ ion from the decomposition of the metastable m_1^+ ion is known as a *metastable peak* and such peaks will be seen in single-focussing sector instruments.

In double-focussing instruments of the reversed geometry type (as used in our laboratory, see Fig. 9 in Chapter 3.5 in Vol. I), the magnetic sector precedes an electric sector. The m_2^+ daughter ions from a decomposition of metastable ions in the first field-free region (in front of the magnetic sector field) will not be present in the conventional mass spectrum recorded with such an instrument (in contrast to the situation in a single-focussing instrument where all ion peaks are present, which sometimes leads to ambiguities due to overlap between the spectra of stable ions and metastable ions). This is due to the fact that the voltage of the electric sector field following the magnetic sector field is coupled to the main accelerating voltage V. Only ions having the full initial energy, qV, will be transmitted and constitute the conventional mass spectrum. If the voltage across the electric sector is changed to a fraction m_2/m_1 of V [51], i.e.

$$V^* = m_2 V/m_1 \tag{3}$$

then daughter ions m_2^+ decaying either in the first field-free region (with the magnetic sector tuned to m^* (Eq. 2)) or in the second field-free region (with the magnetic sector tuned to m_1) will pass the electric sector. In both cases, only peaks due to ions that have decomposed in the field-free regions will be recorded, thus avoiding overlap with ions of the conventional mass spectrum. For more details on the experimental methods used see [51, 61].

2.6.5.2   Metastable Decay Mechanisms

Fragmentation of an excited molecular ion usually occurs immediately after the ionization event (prompt dissociative ionization via repulsive hypersurfaces).

However, in certain cases such a dissociation cannot take place immediately due to obstacles along the possible reaction pathways. Ordinary polyatomic metastable ions can be categorized into three groups [62] corresponding to three *different mechanisms* (storage and disposal of excess energy):

Electronic (forbidden) Predissociation. In this case the radiative decay or the radiationless transition from an excited (bound) state to a repulsive state (leading to instantaneous dissociation) is strongly forbidden by some selection rules. A typical example is the metastable decay of Ar_2^+ (II $(1/2)_u$) via Ar_2^+ $(I(1/2_g)$ into Ar^+ $(^2P_{3/2})$ + Ar with a lifetime of 91 µs [63, 64].

Barrier Penetration. A second possible mechanism giving rise to metastable decay of ions is the excitation to bound levels which are above the dissociation limit of this ion but below the top of a weak dissociation barrier. The quantum mechanical tunnel effect is responsible for the observed metastable decay and the corresponding lifetimes. A particular variant of this mechanism is tunneling through a centrifugal barrier (rotational predissociation). Based on experimental findings [3, 63] and a model calculation [65] it was suggested [3, 63, 65] that tunneling through a rotational barrier is responsible for the slow decay of small Ar cluster ions.

Vibrational Predissociation. If a polyatomic ion is complex enough, the Lissajou motion of an activated ion on its potential hypersurface will be complicated enough to increase the lifetime into the metastable time regime. This process has to be described in the framework of a statistical theory (RRKM or QET [51, 52, 66]), where the unimolecular rate constant k (and other properties such as the release of translational kinetic energy T) are assumed to depend only on the internal energy of the activated ion. The excitation process is assumed to have no influence on the values of k or T ("microcanonical ensemble"). Exceptions to this, however, are known under the name of "isolated state decay" [11, 67]. Vibrational predissociation is thought to be the dominant metastable dissociation mechanism for large cluster ions and several variants of statistical theory have been applied to this rather new field [68, 69]. Moreover, *Scharf* et al. [70] have recently explored by classical molecular dynamics vibrational predissociation induced by exciton trapping in neutral rare gas clusters.

It is now well established that under certain conditions following the ionization of a cluster, metastable decay not only proceeds according to one of the above mentioned (intramolecular) mechanisms, but also via mechanisms only possible in the complex environment of a cluster. To date three of these intermolecular mechanisms are known:

Delayed Internal Ion Molecule Reaction. One of the first examples discussed in this context was the formation of protonated water clusters by *Klots* and *Compton* [46], where they attributed observed metastable fragments to an ion molecule reaction taking place within a water cluster following its ionization. They also reported observing metastable components in the formation of $C_3H_5^+$

following the ionization of the ethylene dimer. Other authors have made similar observations [71]; for instance, *Futrell* et al. [72] have reported that in case of ammonia clusters a rearrangement channel is a relevant factor for metastable dissociation. *Morgan* et al. [73] have recently obtained first evidence for a delayed internal cluster reaction, namely the metastable loss of H_2O following the ionization of methanol clusters via multiphoton ionization. This is in contrast to earlier work where the product ions detected were also present as peaks in the direct ionization spectrum.

Intermolecular Energy Transfer. In this case *vibrational* or *electronic* energy stored in one moiety of a cluster ion is released with a delay into intermolecular motion leading to the evaporation of specific (magic) numbers of monomers. In particular, up to four vibrational quanta may be stored in a nitrogen cluster ion [74–77] leading to a rather peculiar metastable evaporation pattern (see Fig. 4). Another process involving the intermolecular transfer of energy in the metastable time regime has been observed recently for argon cluster [54] and propane cluster [78] ions resulting in the preferential evaporation of specific magic numbers of cluster constituents.

Finally, the observation of giant neutral and ionized metastable He clusters by *Gspann* [79] should be mentioned. Liquid phase phenomena are responsible for the ejection of the observed charged miniclusters. Furthermore, metastable

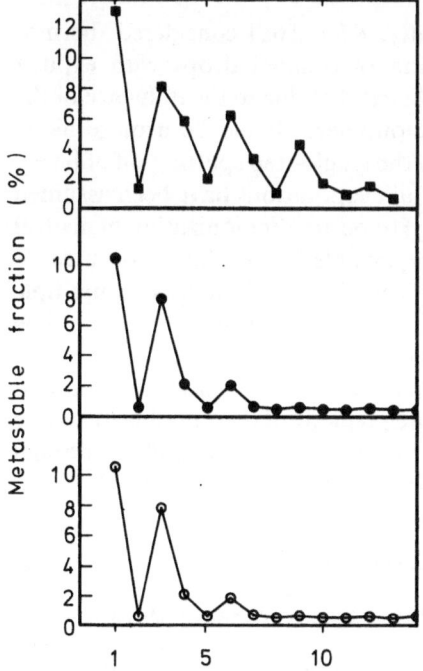

Fig. 4. Metastable fractions as a function of the number of evaporated monomers for the parent cluster ion $(N_2)_{20}^{+*}$. ■, data obtained by *Echt* and co-workers [75], ●, data obtained by *Scheier* and *Märk* [76], ○, calculated results using the reaction rate constants given in [76]

dissociations have been also reported in case of negative cluster ions [80]. A particular variant of a delayed decay in these systems is the occurrence of electron autodetachment in the metastable time regime [80, 81] (see also the recent observation of thermionic electron emission from refractory metal and fullerene clusters [82]).

2.6.5.3 Metastable Decay Processes

Not only can metastable decay be initiated by a number of different mechanisms (see Sect. 2.6.5.2), but the actual decay process can also proceed via a number of different channels, some of which have been predicted/discovered only recently:

Monomer Evaporation. The most common metastable decay process for atomic and molecular cluster ions is monomer evaporation

$$X_n^{+*} \rightarrow X_{n-1}^+ + X \ . \tag{4}$$

The large probability of this process in comparison with other possibilities is most likely due to the importance of the energetics in the decay process, i.e., the predominant fragmentation channel will follow the adiabatic dissociation channel associated with the lowest dissociation energy (see also below). Examples for this behavior have been given by *Brechignac* and coworkers [83] in the case of alkali clusters and by *Märk* [7] for ionized van der Waals clusters.

Sequential Metastable Decay Series. Recently, *Klots* [68] considered thermochemical aspects of the evaporative cooling of isolated drops with explicit reference to van der Waals clusters. He predicted that due to the influence of the surface energy sequential evaporation of monomers should be a more likely cooling process for excited cluster ions than the single-step splitting off of larger fragments (see also Ref. [84]). Such sequential evaporations have been assumed (in molecular dynamics calculations [84, 85]) to occur after ionization of neutral vdW clusters in the ps time regime and be responsible, in part, for the occurrence of magic numbers in mass spectra (see Sect. 2.6.5.5a). Moreover, multiple evaporations (i.e. the loss of more than one monomer) in the metastable time regime (μs) have been observed by numerous authors. However, only recently it became possible to determine the true nature of this process. *Märk* and coworkers [76, 77, 86, 87], using both field-free regions of their double focussing mass spectrometer (Fig. 9 in Chapter 3.5 in Vol. I) as independent observational windows, were able to demonstrate that certain cluster ions (e.g., Ar_n^+, $(N_2)_n^+$ and decay by sequential decay series, e.g.,

$$(N_2)_6^{+*} \rightarrow (N_2)_5^{+*} \rightarrow (N_2)_4^{+*} \rightarrow (N_2)_3^{+*} \rightarrow (N_2)_2^{+*} \rightarrow N_2^+ \tag{6}$$

evaporating a single monomer in each of these successive decay steps.

Single Step Fissioning and Size Specific Neutral Fragment Loss. In contrast to larger Ar cluster ions where the loss of two monomers proceeds via sequential evaporation, the strong metastable decay of Ar_4^{+*} into Ar_2^+ appears to be a single step fissioning process [86, 88]. Moreover, recently this process has been discovered to occur also is larger Ar and Ne clusters and to be initiated by excimer decay [54] (see also decay of hydro carbon cluster ions driven by isomerization reactions [78]).

Similar observations have been reported for Bi and C clusters [89, 90]. In case of carbon clusters *Bowers* and coworkers [90] reported that below C_{30}^+ the loss of C_2 and above C_{30}^+ the loss of C_3 dominates the metastable reactions (see also the result of *Lifshitz* et al. [91]). The detailed results are interpreted in terms of stable structures of both the parent ions and neutral fragments that have been predicted theoretically.

Fission processes in the metastable time regime for multiply charged ions due to Coulomb repulsion have been observed for triply charged molecular cluster ions [92], for doubly charged sodium cluster ions [93] and recently for multiply charged fullerene ions [94] (for more details see Chapter 2.7).

2.6.5.4 Metastable Reaction Kinetics and Energetics

A metastable dissociation reaction of an excited cluster ion X_n^{+*} via

$$X_n^{+*} \xrightarrow{k} X_{n-1}^+ + X + T \qquad (7)$$

may be characterized by the rate constant k (which is the inverse of the mean lifetime τ) and the kinetic energy T released in the decay channel. Reaction (7) constitutes the simplest case of ions being excited to a specific energy E^* and decaying via one reaction channel. Cluster ions produced by electron (or photon) impact ionization of a neutral cluster beam normally comprise, however, a broad range of energies due to the broad range of energies deposited in the ions by the primary ionization process and during the latter dissociation events. Parent ions with different energies will have different decay rate constants and produced fragment ions will receive different internal and kinetic energies. Moreover, in many cases X_n^{+*} may decay by competing reactions and a daughter ion X_{n-1}^+ produced may not be stable and decay again by further decomposition reactions. This situation makes analysis of experimental data very difficult and usually only averaged or relative values for the rate constants and kinetic energy release may be obtained. Nevertheless, valuable information has been compiled recently on these properties, in particular in those studies, where a larger cluster size range has been investigated.

2.6.5.4a) Kinetics

Metastable Fraction as a Function of Cluster Size. In a typical experiment of this kind either the amount of fragment ions $[X_{n-1}^+]_{\Delta t}$ (Fig. 4) or the loss of parent

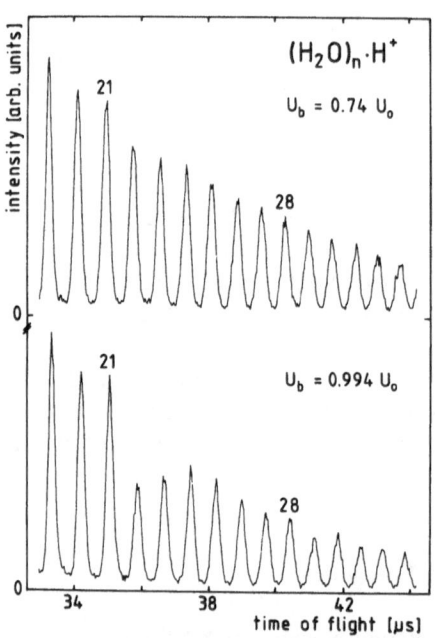

$(H_2O)_n \cdot H^+$

$U_b = 0.74\ U_o$

21

28

$U_b = 0.994\ U_o$

21

28

34   38   42
time of flight [μs]

intensity [arb. units]

Fig. 5. Size distribution of water cluster ions ($H_2O_n \cdot H^+$ ($19 \le n \le 33$) after *Echt* et al. [95]. Top: 4 μs after ion formation from neutral clusters by electron impact ionization. Bottom: appr. 38 μs after ion formation (excluding those ions which have decomposed in the time window between 4 and 38 μs). Similar results have been reported for instance for cesium iodide cluster ion [96]

ions ($[X_n^+(t_1)] - [X_n^+(t_2)]$, e.g. see Fig. 5) originating from a metastable decay in a certain time window ($\Delta t = t_2 - t_1$) is measured. Usually, the amount of excited parent ions at the beginning or end of the time window is not known. Nevertheless, in order to obtain a relative measure for the metastability of cluster ions as a function of cluster size, authors usually divide $[X_{n-1}^+]_{\Delta t}$ ($= [X_n^+(t_1)] - [X_n^+(t_2)]$) by the number of parent ions arriving at the detector ($[X_n^+(t_3)]$). This ratio is called metastable fraction and in its reduced form, i.e. divided by the length of the time window, is called apparent rate constant [88]. Examples for these ratios as a function of cluster size are given in Figs. 6 and 7 for Ar and Xe clusters, respectively. Similar results (also for anions), including also Ne, C, N_2, O_2, CO_2, H_2O, NH_3, C_6H_6, CH_3OH, alkali and metal clusters have been reported in [75–77, 80, 83, 90, 91, 95–104].

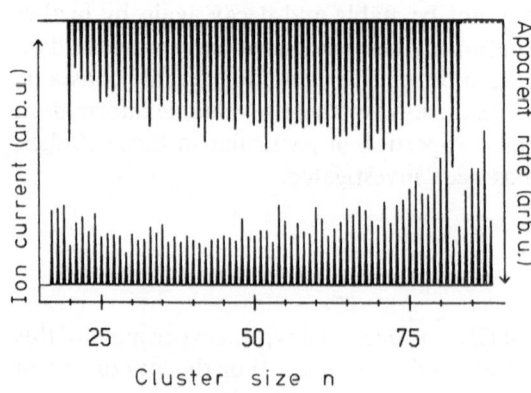

Ion current (arb. u.)

Apparent rate (arb u.)

25   50   75

Cluster size n

Fig. 6. Apparent metastable decay rate for $Ar_n^+ \to Ar_{n-1}^+ + Ar$ and ordinary mass peaks of Ar_n^+ versus n ($20 \le n \le 83$) after [88]. See also similar results for argon cluster ions by [25, 97]

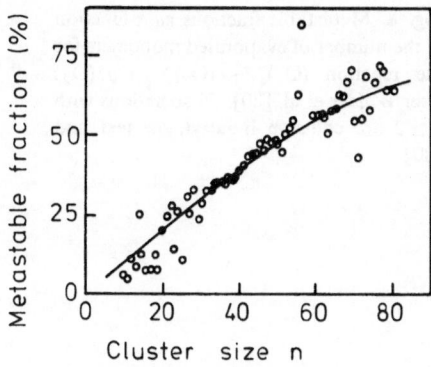

Fig. 7. Comparison of measured [55] and predicted [68] metastable fractions for $Xe_n^+ \rightarrow Xe_{n-1}^+ + Xe$. $C = n \cdot 2.79$

It can be seen that the relative probability for monomer evaporation is increasing in general with cluster size n. This result is in accordance with predictions of *Klots* [68] using the concept of an evaporative ensemble model (EEM) and considering the kinetics of evaporation within the framework of quasiequilibrium theory. According to *Klots* [68] the time dependence of an evaporating parent population, isolated and normalized to unity at time t_0, is given by

$$[X_n^+](t) = 1 - (C/\gamma^2)\ln\{t/[t_0 + (t - t_0)\exp(-\gamma^2/C)]\}, \tag{8}$$

where C is the heat capacity of the cluster (in units of the Boltzmann constant k_B) and γ is the modified Gspann parameter, defined by

$$\gamma = E_{vap}/k_B (TT^*)^{1/2}. \tag{9}$$

It contains the energy of evaporation E_{vap} and a geometric mean of the before-and-after temperatures T and T^*, respectively. The Gspann parameter is very nearly independent of the size and composition of the cluster and on a typical laboratory time scale of tens of µs equal to about 25. The single unknown parameter in the Klots formula is the heat capacity. Choosing plausible values for C there is very good agreement in the general trend between existing experimental data for Cu [99], Ar [88], Xe [55] (see Fig. 7) and NH$_3$ [104] and the predicted curve. It is interesting to note that for certain cluster sizes the experimental values deviate from the predicted curve beyond quoted error bars. The reason for this are additional structural stabilities ("magic numbers") not included in the continuum based model of Klots (see Sect. 2.6.5.5a).

Metastable Fraction as a Function of Evaporated Monomers. Some of these studies on metastable kinetics give also a detailed account of metastable branching ratios for competing reactions. Figures 4 and 8 demonstrate two different cases observed so far. In the case of oxygen cluster ions (Fig. 8) only monomer evaporations ($p = 1$) are induced by a true unimolecular decay [80]. Whereas decays with $p \geq 2$ are caused by collision induced dissociations (via background gas collisions) and the probability that a specific precursor ion $(O_2)_n^{+*}$ is loosing

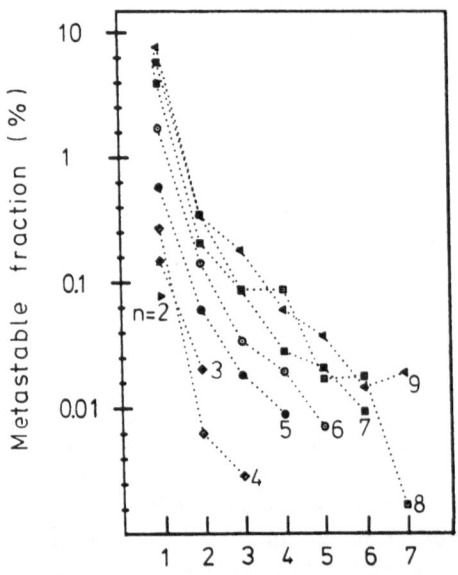

Fig. 8. Metastable fractions as a function of the number of evaporated monomers for the reaction $(O_2)_n^{+*} \rightarrow (O_2)_{n-p}^+ + p \cdot (O_2)$ after *Walder* et al. [80]. Dissociations with $p \geq 2$ are collision induced, see text and [80]

p monomers is decreasing monotonically with increasing p. In the case of nitrogen cluster ions (Fig. 4) the dependence of the metastable fraction on the number of evaporated neutrals exhibits a quasiperiodic pattern of magic losses (in the case of $(N_2)_{20}^+$ with a period of $p = 3$). The occurrence of these magic evaporations has been rationalized in 2.6.5.2.

Metastable Fraction as a Function of Time (Lifetime). As mentioned above, in general a given mass selected cluster ion beam generated in a nozzle expansion source will comprise a broad range of internal energies equal at least to the average energy loss per evaporation [68]. The range of evaporative metastable rate constants thus subtended in the frame of RRKM theory should be large leading to an apparent time dependence (e.g. as shown recently in several experiments using variable time windows [53, 54, 57, 80, 99, 100]) of the rate constants (or lifetime) of such an ion ensemble. This leads according to Eq. (8) to a nonexponential decay, which is in excellent agreement (at least for larger clusters) with experimental findings [57, 88] (see Fig. 9).

It should be noted, however, that apparent metastable rate constants (determined as mentioned above) can only be converted to true metastable rate constants if (i) $X_n^+(t_3)$ can be related to the true number of excited (metastable) parent ions [52, 88] and (ii) the decay is governed by a single decay constant. N_2O trimer ions produced by photoionization were found by Hertel and coworkers [59] to decay with a single metastable rate constant, presumably because these ions were produced with a fairly narrow range of internal energies. Moreover, there exists evidence for a single decay constant in case of certain

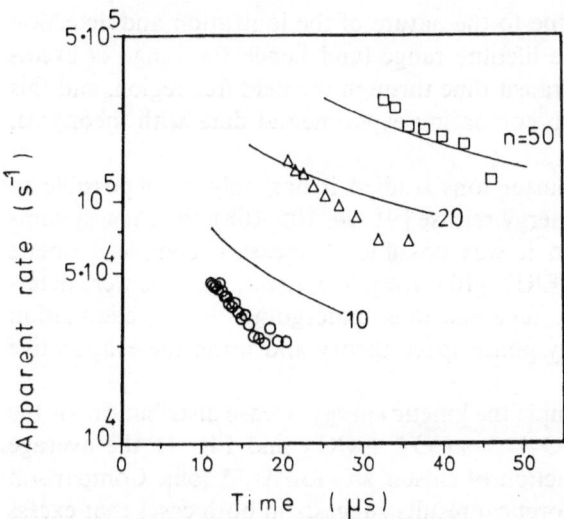

Fig. 9. Apparent metastable decay rate (metastable fraction divided by the length of the time window) as a function of flight time for monomer evaporation of $Ar_{20}^+{}^*$, $Ar_{20}^+{}^*$ and $Ar_{50}^+{}^*$ after Ref. [54, 57, 88]. Also shown (*full lines*) for comparison are the predicted dependencies Eq. (8) with $\gamma = 25$, $C = n \cdot 3.9$ and $t_0 = 10^{-7}$ s

single step fissioning processes in argon and neon cluster ions [54, 86, 87] which are due to dissociations via "isolated electronic state" [67], i.e. excimer states [54].

2.6.5.4b) Kinetic Energy Release

When a cluster ion decays by metastable dissociations, any residual energy in excess of the critical energy of decomposition ε_0, may be partitioned between the various degrees of freedom of the reaction products. Of particular interest in case of cluster ions, is the product translational energy. According to Stace [105] the probability that an ion with excess energy $E^* = E - \varepsilon_0$ (E excitation energy) will give rise to products with a relative kinetic energy ε_t is given within the framework of RRKM-QET by

$$P(E^*, \varepsilon_t) = \rho(E^* - \varepsilon_t) \Big/ \int_0^{E^*} \rho(E^* - \varepsilon_t) d\varepsilon_t \tag{10}$$

where $\rho(E^* - \varepsilon_t)$ is the density of energy states for the transition state at an energy $E^* - \varepsilon_t$. The average value of ε_t is then given by

$$\bar{\varepsilon}_t = \int_0^{E^*} \varepsilon_t P(E^*, \varepsilon_t) d\varepsilon_t . \tag{11}$$

According to a classical consideration by Stace [105] it would be expected that $\bar{\varepsilon}_t$ should decrease as the size of the cluster ion increases (if E^* is constant). As mentioned above, however, metastable cluster ion peaks usually arise not with fixed values of excess energy due to the nature of the ionization and detection process. For a specific ion the lifetime range (and hence the range of excess energies) is determined by its transit time through the field free region, and this fact has to be considered when comparing experimental data with theoretical calculations.

For the majority of the cluster ions studied it has only been possible to measure an averaged kinetic energy release [91, 94, 105–108], for a few systems using suitable instrumentation it was possible to measure complete kinetic energy release distributions (KERD) [103, 109]. Moreover, KERDs were determined recently for metastable fullerene ions undergoing the C_2 elimination reaction and were modelled by phase space theory and using the evaporative ensemble model [110].

Figure 10 shows as an example the kinetic energy release distribution for the metastable decay reaction $(CO_2)_2^{+*} \rightarrow CO_2^+ + CO_2$ and Fig. 11 the average kinetic energy release as a function of cluster size for Ar_n^{+*} ions. Comparison between experimental and theoretical results suggests in both cases that excess energy in the cluster ions is partitioned in a statistical manner. The considerable scatter in the experimental data in Fig. 11 could be again (see also the discussion of Fig. 7) interpreted as being due to structural peculiarities of specific cluster sizes not accounted for in the RRKM description. A similar outstanding discontinuity in the general shape of $\bar{\varepsilon}_t$ versus n has been recently observed by *Lifshitz* et al. [103] for $(NH_3)_n H^+$ clusters. This peculiar behavior has been attributed to the filling of the first solvation shell at $n = 5$ [41]. Moreover, *El-Sayed* and coworkers [111] have recently studied not only the size, but also the temporal dependence of the average kinetic energy release during the evaporation of one and two CsI molecules of sputtered $Cs(CsI)_n^+$ ions. Their results are in reasonable agreement with predictions of the evaporative ensemble model and they prove that the evaporation of two CsI molecules involves a direct fission rather than a sequential process (see also Ref. [112]).

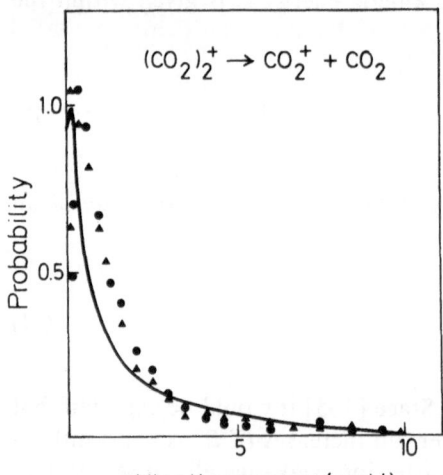

Fig. 10. Kinetic energy release distribution for the metastable decay reaction $((CO_2)_2^{+*} \rightarrow CO_2^+ + CO_2$ after *Illies* et al. [109]. The points are the experimental data and the line is the result of a statistical phase space theory calculation

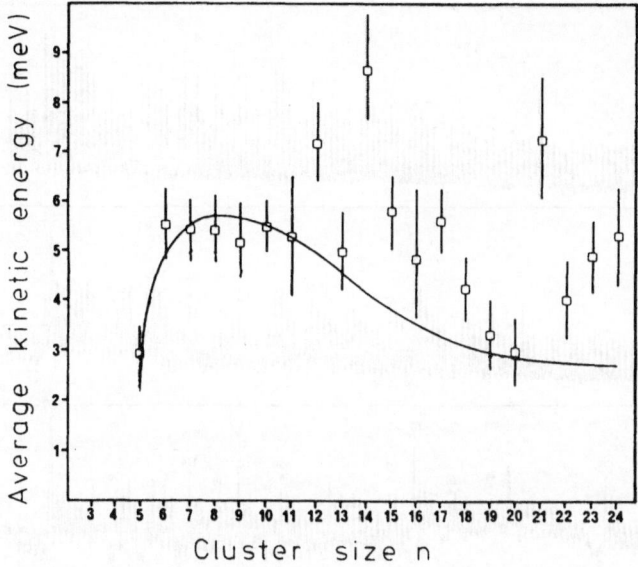

Fig. 11. Average kinetic energy release $\bar{\varepsilon}_t$ as a function of cluster size n for the metastable decay reaction $Ar_n^{+*} \to Ar_{n-1}^+ + Ar$ after *Stace* [105]. Also shown (*solid line*) are the results calculated from a modified RRKM model

2.6.5.5   Related Properties and Applications

2.6.5.5a)   Magic Numbers in Mass Spectra

Self-bound few-body systems ($n < 100$) often show exceptional stability at so-called magic numbers n_m due to closed shell phenomena. The best known examples are electron shells of noble atoms ($n_m = 2, 10, 18, 36, 54$, etc.) and magic numbers in nuclei ($n_m = 2, 8, 20, 28, 50$, etc.). Recently, similar effects have been observed in mass spectra of van der Waals clusters [4, 7, 8], of valence clusters (see 4.4 in Vol. I) and of metallic clusters (see 2.6 in Vol. I).

The first well-resolved mass spectrum with cluster sizes up to $n \sim 150$ was reported for Xe by *Echt* et al. [113] in 1981. This was followed by the determination of a He cluster spectrum (up to $n = 36$) by *Stephens* and *King* [114] and Ar and Kr spectra (up to $n \sim 60$) by *Ding* and *Hesslich* [115] in 1983. Recently, we were able to measure well-resolved Ne (up to $n = 90$) [100] and Ar, Kr and Xe (up to $n \sim 1000$) cluster spectra [116, 117] (see Fig. 12). In all of these spectra, significant magic number effects are visible, i.e. anomalous abundances are present at certain cluster sizes in an otherwise smoothly varying cluster size distribution. The origin of these magic numbers used to be a controversial subject [4], i.e. some investigators interpreted them as being due to special structure and stability of the neutral clusters, whereas others attributed these features to the stability of the cluster ions produced in the course of investigating

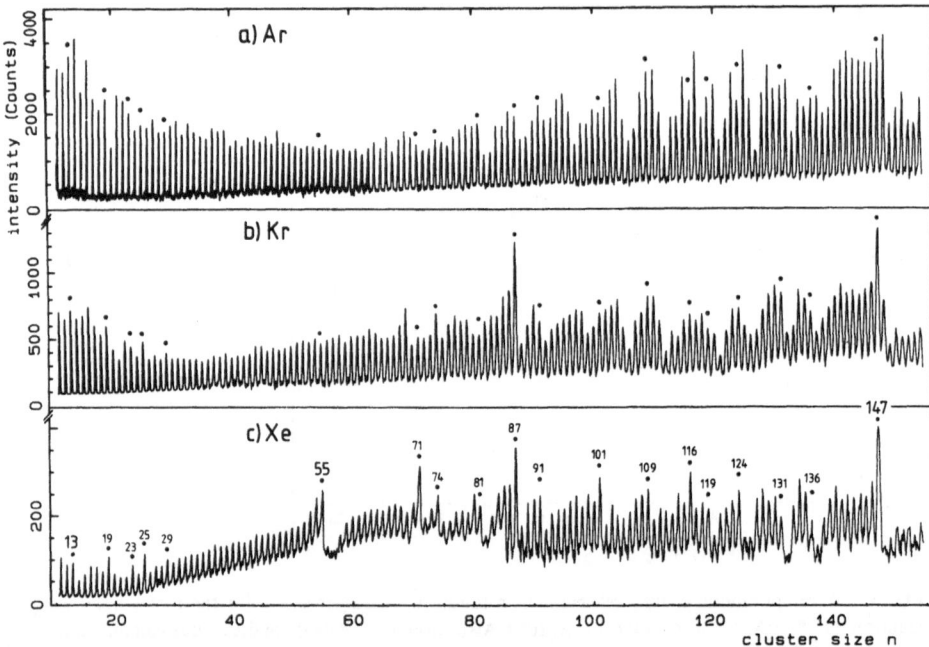

Fig. 12. Mass spectra of Ar, Kr and Xe clusters covering the size range where the 2nd and 3rd icosahedral shells (at $n = 55$ and 147, respectively) are expected to fill after *Miehle* et al. [117]

neutral cluster distributions with mass spectrometry. However, several recent studies of Ne, Ar, Xe, and H_2O [25, 55, 88, 95, 97,100] have shown, beyond any doubt that a least for vdW clusters, the evaporative stabilization after the ionization process in the metastable time regime is the main reason, and the likely origin, for the occurrence of these abundance anomalies.

Figure 13a shows the metastable abundance fraction of Ar_n^+ clusters ($10 \leq n \leq 25$) for the loss of one monomer measured in the first field-free region of a sector field mass spectrometer. The most [25] salient feature to be seen (see also Fig. 6) is the strong metastable fraction for Ar_{20}^+, indicating that this cluster ion is less stable than others under the experimental conditions employed. Figure 13b shows a conventional mass spectrum of Ar clusters obtained under identical experimental conditions. It can be seen that these two distributions are almost mirror like, e.g. the characteristic troughs at $n = 15$, 20, and 24 in the conventional mass spectra (Fig. 13b) correspond to strong metastable fractions in Fig. 13a and, conversely, the characteristic troughs at $n = 14$, 16, 19, and 21 in the metastable abundance fractions correspond to strong mass peaks in Fig. 13b.

For the most striking example of Ar_{19}^+ and Ar_{20}^+, it is possible to explain, with these results in mind, the observed anomalous step in the mass spectrum as being in part due to the strong metastable decay of Ar_{20}^+ during its flight through the first field-free region (i.e. between 8.2 and 22.7 μs after its production by

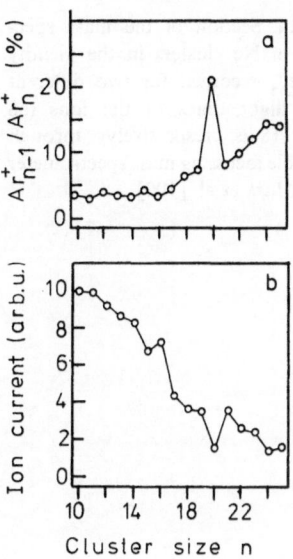

Fig. 13. Metastable fractions and ordinary mass spectrum for Ar_n^+ cluster ions after *Märk* et al. [25]

electron ionization) and in part due to unimolecular decay occurring before entering the first field-free region. From such a comparison (see also similar findings in Ne [100], Xe [55], and H_2O [95]), it follows that abundance anomalies in the conventional mass spectrum of vdW clusters are not due to abundance anomalies in the neutral cluster distribution, as argued previously, but rather evolve from the cluster ion production as a result of the intrinsic stability of the cluster ions produced. This conclusion has recently been confirmed by additonal results, i.e. doubly charged argon clusters have different magic numbers than singly charged argon clusters [19, 118] and non-resonant multiphoton ionization of Xe clusters show the same magic numbers as found in the case of electron impact ionization with no discernible influence of the laser fluence or expansion conditions [53].

In this context, it is worth recalling that *Klots* noted [68] that any ionized cluster of unusual abundance not accompanied by an abnormal lifetime does not necessarily constitute a local stability, but is perhaps only an echo of some neutral precursor. In any case, simple operational tests (see above) may be used to distinguish between ionized and neutral structures, of unusual stability. Moreover, depending on the time scale of the experiment and on the distribution of metastable rate constants of the cluster ions under study, local stabilities in the conventional mass spectrum will only show up at very late times (see also Ref. [119]). A special instructive example is the Ne_{55}^+ ion [100]. Figure 14 shows a section of the Ne cluster spectrum around Ne_{55}^+ for two different total flight times of the ions through a mass spectrometer (e.g. 68 and 118 µs). It can be seen that Ne_{55}^+ (which has a relatively smaller k than its neighbours) is only the dominant ion at long flight times whereas, at short flight times, Ne_{56}^+ is larger than Ne_{55}^+, very likely due to strong production of Ne_{56}^+ in the ion source by

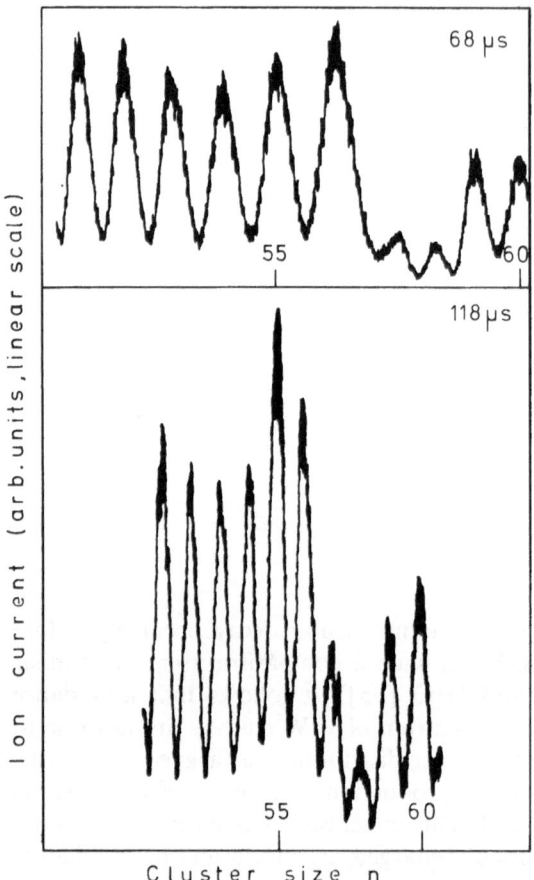

Fig. 14. Section of the mass spectrum of Ne clusters in the vicinity of Ne_{55}^+ recorded for two different total flight times of the ions (68 and 118 µs, respectively) through a double focussing mass spectrometer after *Märk* et al. [100]

fragmentation of Ne_{57}^+ and Ne_{58}^+ (see the large decay rate constants for these ions given in [100]).

The analogy between nuclear and atomic magic numbers prompted earlier scientists to look for an explanation of the nuclear phenomenon that was similar to that used in the atomic case. This was solved by assuming that each nucleon of the nucleus moves in an attractive net potential (Fermi gas model) and that, in addition, each nucleon feels a strong inverted spin orbit interaction. According to recent work, shell models which have worked well for atoms and nuclei arc also applicable to metal clusters (see Chapter 2.6 in Vol. I). Magic numbers of vdW clusters cannot be explained within the framework of this theory because electrons are not delocalized in neutral vdW clusters. Therefore structural arrangement (as is the case for fullerenes; see 4.4 in Vol I) of the cluster constituents has to be considered as a building principle. Hence, it is interesting to note that, in computer simulations of soft spheres (Lennard Jones interaction) and in hard sphere packing models [120], especially stable atomic clusters were obtained for $n = 13, 55, 147, 309$, etc. atoms, their corresponding structure being icosahedral (see also discussion in Ref. [121]). Moreover, clusters containing 19 atoms are also especially stable because the additional 6 atoms arranged in the form of a cap yield a so-called double icosahedron. Similarly, a cap of 16 can be added to the 55-mer, leading to a magic number at $n = 71$ (see Fig. 6). This can be extended further along the same axis of symmetry [122]. As of now, some of the magic numbers corresponding to structures based on icosahedral cores have been observed in all of the rare gases, although there are exceptions, and additional magic numbers are observed which are not accounted for by these considerations.

The well known, but puzzling differences in size distributions (see also Fig. 12) of Kr and Xe clusters disappear beyond $n \sim 130$, while those between Ar and Xe disappear beyond $n \sim 220$. The most pronounced magic numbers in the distribution of large cluster ions occur at $n = 147$ (148 for Ar), 309 and 561, in striking agreement with the number of atoms required to build icosahedral clusters with 3, 4 and 5 complete coordination shells, respectively. For more details see also the discussions in Ref. [7, 8, 117, 123–127].

2.6.5.5b)   Determination of Cluster Ion Binding Energy

Recently, *Engelking* [69] proposed that a modified RRK-QET statistical method may be used to determine the binding energy of a monomer within a cluster ion from the measurement of the metastable lifetime ($\tau = 1/k$) and concomitant average kinetic energy release. If the final product density of states is approximated by a Kassel distribution, the expression for lifetime can be inverted to give the binding energy BE,

$$\text{BE} = 0,5(s-1)\left[C^{1/(s-1)}\bar{\varepsilon}_t^{(s-2)(s-1)} - \bar{\varepsilon}_t\right] \tag{12}$$

where

$$C = v^3 16\pi\mu g\, S/k \tag{13}$$

with v the characteristic mode frequency, μ the reduced mass, g the channel degeneracy, S the geometric cluster cross section, s the number of internal modes. It turns out that stringent demands are placed upon the accurate determination of $\bar{\varepsilon}_t$, but only moderate demands are placed upon the accuracy of the lifetime or other model parameters. Engelking applied this method to data of $(CO_2)_n^+$ and Ar_n^+ reported by Stace [105,106], and *Castleman* and coworkers [108] obtained binding energies of the protonated ammonia ions in very good agreement with reported literature values (see Fig. 15). Moreover, *Stace* and coworkers [97] demonstrated recently that the binding energy is the only important variable in determining the fragmentation pattern of Ar_n^{+*} through the use of RRKM theory. *Brechignac* and coworkers [102] have extended the approach of Engelking to $(Na)_n^+$ cluster ions and *Castleman* and coworkers [128] and *Lezius* et al. [129] the EEM model of Klots [130] to positive and negative cluster ions, respectively.

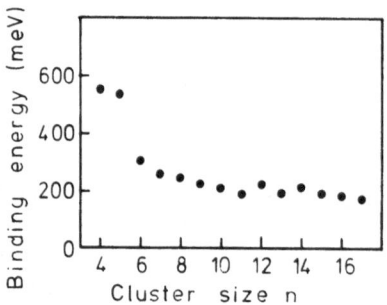

Fig. 15. Binding energies of $(NH_3)_n \cdot H^+$ versus cluster size determined from the measured average kinetic energy release after *Wei* et al. [108]. Note the abrupt decrease in binding energy from $n = 5$–6 which is also reflected in the special abundance of the ammonia pentamer in mass spectra [41] (see also *Lifshitz* et al. [103])

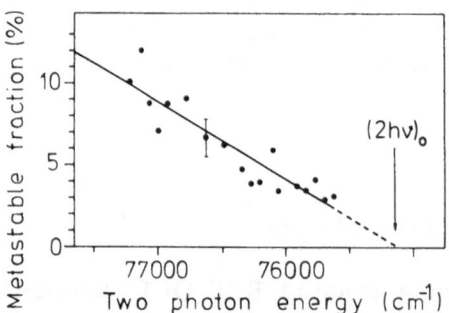

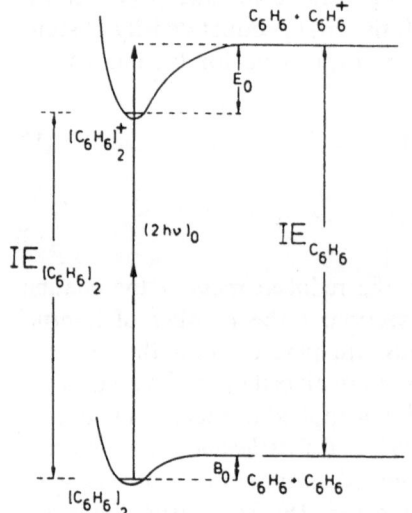

Fig.16. Breakdown graph (*upper part*) for the metastable decay $(C_6H_6)_2^{+*} \rightarrow C_6H_6^+ + C_6H_6$ after *Kiermeier* et al. [101]. The *solid line* is the result of a least-squares fit yielding the threshold energy $(2h\nu)_0$ at the intercept with the base line. From this threshold value the binding energy BE of the dimer ion (E_0) and of the neutral dimer (B_0) can be deduced with help of known ionization energies. This is illustrated in the *lower part* of the figure

It is interesting to note that *Schlag* and coworkers [101] were able to determine the binding energy of the benzene dimer ion from an experimentally determined breakdown graph (metastable fraction versus photon excitation energy, see Fig. 16) and *Foltin* et al. [131] the activation energy of C_{60}^+ for C_2 elimination from a comparison of measured (electron impact) and calculated (RRKM, EEM) breakdown graphs.

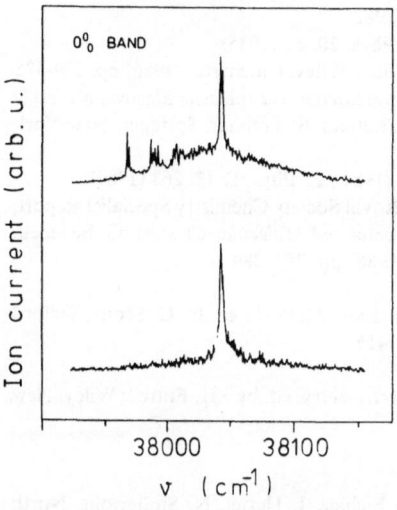

Ion current (arb. u.)

0_0^0 BAND

38000   38100

v (cm^{-1})

Fig. 17. Section of the $S_1 \leftarrow S_0$ spectrum of the benzene dimer obtained by ordinary two photon ionization (*upper part*) and by metastable decay spectrometry (*lower part*) after *Kiermeier* et al. [101]. Whereas the upper spectrum is contaminated by lines from larger clusters, the lower spectrum represents only the pure dimer spectrum

2.6.5.5c)   Fragmentationfree Spectra

Features associated with vdW clusters can be difficult to assign unambiguously in electronic excitation spectra due to overlapping structures originating from absorption by higher vdW clusters. *Schlag* and coworkers [101] presented very recently a method based on the observation of the metastable decay of the corresponding cluster ion of benzene in order to obtain pure intermediate state $(S_1 \leftarrow S_0)$ spectra. Figure 17 shows as an example part of the $S_1 \leftarrow S_0$ spectrum of the benzene dimer recorded by ordinary two photon ionization mass spectrometry and by recording the metastable decay signal as a function of photon energy with a reflectron time of flight mass spectrometer.

2.6.6   Recent Developments

Recently, further interesting research has been carried out in the field of internal and metastable reactions of cluster ions. The results are summarized in several recent reviews [132–138].

Acknowledgements. Work partially supported by the Österreichischer Fonds zur Förderung der Wissenschaftlichen Forschung.

References

1. T.D. Märk: In *Electron–Molecule Interaction and Their Applications*, ed. by L.G. Christophorou, Academic, New York (1984) Vol. 1, pp. 251-334
2. T.D. Märk, G.H. Dunn: *Electron Impact Ionization*, Springer, Wien (1985)

3. K. Stephan, T.D. Märk: Chem. Phys. Lett. **90**, 51 (1982)
4. T.D. Märk, A.W. Castleman Jr.: Adv. Atom. Mol. Phys. **20**, 65 (1985)
5. T.D. Märk: In: *Adv. Mass Spectrom*, ed. by J.F.J. Todd, Wiley, Chichester (1986) pp. 379-395
6. R.G. Keesee, A.W. Castleman Jr., T.D. Märk: In *Swarm Studies and Inelastic Electron Molecule Collisions*, ed. by L.C. Pitchford, B.V. McKoy, A. Chutjian, S. Trajmar, Springer, New York (1987) pp. 351–366
7. T.D. Märk: Int. J. Mass Spectrom. Ion Proc: **79**, 1 (1987); Z. Phys. **D 12**, 263 (1989)
8. A.J. Stace: In *Mass Spectrometry*, ed. by M.E. Rose, Royal Society Chemistry Specialist Report, London (1987) Vol. 9, pp. 96–121; O. Echt: In *Elemental and Molecular Clusters* (G. Benedek, T.P. Martin, G. Pacchioni, Eds.) Springer, Berlin (1988) pp. 263–284
9. U. Buck: J. Phys. Chem. **92**, 1023 (1988)
10. M. Kappes, S. Leutwyler: In *Atomic and Molecular Beam Methods*, ed. by G. Scoles, Oxford University Press, New York (1988) Vol. 1, pp. 380–415
11. C. Lifshitz: Adv. Mass Spectrom. .7A, 3 (1978)
12. T.D. Märk: In *Gaseous Ion Chemistry and Mass Spectrometry*, ed. by J.H. Futrell, Wiley, New York (1986) pp. 61–93
13. P. Dehmer: J. Chem. Phys. **76**, 1263 (1982)
14. T.D. Märk: Europhys. Conf. Abstr. **6 D**, 29 (1982)
15. H. Haberland: In Proc. 13th ICPEAC, ed. by J. Eichler, I. Hertel, N. Stolterfoht, North Holland, Amsterdam (1984); Surf. Science **156**, 305 (1985)
16. W. Henkes, F. Mikosch: Int. J. Mass Spectrom. Ion Proc. **13**, 151 (1974)
17. J. Gspann, H. Vollmar: J. Chem. Phys. **73**, 1657 (1980)
18. F. Bottiglioni, J. Coutant, M. Fois: Phys. Rev. A **46**, 1830 (1972)
19. M. Lezius, T.D. Märk: Chem. Phys. Lett. **155**, 496 (1989)
20. T.E. Gough, R.E. Miller: Chem. Phys. Lett. **87**, 280 (1982); J. Geraedts, S. Stolte, J. Reuss: Z. Phys. A **304**, 167 (1982)
21. D.R. Worsnop, S.J. Buelow, D.R. Herschbach: J. Phys. Chem. **88**, 4506 (1984)
22. E. Holub–Krappe, G. Ganteför, G. Bröker, A. Ding: Z. Phys. **D 10**, 319 (1988); L. Cordis, G. Ganteför, J. Heßlich, A. Ding: Z. Phys. **D 3**, 323 (1986)
23. H. Helm, K. Stephan, T.D. Märk: Phys. Rev. A **19**, 2154 (1980)
24. U. Buck, H. Meyer: J. Chem. Phys. **84**, 4854 (1986)
25. T.D. Märk, P. Scheier, K. Leiter, W. Ritter, K. Stephan, A. Stamatovic: Int. J. Mass Spectrom. Ion Proc. **74**, 281 (1986)
26. P.M. Dehmer, S.T. Pratt: J. Chem. Phys. **76**, 4804 (1982) **77**, 4804 (1982); E. Rühl, B. Brutschy, H. Baumgärtel: Chem. Phys. Lett. **157**, 379 (1989)
27. C.Y. Ng: Adv. Chem. Phys. **52**, 263 (1983)
28. P.M. Dehmer, E.D. Poliakoff: Chem. Phys. Lett. **77**, 326 (1981)
29. Y. Ono, S.H. Linn, H.F. Prest, M.E. Gress, C.Y. Ng: J. Chem. Phys. **74**, 1125 (1981)
30. H.P. Birkhofer, H. Haberland, M. Winterer, D. Worsnop: Ber. Bunsenges. Physik. Chem. **88**, 207 (1984)
31. H.R. Siddiqui, D. Bernfeld, P.E. Siska: J. Chem. Phys. **80**, 567 (1984)
32. W. Kamke, B. Kamke, H.U. Kiefl, I.V. Hertel: Chem. Phys. Lett. **122**, 356 (1985)
33. P.D. Dao, A.W. Castleman Jr.: J. Chem. Phys. **84**, 1435 (1986)
34. B. Kamke, W. Kamke, R. Herrmann, I.V. Hertel: Z. Phys. **D 11**, 153 (1989)
35. E.D. Poliakoff, P.M. Dehmer, J.L. Dehmer, R. Stockbauer: J. Chem. Phys. **76**, 5214 (1982)
36. K. Norwood, J.H. Guo, G. Luo, C.Y. Ng: Chem. Phys. **129**, 109 (1989)
37. A.J. Stace, D.M. Bernhard: Chem. Phys. Lett. **146**, 531 (1988) and earlier references therein
38. C.E. Klots: Radiat. Phys. Chem. **20**, 51 (1982)
39. A.J. Stace, A.K. Shukla: J. Phys. Chem. **86**, 157 (1982); K. Stephan, J.H. Futrell, K.I. Peterson, A.W. Castleman Jr., H.E. Wagner, N. Djuric, T.D. Märk: Int. J. Mass Spectrom. Ion Phys. **44**, 167 (1982); C.Y. Ng, D.J. Trevor, P.W. Tiedemann, S.T. Ceyer, P.L. Kronebusch, B.H. Mahan, Y.T. Lee: J. Chem. Phys. **67**, 4235 (1977)
40. M.T. Coolbaugh, W.R. Pfeiffer, J.F. Garvey: Chem. Phys. Lett. **156**, 19 (1989) and references herein

41. O. Echt, S. Morgan, P.D. Dao, R.J. Stanley, A.W. Castleman Jr.: Ber. Bunsenges. Physik. Chem. **88**, 217 (1984)
42. V. Hermann, B.D. Kay, A.W. Castleman Jr.: Chem. Phys. **72**, 185 (1982)
43. D.M. Bernard, A.J. Stace: Int. J. Mass Spectrom. Ion Proc. **84**, 215 (1988)
44. U. Buck, C. Lauenstein, H. Meyer, R. Sroka: J. Phys. Chem. **92**, 1916 (1988)
45. B. Brutschy: J. Phys. Chem. **94**, 8637 (1990)
46. C.E. Klots, R.N. Compton: J. Chem. Phys. **69**, 1636, 1644 (1978)
47. T.D. Märk: Int. J. Mass Spectrom. Ion Proc. **107**, 143 (1991)
48. E. Illenberger: Chem. Rev. **92**, 1589 (1992)
49. A.J. Stace, A.K. Shukla: Int. J. Mass Spectrom. Ion Proc. **36** 119 (1980)
50. K. Stephan, T.D. Märk: Chem. Phys. Lett. **87**, 226 (1982)
51. R.G. Cooks, J.H. Beynon, R.M. Caprioli, G.R. Lester: *Metastable ions*, Elsevier, Amsterdam (1973)
52. K. Levsen: *Fundamental aspects of organic mass spectrometry*, Verlag Chemie, Weinheim (1978)
53. O. Echt, M.C. Cook, A.W. Castleman Jr.: Chem. Phys. Lett. **135**, 229 (1987)
54. M. Foltin, G. Walder, S. Mohr, P. Scheier, A.W. Castleman Jr., T.D. Märk: Proc. Z. Phys. D. **20**, 157 (1991); M. Foltin, G. Walder, A.W. Castleman Jr, T.D. Märk: J. Chem. Phys. **94**, 810 (1991); M. Foltin, T.D. Märk: Chem. Phys. Lett. **180**, 317 (1991)
55. D. Kreisle, O. Echt, M. Knapp, E. Recknagel: Phys. Rev. A **33**, 768 (1986)
56. P.G. Lethbridge, A.J. Stace: J. Chem. Phys. **91**, 7685 (1989)
57. Y. Ji, M. Foltin, C.H. Liao, T.D. Märk: J. Chem. Phys. **96**, 3624 (1992)
58. T. Leisner, O. Echt, D. Kreisle, E. Recknagel: Int. J. Mass Spectrom. Ion Proc. **87**, R19 (1989)
59. W. Kamke, B. Kamke, H.U. Kiefl, I.V. Hertel: J. Chem. Phys. **84**, 1325 (1986)
60. G. Romanowski, R.H. Gabling, K.P. Wanczek, Int. J. Mass Spectrom. Ion Proc. **71**, 119 (1986)
61. H.J. Neusser: Int. J. Mass Spectrom. Ion Proc. **79**, 141 (1987)
62. G. Herzberg: *Molecular Spectra and Molecular Structure*, Van Nostrand, Princeton (1967)
63. K. Stephan, A. Stamatovic, T.D. Märk: Phys. Rev. A**28**, 3105 (1983)
64. K. Norwood, J.H. Guo, C.Y. Ng: J. Chem. Phys. **90**, 2995 (1989)
65. E.E. Ferguson, C.R. Albertoni, R. Kuhn, Z.Y. Chen, R.G. Keesee, A.W. Castleman Jr.: J. Chem. Phys. **88**, 6335 (1988)
66. J.C. Lorquet: Org. Mass Spectrom. **16**, 469 (1981)
67. C.L. Lifshitz: J. Phys. Chem. **87**, 2304 (1983)
68. C.E. Klots: J. Phys. Chem. **92**, 5864 (1988); J. Chem. Phys. **83**, 5854 (1985)
69. P.C. Engelking: J. Chem. Phys. **87**, 936 (1987)
70. D. Scharf, J. Jortner, U. Landmann: Chem. Phys. Lett. **126**, 495 (1986)
71. A.J. Stace, A.K. Shukla: J. Am. Chem. Soc. **104**, 5314 (1982)
72. J.H. Futrell, K. Stephan, T.D. Märk: J. Chem. Phys. **76**, 5893 (1982)
73. S. Morgan, A.W. Castleman Jr.: J. Am. Chem. Soc. **109**, 2667 (1987)
74. T.F. Magnera, D.E. David, J. Michl: Chem. Phys. Lett. **123**, 327 (1986)
75. T. Leisner, O. Echt, O. Kandler, X.J. Yan, E. Recknagel, Chem. Phys. Lett. **148**, 386 (1988)
76. P. Scheier, T.D. Märk: Chem. Phys. Lett. **148**, 393 (1988)
77. G. Walder, C. Winkler, T.D. Märk: Chem. Phys. Lett. **157**, 224 (1989)
78. M. Foltin, V. Grill, T. Rauth, Z. Herman, T.D. Märk: Phys. Rev. Lett. **68**, 2019 (1992)
79. J. Gspann: Surf. Science **106**, 219 (1981)
80. G. Walder, D. Margreiter, C. Winkler, A. Stamatovic, Z. Herman, T.D. Märk: Faraday Trans. Chem. Soc. **86**, 2395 (1990); G. Walder, D. Margreiter, C. Winkler, V. Grill, T. Rauth, P. Scheier, A. Stamatovic, Z. Herman, M. Foltin, T.D. Märk: Z. Phys. D. **20**, 201 (1991)
81. M. Knapp, O. Echt, D. Kreisle, E. Recknagel: J. Phys. Chem. **91**, 2601 (1987)
82. G. Walder, O. Echt: Int. J. Mod. Phys. B**6**, 3881 (1992) and references therein
83. C. Brechignac, P. Cahuzac, J.P. Roux, D. Pavolini, F. Spiegelmann: J. Chem. Phys. **87**, 5694 (1987)
84. E.E. Polymeropoulos, S. Löffler, J. Brickmann: Z. Naturforschg. **40a**, 516 (1985)
85. J.M. Soler, J.J. Saenz, N. Garcia, O. Echt: Chem. Phys. Lett. **109**, 71 (1984); J.J. Saenz, J.M. Soler, N. Garcia: Chem. Phys. Lett. **114**, 15 (1985)

86. P. Scheier, T.D. Märk: Phys. Rev. Lett. **59**, 1813 (1987)
87. P. Scheier, A. Stamatovic, T.D. Märk, J. Chem. Phys. **89**, 295 (1988)
88. P. Scheier, T.D. Märk: Int. J. Mass Spectrom. Ion Proc. **102**, 19 (1990)
89. M.M. Ross, S.W. McElvany: J. Chem. Phys. **89**, 4821 (1988)
90. P.P. Radi, T.L. Bunn, P.R. Kemper, M.E. Molchan, M.T. Bowers: J. Chem. Phys. **88**, 2809 (1988)
91. C. Lifshitz, T. Peres, S. Kabikia, I. Agranat: Int. J. Mass Spectrom. Ion Proc. **82**, 193 (1988); C. Lifshitz, T. Peres, I. Agranat: Int. J. Mass Spectrom. Ion Proc. **93**, 149 (1989)
92. D. Kreisle, O. Echt, M. Knapp, E. Recknagel, K. Leiter, T.D. Märk, J.J. Saenz, J.M. Soler: Phys. Rev. Lett. **56**, 1551 (1986); K. Leiter, D. Kreisle, O. Echt, T.D. Märk: J. Phys. Chem. **91**, 2583 (1987)
93. C. Brechignac, P. Cahuzac, F. Carlier, M. de Frutos: Phys. Rev. Lett. **64**, 2893 (1990)
94. P. Scheier, B. Dünser, T.D. Märk: to be published (1994)
95. O. Echt, D. Kreisle, M. Knapp, E. Recknagel: Chem. Phys. Lett. **108**, 401 (1984)
96. W. Ens, R. Beavis, K.G. Standing: Phys. Rev. Lett. **50**, 27 (1983); J.E. Campana, B.N. Green: J. Am. Chem. Soc. **106**, 531 (1984); I. Katakuse, H. Nikabushi, T. Ichihara, T. Sakurai, T. Matsuo, H. Matsuda: Int. J. Mass Spectrom. Ion Proc. **62**, 17 (1984)
97. A.J. Stace, C. Moore: Chem. Phys. Lett. **96**, 80 (1983); P.G. Lethbridge, A.J. Stace: J. Chem. Phys. **89**, 4062 (1988)
98. I. Katakuso, T. Ichihara, Y. Fujita, T. Matsuo, T. Sakurai, H. Matsuda: Int. J. Mass Spectrom. Ion Proc. **67**, 229 (1985)
99. W. Begemann, K.H. Meiwes-Broer, H.O. Lutz: Phys. Rev. Lett. **56**, 2248 (1986); W. Begemann, S. Dreihöfer, K.H. Meiwes-Broer, H.O. Lutz: Z. Phys. D **3**, 183 (1986)
100. T.D. Märk, P. Scheier: Chem. Phys. Lett. **137**, 245 (1987); J. Chem. Phys. **87**, 1456 (1987)
101. A. Kiermeier, B. Ernstberger, J. Neusser, E.W. Schlag: J. Phys. Chem. **92**, 3785 (1988); Z. Phys. D **10**, 311 (1988)
102. C. Brechignac, P. Cahuzac, J. Leygnier, J. Weiner: J. Chem. Phys. **90**, 1492 (1989)
103. C. Lifshitz, F. Luage: J. Phys. Chem. **93**, 5633 (1989); M. Jraqi, C. Lifshitz: Int. J. Mass Spectrom. Ion Proc. **88**, 45 (1989)
104. A.W. Castleman Jr., W.B. Tzeng, S. Wei, S. Morgan: J. Chem. Soc. Faraday Trans. **86**, 2417 (1990)
105. A.J. Stace: J. Chem. Phys. **85**, 5774 (1986)
106. A.J. Stace, A.K. Shukla: Chem. Phys. Lett. **85**, 157 (1982)
107. S.K. Cole, K. Liu: J. Chem. Phys. **89**, 780 (1988)
108. S. Wei, W.B. Tzeng, A.W. Castleman Jr.: J. Chem. Phys. **92**, 332 (1990)
109. A.J. Illies, M.F. Jarrold, L.M. Bass, M.T. Bowers: J. Am. Chem. Soc. **105**, 5775 (1983)
110. P.P. Radi, M.T. Hsu, M.E. Rincon, P.R. Kemper, M.T. Bowers: Chem. Phys. Lett. **174**, 223 (1990); P. Sandler, C. Lifshitz, C.E. Klots: Chem. Phys. Lett. **200**, 445 (1992)
111. H.J. Hwang, D.K. Sensharma, M.A. El-Sayed: Chem. Phys. Lett. **160**, 243 (1989); Phys. Rev. Lett. **64**, 808 (1990)
112. T. Drewello, R. Herzschuh, J. Stach: Z. Phys. **D28**, 339 (1993)
113. O. Echt, K. Sattler, E. Recknagel. Phys. Rev. Lett. **47**, 1121 (1981)
114. P.W. Stephens, J.G. King: Phys. Rev. Lett. **51**, 1538 (1983)
115. A. Ding, J. Hesslich: Chem. Phys. Lett. **94**, 54 (1983)
116. P. Scheier, T.D. Märk: Int. J. Mass Spectrom. Ion Proc. **76**, R11 (1987)
117. W. Miehle, O. Kandler, T. Leisner, O. Echt: J. Chem. Phys. **91**, 5940 (1990)
118. P. Scheier, T.D. Märk: Chem. Phys. Lett. **136**, 423 (1987)
119. R. Caseto, J.M. Soler: J. Chem. Phys. **95**, 2927 (1991)
120. A.L. Mackay: Acta Cryst. **15**, 916 (1962); M.R. Hoare: Adv. Chem. Phys. **40**, 49 (1979)
121. J.C. Phillips: Chem. Phys. Rev. **86**, 619 (1986)
122. O. Echt, A. Reyes Flotte, M. Knapp, K. Sattler, E. Recknagel: Ber. Bunsenges. Phys. Chem. **86**, 860 (1982)
123. I.A. Harris, K.A. Norman, R.V. Mulhern, J.A. Northby: Chem. Phys. Lett. **131**, 316 (1986); J.A. Northby: J. Chem. Phys. **87**, 6166 (1987)

124. J.Heßlich, P.J. Kuntz: Z. Phys. D 2, 251 (1986)
125. M. Amarouche, G. Durand, J.P. Malrieu: J. Chem. Phys. 88, 1010 (1988); Z. Phys. D 8, 289 (1988)
126. H.U. Böhmer, S.D. Peyerimhoff: Z. Phys. D 8, 91 (1988)
127. P.J. Kuntz, J. Valldorf: Z. Phys. D 8, 195 (1988)
128. S. Wei, Z. Shi, A.W. Castleman Jr.: J. Chem. Phys. 94, 8604 (1991)
129. M. Lezius, T. Rauth, V. Grill, M. Foltin, T.D. Märk: Z. Phys. D24, 289 (1992)
130. C.E. Klots: Z. Phys. D21, 335 (1991)
131. M. Foltin, M. Lezius, P. Scheier, T.D. Märk: J. Chem. Phys. 98, 9624 (1993)
132. J.F. Garvey, W.R. Peifer, M.T. Coolbaugh: Acc. Chem. Res. 42, 48 (1991)
133. T.D. Märk: In Nuclear Physics Concepts in the Study of Atomic Cluster Physics, ed. R. Schmidt, H.O. Lutz, R. Dreizler, Springer, Berlin (1992) pp. 83–92
134. A.J. Stace: Org. Mass Spectrom. 28, 3 (1993)
135. W. Kamke: In Cluster Ions, ed. C.Y. Ng, T. Baer, I. Powis, John Wiley, New York (1993) pp. 1–119
136. J.A. Syage: In Ultrafast Spectroscopy in Chemical Systems, ed. J.D. Simon, Kluwer, in print (1993)
137. C. Lifshitz: Mass Spectrom. Reviews, in print (1993/94)
138. T.D. Märk: In Linking the Gaseous and Condensed Phases of Matter: The Behaviour to Slow Electrons, ed. L.G. Christophorou, W.F. Schmidt, E. Illenberger, NATO ASI Series, Plenum, New York, in print (1994); Physica Scripta, in print (1994)

2.7   Multiply Charged Clusters

O. Echt and *T.D. Märk*

2.7.1   Introduction

All the available experimental and theoretical evidence indicates that singly charged homogeneous clusters, P_n^+, at zero temperature, are thermodynamically stable with respect to dissociation into any pair of fragments, for any size and any material. This stability arises from the fact that the long-range interaction between fragments P_x^+ and P_{n-x} is always attractive, and because the interaction (being dominated by ion-dipole or ion-induced dipole forces) decreases relatively slowly with increasing separation r. In contrast, the long-range interaction between a pair of charged fragments from a multiply charged cluster, P_n^{z+}, is always repulsive.

Depending on the balance between repulsive and attractive forces, hence depending on cluster size, charge state, and material, a multiply charged cluster may be thermodynamically ("intrinsically") stable, kinetically stable ("metastable"), or unstable. In the first case, fission of P_n^{z+} would be endothermic for all possible reaction channels, whereas in the second case, at least one channel satisfying charge and mass conservation would be exothermic, but spontaneous fissioning would be impeded by a barrier. For an unstable cluster, fission can proceed without barrier for at least one reaction channel, and the life-time of the parent cluster would be much less than the time required for mass spectrometric detection.[1]

Generally, for a given charge state and material, we expect a transition to occur from unstable (for small size n) over kinetically stable to, ultimately,

---

[1] Our definitions in this paragraph, which will be used throughout this Chapter, characterize the energy of a cluster P_n^{z+} in its (global or local) rovibronic ground state, relative to that of a pair of ground state fragments at infinite separation and zero kinetic energy. More formal definitions will be introduced in Chap. 2.7.3. The terms *thermodynamically stable* and *kinetically stable* (or *metastable*) are borrowed from thermodynamics. None of these clusters would exist in true thermodynamic equilibrium, whatever the temperature may be, and their lifetime "in vacuum" crucially depends on their vibrational excess energy, in addition to their size n and to the height of the barrier protecting them from fissioning. The lifetime of kinetically stable, very cold clusters may ultimately be restricted by quantum mechanical tunneling, especially for $n = 2$. We will use the term *stable* for cluster ions which are either thermodynamically or kinetically stable. Some authors use a different terminology. In mass spectrometry, for example, "stable" and "metastable" usually characterize the effective lifetime of excited ions, cf. Chap. 2.6.

thermodynamically stable. The first of these transitions defines the *critical size* $n_c(z)$, setting a definite lower limit to the observability of P_n^{z+}. Early mass spectra of doubly charged Pb, NaI and Xe clusters did, indeed, display lower size limits[2] (*appearance sizes* n_2) ranging from 21 for NaI to 53 for Xe [1]. During the past decade, a large number of appearance sizes of weakly bound (van der Waals or hydrogen bonded) clusters have been determined, and their correlation with bulk properties is well established. For other materials, the subject of appearance size is more controversial. In some cases, doubly charged clusters of all sizes may be thermodynamically or, at least, kinetically stable, and their observability will crucially depend on the amount of excess energy introduced upon formation. For example, the appearance of C_n^{2+}, $n \geq 3$, was reported as early as 1959 [2].

Historically, one of the first discussions of the stability of multiply charged clusters ("droplets") was published by Rayleigh. In his words: "In consequence of electrical repulsion, a charged spherical mass of liquid, unacted upon by other forces, is in a condition of unstable equilibrium . . . . When (the charge) is great, the spherical form is unstable . . . ; the liquid is thrown out in fine jets, whose fineness, however, has a limit" [3]. This description points to another topic of current interest: The size distribution of fragments from multiply charged clusters. Part of the interest arises from the formal relationship between fission of atomic clusters and fission of atomic nuclei ("nucleonic clusters") [4–6].

This Chapter will discuss the topics mentioned above in detail, but no attempt will be made to completely cover the vast literature. The literature up to 1987 has been discussed in detail in several reviews [7–10].

At the end of this Introduction we briefly point out two other areas of active research: (1) The observation of highly charged, very large clusters or molecules is not uncommon. The occurrence of multiply charged nitrogen and hydrogen clusters, with z ranging up to five or more, was reported two decades ago [11], and multiply charged macromolecules have been detected at an even earlier date [12]. Doubly and triply charged carbon clusters, containing hundreds of atoms, may be formed by thermionic emission [13]. The production of copious amounts of highly charged molecules by plasma desorption [14] or by the electrospray technique [15] finds application in the analysis of molecules of biological importance [16]. The mass of these particles may be as great as 130 000 amu, but their high charge state (being as high as $z = 45$) makes possible their detection in commercial mass spectrometers of limited mass-to-charge range. (2) Sufficiently large particles will be able to bind several excess electrons; these species have, indeed, been observed [17]. In this context it is also interesting to mention studies on the electrohydrodynamic instability of multielectron bubbles in liquid helium [18]. Small dinegative clusters or molecules, however, are prone to rapid autodetachment. Several early reports of doubly charged atomic anions remained unconfirmed [19]. Recently, however, evidence for the

---

[2] Some authors assign the term "critical size" to the appearance size n_z or to the transition towards thermodynamic stability which we shall denote $n_t(z)$.

existence of small dinegative clusters, $(O_2)_n^{2-}$ with $n = 3, 5, 7$ [20] and C_n^{2-} with $n \geq 7$ [21] has been presented.

2.7.2   Formation of Multiply Charged Clusters

2.7.2.1   Mass Spectrometric Identification

Mass spectrometers analyze the mass-to-charge ratio m/z of ions; they cannot directly identify the charge state z. Only those members of a series of z-fold charged elemental clusters, P_n^{z+}, can be identified readily for which n/z is non-integral. Even so, care has to be taken to distinguish between what appears to be a series of, say, doubly charged odd sized clusters, and a series of contaminated singly charged clusters coinciding in the m/z ratio. This may be accomplished by an isotope analysis, but few instruments provide the required mass resolution of $\Delta m < 1$ amu (amu = atomic mass unit) for large clusters. Alternatively, the appearance energy of the ions in question should be analyzed (cf. 2.7.2.5).

Once the existence of z-fold charged clusters with nonintegral n/z has been established for $n \geq n_z$, the existence of the "hidden" cluster ions, having integral n/z, is often taken to be granted above the appearance size n_z. This conclusion is reasonable for, say, van der Waals clusters, but it may be in error for alkali clusters where stability is controlled by strong odd-even effects, cf. Vol. 1, Chaps. 2.6 and 4.1. To be sure, one should identify a stepwise abundance increase of the "main" peaks (n/z = integer) above n_z/z, arising from the mass spectral coincidence of $P_{n/z}^+$ and P_n^{z+}. Ultimately, a detailed study of the ion abundance on electron/photon energy or photon fluence may be required. Fig. 1 displays mass spectra of argon clusters, produced by electron impact ionization at 400 eV [7]. Note the steep rise in the yield of Ar_n^{2+} (odd n) above the appearance size $n_2 = 91$ (the insert [22] provides a more detailed view of this region). A concomitant increase in the abundance of the "main" peaks (n/z = integer) is also apparent, even though the ordinate is logarithmic.

More direct evidence for the existence of multiply charged clusters with integral n/z may be obtained from a mass analysis of their isotopic composition, provided the monomer features at least two natural isotopes of significant abundance. In the most favorable case, clusters with fractional mass-to-charge ratios (in units of amu/e, e = elementary charge unit) are identified. Otherwise, an identification of all possible isotopic combinations may still be accomplished for a given n. For example, the existence of long-lived doubly charged dimers including He, N, Cl, and Mo, has been verified by isotope analysis.

In case of molecular clusters which undergo prompt intramolecular fragmentation[3] upon ionization (cf. Chap. 2.7.2.6), multiply charged clusters

---

[3] Throughout this Chapter we adopt, with few exceptions, a notation that merely specifies the net composition of the cluster ions. E.g., we shall write $(H_2O)_n \cdot H_2^{2+}$ rather than $(H_2O)_{n-2} \cdot (H_3O^+)_2$. The latter notation might be more informative, but it is more clumsy and, often, speculative (see Chap. 2.7.2.6)

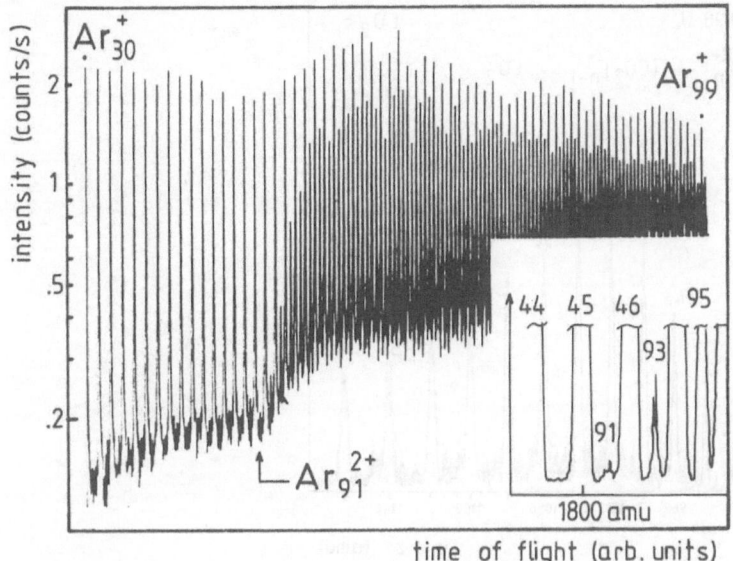

Fig. 1. Mass spectrum of argon clusters, produced by electron impact ionization, featuring doubly charged clusters starting at $n_2 = 91$ (adapted from [7, 22])

(e.g., $(SO_2)_n \cdot O^{2+}$) may be distinguished from singly charged clusters for all n, even if unit mass resolution is not achieved.

Alternatively, a low-resolution study of the metastable decay of cluster ions may provide unique evidence for the occurrence of clusters with integral n/z, $z > 1$. For a given z, the reaction $P_n^{z+} \rightarrow P_{n-1}^{z+} + P$ can be studied without interference from parent clusters having the same size-to-charge ratio, but a lower charge state z. As an example, Fig. 2 displays the yield of product ions for this reaction in case of $(CO_2)_n^{2+}$ [23]. Mass peaks are labelled by their parent size n. The spectrum provides evidence for the existence of long-lived $(CO_2)_n^{2+}$, $n \geq 44$ (all n), complementing the information from direct mass spectra which reveal the existence of these ions for $n \geq 45$ (odd n).

Along the same line, the existence of long-lived Au_2^{2+} [24] and Nb_2^{2+} [25] has been established by detecting their singly charged counterparts arising from charge-exchange reactions.

An interesting way of distinguishing between ions of different charge states has been reported by *Pfau* and *Sattler* [26]. The abundance of Pb_n^{3+} relative to that of singly and doubly charged clusters having nearly the same n/z was found to be greatly enhanced if the discriminator following the preamplifier in the ion detection circuit, operating in the ion counting mode, was set to high values. This result indicates that triply charged lead clusters generate, on an average, more secondary electrons at the conversion dynode, as to be expected from

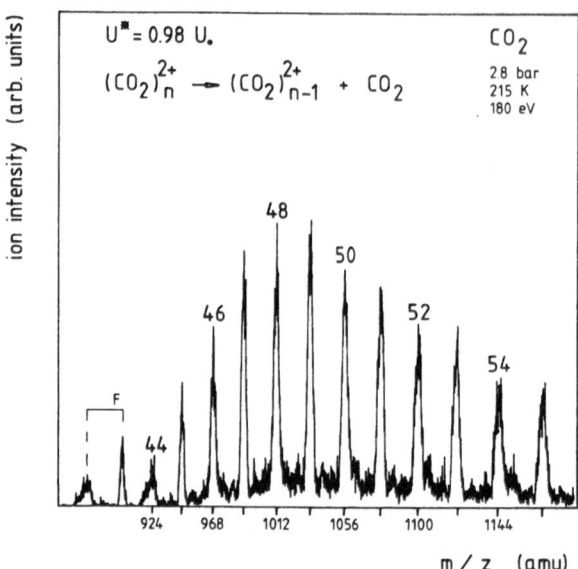

Fig. 2. Odd and even sized, doubly charged CO_2 cluster ions, arising from metastable precursors $(CO_2)_n^{2+}$ by loss of one monomer, in the size range size $44 \leq n \leq 55$ (peaks designated F are likely due to metastable decay of singly charged fragment ions) [23]

considerations of the mechanisms of kinetic and potential electron emission (Vol. 1. Chap. 3.2, and [27]).

2.7.2.2 Very Small Clusters

For a given charge state z, repulsive interaction arising from the positive net charge density of the cluster will generally decrease with increasing size n, hence we expect a transition to occur from unstable to kinetically stable and, ultimately, to thermodynamically stable. Per definition, the first of these transitions occurs at the "critical size"[2] $n_c(z)$, while the second transition defines $n_t(z)$. However, the scientific literature abounds with reports on the observation of doubly charged dimers and trimers. Is it possible that those clusters are stable down to $n = 2$ or 3? (The stability is trivial for $n = 1$ in case of atomic clusters).

For a crude estimate, we note that the potential energy of two elementary point charges in vacuum, at a distance d, is $E = e^2/4\pi\varepsilon_0 d = 1.44 \text{ eV} \cdot (d/\text{nm})^{-1}$ (referenced to infinetely separated charges e). Bond strengths of neutral, covalently bound dimers are typically 1 to 5 eV, hence P_n^{2+} may indeed be kinetically or even thermodynamically stable down to $n = 2$ for selected materials. Generally, however, the bond strength of P_2 is of little value in estimating the stability of P_2^{2+}, because the "missing electrons" may come from either bonding, antibonding, or nonbonding orbitals. The most striking example is

He_2^{2+}, which was predicted to be kinetically stable as early as 1933 by Pauling [28, 29]. The barrier towards dissociation was calculated to be 1.4 eV, which contrasts with a well depth of a few meV for the potential of He_2. Preparation of long-lived He_2^{2+} is difficult, because its local minimum is 10 eV above the dissociation asymptote (owing to its short bond length of 0.075 nm), but it has been identified mass spectrometrically [30]. We expect somewhat larger, doubly charged He clusters to be unstable, because neutral atoms, separating the two positive charges, will merely interact via van der Waals forces, with modifications being due to induced dipoles. These qualitative arguments are corroborated by a theoretical study of He_3^{2+}, which turns out to be stable in C_{2v} symmetry, but unstable upon symmetry breaking [31].

More generally, there will be a class of materials that feature a well defined critical size $n_c(z)$ well above $n = 2$. This class is likely to comprise most van der Waals bound materials for $z \geq 2$, and many other materials for charge states higher than $z = 2$. However, excluding species such as N_2^{2+} or O_2^{2+} [32] from our list of clusters, helium presents the only firm evidence that an additional transition, from kinetically stable to unstable at, or just above, $n = 2$, can indeed occur.

In the other extreme, for strongly bound materials, cluster ions P_n^{z+} may be stable (kinetically or thermodynamically) down to $n = 2$ or 3. The following species, being identified mass spectroscopically, may be tentatively assigned to this class: Mo_2^{2+} [33], Au_n^{2+} ($n = 2$–5) [24, 34], Nb_n^{2+} ($n = 2, 3, 5, 7, \ldots$) [25], Ge_n^{2+} ($n = 2$–10) [35], B_2^{2+}, C_n^{2+} ($n = 2, 3, 5, 7, \ldots$) [36, 37], Mg_n^{2+} ($n = 2, 3$) [38], Sb_n^{2+} ($n = 2, 3$) [39, 40]. The observation of doubly charged mixed dimers, which are identified mass spectrometrically much more easily, is also noteworthy, e.g. $AuSb^{2+}$, $PtSb^{2+}$ [39], $LaFe^{2+}$ [41], $AuSi^{2+}$ [42], PtP^{2+} [43], and AsP^{2+} [44]. In fact, a large number of chemically bound two-atomic molecules (e.g. CO) have been observed in the doubly charged state. Back to homogeneous clusters, doubly charged trimers of the following elements have been observed: Pb ($n \geq 3, n$ odd) [45, 46], Sn ($n = 3, 5, 7, 9$) [47], P [44, 48], As [49], and Ni, W, Si, Bi ([50], and references therein). On the other hand, the absence of Ag_3^{2+}, Cu_3^{2+}, Ga_3^{2+}, and In_3^{2+} under similar experimental conditions is noteworthy ([50], and references therein).

Most of the studies quoted above have employed field evaporation or closely related techniques (liquid metal ion source, laser-assisted field evaporation, etc.). These techniques are known to produce multiply charged ions in copious amounts, probably due to postionization of field-evaporated singly charged ions [51–54]. This process results in large kinetic energy spreads which degrade the resolution of mass spectra, usually precluding information on the presence or absence of even-sized doubly charged clusters. Furthermore, the abundance of cluster ions falls off rapidly with size. Hence, for many of the elements listed above, we do not know for sure if all doubly charged clusters, down to the dimer, are stable. The experimental evidence, listed above, is most convincing in case of germanium, niobium and, perhaps, carbon; while lead and tin are possibly stable for all n starting at $n = 3$ (the calculated fission barrier of Pb_3^{2+} amounts to only

0.13 eV [55]). On the other hand, data for Au_n^{2+} do not rule out a stability gap at $n = 7$ [24, 56]. Moreover, the observation of Au_2^{2+} is interesting in its own right, because of the closed-shell configuration of the separated ions [57, 58].

The existence of a stability gap is also conceivable for materials such as P, As, and Sb, which have a relatively low bulk cohesive energy. This is even more likely for clusters in higher charge states. For example, mass spectrometric evidence for S_2^{3+} [59], W_2^{3+}, Mo_2^{3+}, Re_2^{3+}, Ir_2^{3+} [54], Au_4^{3+} [24], Sn_4^{3+} [47], Ge_4^{3+}, Ge_9^{4+} [46], and Sb_5^{4+} [40] has been reported.

Concerning doubly charged clusters, ab-initio studies indicate that dimers of many, though not all, metallic or semiconducting elements are (kinetically) stable. As an illustration, Fig. 3 displays the interaction energies of several transition metal dimers as functions of the internuclear separation [60]. The interaction is purely repulsive for Ni, Co, and (not shown) Pd and Pt, while local minima, being as deep as 2 eV in case of Mo, occur for several other elements (V, Cr, Fe, Mo, W). None of these species, however, is thermodynamically stable. The trends in stability may be rationalized in a simple bonding-antibonding picture, but there is no direct correlation with the bond strengths of the neutral dimers. It is noteworthy that Ni_3^{2+} has been observed (see above), even though the potential curve of Ni_2^{2+} in Fig. 3 is dominated by Coulomb repulsion for all distances.

Several theoretical studies of doubly charged dimers have been devoted to Be [61, 62], Hg [63–65], and Mg [65, 66, 100]; all of them are found to be kinetically stable. Small neutral clusters of these elements are rather weakly bound, because of the closed-shell nature of the atoms. However, there is no firm evidence that P_n^{2+} destabilizes beyond $n = 2$: Firstly, Hg_n^{2+} has been observed

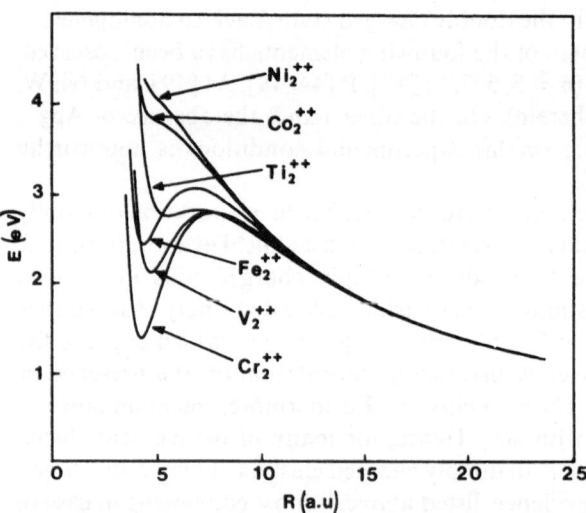

Fig. 3. Calculated interaction energies of doubly charged transition metal dimers as functions of the interparticle separation [60]

down to $n = 5$ (n = odd) [67]. Secondly, Mg_n^{2+} and Be_n^{2+} have been calculated to be (kinetically) stable for $n = 2$ up to $n = 7$ and 5, respectively. Mg_3^{2+} is protected from dissociation by a 0.2 eV barrier, while barriers are on the order of 1 eV for Be_n^{2+}. Mg_7^{2+} and Be_5^{2+} are on the verge of thermodynamic stability [62, 66, 100]. These studies of the multidimensional energy surfaces provide insight into the problems of forming long-lived small multiply charged clusters from neutral precursors; this point will be pursued further in Chap. 2.7.3.

Finally, there is a third class of small, stable, multiply charged clusters: In a mixed doubly charged cluster, the net charge (i.e. the two holes) may be localized at one atom or molecule if the ionization energies of the atoms are greatly disparate. The configuration $A^{2+} \cdot B_n$, e.g., will be kinetically stable with respect to dissociation if it is energetically lower than the configuration $A^+ \cdot B_{n-1} \cdot B^+$, which requires that the second ionization energy of A is lower than the first one of B (this criterion should not be taken too literally; solvation energies, e.g., also contribute to the energy balance). This mechanism is likely to be responsible for the existence of long-lived $Ne \cdot Xe^{2+}$, $Ar \cdot Xe^{2+}$ [68], $Ne \cdot Kr^{2+}$ [69], and of $Ar_n \cdot X^{2+}$ ($n \geq 1$, X = 1,3-butadiene or 1,3,5-hexatriene) [70]. An illuminating discussion of the energy balance in inhomogeneous doubly charged species is to be found in [71].

2.7.2.3 Appearance Sizes

Ionization of free clusters in a crossed-beam experiment by electrons or photons results in the formation of long-lived multiply charged clusters if their size n is sufficiently large, and if the energy and/or flux of the ionizing particles is sufficiently high. In some cases, the abundance of multiply charged species rises steeply above the appearance size n_z, and its value is not significantly affected by experimental parameters. In this paragraph, we shall compile values of n_z determined in crossed-beam experiments. The compilation is not meant to be exhaustive; references to metal halides, metal oxides and some other mixed systems may be found in [8].

2.7.2.3a) Van der Waals and Hydrogen-Bound Clusters

A wealth of data is available for weakly bound systems, partly because the formation of intense, continuous beams of neutral clusters by adiabatic expansion is readily available. In Fig. 4 we display appearance sizes of doubly charged clusters, plotted versus the quantity $1/T_c \cdot v^{1/3}$. T_c denotes the critical temperature, and the molecular volume v is calculated in the continuum approximation from the density of the bulk liquid at the boiling point [79] (in other words, v would equal the volume of the Wigner–Seitz cell in case of monovalent metals).

The data displayed in Fig. 4 suggest the following scaling law:

$$n_2 \cdot T_c \cdot v^{1/3} = \text{constant} \tag{1}$$

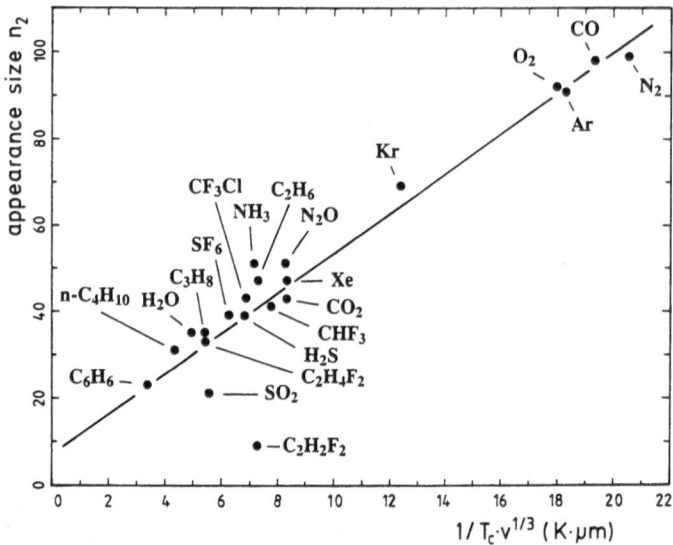

Fig. 4. Appearance sizes of doubly charged van der Waals and hydrogen-bonded clusters, plotted vs. $1/T_c v^{1/3}$ (T_c = critical temperature, v = molecular volume). The line represents a least squares fit. The data are those compiled in [72], with the exceptions of: Kr [13], Xe [74], SO_2 [75], C_2H_6 [76], $C_2H_2F_2$ [77] ,C_3H_8 [76] . Not included is the value $n_2 = 3$, recently reported for $C_6F_2H_4$ [78]

This kind of scaling law can indeed be derived for $n_t(2)$ within a very simple liquid drop model [7].[4] Assume that the cluster of size $n_t(2)$, or diameter d_2,[5] is on the verge of thermodynamic stability. If this cluster undergoes symmetric fission, the diameter of its products will decrease by 21%, while the sum of their surface energies will increase by 26%, assuming that the surface tension remains constant, and that the total volume and volume energy are conserved. Hence, the surface energy will increase by $c \cdot \sigma \cdot d_2^2$, where c has the same value for all materials. On the other hand, upon fission, the cluster gets rid of its Coulomb energy $c' \cdot e^2/d_2$. Equating these terms, and replacing d_2 by $c'' \cdot (n_t(2) \cdot v)^{1/3}$, we arrive at

$$n_t(2) \cdot v \cdot \sigma = \text{constant} \tag{2a}$$

Technically, surface tensions are often quoted at or near room temperature. These are rather useless for testing the above scaling laws, because σ is temperature dependent, approaching zero at the critical temperature. We avoid this

---

[4] In this section we shall occasionally discuss n_z, $n_c(z)$ and $n_t(z)$ next to each other. These are, of course, distinct quantities, with the only obvious relations being $n_c(z) \leq n_z$ and $n_c(z) < n_t(z)$. However, data to be presented in this section as well as theoretical considerations to be discussed later (Chap. 2.7.3.2), indicate that the critical size, the appearance size, and $n_t(z)$, are correlated more closely. Equation (2a), e.g., which is derived for $n_t(2)$, will be shown to also hold for $n_c(2)$ (Eq. 10).

[5] The radius r is related to the number of cluster constituents, n, by $r^3 = n \cdot r_0^3$, or by $4\pi r^3/3 = n \cdot v$, where r_0 is the effective radius of the monomer, and v its effective volume, usually calculated from the density of the bulk phase.

problem by substituting $E_{coh} \cdot v^{-2/3}$ for σ, where E_{coh} is the cohesive energy per monomer at $T = 0$ K. As a further simplification, we replace E_{coh} by the critical temperature T_c (or, alternatively, by the boiling temperature at standard pressure), because these quantities scale as E_{coh} according to the law of corresponding states [80]. Thus, the scaling law (1) is obtained for the quantity $n_1(2)$.[4] Note that the equations above do not invoke the dielectric constant ε. In fact, introduction of the static dielectric constant into the scaling law would significantly deteriorate the agreement whenever the value of ε is large (which applies to H_2O, H_2S, NH_3, and C_2H_6O, in particular). This indicates that the use of the static dielectric constant is inappropriate, presumably because the proximity of the surface, and of other charges, hinders alignment of solvent molecules around the ions [72].

Appearance sizes can be deduced from mass spectra in a straightforward manner, but a few observations are worth being mentioned:

(1) The relative yield of doubly charged clusters may be huge. Slightly above n_2, it is sometimes as large as that of singly charged clusters of the same size, hence exceeding the relative yield of doubly charged atoms by one to two orders of magnitude.

(2) No systematic influence of cluster source conditions, electron energy, or electron flux, on n_2 has been observed. Of course, the abundance of P_n^{2+} rises steeply beyond the appearance size, but not infinitely steep. Hence, improvements in mass spectrometric resolution, dynamic range, and absolute ion abundance, will tend to decrease the experimental size limit. Over the past decade, however, improved techniques have lowered reported appearance sizes by a mere 4% in case of CO_2 [81, 82], and by 11% in case of Xe [1, 74]. Furthermore, identical appearance sizes have been reported for doubly charged benzene clusters under electron impact and multiphoton ionization [83, 84].

(3) The appearance size is independent of the exact stoichiometry of doubly charged molecular clusters [75, 85–89]. $(N_2)_n^{2+}$ and $(N_2)_n N^{2+}$, e.g., feature a common appearance size of 99 [85]. (Depending on the definition of cluster size n for fragmented molecular clusters, a small shift by $\Delta n = \pm 1$ may be deduced from the data [88]). This indicates negligible effect of the nature of charge carrier(s) on the stability of multiply charged clusters.

Appearance sizes of triply and quadruply charged clusters have been reported for several weakly bound systems. In most cases, these clusters were not mass-resolved, their existence was inferred from a stepwise increase of ion abundance with increasing mass. Hence, those appearance sizes are not exactly known (uncertainty $\approx 10\%$). Nevertheless, according to Lezius et al. [73], the available data of n_3 closely follow a line through the origin if they are plotted versus $1/T_c \cdot v^{1/3}$. Taken together with the fact that a similar scaling law holds for n_2 (see Fig. 4), this implies that n_3/n_2 assumes a constant value. In Fig. 5 we present all experimentally known values of n_3 and n_4, normalized to n_2. The average of the former ratio is 2.22, while that of the latter is 4.08, with standard deviations of 7 and 10%, respectively.

The stability limit of a charged liquid drop with respect to small deformations is, indeed, predicted to scale as

$$n_c(z) \cdot v \cdot \sigma = z^2 \cdot \text{constant} \tag{2b}$$

(cf. Chap. 2.7.3.2, Eq. (10)), i.e. the relation

$$n_c(2):n_c(3):n_c(4) = 1:2.25:4.0 \tag{3}$$

is predicted in excellent agreement with the experimental ratio of appearance sizes, $1:2.22:4.08$. A slightly different ratio, $1:2.3:3.7$, was predicted for $n_t(z)^4$ on the basis of a continuum approach where the charges were assumed to be localized, and fission was assumed to proceed symmetrically [90].

2.7.2.3b)   Metal Clusters

According to a model suggested by *Delley*, which focusses on the competition between evaporation of neutral and charged monomers, a ratio $n_c(2):n_c(3):n_c(4) = 1:2.8:5.2$ is to be expected for metal clusters [91]. Other predictions will be mentioned in Chap. 2.7.3.1, but a comparison with experimental data is not warranted in view of their large scatter (Fig. 5). What is the origin of these fluctuations?

Figure 6 displays a mass spectrum of cesium clusters [92]. This spectrum is rather exceptional among metal clusters, in that it features an abrupt onset of doubly charged clusters at $n_2 = 19$. A more typical case is displayed in Fig. 7, where the abundance of doubly and triply charged silver clusters rises gradually beyond their appearance size $n_2 = 9$ (not shown) and $n_3 = 37$, respectively [93]. The poor reproducibility of these values is apparent from Table 1 which compiles appearance sizes of multiply charged metal clusters produced by

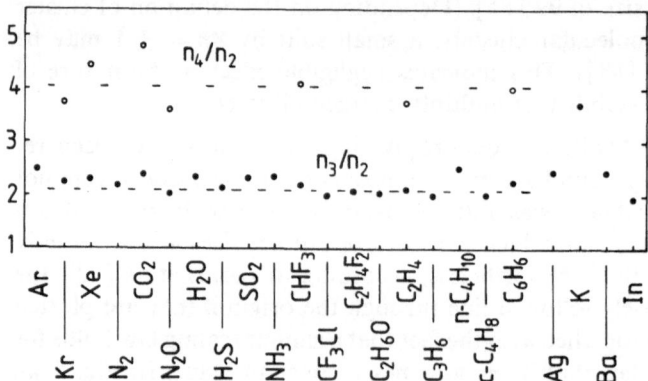

Fig. 5. Ratio of appearance sizes n_3/n_2, and n_4/n_2, determined in crossed-beam experiments, for weakly bound clusters and for a few metallic systems. For references, see captions of Fig. 4 and of Table 1

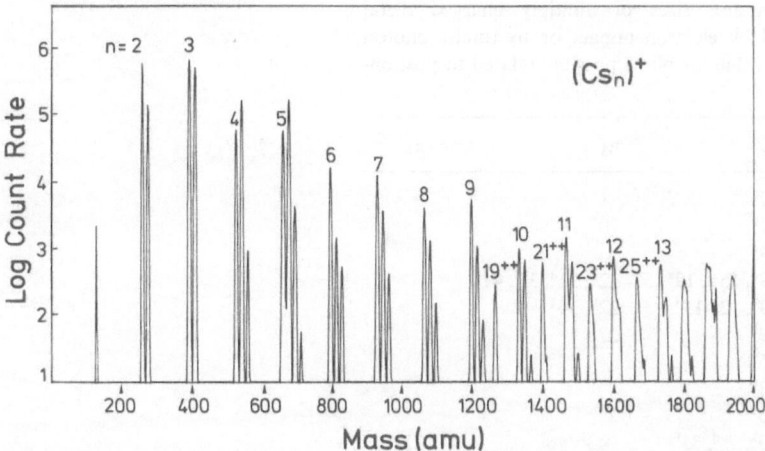

Fig. 6. Mass spectrum of cesium clusters, produced by electron impact ionization at 70 eV. Doubly charged clusters occur at and beyond $n_2 = 19$ (unlabelled peaks are due to lightly oxidized clusters) [92]

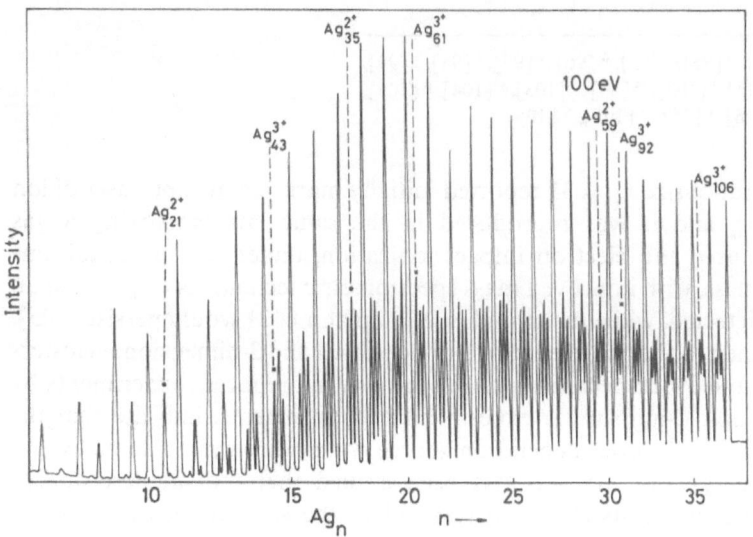

Fig. 7. Mass spectrum of silver clusters, produced by electron impact ionization at 100 eV. Doubly charged clusters occur over the full mass-to-charge range, triply charged clusters start at $n_3 = 37$. Abundance anomalies in the distribution of doubly and triply charged ions have been labelled [93]

electron impact or (multi-)photon ionization. The discrepancy is even larger if we recall that for some of the elements listed in Table 1, field-evaporation or related techniques result in the formation of doubly charged dimers or trimers (Chap. 2.7.2.2). We also mention a report on Na_7^{2+}, produced in a liquid metal ion source [109]. The following discussion will be restricted to lead, because this system has been investigated most thoroughly.

Table 1. Appearance sizes of multiply charged metal clusters, ionized by electron impact or by (multi-)photon ionization. The value for Nb is possibly related to postionization of Nb_n^+

Metal	n_2	n_3	n_4
Na	27[a]		
K	19[b]	70[b]	
Cs	19[c]		
Ag	9[d], 15[ef], 19[g]	22[k], 31[d], 37[i], 41[e]	
Au	9[k], 13[i], 15[j]	22[k], 34[j]	
Ba	9[l]	22[l]	
Hg	5[m]		
Al	13[n]		
In	15[o]	29[o]	
Pb	3[k], 7[pq], 31[r]	43[k], 46[s]	73[s]
Sb	29[s]		
Bi	5[k], 27[s]	38[k]	
Nb	2[t]		
Mo	19[u]		
Ta	11[v]		
W	7[u]		

References: [a] [94], [b] [95], [c] [92], [d] [96], [e] [97], [f] [98], [g] [99], [i] [93], [j] [101], [k] [45], [l] [102], [m] [67], [n] [103], [o] [104], [p] [105], [q] [106], [r] [1], [s] [26], [t] [25], [u] [107], [v] [108]

The appearance size $n_2 = 31$ reported initially marks an abrupt onset of ion abundance [1], and it was reproduced in the same lab, employing a gas aggregation source and electron impact ionization, under various conditions [26]. Nevertheless, with improved mass spectrometric techniques Pb_2^{2+} as small as $n = 7$ was detected, while the abundance jump at $n = 31$ would persist [105]. This was assigned to the occurrence of 1-dimensional and 3-dimensional clusters beyond $n = 7$ and 31, respectively [110]. However, more recent experiments by two other groups, using similar experimental arrangements, indicate that the minimum size of Pb_n^{2+} does depend on source conditions and, to some extent, on electron energy [45, 106]. It has been argued that the smallest doubly charged clusters in the spectrum correspond to the smallest neutral clusters generated in the source, provided the sensitivity of the mass spectrometer is sufficiently high [106]. Measurements under widely different electron energies support this model, but it fails to explain the nature of the persistent onset at $n = 31$.

Anticipating the discussion in Sect. 2.7.3.2, we tentatively interpret the observations as follows: Pb_n^{2+} is metastable down to the trimer, but its barrier towards fission is below the activation energy for the competing reaction $Pb^{2+} \rightarrow Pb_{n-1}^{2+} + Pb$ as long as $n < 31$. If small, cold neutral clusters $n_x \leq n < 31$ exist in the beam, and if ionization proceeds without introducing a significant amount of excess energy into the system, long-lived doubly charged clusters as small as n_x may be observed. Under more violent ionizing conditions, doubly charged clusters below $n = 31$ will be depleted due to rapid fission, hence n_x is still a valid lower size limit. Above $n = 31$, however, metastable monomer evaporation is a viable path, and any neutral precursor of sufficient size can contribute to the yield of doubly charged clusters just above $n = 31$, whatever the value of n_x.[6]

Is this (rather speculative) model applicable to other elements listed in Table 1? We shall return to this point in Sect. 2.7.3.2, but we remark that no other system exhibits mass spectral features similar to those of lead (which may be due to lack of adequate data). It is also interesting to note that the appearance size of triply charged lead clusters has, until now, been reproduced remarkably well [45].

2.7.2.4 Magic Numbers

Abundance anomalies in size distributions of clusters have helped unraveling atomic or electronic structures. It is of interest to compare magic numbers in spectra of multiply charged clusters with those of singly charged ones, even though such a study is hampered by the "non-observability" of multiply charged elemental clusters having integral n/z values. Does the presence of an additional electronic hole alter the structure?

Apparently it does not in the case of calcium oxide clusters. Both, $[Ca(CaO)_n]^+$ and $[Ca(CaO)_n]^{2+}$, feature unusually weak lines at $n = 13, 24, 32$, 38, 40, 50, and 64, which can be assigned to rectangular clusters with rock salt structure plus one extra ion pair [114]. In either case, the hole(s) would be located at just one calcium ion.

Barium clusters feature identical magic numbers for $z = 1$ and 2 [102], even if the clusters are slightly oxidized [114]. The numbers ($n = 13, 19, 23, 26$, 29, . . . barium atoms) suggest that a close packed, non-crystalline (icosahedral) structure is preferred, independent of the charge state and of oxygen content.

Rare gas clusters are prime candidates for icosahedral structures. Recently, a complete distribution of Ar_n^{2+} has been reported [115]. Characteristic differences in the electron energy dependence and spatial distribution of Ar_n^+ and odd-sized Ar_n^{2+} allowed to extract the yield of even-sized Ar_n^{2+}. Fig. 8 compares the abundance of singly and doubly charged clusters. Pronounced minima in the distribution of Ar_n^+, in the size range $n = 90$–150, are not directly mirrored in the distribution of doubly charged clusters. However, as pointed out by the authors, the latter distribution does feature a series of minima being displaced to smaller n values by 4 units in comparison to singly charged ions. An explanation for this

---

[6] An interesting parallel occurs in negatively water clusters: it has been suggested that the apearance of a dramatic, easily reproduced increase in the ion abundance above $n \approx 11$ relates to competition between monomer evaporation and electron detachment [111, 112]. Here, too, smaller ions can be observed under suitable source conditions, but the step in the spectrum at $n \approx 11$ usually persists [113].

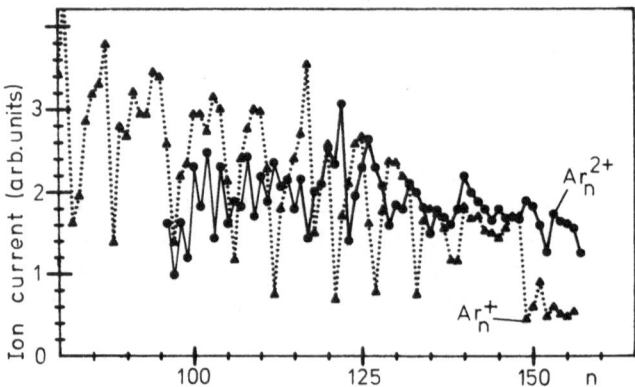

Fig. 8. Size distribution of singly charged (*triangular symbols, dotted line*) and doubly charged (*dots, solid line*) argon clusters [115]

shift has not yet been proposed. It is also of interest to note that a particularly prominent magic number in xenon spectra at $n = 55$, assigned to a complete-shell icosahedron, also shows up in distributions of Xe_n^{2+}, while other magic numbers (e.g. at 71, 81, 87, which are related to capped icosahedra) do not [74].

Given the strong perturbation of a rare gas cluster by an electronic hole (see Vol. 1. Chap. 4.6) we should, in fact, not expect any direct similarity among features in singly and doubly charged argon or xenon clusters. In the other extreme, for clusters with free-electron like valence electrons, the number of electrons rather than the atomic structure (and, hence, the total number of ionic cores) controls the relative stability (Vol. 1. Chap. 2.6 and 4.1). This conclusion, initially drawn from distributions of neutral and singly charged clusters, finds further support by recent studies of doubly and triply charged silver, gold, and lead clusters [93, 97, 116]. For example, the size distribution of Ag_n^{2+} and Ag_n^{3+}, presented in Fig. 7, exhibits abundance anomalies which are fully consistent with electronic shell closings at electron number $n_e = n - z = 20$ and 58, in accord with the spherical jellium model.

2.7.2.5 Ionization Mechanism

The absence of small doubly charged van der Waals clusters and its explanation in terms of a liquid drop model (see 2.7.3) implies that, for instance, stable doubly charged homogeneous clusters, P_n^{2+}, consist of two P^+ ions and $(n - 2)$ neutral constituents P. The formation of multiply charged clusters was originally assumed to follow a mechanism known to occur in ordinary molecules [117]. A core electron of a particular target atom is ejected by the ionizing agent (an electron or photon), and the subsequent electronic relaxation frees a second electron from the same atom (Auger effect), thus producing a $P^{2+} \cdot P_{n-1}$ ion. In a cluster this initial ionization process may be followed by a rapid charge transfer via electron hole recombination, leading to the migration of the positive charges to opposite sides of the cluster, giving an ion $P^+ \cdot P_{n-2} \cdot P^+$ [1]. However, a molecular dynamics simulation has shown that an ion produced in this way experiences a large temperature rise which leads to its immediate break up into singly charged fragments [118]. Based on these simulations, Gay and Berne suggested as a more likely process multiple inelastic electron scattering within the cluster leading to the production of two singly charged ions at different sites.

Experiments to determine the true nature of the formation mechanism were carried out by *Märk* and coworkers for the case of electron impact [73, 74, 85, 86, 119, 120] and by *Brechignac* and coworkers for photoionization [67]. Fig. 9 shows relative electron impact ionization cross section functions near threshold for the production of singly, doubly and triply charged Ar cluster ions. Also shown, for comparison and calibration purposes, are cross section functions of the respective monomer ions. The onsets of singly charged clusters (see inset) is shifted slightly below the appearance energy of Ar^+ due to solvation effects. The onsets of Ar_n^{z+} ($z = 2, 3$), however, are shifted way below the onsets of the respective monomer ions. These large red shifts, observed also for other van der Waals clusters, cannot be solely attributed to solvation effects. Therefore, the result that the appearance energy of doubly charged cluster ions is approximately two times the appearance energy of singly charged cluster ions, constitutes direct evidence for the occurrence of two sequential single ionization processes during the formation of doubly charged cluster ions (see Fig. 10). The linear threshold law observed for the doubly charged cluster ions (see Fig. 9), the large abundance of doubly charged cluster ions (see Fig. 8), and the linear dependence on electron current are additional arguments in favor of this

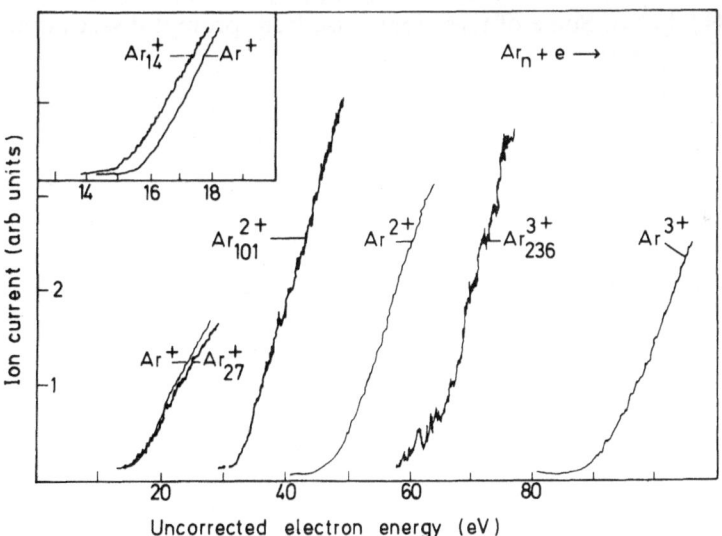

Fig. 9. Ion current versus electron energy close to threshold for the production of Ar^+, Ar^{2+}, Ar^{3+}, Ar_{14}^+, Ar_{27}^{2+}, Ar_{101}^{2+}, and Ar_{236}^{3+} by electron impact ionization of a supersonic argon cluster beam [119]

mechanism. Whereas, in case of electron impact ionization either of the two secondary electrons may contribute to this two-step process, in case of photoionization the second step is carried out by the photoelectron [67] leading to the ejection of two electrons from the cluster. This latter process has been observed unambigously in studies of potassium solids [121] and rare gas solids [122].

In either case (electron or photon ionization), the multiple ionization in van der Waals clusters, or van der Waals-type systems (e.g., small Hg clusters), are believed to occur on the time scale of electronic motion, within 10^{-15} s. The situation in case of metallic clusters is not yet clear, even though ionization cross section functions near threshold have recently been reported for lead [106, 123], silver and gold [124], and potassium ([95], cf. Vol. 1. Chap. 4.1).

Evidence for another interesting ionization mechanism has been reported by *Whetten* and coworkers [84]. They observe extremely efficient double ionization of large benzene clusters by multiphoton absorption under conditions where the energy of a single photon (6.4 eV) is well below the first ionization energy of the cluster. The strong size dependence of this enhancement, and the dependence of the ion yield on laser fluence, is interpreted in terms of exciton annihilation, i.e. fusion of several electronic excitations at different molecules within the same cluster causes the ejection of 2 electrons.

2.7.2.6 Nonstoichiometric Multiply Charged Molecular Clusters

Singly charged molecular clusters usually come in several different stoichiometries.[3] For example, sulfur dioxide $(P = SO_2)$ gives rise to 4 different homologous series: P_n^+, $P_n \cdot O^+$, $P_n \cdot S^+$, and $P_n \cdot SO^+$ [75], while six homologous series have been reported for ammonia clusters $(P = NH_3)$: P_n^+, $P_n \cdot H^+$, $P_n \cdot H_2^+$, $P_n \cdot N^+$, $P_n \cdot NH^+$, $P_n \cdot NH_2^+$ [125]. Some of these ions arise from prompt dissociation

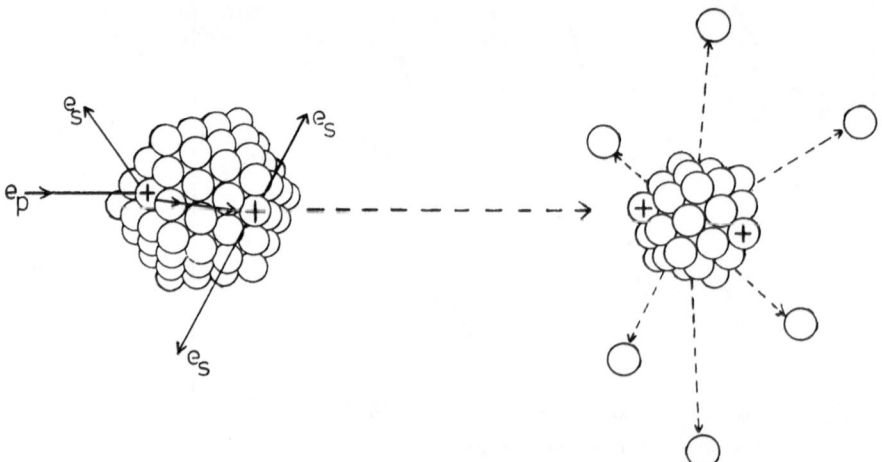

Fig. 10. Schematic representation of the sequential two-step process leading to the production of doubly charged van der Waals cluster ions. e_p = primary electron, e_s = scattered electron

during the primary ionization of a single molecule in the cluster, while others are formed by ion-molecule reactions within the cluster (cf. Chaps. 2.6.3, 2.6.4). Their relative abundance will, in general, depend on the nature of the ionizing particle (electron or photon) and their energy, especially in case of prompt dissociation. In mass spectrometric studies of multiply charged clusters, the occurrence of more than one homologous series per charge state is a nuisance, unless the resolving power and dynamic range of the mass spectrometer is exceptionally high. In that case, however, their analysis offers valuable information.

Identification. Even-sized atomic clusters P_n^{2+} are normally undetectable, because of their coincidence with the peak arising from $P_{n/2}^+$. Intramolecular fragmentation, however, offers the possibility of unambiguously identifying doubly charged clusters for odd and even n. One example, $(CaO)_n \cdot Ca^{2+}$, arising from electron-impact ionization of calcium oxide clusters [114], has been mentioned previously (2.7.2.4), the homologous series $(CO)_n \cdot O^{2+}$, arising from electron impact ionization of carbon monoxide clusters, presents another one [87]. In both cases, singly charged clusters, whatever their stoichiometry, cannot give rise to peaks at the same mass-to-charge-ratio. Water clusters present a somewhat different case [88, 126]: Following electron impact ionization, only one form of singly charged clusters is observed[7], namely the protonated series $P_n \cdot H^+$, while doubly charged clusters come in two forms, $P_n \cdot H_2^{2+}$ and P_n^{2+} (cf. below). Hence, the latter series is free from contamination by singly charged clusters.

Ionization mechanism and charge carriers. The ionization mechanism as well as the charge carrier may be inferred from a detailed comparison of the stoichiometries of singly and multiply charged clusters. Let us assume that singly charged clusters come in the following three forms: P_n^+, $P_n \cdot X^+$, and $P_n \cdot Y^+$. If, as argued above, doubly charged clusters are formed by two sequential single ionization events, 6 different forms (P_n^{2+}, $P_n^+ \cdot X^+$, $P_n \cdot X^+ X^+$, etc.) will be generated. This conclusion should be valid independent of the formation mechanism of the fragments X^+ and Y^+ (prompt dissociative ionization versus ion-molecule reactions). *Märk* and coworkers have, indeed, reported that all four different forms of singly charged sulfur dioxide clusters, mentioned above, have doubly charged analogs [75]. Likewise, all possible combinations of the dominant charge carriers observed in mass spectra of singly charged ammonia clusters (stoichiometry P_n^+, $P_n \cdot H^+$, $P_n \cdot H_2^+$, $P_n \cdot NH_2^+$ [125]) appear to occur in mass spectra of doubly charged clusters [89]. However, some combinations of these ions coincide in mass, hence they cannot be identified unambiguously, except by investigating the electron energy dependence close to threshold [120]. The presence of some other charge carriers, known to exist in singly charged clusters at lower abundance, could not be confirmed (also see [129, 130]).

A recent high-resolution study of water clusters suggests that sequential ionization may not be the only process at elevated electron energies (70 eV)

[126]. Singly charged clusters appear in only one form, $P_n \cdot H^+$.[7] The corresponding forms $P_n \cdot H_2^{2+}$ and $P_n \cdot H_3^{3+}$ are, indeed, observed, indicating that ionization proceeds by three sequential steps. But, in addition, the following homologous series are observed: P_n^{2+}, $P_n \cdot H^{3+}$, and $P_n \cdot OH^{3+}$. The following model accounts for these facts: Double ionization of a single molecule in the cluster may occur. If the products $H^+ + OH^+$ are caged within the cluster, doubly charged clusters of the apparent stoichiometry P_n^{2+} are observed. Similarly, other stoichiometries can be explained by a combination of double ionization with sequential ionization, or of two double ionization events. For example, the ion $P_n \cdot OH^+$ is believed to be a product of double ionization followed by loss of H^+.

Verification of this intriguing model, and of several other interpretations mentioned earlier, could probably be accomplished by a detailed analysis of the dependence of the ion intensity on electron energy. At present, however, hardly any mass spectrometer qualifies for such an ambitious study.

2.7.3 Stability and Fragmentation of Multiply Charged Clusters

A number of recent studies have explored the dissociation dynamics of multiply charged clusters. These are controlled by several factors: energy difference between ground states of products (at infinite separation) and reactant, activation barriers for the various reaction channels, excess energy in the precursor cluster, and, at least for small systems, the detailed geometric configuration in which the precursor is initially prepared. As an illustration, Fig. 11 [66] presents a schematic view of the energy surface of Mg_3^{2+} in its electronic ground state as functions of interatomic distance $r = r_{AB} = r_{BC}$, and bond angle $\sphericalangle ABC$. Based on our definition,[1] Mg_3^{2+} is metastable. A fission barrier of 0.35 eV stabilizes its local ground state configuration, which is linear. A long-lived doubly charged trimer might be formed by vertical ionization of the relaxed singly charged trimer, which has a similar ground state structure. However, vertical double ionization of the neutral trimer (equilateral triangle) would result in a highly excited species which would immediately undergo fission without visiting its local ground state.

Similar features have emerged from other theoretical studies of small clusters [62, 100, 131]. It may appear that without detailed knowledge of the multidimensional energy surfaces of P_n^{2+} and its singly charged or neutral precursors, little can be said about the fragmentation dynamics. However, for large clusters, the situation will possibly simplify if the excess energy in the newly formed multiply charged species is dissipated into nuclear motion before fission occurs. The following quantities will be referred to throughout the rest of this chapter:

---

[7] Unprotonated cluster ions $(H_2O)_n^+$ have been observed in experiments exploiting ionization techniques other than electron impact ionization [127, 128].

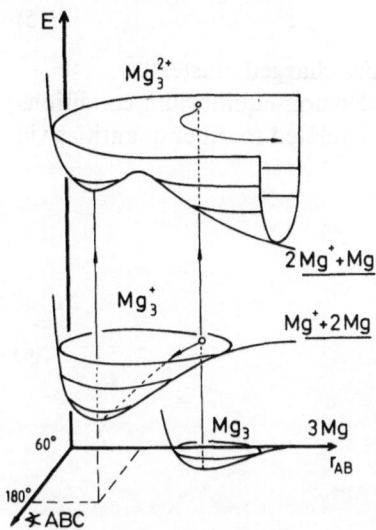

Fig. 11. Schematic view of the potential energy surface of Mg_3^{2+}, and the points on this surface that are reached upon vertical ionization from Mg_3 and Mg_3^+, respectively (after [66])

Fission may be described by the reaction

$$P_n^{z+} = P_x^{z'+} + P_{n-x}^{(z-z')+} - E_{tn}^{z+}(x, z') \tag{4}$$

if we restrict our discussion to reactions that produce only two products. The heat of reaction (4) is positive for endothermic reactions. P_n^{z+} is *thermodynamically stable* if and only if the heat of reaction is positive for all $x < n$, $z' < z$. Its minimum value (being associated with the channel that maximizes the exothermicity) will be written E_{tn}^{z+}. The minimum size of a thermodynamically stable cluster will be written $n_t(z)$. Likewise, P_n^{z+} is *metastable* if and only if the fission barrier $E_{bn}^{z+}(x, z')$ is positive for all $x < n$, $z' < z$. Its minimum value will be written E_{bn}^{z+}. By definition, this quantity is non-positive below the critical size $n_c(z)$. The size distribution of fission products is likely to be controlled by the x-dependence of the barrier, which may qualitatively differ from the x-dependence of the heat of reaction (4).

Under basically all circumstances, ejection of neutral atoms or neutral clusters from P_n^{z+} will be endothermic, and it will not lower the repulsive Coulomb energy. Nevertheless, these reactions may compete with fission in some cases. The activation energy of the most important of these, loss of monomers, will be designated D_n^{z+}. (Loss of a monomer probably proceeds barrierless, hence we expect $D_n^{z+} = E_{tn}^{z+}(x = n - 1, z' = z)$).

2.7.3.1   Thermodynamically Stable Clusters

2.7.3.1a)   Experiment

Thermodynamic stability is neither a required nor a sufficient condition for the observability of P_n^{z+}. Accordingly, little is known about the heat of the reaction

$$P_n^{2+} = P_x^+ + P_{n-x}^+ - E_{tn}^{2+}(x, z' = 1) \tag{5}$$

(this section will be restricted to fission of doubly charged clusters).

A direct determination of $E_{tn}^{2+}(x, z' = 1)$ under non-equilibrium conditions would be exceedingly difficult, but this quantity is related to other quantities via the following thermodynamic cycle:

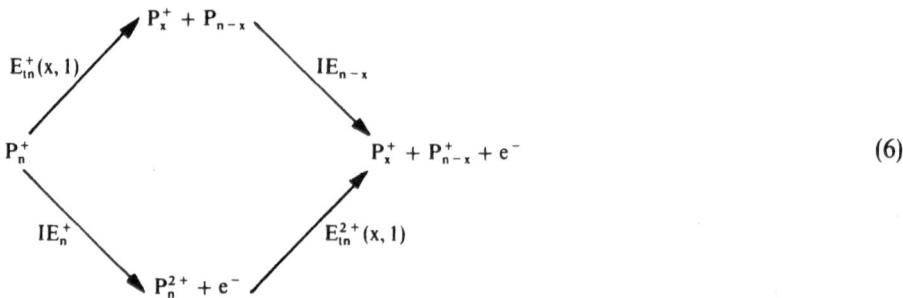

$$(6)$$

where IE_n^{z+} denotes the (adiabatic) ionization energy of P_n^{z+}. From the thermodynamic cycle (6), the minimum size $n_t(2)$ may be obtained if the binding energies of all neutral and singly charged clusters up to and including $n_t(2)$, plus their (adiabatic) ionization energies, are known. The thermodynamic stabilities of Ge_n^{2+} [35] and of K_n^{z+} ($z = 2, 3$) ([95], cf. Vol. 1. Chap. 4.1) have been analyzed in this way. We shall briefly discuss the former of these studies.

A liquid–metal ion source generates Ge_n^{2+} as small as $n = 2$. Hence, all these ions are stable. A mass-selected beam is passed through a collision cell at low, variable energy. The occurrence of the charge exchange reaction

$$Ge_n^{2+} + Xe \rightarrow Ge_n^+ + Xe^+ \tag{7}$$

is monitored in a second mass filter. Repeating this experiment with various collision gases of different ionization energies, one can bracket the electron affinity EA of Ge_n^{2+}, which equals the ionization energy of Ge_n^+ if charge exchange is adiabatic, i.e. if no significant structural rearrangement follows the capture of an electron. The other quantities involved in (6) were taken from the literature: The binding energies of neutral clusters ($n \leq 7$) have been determined from the partial pressures in a Knudsen cell, while $IE(Ge_n)$ and the binding energies of Ge_n^+ were derived from (suitably scaled) theoretical studies of silicon clusters.

From the cycle (6) we see that Ge_n^{2+} is thermodynamically stable (i.e. $E_{tn}^{2+}(x, 1) > 0$ for all x) if and only if

$$EA_t(Ge_n^{2+}) < E_{tn}^+(x, 1) + IE_{n-x} \text{ for all } x \tag{8}$$

Figure 12 compares the minimum value of the right-hand-side of Eq. (8) (solid curve) with the experimentally determined electron affinity (hatched bars). Ge_n^{2+} appears to be thermodynamically stable for $n \geq 4$ or 5, while the dimer and trimer violate Eq. (8), i.e. they are metastable. However, systematic errors are not included in Fig. 12. These may arise from uncertainties in the calculated

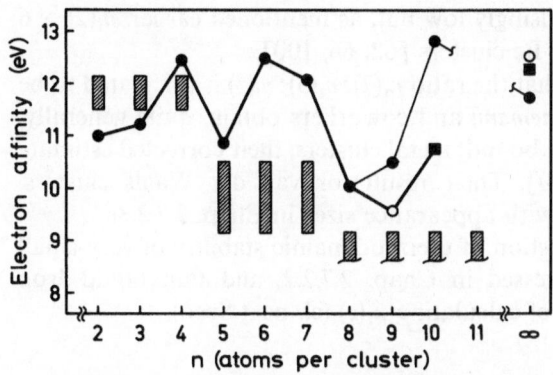

Fig. 12. Experimentally determined electron affinities of Ge_n^{2+} (*hatched boxes*), and derived upper boundaries for the electron affinities beyond which the clusters would be metastable with respect to loss of Ge^+ (*solid circles*), or loss of Ge_2^+ (*open circle*), or loss of Ge_4^+ (*solid square*) [35]

values and the scaling procedure, the Knudsen data, and a non-adiabatic charge exchange process. $n_t(2) = 6$ represents a more conservative estimate for the smallest, thermodynamically stable, doubly charged germanium cluster [35].

2.7.3.1b)   Theory

The thermodynamic stability of a cluster P_n^{z+} is obtained from the heat of reaction (4).

Hence, the minimum size $n_t(z)$ of a thermodynamically stable cluster is obtained if the binding energies of all clusters $n < n_t(z)$ for all charge states $z' < z$ are known. The task is significantly simplified if it is *assumed* that emission of a singly charged monomer is thermodynamically favored. This particular channel minimizes the increase in surface area and was suggested to be favored in case of metal clusters [90, 91, 132, 133].[8]

However, if the size x, $n - x$ of the products is systematically varied, it becomes clear that electronic shell effects in monovalent metal clusters strongly favor fragments containing a "magic" number of valence electrons ($n_e = 2, 8, 20, \ldots$), and the trend towards extremely asymmetric fission is lost [134–138]. Moreover, shell effects cause an oscillatory pattern in the heat of reaction E_{tn}^{2+}, and P_n^{2+} might experience several islands of thermodynamic stability with increasing n. However, as known from a comparison of experimentally and theoretically determined ionization energies, the jellium model, at least in its simpler versions, seriously overestimates shell effects, and these oscillations may be unreal.

Disregarding the oscillations, the size limit $n_t(2)$ for sodium and cesium clusters is calculated to be around 25 [132, 135, 139], and about twice as large for Li [136]. In comparison, the experimental value of germanium clusters,

---

[8] The situation will be different for van der Waals clusters, since their effective surface tension is strongly size dependent due to solvation effects.

$n_t(2) = 6$ [35] may appear surprisingly low but, as mentioned earlier, $n_t(2) \approx 6$ was also calculated for Mg and Be clusters [62, 66, 100].

It is also worth mentioning that the ratio $n_t(2):n_t(3):n_t(4)$ is calculated to be $1:4.7:14$ for cesium [132]. *Bennemann* and coworkers obtain, quite generally, $1:8:27$ for large (i.e. very weakly bound) metal clusters, their corrected estimate for lead clusters is $1:4:10$ [90]. Their result for van der Waals clusters, $1:2.3:3.7$ had been compared with appearance sizes in Chap. 2.7.2.3a.

Finally we note that the question of thermodynamic stability of very small clusters has already been addressed in Chap. 2.7.2.2, and that liquid-drop models provide another means of calculating $n_t(z)$ (cf. next Sect.).

2.7.3.2   Metastable Clusters

In this section we shall address the following questions and topics:

- Approaches towards calculating critical sizes
- Dependence of fission barriers on cluster size
- How close are appearance sizes n_z to the stability limit $n_c(z)$?
- Which mechanisms ensure reproducibility of appearance sizes under widely different experimental conditions?
- Fragmentation channels of multiply charged clusters under various scenarios:
  Fragmentation immediately after ionization
  Unimolecular dissociation after energy relaxation
  Fragmentation of long-lived ions after renewed excitation

As far as theoretical studies are concerned, we shall focus on continuum approaches which can, hopefully, be generalized (specific ab-initio studies of small systems have been discussed in Sect. 2.7.2.2). We shall not attempt to discuss metallic and van der Waals systems separately, partly because these systems behave quite similarly under certain aspects, and partly because of the paucity of studies.

2.7.3.2a)   Onset of Instability in Highly Charged Droplets

The stability of highly charged droplets is of relevance to thunderheads, liquid-metal ion sources, and nuclear physics. Accordingly, a large number of theoretical reports have addressed this phenomenon ever since the seminal paper by *Rayleigh* was published in 1882 [3]. After the discovery of nuclear fission, this phenomenon received renewed interest ([4], and references therein). The discovery of cluster fission has, once again, stimulated interest in this field [140–143]. All these studies share the same basic approach. Matter as well as the (net) charge of aggregates are treated as a continuum. In the absence of quantal shell effects, the equilibrium shape of a weakly charged droplet is a sphere of radius r, because the surface energy has its minimum value in this case. The stability of the sphere with respect to small deformations is analyzed by expanding the surface in spherical harmonics (specified by an integer $\mu \geq 2$). A particular mode is stable if (in SI units)

$$r^3 > \frac{(z \cdot e)^2 \cdot f}{4\pi\varepsilon_0 \cdot 2\pi\sigma \cdot (\mu + 2)} \tag{9}$$

where σ is the surface tension, and the numerical factor f is 1/2 if the charge is distributed on the surface[9] (a metallic droplet, treated by *Rayleigh* [3]), or 3/5 if the charge is homogeneously distributed throughout the volume (a nucleus, treated by *Bohr* and *Wheeler* [4]). The droplet is stable (i.e., at least metastable) if and only if it is stable with respect to all possible deformations, i.e. for all μ. In other words, the critical radius is given by

$$r_c^3 = \frac{(z \cdot e)^2 \cdot f}{4\pi\varepsilon_0 \cdot 8\pi\sigma} \tag{10}$$

This can be easily rewritten in terms of $n_c(z)$,[5] and is seen to agree with our previous equations 2a and 2b. At this critical size, the droplet becomes unstable with respect to its fundamental mode ($\mu = 2$, quadrupolar deformation), while still being stable with respect to higher modes.

If one analyzes the stability of the droplet with respect to large deformations, the model also predicts the size $n_t(z)$ where the cluster becomes thermodynamically stable [4, 141, 142]. It is convenient to introduce the so-called fissility parameter

$$x = \frac{z^2/n}{z'^2/n_c(z')} \tag{11}$$

where the denominator involves the charge and the critical size of a z'-fold charged cluster of a given material. P_n^{z+} is thermodynamically stable if

$$x < 0.351 \tag{12}$$

For $z = z'$ we simply obtain

$$n_t(z) = n_c(z)/0.351 \tag{13}$$

Hence, using the bulk surface tension and the specific volume, we can easily estimate the limiting size where a cluster is on the verge of thermodynamic stability.[9] For the alkali metals these values agree with more sophisticated approaches (cf. 2.7.3.1b) within a factor of 2. Keeping in mind that surface tensions are temperature dependent,[10] and recalling the general problem of applying surface tensions to small aggregates, this agreement is more than fair.

---

[9] In a non-spherical metallic droplet, the surface charge will be non-uniform. This invalidates Eqs. 12–14 [144]

[10] In our analysis we used, for the sake of consistency, values close to the boiling point of the corresponding materials.

The equations quoted above are appropriate for metal clusters (using $f = 1/2$), because the charge will be delocalized. Recent studies have aimed at incorporating corrections due to electronic shell effects, under the simplifying assumption of pure quadrupolar deformation [141, 143]. Concerning van der Waals clusters, the assumption of a continuous charge distribution is not warranted, especially if z is small. *Soler* and coworkers have modified *Rayleigh*'s liquid drop model [145]. They calculate the total energy and, hence, the fission barrier, of doubly charged xenon clusters with two localized charges as a function of cluster shape. The critical size is but one outcome of their study, we shall discuss others further below.

Another model has been introduced by *Soler* and coworkers which does not follow *Rayleigh*'s approach [72]. This work seeks to determine fission barriers of multiply charged van der Waals clusters by comparing the energy of the spherical parent cluster with the total energy (self-energy plus repulsive cluster–cluster interaction) of a pair of spherical fragments just touching each other. At first sight this may appear crude. However, under certain conditions the critical shape (i.e. the shape of the cluster when it reaches the fission barrier) will indeed closely resemble that particular configuration, i.e. there will be a narrow neck of matter connecting the two emerging fragments [4, 5]. Critical sizes calculated within this model agree surprisingly well with appearance sizes, even though the model treats the cluster–cluster interaction in a very simplified manner, and it assumes a continuous charge distribution [72]. A similar approach towards estimating fission barriers has been applied to Na_n^{2+} [139]. The energy of the two fragments at infinite separation is calculated within the spherical jellium model (excluding shell corrections), whereas the energy of the two spherical fragments at small separation is calculated by a deformed self-consistent extended Thomas–Fermi model. Barriers have been reported for two competing decay channels, loss of Na and of Na^+, but not for any other fission channel.

2.7.3.2b)   Fission Barriers, Appearance Sizes, and their Relation to Critical Sizes

Experimentally determined appearance sizes are upper limits to critical sizes — but can we say more about the correlation between these two quantities? An answer might be sought by simply comparing (calculated) critical sizes with (experimental) appearance sizes. Fig. 13 presents $n_c(2)$, calculated from Eq. (10), and n_2 for a variety of materials. In spite of our earlier caveats about the general validity of Eq. 10, we have applied it (using $f = 1/2$) to metallic as well as to van der Waals systems.[11] We have consistently used values of the surface tension at

---

[11] Rayleigh himself calculated the critical size of water clusters by applying Eq. (10) (using $f = 1/2$), without taking into account the dielectric constant of water.

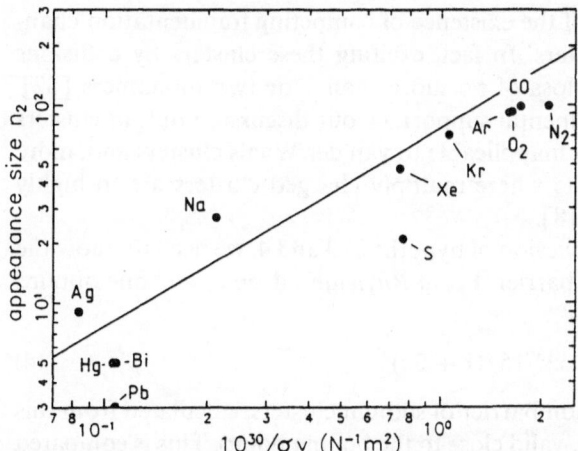

Fig. 13. Comparison of critical size $n_c(2)$, calculated from Rayleigh's liquid drop model (*solid line*, Eq. 10), and experimentally determined appearance sizes n_2 (*full dots*). σ = surface tension, v = molecular volume. For references to experimental values see captions of Fig. 4 and of Table 1

or very close to the boiling point of the material in question. Appearance sizes are from Fig. 4 and Table 1 (taking the minimum of the reported values in each case). The ratio $n_2/n_c(2)$ is close to 0.7 for van der Waals systems, while it fluctuates roughly between 0.5 and 2 for more tightly bound systems. The ratio should, of course, never be less than one, hence Fig. 13 indicates that Eq. 10 is not sufficiently accurate.[12] We need to find the true ratio by some other means.

The appearance size is probably controlled by either one of the following mechanisms:

1) n_2 depends on the size dependence of the fission barrier E_{bn}^{2+} and on the total excess energy in the cluster immediately after ionization
2) n_2 corresponds to the smallest neutral clusters generated in the source [45, 106]
3) At n_2, E_{bn}^{2+} equals approximately the dissociation energy D_n^{z+} [72, 94, 147, 148].
4) At n_2, E_{bn}^{2+} equals approximately the potential energy (or the "directed energy") in the compression mode immediately after ionization [118, 145].

Hypothesis 1 and 2 postulate, in contrast to hypothesis 3 and 4, that the appearance size strongly depends on experimental conditions. Hypothesis 1 can be abandoned, even though it is in line with general findings pertaining to prompt and delayed fragmentation of singly charged clusters. Those studies are

---

[12] This discrepancy would be less serious if we would have used surface tension measured well below the boiling point. Often, values for σ are taken from the melting point. This increases σ, and, hence, the ratio n_2/n_c, but the procedure invalidates a comparison of different materials, because the most likely effective cluster temperature scales as T_{boil}, not as T_{melt} [146].

simply not relevant, because of the existence of competing fragmentation channels in multiply charged clusters. In fact, exciting these clusters by collisions leads to fission, preceded by loss of no more than 1 or two monomers [82]. Hypothesis 2 has some experimental support, cf. our discussion of lead clusters in Sect. 2.7.2.3b. However, it is inapplicable to van der Waals clusters and, more generally, to those experiments where multiply charged clusters are in highly excited (vibrational) states [118].

Before proceeding to a discussion of hypothesis 3 and 4, we need to know the size dependence of the fission barrier. From *Rayleigh*'s drop model one obtains [142][9]

$$E_b^{z+}(r) = 98 \cdot 4\pi r^2 \cdot \sigma \cdot (1 - x)^3/15 \cdot (1 + 2x)^2 \tag{14}$$

Figure 14 displays the fission barrier of sodium clusters, calculated from this relation (using $\sigma = 0.120$ N/m, valid close to the boiling point). This is compared with experimental values of the dissociation energy of singly charged sodium clusters [149].[13] The calculated fission barrier rises slowly close to $n_c(z) \approx 16$, it approaches the experimental dissociation energy of Na_n^+ (which is assumed to equal that of Na_n^{2+}, apart from shell effects) at $n \approx 32$. The latter value is close to the appearance size $n_2 = 27$ [94]. The agreement may be, to some extent, fortuitous,[14] but the general rationale behind hypothesis 3 is as follows: Under typical conditions (and certainly for non-seeded supersonic beams), neutral clusters are "boiling hot", and the ionization process adds to the excess energy. Hence all cluster ions undergo extensive fragmentation along the kinetically preferred channel before mass analysis can be accomplished [150]. The fission barrier E_{bn}^{z+} is much more strongly size dependent than the dissociation energy D_n^{z+}. Hence these two quantities will be numerically equal somewhere above $n_c(z)$, and a cluster which has boilt down to this point and is still sufficiently excited, will undergo fission. Anyhow, cluster ions of a given charge cannot shrink below this point by evaporation. The total excess energy in the newly formed ions is immaterial, provided the neutral clusters in the beam cover the relevant size range.

Figure 14 suggests $n_2/n_c(2) \approx 2$ for sodium, while a comparison with recent ab-initio calculations[14] suggests an even larger ratio. On the other hand, the above-mentioned analysis of van der Waals clusters by *Soler* and coworkers suggests a ratio significantly closer to 1, mainly because the calculated slope of the fission barrier is non-zero at the critical size [72, 145].

Theoretical studies of doubly charged van der Waals systems with proper incorporation of localized charges support hypothesis 4, essentially because in

---

[13] This kind of presentation was proposed by Saunders [148]. His plot, however, was based on an equation different from our Eq. 14, and it applied the surface tension of sodium at its melting temperature, which is 70% higher than that at T_{boil}.

[14] Two recent theoretical studies conclude that $n_c(2)$ is a factor two smaller than our value [138, 139]. Also, they indicate that the fission barrier rises much more steeply beyond n_c than predicted by the liquid-drop model.

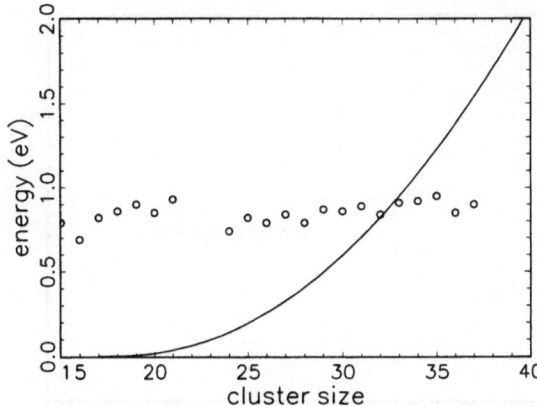

Fig. 14. Height of fission barrier of doubly charged sodium clusters, calculated from Eq. 14 (*solid line*), and experimentally determined [149] dissociation energies (*open circles*)

this case the equilibrium shape is non-spherical [118, 145]. Introducing, say, two charges (holes) into a neutral, relaxed cluster, followed by rapid charge hopping, will result in a cluster ion which features a well-defined potential energy with respect to its non-spherical equilibrium shape.[15] A molecular-dynamics simulation of Xe_n^{2+} [118] illustrates this nicely (Fig. 15). The energy in the compression mode drives Xe_{51}^{2+}, but not Xe_{55}^{2+}, over its fission barrier within half a vibrational period of its collective mode.

Will the 55-mer survive forever? The collective oscillation is strongly damped, hence the 55-mer can undergo fission at some later time only if it concentrates enough of its randomized energy back into a mode suitable for fission. According to *Gay* and *Berne* [118] this will eventually happen, because the random energy is huge compared to the fission barrier. However, one should be cautious about this point: If the fission barrier is larger than the dissociation energy (and the studies [118, 145] indicate that this holds true at the size under consideration), evaporation of monomers may occur at a much higher rate, ultimately producing doubly charged clusters smaller than those which did not survive the first vibrational period of the collective mode. Hence, while mechanism (4) may be decisive immediately after ionization, mechanism (3) would control the appearance size on the time scale of mass spectrometric detection.

Experimentally, the situation is not clear for van der Waals clusters: If mechanism (3) is relevant, delayed unimolecular fission of clusters at or just above the appearance size should be observable. This phenomenon has been reported (and will be discussed later) for triply and quadruply charged CO_2 and C_2H_4 clusters [72, 147] and for triply charged ammonia clusters [151]. However, as shown in Fig. 2, their doubly charged counterparts emit neutral monomers, rather than undergoing fission [23, 72, 151]. In other words, these evaporating clusters have not yet reached the size where fission would be

---

[15] There will also be excess energy due to ion solvation, dimer-ion formation, etc., which will depend on the details of the primary ionization event, but this energy will be distributed at random over many modes.

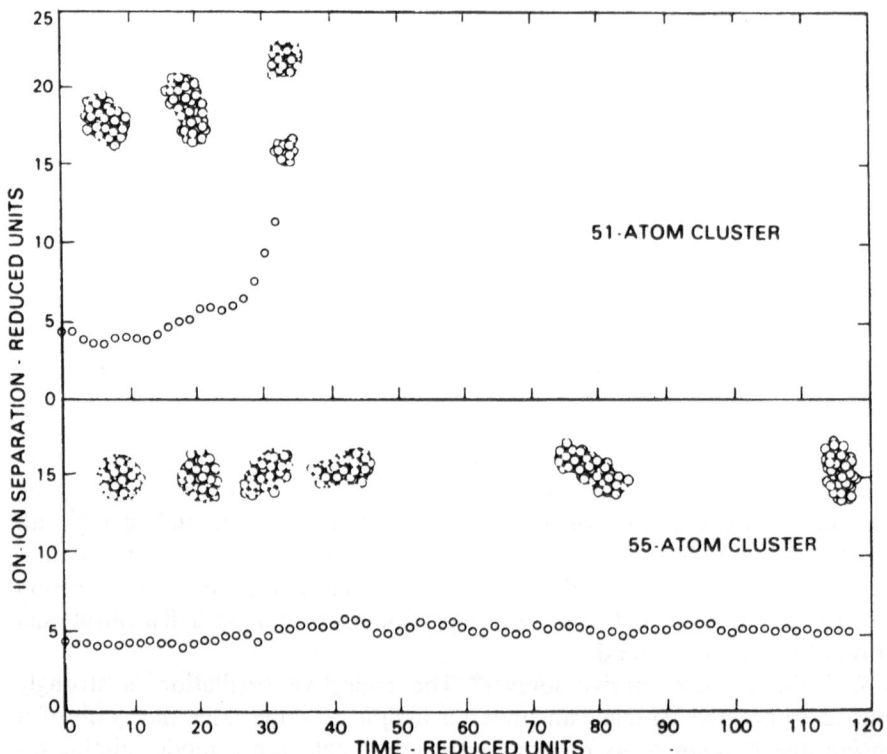

Fig. 15. Ion-ion separation as a function of time for Xe_{51}^{2+} and Xe_{55}^{2+}, after introducing two charges at opposite sides into a relaxed neutral cluster at $t = 0$ (classical molecular dynamics simulation, 1 reduced time unit = 2.76 ps). Xe_{51}^{2+} fractures rapidly, because the initial energy in the compression coordinate exceeds the fission barrier [118]

kinetically favored over evaporation.[16] According to a preliminary report by *Stace*, there is evidence for unimolecular fission of doubly charged argon clusters into equally sized fragments, but these observations require further confirmation [152, 153].

2.7.3.2c)   Size Distribution of Fission Products

Is the size distribution of fission fragments symmetric or asymmetric? Does fission produce fragments of a specific size? Our experimental knowledge is scarce, and universal trends have not yet been established. In a simple view, the size distribution may be predicted if the critical shape of the cluster (its shape in the transition state) is known, i.e. if one knows the reaction channel that

---

[16] Experiments probing the transition from evaporation to fission will be discussed in Sect. 2.7.3.2c.

minimizes the fission barrier. Some studies have adopted this approach (see below) but, unfortunately, it may be inadequate. For example, a charged liquid drop slightly below the critical size is unstable with respect to just one particular mode, the quadrupolar deformation (Eq. 9). This wrongly suggests that the droplet will fission symmetrically, into two fragments of equal size. However, once the metallic droplet is deformed, the excess charge is no longer homogeneously distributed over the surface, and higher modes may also become unstable [154]. More generally, once a metastable cluster has successfully passed over the fission barrier by virtue of its internal vibrational excess energy, the further evolution of its shape until the instant of scission will depend on its potential energy with respect to all possible deformations. Even complete knowledge of the multidimensional energy surface, with its possible existence of secondary minima, barriers, and bifurcation points, may not be sufficient: Because of its inertia, the cluster will not necessarily slide downhill along the path of steepest descent. These details are essential for an understanding of size distributions of fission fragments from radioactive nuclei [5, 6, 155]; they are also likely to be important in case of simple metal clusters where electronic shell effects cannot be ignored.

The potential energy of multiply charged sodium clusters has been calculated (including electronic shell corrections) by *Sugano* and coworkers [141]. Unfortunately, their analysis is restricted to deformations that are symmetric under inversion. These are unlikely to be essential for the process of fissioning.

The most adequate theoretical analysis of atomic clusters undergoing unimolecular fission is by molecular dynamics. In the liquid-drop language, this approach does not impose any restrictions on the geometry of the deformed cluster, it incorporates the effects of random (vibrational) energy and of "directed" energy (inertia, viscous effects, etc.). An early classical molecular dynamics simulation of Xe_n^{2+} by *Gay* and *Berne* [118] has already been mentioned in the previous section. Xe_{51}^{2+} was found to fracture rapidly into Xe_{36}^+ and Xe_{15}^+, i.e. mildly asymmetric. The significance of this result is uncertain, only one run was reported. *Landman* and coworkers have performed a local-spin-density molecular dynamics study of Na_n^{2+} ($n = 8$, 10, 12) [138]. These systems are metastable, the fission barrier is highest (0.7 eV) for the 10-mer, which has a closed electronic shell. This cluster, if thermally excited, undergoes fission into $Na_3^+ + Na_7^+$. This particular channel also maximizes the exothermicity of the reaction, because the singly charged trimer has a closed electronic shell [135, 138]. An interesting result of the molecular dynamics simulation is that the fission barrier features a double hump,[17] which is related to structural isomerization of the fragments prior to the final act of scission.

Experimental evidence for preferential fission of doubly charged clusters of monovalent metals into "magic" fragments exists, indeed. *Katakuse* and coworkers have analyzed the unimolecular decay of Ag_n^{2+} ($n = 12–22$), gener-

---

[17] Double-humped barriers are known to be of prime importance for an understanding of low-energy fission of actinide nuclei [6].

ated by ion bombardment of silver under vacuum [156]. For technical reasons, only "heavy" fragments Ag_m^+, $m > n/2$, could be detected (lighter fragments are buried under the background of decaying singly charged cluster ions). Up to $n = 16$, and for $n = 18$, the fragment of size $m = n - 3$ is dominant; for some parent ions, Ag_9^+ is another preferred fragment. In other words, fission into the closed-shell clusters Ag_3^+ and Ag_9^+ is strongly preferred.[18] For heavier clusters Ag_n^{2+} ($n = 17$, 19–22), the size of the most intense heavy fragments is but slightly larger than $n/2$. In some cases, either Ag_9^+ or Ag_{n-9}^+ is a dominant fragment.

Unimolecular decay of doubly charged sodium clusters has been analyzed in the vicinity of the appearance size, $n_2 = 27$ ([94], also see Vol. 1. Chap. 4.1). For $27 \leq n \leq 31$, two reaction channels, evaporation of neutrals and fission, co-exist. For Na_{26}^{2+} and Na_{24}^{2+}, which can be identified by detection of their fragments fission is dominant. This trend is, of course, to be expected, because the fission barrier decreases with decreasing size. The size of the heavy charged fragments is 1, 3, 5, 7, or, rarely, 9 units less than that of their parents. Na_{n-3}^+ is not particularly abundant, but a detailed analysis is hampered by the low mass resolution. A search for the light fission fragments remained without success. In addition to the background problem mentioned previously, this failure is very likely due to the kinetic energy that is imparted to the recoiling fission fragments, dependent on their relative masses. The energy release is expected to be of the order of 1 eV, this will significantly increase the divergence (in the laboratory system) of the beam of the light ions [72, 94, 147].

Finally, evidence for preferential fission into magic products comes from an investigation of collision-induced dissociation of doubly charged gold clusters. In this case, heavy as well as light fragments could be observed [56]. The size distribution displayed in Fig. 16 was obtained by colliding Au_{15}^{2+} at a kinetic energy of 150 eV (top) and 300 eV (bottom) with krypton gas, corresponding to a maximum energy transfer of 4.5 and 9 eV in the center of mass. The intense peak at $n/z = 7.5$ is the unfragmented parent, the intense peak to the left is Au_{14}^{2+}, arising from evaporative decay, with a possible, but less likely, contribution from the fission fragment Au_7^+. Two fission channels appear to co-exist, one of them generating $Au_3^+ + Au_{12}^+$ and $Au_3^+ + Au_{11}^+ + Au$, the other generating $Au_5^+ + Au_9^+ + Au$. In either case, one of the fragments is magic. Here we have assumed that a neutral atom is emitted during or after fission of Au_{15}^{2+} but, in fact, one cannot rule out that this parent ion emits an atom, hence destabilizes, and fissions thereafter. The observation of Au_{14}^{2+} and, in fact, Au_{13}^{2+}, points to the importance of this channel.

It is difficult to assess the effect of elevated collision energies from Fig. 16. However, it has been demonstrated that the branching ratio between the two

---

[18] Note that the experiment analyzes the size of fragments typically 10^{-5} s after the reaction, whereas molecular dynamics reveals the size of fragments within, say, 10^{-11} s. If the emitted, and eventually detected, heavy fragment undergoes delayed evaporation, its size will be underestimated by the experiment, hence the size of the light fragment (at the instant of scission) will be overestimated.

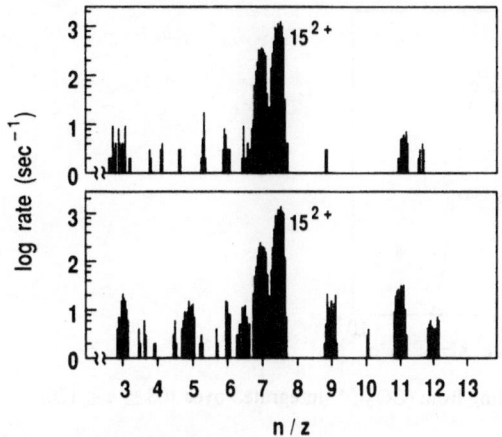

log rate (sec^{-1})

15^{2+}

15^{2+}

3 4 5 6 7 8 9 10 11 12 13

n / z

Fig. 16. Product ions from collision-induced dissociation of Au_{15}^{2+} at 150 eV (*top*) and 300 eV (*bottom*) [56]

competing fission channels of Au_5^{2+} into either a (magic) trimer plus a dimer, or a tetramer plus a monomer, changes from 1 at low collision energy to essentially zero if the center-of-mass energy approaches 10 eV [34].

Fission is the dominant decay channel of triply charged CO_2, NH_3, and C_2H_4 clusters, and of $(C_2H_4)_n^{4+}$, in the vicinity of the respective appearance sizes [23, 72, 147, 151]. For example, the appearance size of $(CO_2)_n^{3+}$ is $n_3 \approx 108$. The effective rate for (spontaneous) fission is maximum at this size, it decreases by at least 2 orders of magnitude upon increasing n by ≈ 15 units (individual sizes could not be resolved in this study) [72]. The probability for the reaction

$$(CO_2)_n^{3+} \rightarrow (CO_2)_m^{2+} + (CO_2)_x^+ + y \cdot (CO_2)$$ (15)

where $m + x + y = n$, is displayed in Fig. 17. The signal has been integrated over the range $108 \leq n \leq 120$, and the abscissa shows the relative fragment size, m/n. Due to previously mentioned interference problems, only "heavy" fragments of size $m/n > 2/3$ could be analyzed. The size distribution is strikingly narrow and asymmetric. It peaks at 92%, i.e. the (undetectable) singly charged fragment ion carries no more than $\approx 8\%$ of the parent mass. Experiments on triply charged ammonia clusters, with definite mass identification of fragment ions, lead to similar results [151], while the size distribution of doubly charged fragments from $(C_2H_4)_n^{3+}$ is considerably less asymmetric [72].

The previously mentioned tangent-sphere model was used to calculate the fission barrier versus fragment size for van der Waals clusters [72]. The critical size of $(CO_2)_n^{3+}$ is calculated to be $n_c(3) = 110$. The dependence of the fission barrier on fragment size is very pronounced. The barrier is minimum (zero in this case) for fission into a doubly charged cluster carrying 95% of the parent mass.

Another important piece of information lies in the kinetic energy released upon fission. *Stace* and coworkers have analyzed the size distribution and kinetic energy distribution of collision-induced fragment ions originating from $(CO_2)_n^{2+}$ [82] and $(C_6H_6)_n^{3+}$ [157] slightly above their appearance sizes

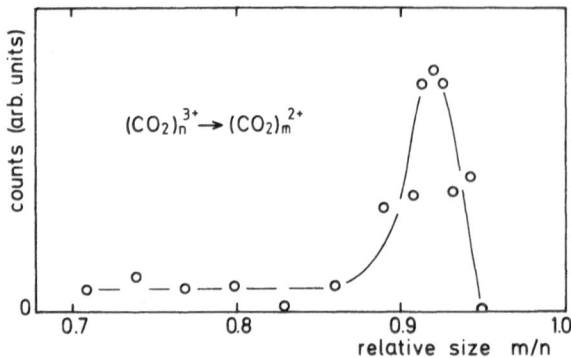

Fig. 17. Size distribution of $(CO_2)_m^{2+}$ originating from $(CO_2)_n^{3+}$ (integrated over $108 \leq n \leq 120$) by unimolecular decay (after [72, 147])

($n_2 = 43$ and $n_3 = 52$, respectively). In these studies, the kinetic energy of the parent ions, colliding with a stationary gas (air), is 6–8 keV per charge state, equivalent to a maximum energy transfer of more than 100 eV in the center-of-mass system. The size distribution of fragments $(CO_2)_m^+$ is very broad, and none of them is larger than $n/2$. This implies that several neutral monomers are emitted during or shortly after fission. Likewise, doubly charged fragments from $(C_6H_6)_n^{3+}$ typically carry 50 to 60% of the parent mass. Because of their high excitation energy, many of them continue to undergo fission into singly charged fragments, generating a very broad size distribution. The difference between these results and the previously mentioned spontaneous fission of $(CO_2)_n^{3+}$ is probably not suprising if we recall the dramatic effect of a ten-fold smaller collision-energy on the branching ratio of fissioning Au_3^{2+} [34].

The total kinetic energy released upon fission of $(CO_2)_n^{2+}$ and $(C_6H_6)_n^{3+}$ is found to be 0.35 eV and ≈ 0.6 eV, respectively [82, 157]. These values are significantly lower than those obtained from a crude estimate involving two point charges at a distance equivalent to the diameter of the precursor cluster [72]. A possible explanation is that the repulsive Coulomb energy is transferred into vibrational energy [82]. In the language of the liquid-drop model, applied to fissioning nuclei, this is a well-known effect: Not all of the potential energy difference between the droplet in its critical shape, and the infinitely separated fragments, is converted into kinetic energy because of viscous effects [5]. In the extreme case, the repulsive energy being released is dissipated until scission occurs. For atomic clusters, however, the point of scission is ill-defined, because the range of repulsive and attractive forces is not greatly disparate. It is also interesting to note that *Bowers* and coworkers have determined the kinetic energy released upon fission of Nb_2^{2+} and of Nb_3^{2+} [25]. Their values of 3.1 and 2.5 eV correspond to a charge separation distance in the parent ions of 0.43 and 0.57 nm, respectively. These values are reasonable if one assumes that they relate to the potential energy of the precursor ions at the instant of their passage over the fission barrier.

2.7.4 Outlook

A decade after the discovery of well-defined lower size limits in mass spectra of multiply charged clusters [1], some of their features are still poorly understood. We have discussed possible correlations between the (experimental) appearance size and the (uniquely defined, but experimentally unknown) critical size. Proofs for the correctness of the proposed models are eagerly awaited. The dynamics of metastable multiply charged clusters has been probed in some experiments, but the internal excitation energies of these species can only be estimated. Hence, we have no experimental data concerning the height of fission barriers versus cluster size. Future experiments should, perhaps, involve cold (or, at least, thermalized) cluster ions, subsequently excited by the absorption of defined energy quanta (photons). Recent experiments on singly charged clusters demonstrate that reliable dissociation energies can be obtained from a state-of-the art analysis of their decay dynamics [158], similar experiments are also feasible for multiply charged clusters.

In our discussion we have frequently resorted to the simple liquid-drop model. This is certainly inadequate for doubly charged clusters containing just a few atoms, but even tightly bound systems (e.g., refractory metals) will feature large critical sizes if the charge state of the cluster is very high.

The kinetic energy released upon fission provides important information [25, 82, 157], but the energetics can be unravelled only if all fragments, and their kinetic energies, are detected in coincidence. Analysis of just one fragment is sufficient only if no more than two fragments are produced per event. It would be intriguing to measure the time-dependent probability for emission of neutral monomers during and after induced fission.

Interpretation of these data in terms of other cluster properties (fission barriers, cluster geometry at the transition state, etc.) would greatly benefit from ab-initio calculations and molecular dynamics simulations [62, 118, 131, 138] of the system under investigation. As far as very small clusters are concerned, detailed predictions of these studies should be verified by experiments involving near-threshold ionization of size selected, singly charged cluster ions and analysis of possible fragment ions, including their kinetic energies. Finally, an attempt to search for fusion of singly charged clusters in colliding beams at low, controlled kinetic energies might be worthwile.

2.7.5 Recent Developments

The stability of doubly charged post-transition metal dimers and of very small (para-$C_6F_2H_4)_n^{2+}$ has been calculated [159]. Fission channels of doubly charged monovalent metal clusters have been measured [160] and, with shell corrections, calculated [161]. Dissociation experiments on Au_2^{2+} have been discussed in the context of the liquid-drop model [162]. Fission possibly competes with electron detachment from $(Na_n)^{2-}$ [163]. Appearance sizes of

$(Na_n)^{z+}$ $(z \leq 7)$ have been determined through evaporative shrinking [164]. Fragment sizes have been measured for small, unstable, doubly charged van der Waals clusters [165]. An island of stability exists for doubly charged alkali halide clusters at $n \approx 13$ [166]. $(C_{60})^{z+}$ is kinetically stable up to, at least, $z = 6$; ionization energies and fission reactions have been determined [167]. Several other contributions may be found in the Proceedings of two recent meetings [168].

References

1. K. Sattler, J. Mühlbach, O. Echt, P. Pfau, E. Recknagel: Phys. Rev. Lett. **47**, 160 (1981)
2. E. Dörnenburg, H. Hintenberger: Z. Naturf. **14a**, 765 (1959)
3. Lord Rayleigh: Phil. Mag. **14**, 185 (1882)
4. N. Bohr, J.A. Wheeler: Phys. Rev. **56**, 426 (1939)
5. U. Brosa, S. Grossmann, A. Müller: Phys. Rep. **197**, 167 (1990)
6. S. Bjørnholm, J.E. Lynn: Rev. Mod. Phys. **52**, 725 (1980)
7. O. Echt: In *Electronic and Atomic Collisions*, H.B. Gilbody, W.R. Newell, F.H. Read, A.C.H. Smith, eds., Elsevier 1988, p. 719
8. O. Echt: In *Physics and Chemistry of Small Clusters*, Vol. 158 of NATO ASI Series B, ed. by Jena, P., Rao, B.K., Khanna, S.N., Plenum, New York, 1987, p. 623
9. T.D. Märk: Int. J. Mass Spectrom. Ion Proc. 79, 1 (1987)
10. T.D. Märk: In *Book of Invited Papers*, 6th Symp. Elem. Proc. Reak. Nizk. Plazma, Jelsava-Hradok, CSSR, 1986, 142
11. H. Henkes, G. Isenberg: Int. J. Mass Spectrom. Ion Proc. **5**, 249 (1970); J. Gspann, K. Körting: J. Chem. Phys. **59**, 4726 (1973)
12. M. Dole, L.L. Mack, R.L. Hines, R.C. Mobley, L.D. Ferguson, M.B. Alice: J. Chem. Phys. **49**, 2240 (1968)
13. S. Maruyama, M.Y. Lee, R.E. Haufler, Y. Chai, R.E. Smalley: Z. Phys. D **19**, 409 (1991)
14. B.U.R. Sundqvist: Nucl. Instr. Meth. B **48**, 517 (1990)
15. C.K. Meng, M. Mann, J.B. Fenn: Z. Phys. D **10**, 361 (1988)
16. J.B. Fenn, M. Mann, C.K. Meng, S.F. Wong, C.M. Whitehouse: Science **246**, 64 (1989)
17. J.J. Wu, R.J. Miller: J. Appl. Phys. **67**, 1051 (1990)
18. U. Albrecht, P. Leiderer: Europhys. Lett. **3**, 705 (1987)
19. K.H. Chang, R.D. McKeown, R.G. Milner, J. Labrenz: Phys. Rev. A **35**, 3949 (1987)
20. K. Leiter, W. Ritter, A. Stamatovic, T.D. Märk: Int. J. Mass Spectrom. Ion Proc. **68**, 341 (1986)
21. S.N. Schauer, P. Williams, R.N. Compton: Phys. Rev. Lett. **65**, 625 (1990); P.A. Limbach, L. Schweikhard, K.A. Cohen, M.T. McDermott, A.G. Marshall, J.V. Coe; J. Am. Chem. Soc. **113**, 6795 (1991)
22. P. Scheier, T.D. Märk: J. Chem. Phys. **86**, 3056 (1987)
23. K. Leiter, D. Kreisle, O. Echt, T.D. Märk: J. Phys. Chem. **91**, 2583 (1987)
24. W.A. Saunders: Phys. Rev. Lett. **62**, 1037 (1989)
25. P.P. Radi, G. von Helden, M.T. Hsu, P.R. Kemper, M.T. Bowers: Chem. Phys. Lett. **179**, 531 (1991)
26. P. Pfau: Ph.D. Thesis, University of Konstanz, 1984 (unpublished), also see: K. Sattler, Surface Science **156**, 292 (1985)
27. M. Delaunay, M. Fehringer, R. Geller, D. Hitz, P. Varga, H. Winter: Phys. Rev. B **35**, 4232 (1987); F. Aumayr, T.D. Märk, H. Winter: Int. J. Mass Spectrom. Ion Proc. **129**, 17 (1993)
28. L. Pauling: J. Chem. Phys. **1**, 56 (1933)
29. C.A. Nicolaides: Chem. Phys. Lett. **161**, 547 (1989)
30. M. Guilhaus, A.G. Brenton, J.H. Beynon, M. Rabrenovic, P. von Rague Schleyer: J. Phys. B **17**, L605 (1984)
31. P. Valtazanos, C.A. Nicolaides: Chem. Phys. Lett. **172**, 254 (1990)

32. K.P. Huber, G. Herzberg: Molecular Spectra and Molecular Structure Vol. 4, Van Nostrand, 1979
33. T.T. Tsong: J. Chem. Phys. **85**, 639 (1986)
34. W.A. Saunders, S. Fedrigo: Chem. Phys. Lett: **156**, 14 (1989)
35. W.A. Saunders: Phys. Rev. B **40**, 1400 (1989)
36. A. Galindo-Uribarri, H.W. Lee, K.H. Chang: J. Chem. Phys. **83**, 3685 (1985)
37. I. Cornides, L. Morvay: Mass Spectroscopy **31**, 81 (1983)
38. Y. Saito, T. Ishida, T. Noda: J. Am. Soc. Mass Spectrom **2**, 76 (1991)
39. T. Ishitani, K. Umemura, Y. Kawanami: Surf. Sci. **218**, 259 (1989)
40. K. Gamo, Y. Ukegawa, Y. Inomoto, Y. Ochiai, S. Namba: J. Vac. Sci. Technol., **19**, 1182 (1981)
41. Y. Huang, B.S. Freiser: J. Am. Chem. Soc. **110**, 4434 (1988)
42. F. Machalett, R. Mühle, I. Stiebritz: J. Phys. D **20**, 1417 (1987)
43. T. Ishitani, K. Umemura, Y. Kawanami: J. Appl. Phys. **61**, 748 (1987)
44. G.P. Schwartz, V.E. Bondybey, J.H. English, G.J. Gualtieri: Appl. Phys. Lett. **42**, 952 (1983)
45. W. Schulze, B. Winter, I. Goldenfeld: Phys. Rev. B **38**, 12937 (1988)
46. J. Van de Walle, P. Joyes: Phys. Rev. B **32**, 8381 (1985)
47. Y. Saito, T. Noda: Z. Phys. D **12**, 225 (1989)
48. M. Tomita, T. Kuroda: Surf. Sci. **201**, 385 (1988)
49. O. Nishikawa, E. Nomura, H. Kawada, K. Oida: J. de Phys. **47** Colloque C2, 297 (1986)
50. W. Drachsel, Th. Jentsch, K.A. Gingerich, J.H. Block: Surf. Science **156**, 173 (1985)
51. J. Van de Walle, P. Joyes: Z. Phys. D **12**, 221 (1989)
52. S. Papadopoulos: J. Phys. D **20**, 530 (1987)
53. D.R. Kingham, L.W. Swanson: Vacuum **34**, 941 (1984)
54. G.L. Kellogg: Surf. **120**, 319 (1982); Phys. Rev. B **24**, 1848 (1981)
55. S.L. Reindl, G.M. Pastor, K.H. Bennemann: Z. Phys. D **20**, 133 (1991); Phys. Rev. Lett. **67**, 1250 (1991)
56. W.A. Saunders: Phys. Rev. Lett. **64**, 3046 (1990)
57. Y. Li, S.N. Khanna, P. Jena: Phys. Rev. Lett. **64**, 1188 (1990)
58. S. Mukherjee, G.M. Pastor, K.H. Bennemann: Z. Phys. D **20**, 131 (1991); Phys. Rev. B **42**, 5327 (1990)
59. L. Morvay, I. Cornides: Int. J. Mass Spectrom. Ion Proc. **62**, 263 (1984)
60. F. Liu, M.R. Press, S.N. Khanna, P. Jena: Phys. Rev. Lett. **59**, 2562 (1987)
61. C. Bauschlicher, Jr., M. Rosi: Chem. Phys. Lett. **165**, 501 (1990)
62. S.N. Khanna, F. Reuse, J. Buttet: Phys. Rev. Lett. **61**, 535 (1988)
63. D. Strömberg, U. Wahlgren: Chem. Phys. Lett. **169**, 109 (1990)
64. R.P. Neisler, K.S. Pitzer: J. Phys. Chem. **91**, 1084 (1987)
65. G. Durand, F. Spiegelmann, A. Bernier: J. Phys. B **20**, 1161 (1987)
66. G. Durand, J.-P. Daudey, J.-P. Malrieu: J. de Phys. **47**, 1335 (1986); J.-P. Malrieu: In *Physics and Chemistry of Small Clusters, Vol. 158 of NATO ASI Series B*, ed. by P. Jena, B.K. Rao, S.N. Khanna, Plenum, New York, 1987, p. 383
67. C. Bréchignac, M. Broyer, Ph. Cahuzac, G. Delacretaz, P. Labastie, L. Wöste: Chem. Phys. Lett. **133**, 45 (1987); ibid, **118**, 174 (1985)
68. H. Helm, K. Stephan, T.D. Märk, D.L. Huestis: J. Chem. Phys. **74**, 3844 (1981)
69. K. Stephan, T.D. Märk, H. Helm: Phys. Rev. A **26**, 2981 (1982)
70. N.G. Gotts, A.J. Stace: Int. J. Mass Spectrom. Ion Proc. **102**, 151 (1990)
71. P.M.W. Gill, L. Radom: Chem. Phys. Lett. **136**, 294 (1987)
72. O. Echt, D. Kreisle, E. Recknagel, J.J. Saenz, R. Casero, J.M. Soler: Phys. Rev. A **38**, 3236 (1988)
73. M. Lezius, P. Scheier, A. Stamatovic, T.D. Märk: J. Chem. Phys. **91**, 3240 (1989)
74. P. Scheier, G. Walder, A. Stamatovic, T.D. Märk: J. Chem. Phys. **90**, 4091 (1989)
75. P. Scheier, G. Walder, A. Stamatovic, T.D. Märk: J. Chem. Phys. **90**, 1288 (1989)
76. T. Rauth: Diploma Thesis, Universität Innsbruck, 1992 (unpublished)
77. W.R. Pfeifer, J.F. Garvey: J. Chem. Phys. **91**, 1940 (1989)
78. S. Martrenchard, C. Jouvet, C. Lardeux-Dedonder, D. Solgadi: J. Chem. Phys. **94**, 3274 (1991)
79. *Sondergase*, Linde GmbH, 1983; *Handbook of Chemistry and Physics*, 68th ed., CRC Press, 1987-88

80. F. C. Andrews: *Thermodynamics: Principles and Applications*, Wiley & Sons, 3rd ed., 1983
81. O. Echt, K. Sattler, E. Recknagel: Phys. Lett. **90A**, 185 (1982)
82. N.G. Gotts, A.J. Stace: Phys. Rev. Lett. **66**, 21 (1991)
83. A.J. Stace, D.M. Bernard, J.J. Crooks, K.L. Reid: Mol. Phys. **60**, 671 (1987)
84. M.Y. Hahn, K.E. Schriver, R.L. Whetten: J. Chem. Phys. **88**, 4242 (1988)
85. P. Scheier, A. Stamatovic, T.D. Märk: J. Chem. Phys. **88**, 4289 (1988)
86. P. Scheier, A. Stamatovic, T.D. Märk: Chem. Phys. Lett. **144**, 119 (1988)
87. O. Kandler, T. Leisner, O. Echt, E. Recknagel: Z. Phys. **D 10**, 295 (1988)
88. A.J. Stace: Phys. Rev. Lett. **61**, 306 (1988)
89. M.T. Coolbaugh, W.R. Pfeifer, J.F. Garvey: Chem. Phys. Lett. **156**, 19 (1989)
90. D. Tomanek, S. Mukherjee, K.H. Bennemann: Phys. Rev. **B 28**, 665 (1983)
91. B. Delley: J. Phys. **C 17**, L551 (1984)
92. T.P. Martin: J. Chem. Phys. **81**, 4426 (1984)
93. I. Rabin, C. Jackschath, W. Schulze: Z. Phys. **D 19**, 153 (1991)
94. C. Bréchignac, Ph. Cahuzac, F. Carlier, M. de Frutos: Phys. Rev. Lett. **64**, 2893 (1990)
95. C. Bréchignac, Ph. Cahuzac, F. Carlier, J. Leygnier: Phys. Rev. Lett. **63**, 1368 (1989)
96. W. Schulze, B. Winter, J. Urban, I. Goldenfeld: Z. Phys. **D 4**, 379 (1987)
97. O. Kandler, K. Athanassenas, O. Echt, D. Kreisle, T. Leisner, E. Recknagel: Z. Phys. **D 19**, 151 (1991)
98. P. Fayet, D. Kreisle, L. Wöste: unpublished results
99. A. Hoareau, P. Melinon, B. Cabaud: J. Phys. **D 18**, 1731 (1985)
100. F. Reuse, S.N. Khanna, V. de Coulon, J. Buttet: Phys. Rev. **B 41**, 11 743 (1990); G. Durand: J. Chem. Phys. **91**, 6225 (1989)
101. H. Weidele: Diploma Thesis, University of Konstanz, 1991 (unpublished)
102. D. Rayane, P. Melinon, B. Cabaud, A. Hoareau, B. Tribollet, M. Broyer: Phys. Rev. **A 39**, 6056 (1989)
103. M.F. Jarrold, J.E. Bower, J.S. Kraus: J. Chem. Phys. **86**, 3876 (1987)
104. D. Rayane, P. Melinon, B. Cabaud, A. Hoareau, B. Tribollet, M. Broyer: J. Chem. Phys. **90**, 3295 (1989)
105. P. Pfau, K. Sattler, R. Pflaum, E. Recknagel: Phys. Lett. **104A**, 262 (1984)
106. A. Hoareau, P. Melinon, B. Cabaud, D. Rayane, B. Tribollet, M. Broyer: Chem. Phys. Lett. **143**, 602 (1988)
107. K. Athanassenas, T. Leisner, O. Kandler, O. Echt, D. Kreisle, E. Recknagel: In *Proc. Symp. on Atomic and Surface Physics*, Obertraun, Austria, 1990, ed. by T.D. Märk, F. Howorka, p. 283
108. D. Kreisle: unpublished results
109. Y. Saito, K. Minami, T. Ishida, T. Noda: Z. Phys. **D 11**, 87 (1989)
110. S. Mukherjee, D. Tomanek, K.H. Bennemann: Chem. Phys. Lett. **119**, 241 (1985)
111. M. Knapp, O. Echt, D. Kreisle, E. Recknagel: J. Phys. Chem. **91**, 2601 (1987)
112. L.A. Posey, M.A. Johnson: J. Chem. Phys. **89**, 4807 (1988)
113. H. Haberland, C. Ludewigt, H.G. Schindler, D.R. Worsnop: Surf. Sci. **156**, 157 (1985)
114. T.P. Martin, T. Bergmann: J. Chem. Phys. **90**, 6664 (1989)
115. M. Lezius, T.D. Märk: Chem. Phys. Lett. **155**, 496 (1989)
116. I. Rabin, W. Schulze, B. Winter: Phys. Rev. **B 40**, 10282 (1989)
117. T.D. Märk: In: *Electron-Molecule Interactions and their Application*, Vol. 1, ed. by Christophorou, L.G., Academic Press, 1984, p. 251
118. J.G. Gay, B.J. Berne: Phys. Rev. Lett. **49**, 194 (1982)
119. P. Scheier, T.D. Märk: Chem. Phys. Lett. **136**, 423 (1987)
120. P. Scheier, G. Walder, A. Stamatovic, T.D. Märk: Chem. Phys. Lett. **150**, 222 (1988)
121. C. Gazier, J.R. Prescott: Phys. Lett. **32A**, 425 (1970)
122. H.W. Biester, M.J. Besnard, G. Dujardin, L. Hellner, E.E. Koch: Phys. Rev. Lett. **59**, 1277 (1987)
123. I. Rabin: Ph.D. Thesis, Freie Universität Berlin, 1990 (unpublished)
124. I. Rabin, W. Schulze: Chem. Phys. Lett. **201**, 265 (1993)
125. W.R. Pfeifer, M.T. Coolbaugh, J.F. Garvey: J. Phys. Chem. **93**, 4700 (1989)
126. A.J. Stace: Chem. Phys. Lett. **174**, 103 (1990)

127. H. Haberland, H. Langosch: Z. Phys. **D 2**, 243 (1986)
128. H. Shinohara, N. Nishi, N. Washida: J. Chem. Phys. **84**, 5561 (1986)
129. T.D. Märk: Chem. Phys. Lett. **163**, 461 (1989)
130. M.T. Coolbaugh, W.R. Pfeifer, J.F. Garvey: Chem. Phys. Lett. **164**, 441 (1989)
131. R.N. Barnett, U. Landman, A. Nitzan, G. Rajagopal: J. Chem. Phys. **94**, 608 (1991)
132. C. Baladrón, J.M. López, M.P. Iñiguez, J.A. Alonso: Z. Phys. **D 11**, 323 (1989)
133. I.T. Iakubov, A.G. Khrapak, L.I. Podlubny, V.V. Pogosov: Sol. State Comm. **53**, 427 (1985)
134. L.C. Balbás, A. Rubio, J.A. Alonso, G. Borstel: Chem. Phys. **120**, 239 (1988)
135. M.P. Iñiguez, L.C. Bálbas, J.A. Alonso: Physica **147B**, 243 (1987); M.P. Iñiguez, J.A. Alonso, M.A. Aller, L.C. Bálbas: Phys. Rev. **B 34**, 2152 (1986)
136. B.K. Rao, P. Jena, M. Manninen, R.M. Nieminen: Phys. Rev. Lett. **58**, 1188 (1987)
137. M. Nakamura: Z. Phys. **D 19**, 149 (1991)
138. R.N. Barnett, U. Landman, G. Rajagopal: Phys. Rev. Lett., **67**, 3058 (1991); ibid, **69**, 1472 (1992); P. Jena, S.N. Khanna, C. Yannouleas: Phys. Rev. Lett. **69**, 1471 (1992)
139. F. Garcias, J.A. Alonso, J.M. López, M. Barranco: Phys. Rev. **B 43**, 9459 (1991)
140. Y. Ishii: Sol. State Comm. **61**, 227 (1987)
141. M. Nakamura, Y. Ishii, A. Tamura, S. Sugano: Phys. Rev. **A 42**, 2267 (1990)
142. E. Lipparini, A. Vitturi: Z. Phys. **D 17**, 57 (1990)
143. M. Nakamura, Y. Ishii, A. Tamura, S. Sugano: Z. Phys. **D 19**, 145 (1991)
144. R. Schmidt: private communication
145. R. Casero, J.J. Saenz, J.M. Soler: Phys. Rev. **A 37**, 1401 (1988)
146. C.E. Klots: J. Phys. Chem. **92**, 5864 (1988)
147. D. Kreisle, O. Echt, M. Knapp, E. Recknagel, K. Leiter, T.D. Märk, J.J. Saenz, J.M. Soler: Phys. Rev. Lett. **56**, 1551 (1986)
148. W.A. Saunders: Phys. Rev. Lett. **66**, 840 (1991)
149. C. Brechignac, Ph. Cahuzac, J. Leygnier, J. Weiner: J. Chem. Phys. **90**, 1492 (1989)
150. J.M. Soler, J.J. Saenz, N. Garcia, O. Echt: Chem. Phys. Lett. **109**, 71 (1984)
151. D. Kreisle, K. Leiter, O. Echt, T.D. Märk: Z. Phys. **D 3**, 319 (1986)
152. A.J. Stace: J. Chem. Soc. Faraday Trans. **86**, 2546 (1990)
153. A.J. Stace: private communication
154. A.I. Grigor'ev: Sov. Phys. Tech. Phys. **30**, 736 (1985)
155. A. Sandulescu: J. Phys. **G 15**, 529 (1989)
156. I. Katakuse, H. Ito, T. Ichihara: Int. J. Mass Spectrom. Ion Proc. **97**, 47 (1990); Z. Phys. **D 20**, 101 (1991)
157. N.G. Gotts, A.J. Stace: J. Chem. Phys. **95**, 6175 (1991)
158. U. Ray, M.F. Jarrold, J.E. Bower, J.S. Kraus: J. Chem. Phys. **91**, 2912 (1989)
159. G. Ortiz, P. Ballone: Phys. Rev. **B 44**, 5881 (1991); S. Martrenchard-Barra, C. Jouvet, C. Lardeux-Dedonder, D. Solgadi: Chem. Phys. Lett. **215**, 291 (1993)
160. C. Bréchignac, Ph. Cahuzac, F. Carlier, J. Leygnier, A. Sarfati: Phys. Rev. **B 44**, 11 386 (1991)
161. H. Koizumi, S. Sugano, Y. Ishii: Z. Phys. **D 28**, 223 (1993); J.M. López, J.A. Alonso, F.F. Garcias, M. Barranco: Ann. Physik **1**, 270 (1992); Comm. At. Mol. Phys.: in print (1994)
162. W.A. Saunders: Phys. Rev. **A 46**, 7028 (1992)
163. C. Yannouleas, U. Landman: Phys. Rev. **B 48**, 8376 (1993); Chem. Phys. Lett. **210**, 437 (1993)
164. U. Näher, H. Göhlich, T. Lange, T.P. Martin: Phys. Rev. Lett. **68**, 3416 (1992); Chem. Phys. Lett. **196**, 113 (1992)
165. E. Rühl: Ber. Bunsenges. Phys. Chem. **96**, 1172 (1992), and reference therein
166. X. Li, R.L. Whetten, Chem. Phys. Lett. **196**, 535 (1992)
167. S. Petrie, J. Wang, D.K. Bohme, Chem. Phys. Lett. **204**, 473 (1993); P. Scheier, R. Robl, B. Schiestl, T.D. Märk, Chem. Phys. Lett., in print (1994); P. Scheier, B. Dunser, T.D. Märk, to be published (1994)
168. *Clustering Phenomena in Atoms and Nuclei*, ed. by M. Brenner, T. Lönnroth, F.B. Malik, Springer Series in Nuclear and Particle Physics, 1992; *Nuclear Physics Concepts in the Study of Atomic Cluster Physics*, ed. by R. Schmidt, H.O. Lutz, R. Dreizler, Lecture Notes in Physics vol. 404, Springer, 1992

2.8 Chemistry with Neutral Metal Clusters

S.J. Riley

2.8.1 Introduction – Clusters and Heterogeneous Catalysis

This chapter will cover the chemical reactions of isolated metal clusters with small molecules. We will consider *adsorption* reactions, in which the molecules (or perhaps their decomposition fragments) remain bound to the cluster surface, as opposed to being incorporated into the cluster to form an entirely new chemical species. Thus this cluster chemistry is essentially surface chemistry, and we have the enormous field of single crystal surface science studies to compare and contrast to. Borrowing from that field, we will sometimes use its terminology, calling, for example, the reactions *chemisorptions* and the bound molecules *adsorbates*. In spite of this commonality, we will see that clusters can behave in ways quite different from bulk metal surfaces, reflecting the fact that the surface of a small cluster can present to the adsorbing molecule an arrangement of metal atoms quite unlike any found on a single crystal surface. Furthermore, this arrangement can vary considerably with cluster size, resulting in a strong dependence of chemical properties on cluster size.

We will focus here on clusters of the transition metals. (See Vol. I, Chap. 4.3 for a discussion of the physical properties of transition metals.) These metals, located in the middle of the periodic table, are defined as elements with partially filled d shells. Since the d orbitals have considerable spatial extent, the d electrons are often involved in transition metal bonding. As a result, the chemical properties of transition metals are very sensitive to the number and arrangement of the d electrons, and can vary substantially across the transition series. This "chemical diversity" of the transition metals leads to some fascinatingly different behavior, both for single metal atoms or ions and for the surfaces of bulk metals. When this variety is combined with the special property of clusters – the possibility of dramatic cluster size dependences – we can expect to find some very intriguing chemical behavior, as indeed we do.

These studies are motivated by the desire to understand the fundamental properties of heterogeneous catalysts – materials, made from transition metals, that are extensively used in industry for the acceleration of important chemical reactions such as petroleum refining and plastics production. In a typical industrial reactor, reactant gases are passed through a bed containing finely divided metal particles, and the desired reaction products are separated from the gas stream. The actual chemical transformations, the succession of bond breakings and bond formations, occur at the surface of the metal particles. In some cases, the activity of the catalyst preparation is dependent only on total metal surface area, so that the smaller the particles the greater the activity per unit weight. Such reactions are referred to as *structure insensitive*. In other cases, the dependence of activity on particle size is more complex. In these *structure sensitive* reactions, it may be that only a few particular particle sizes are contributing to the catalytic activity of the catalyst preparation.

Despite a long history of heterogeneous catalysis, the actual processes that occur on a catalyst are only poorly understood. The metal surface usually lowers the activation barriers for chemical bond breakage, and the binding of intermediate chemical species to the surface helps to stabilize them so that they can go on to form products. But in most cases the microscopic details of the reaction sequence and the nature of the chemically active surface binding sites are not known. The important parameters, such as site geometry and binding energies and the chemical identities of intermediate species, are only slowly being discovered. Much of our understanding of metal surface chemistry comes from studies of single crystals in high vacuum, an environment that is quite different from that found in catalytic reactors.

Studies of clusters of transition metal atoms offer an obvious hope for gaining insight into metal-catalyzed chemistry. A cluster is a (very) small metal particle, and is likely to possess the special reactive sites that give catalyst particles their chemical activity. By studying, in an inert gas stream, association reactions of metal clusters with small molecules, we can more closely approximate the conditions of actual heterogeneous catalysis. If we can understand the special nature of metal particle surfaces that gives them their catalytic activity, then we might in the future hope to design more efficient catalysts.

2.8.2   Experimental

Given the highly refractory nature of the transition metals, the most convenient technique for cluster generation is via laser vaporization (see Vol. I, Chap. 3.1). Chemical reactions are studied in a flow-tube reactor (FTR), which is usually just an extension of the cluster growth region. At a point downstream of the metal target where cluster growth has terminated, reagent gases are introduced into the flow tube, and reactions with the clusters occur. An additional requirement for careful, quantitative chemistry studies is that the flows of inert carrier and reagent gases be continuous, so that experimental conditions such as reaction temperature and pressure can be varied and measured reproducibly. Carrying out the reactions in a constant pressure of inert gas (typically 3 kPa of helium) allows association reactions to be studied, since the heat of reaction can be readily dissipated by collisions with the carrier gas. It also assures that the clusters are thermalized, i.e., reach the ambient temperature of the surroundings, before they leave the flow tube. Another important feature is the provision of more than one point along the flow tube where reagent gas can be added, so that the interaction time between clusters and reagent gas can be varied.

Following reaction, the clusters and their reaction products exit the flow tube via a nozzle and are formed into a molecular beam, which transports them to a detector. The most common detection scheme is laser ionization and time-of-flight (TOF) mass analysis (see Vol. I, Chap. 3.5). The principal experimental procedure is to monitor changes in the mass spectra, hence changes in the distribution of reaction products, as FTR conditions such as temperature, pressure and time are varied. As will be shown here, a careful analysis of the observed changes can provide quite detailed information about such properties as reactivity, adsorbate binding capacity and energy, and cluster structure.

2.8.3   Adsorbate Uptake – The Path to Coverage

A convenient way to map out the interaction between a metal cluster and a particular adsorbate molecule is to systematically measure the coverage of the cluster, i.e., the number of molecules bound to the cluster, as a function of the partial pressure of the reagent gas in the FTR [1]. In general, as the pressure is increased, the coverage of the cluster increases. Figure 1 shows such data for the reaction of Fe_{61} with ammonia, NH_3, to produce $Fe_{61}(NH_3)_m$ [2]. What is

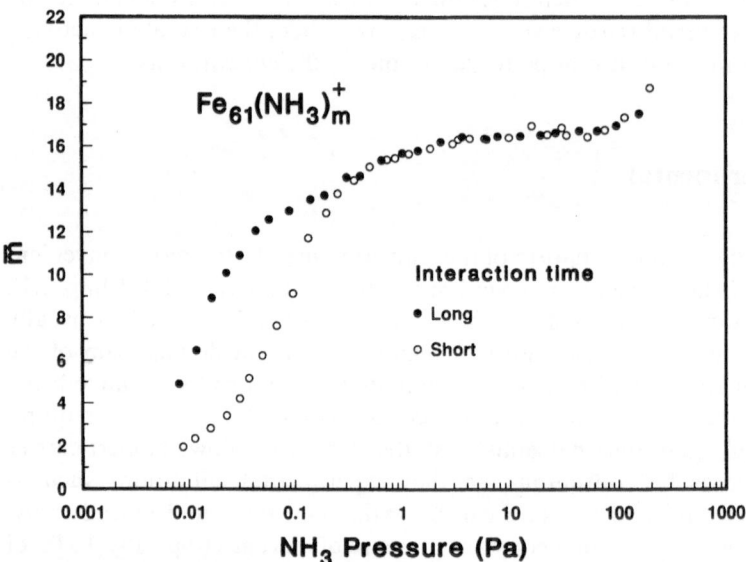

Fig. 1. The average number $\bar{m}$ of NH_3 molecules bound to Fe_{61} vs NH_3 partial pressure in the flow-tube reactor. *Filled symbols* are for a long interaction time, *open symbols* for a short interaction time. Adapted from [2]

plotted here is $\bar{m}$, the average number of NH_3 molecules bound to the cluster. $\bar{m}$ is calculated for each NH_3 pressure from the relative heights in the mass spectrum of the product peaks. To fully map out the uptake pattern it is necessary to vary the NH_3 pressure by over four orders of magnitude, so the pressure scale is logarithmic. Data are shown for conditions in which the NH_3 interacts for a fairly long time (~ 1 ms) and for a short time (~ 0.2 ms). The uptake data can conveniently be divided into four principal regions.

Below about 0.2 Pa, the uptake pattern is characterized by a time dependence: for a given NH_3 partial pressure, a longer interaction time leads to higher coverage. In general, a reaction whose products depend on time is a *kinetically controlled* reaction. Above 0.2 Pa, the extent of coverage loses its time dependence. A reaction independent of time is an *equilibrium* reaction. This conversion from kinetically controlled to equilibrium conditions is caused by a very important aspect of NH_3 binding to iron clusters (or, for that matter, to any iron surface): the binding energy decreases with increasing coverage. For low coverage, the binding energy is sufficiently high that the average (statistical) lifetime for desorption of a thermalized NH_3 molecule is longer than the time the clusters spend in the FTR. Thus if an NH_3 molecule reacts with a cluster, it stays on the cluster, and the number of molecules that end up bound to the cluster depends on the number of cluster-NH_3 collisions, which for a given NH_3 pressure depends on interaction time. Eventually, for increasing coverage, the binding energy becomes so low that the lifetime for NH_3 desorption is less than the time spent in the FTR, so that NH_3 molecules are continually adsorbing and desorbing from the cluster. This, of course, is just a microscopic picture of a reaction at equilibrium: the rate of the forward reaction (adsorption) equals the rate of the reverse (desorption). The extent of coverage is no longer time dependent, but is governed by the standard free energy of adsorption through the equilibrium constant (see Sect. 5 below).

In the kinetically controlled region, the NH_3 uptake pattern is fairly steep. Under equilibrium conditions, the uptake is considerably slower, in many cases approaching a nearly linear dependence on log NH_3 pressure. This region of the uptake plot is quite analogous to an equilibrium adsorption isotherm, a concept often used in surface science to quantify adsorbate binding energies [3]. A linear dependence of $\bar{m}$ on log pressure implies a linear dependence of the free energy of adsorption on coverage.

At still higher pressure the uptake plot often shows a leveling off – a region where $\bar{m}$ is independent of pressure. Such a *plateau* generally indicates that a certain number of binding sites (16 in the case of Fe_{61}) have been filled, and that the *next* site to be occupied has substantially lower binding energy. We say that in this pressure region the cluster has been *saturated* with chemisorbed molecules. By identifying the number of molecules that saturate a given cluster we learn how many binding sites the cluster has, which can provide important information about cluster structure, as we will see in Sect. 7. The final upturn in the $\bar{m}$ plot above 100 Pa represents physisorption – the beginning of the formation of a second layer of ammonia molecules on the cluster. At these very high ammonia pressures there is substantial cooling in the expansion out of the nozzle at the end of the FTR. This, coupled with increased cluster–ammonia collisions in the expansion, leads to physisorption.

The uptake plot, then, provides us with a road map of a given cluster's interaction with a given adsorbate molecule. The three distinct regions, corresponding to kinetics, equilibrium, and saturation, will now be treated in more detail in the next three sections of this chapter.

2.8.4   Kinetics – Strong Cluster Size Dependence and the Approach to Bulk

From the earliest days of the study of transition metal cluster chemistry it was recognized that reactivity, i.e., the rate at which clusters react with simple molecules, can be a strong function of cluster size. Perhaps the prototypical example of this is provided by the reaction of iron clusters with hydrogen (H_2). The simple association reaction

$$Fe_n + H_2 \rightarrow Fe_n H_2$$

should have the second order rate equation

$$\frac{d[Fe_n]}{dt} = -k[Fe_n][H_2] \ ,$$

where $[x]$ is the concentration of species x and k is the rate constant. If we assume that $[H_2]$ is constant, then we get a pseudo-first-order rate equation that integrates to

$$\ln[Fe_n] - \ln[Fe_n]_0 = -k \Delta t [H_2] \ ,$$

where Δt is the interaction time. The advantage of continuous gas flows in the FTR is that the concentration of the gaseous reagent can easily be calculated from relative flows and the total pressure. It is also possible to accurately determine Δt. $[Fe_n]$ is assumed proportional to the iron intensity in the mass spectrum (see Vol. I, Chap. 3.5 for a discussion of this). Thus absolute rate constants can be determined from the slopes of plots of $\ln[Fe_n]$ vs. $[H_2]$.

Such absolute pseudo-first-order rate constants for this reaction are shown in Fig. 2 [4, 5]. As can be seen, the rate constant varies by orders of magnitude with cluster size, increasing, for example, by a factor of 130 between Fe_8 and Fe_9, and a factor of 60 between Fe_{18} and Fe_{19}. Similar variations have been seen for reactions of hydrogen with cobalt, vanadium and niobium clusters, and smaller variations for nickel clusters [6]. Reactions of niobium and cobalt clusters with nitrogen show rate constant variations similar to those seen for hydrogen [6]. In other cases, there is little dependence of reactivity on cluster size. For example, an extensive study of the reaction of carbon monoxide with clusters of 14 transition metals showed that reactivity varied by no more that a factor of two or three [7].

Figure 2 also shows the ionization potentials (IPs) of bare iron clusters [8]. It has been noted [4, 9] that there appears to be a rough correlation between IPs and reactivity: a decrease in IP from one cluster size to the next is often accompanied by an increase in reactivity. There is an appealing interpretation of this correlation. As has been shown by isotope exchange studies [5], hydrogen dissociatively chemisorbs on iron clusters, i.e., binds as H atoms. Any barrier to the reaction would most likely involve breaking the H–H bond. The lower a cluster's IP, the energetically more favorably it can transfer an electron to the H_2. Since this electron would occupy an antibonding orbital in the H_2, this should weaken the H–H bond, facilitating reaction. A more careful analysis of the data, however, shows that the correlation with IP is not perfect, and that the origin of size-dependent reactivity must be more complex.

Also shown in Fig. 2 is the cluster size dependence of the equilibrium constant (see Sect. 5 for more on equilibrium constants) for the reaction of iron clusters with heavy water D_2O [10]. It shows a remarkable correlation with the reactivity data. There is no *a priori* reason for an equilibrium constant, which is essentially a thermodynamic property, to correlate with a rate constant, which is a kinetic parameter. There must be some other underlying property of the cluster that is governing its chemistry, as well as such physical properties as IP. The best candidate is cluster *structure*, i.e., the way the atoms pack together. This

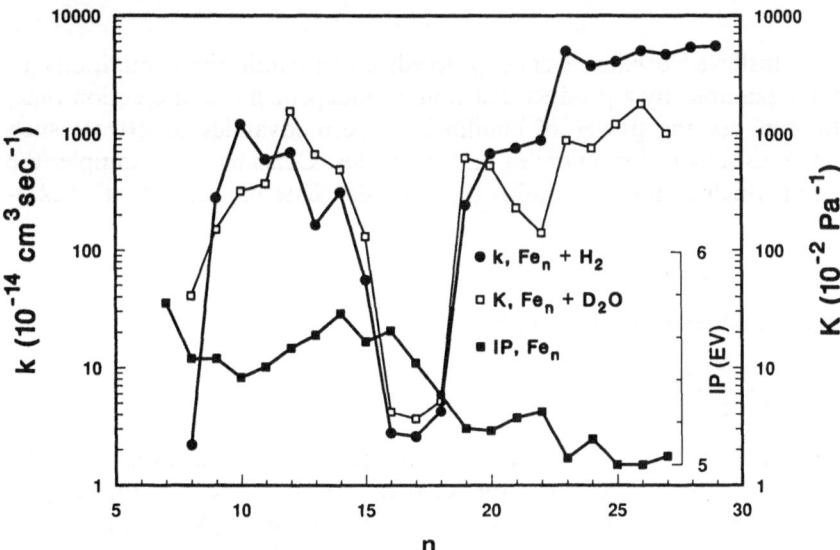

Fig. 2. The dependence on iron cluster size of reactivities with H_2 (*k, left scale, filled circles*), equilibrium constants for binding D_2O (*K, right scale, open squares*), and ionization potentials (*IP, inset scale at right, filled squares*). k values are from [5], K values from [10], IP values from [8]

and other data [11] strongly suggest that in the fairly narrow size range of 8–25 atoms, iron clusters change structure numerous times. These observations are essentially an indication that this reaction is a cluster analog of a *structure sensitive* one. While we may not know with certainty what the cluster structures are in this size region, we see that the microscopic arrangement of metal atoms on the cluster surface can have a dramatic influence on cluster reactivity. We will return to considerations of cluster structure in Sect. 7.

A question often asked in cluster research is "When (i.e., at what size) does a cluster begin to behave like bulk metal?" The reactivity data in Fig. 2 show a final jump between Fe_{22} and Fe_{23}, then no further big changes for larger clusters. (Rate constants have been measured up to Fe_{62} [4].) If the rate constant for Fe_{23} is compared to the hard sphere collision rate, we find that approximately 5% of the collisions of Fe_{23} (and larger clusters) with hydrogen lead to reaction. 5% is a typical figure for the sticking probability of hydrogen on bulk iron surfaces [12]. Thus we can say that, as far as reaction with hydrogen is concerned, iron clusters behave like bulk metal abruptly at 23 atoms. Of course, the answer to the question above will depend strongly on the property being asked about. Transition metal cluster ionization potentials, for example, approach their bulk values very slowly with size (see Vol. I, Chap. 4.3).

2.8.5 Equilibrium – The Thermodynamics of Adsorbate Binding

If a given cluster-adsorbate reaction proceeds under equilibrium conditions (as shown, for example, by a product distribution independent of interaction time) then we can use the power of equilibrium thermodyamics to extract such parameters as adsorption energies and entropies. Consider, for example, the reaction in which a cluster M_n adds the mth adsorbate molecule A at equilibrium:

$$M_n A_{m-1} + A \rightleftharpoons M_n A_m \ .$$

The equilibrium constant is given by

$$K_{eq} = \frac{[M_n A_m]}{[M_n A_{m-1}][A]} \ .$$

K_{eq} is related to the standard free energy of the reaction ΔG^0 by $\Delta G^0 = -RT \ln(K_{eq})$, where R is the ideal gas constant and T is the temperature. Finally, the standard enthalpy change ΔH^0 and entropy change ΔS^0 are determined from $\Delta G^0 = \Delta H^0 - T \Delta S^0$. If we assume that over a fairly narrow temperature range ΔH^0 and ΔS^0 are independent of temperature, then a plot of $R \ln(K_{eq})$ vs. $1/T$ (a van't Hoff plot) will have a slope equal to $-\Delta H^0$ and an intercept equal to ΔS^0. To calculate K_{eq}, we assume, as before, that the relative cluster densities are proportional to the relative ion signals in the mass spectrum.

Figure 3 shows van't Hoff data for the reactions in which Co_{71} and Co_{55} add their 12th NH_3 molecule. The linearity of the data supports the assumptions outlined above. The resulting ΔH^0 and ΔS^0 values, determined from linear least squares fits to the data, are shown in the figure. $-\Delta H^0$ is essentially the cluster-adsorbate binding energy, and we see that it is somewhat higher for Co_{71} than for Co_{55}. Similar results for other clusters, including nickel clusters, show that, as claimed above for iron clusters, the binding energy for NH_3 decreases as cluster coverage increases. Changes in binding energy (or in equilibrium constant) with cluster size can be used to monitor structural changes, as illustrated in Sect. 4 for the iron cluster-D_2O reaction.

Perhaps the most surprising result from these studies is the magnitude of the entropy change. Typically, for the condensation of a gas phase molecule onto a solid the entropy change is dominated by the loss of three translational degrees of freedom, and can be estimated by statistical mechanical calculation to be around -140 J/mol K. We consistently see smaller entropy changes for the adsorption of NH_3 on iron, cobalt, and nickel clusters. Also, there is usually

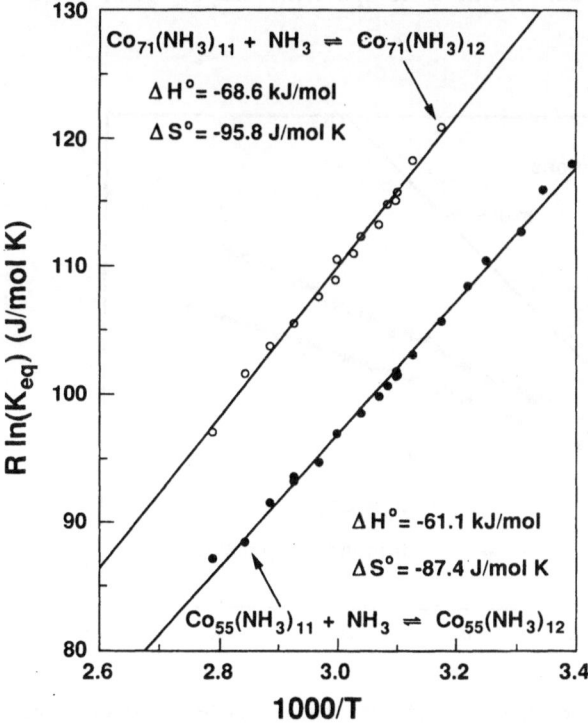

Fig. 3. van't Hoff plots for the reactions in which Co_{55} (*open circles*) and Co_{71} (*filled circles*) add a twelfth NH_3 molecule. The straight lines are linear least squares fits to the data, with the resulting ΔH^0 and ΔS^0 values indicated

a good correlation between ΔS^0 and ΔH^0 values. It would appear that the chemisorbed NH_3 remains fairly loosely coupled to the cluster surface and may, in fact, be practically mobile, so that it retains much of the "translational" freedom it had in the gas phase. The correlation with ΔH^0 is consistent with this, since the lower the binding energy the more loosely coupled to the cluster the adsorbed molecule would be.

2.8.6 Saturated Compositions –
The Number and Nature of Adsorption Sites

As was illustrated in Fig. 1, for sufficiently high reagent pressure in the FTR a cluster can become saturated with adsorbate molecules. From the maximum number of reagent molecules that bind to the cluster, something can be learned about the number and nature of the binding sites. Figure 4 shows the compositions of nickel clusters saturated with deuterium, and of iron clusters saturated with hydrogen or deuterium [13]. Since the hydrogen/deuterium binds to these clusters on the surface, it is convenient to compare the number of adsorbate

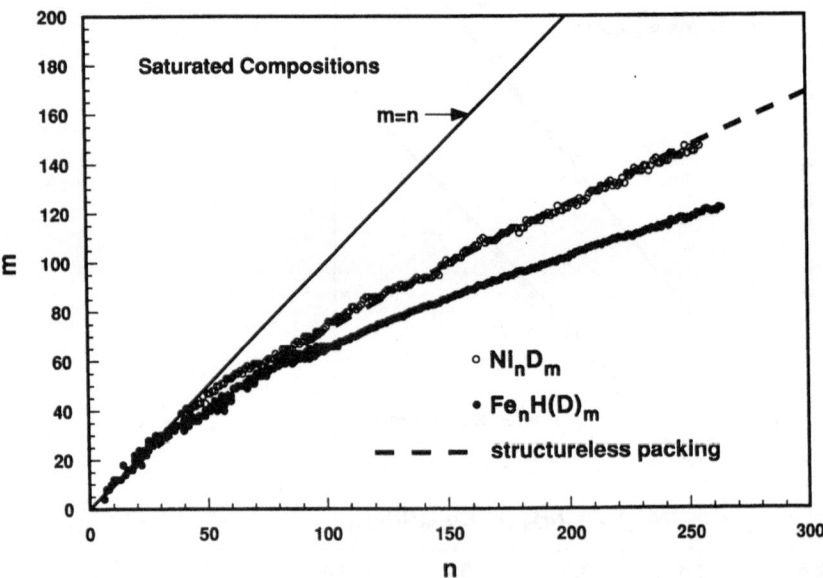

Fig. 4. The compositions of nickel clusters saturated with deuterium (*open circles*) and of iron clusters saturated with hydrogen or deuterium (*filled circles*). Iron cluster hydrides are used for $n = 6$ to 131, deuterides for larger clusters. The solid line shows the stoichiometric behavior, the dashed line the results of the structureless packing model (see text) that estimates the number of metal atoms on a cluster's surface. Adapted from [13]

atoms with some measure of the number of surface metal atoms. The measure shown, as a dashed line, is the structureless packing model [14], which says that the ratio of the number of surface atoms to the total number n of metal atoms is given by

$$1 - [1 - (4\pi/3n)^{1/3}]^3 .$$

This model essentially assumes that the thickness of the surface layer of metal atoms is equal to the cube root of the atomic volume v, that n equals the volume of the cluster divided by v, and that the cluster is spherical.

As seen in Fig. 4, nickel clusters in the 40- to 260-atom size range consistently bind more hydrogen (deuterium) atoms than iron clusters. In fact, when compared to the structureless packing model, nickel clusters are "covered" with hydrogens at saturation, in that there are as many hydrogen atoms as surface nickel atoms. Iron clusters are less than fully covered, and show a decreasing fractional coverage with increasing size, going from 1.0 at Fe_{40} to 0.8 at Fe_{250}. This difference between iron and nickel is also seen in a fundamental catalytic process, the reduction of CO by hydrogen (the Fischer–Tropsch synthesis). The use of nickel catalysts typically results in the production of methane (CH_4), while iron catalysts often produce higher hydrocarbons, including long-chain (many carbon atoms) waxes [15]. An obvious rationale for this follows from the cluster results. On the fully hydrogenated nickel catalyst the CO molecules will be surrounded by hydrogens, and only C–H bond formation is possible, resulting in CH_4. On the less than fully covered iron surface, however, there is room for two CO molecules to bind to adjacent sites, allowing the formation of C–C bonds and the production of higher hydrocarbons. This is an example of how the study of fundamental chemical properties of isolated transition metal clusters may enhance our understanding of actual heterogeneous catalytic processes.

Saturated composition studies of smaller neutral and ionized clusters have been reported for deuterium on nickel, platinum, and rhodium [16]. Unlike the larger clusters, the smallest clusters show abnormally large deuterium uptake, with as many as eight deuterium atoms per metal atom. Such results raise questions about conventional methods used to determine the dispersion (the effective metal surface area, usually reported in m^2/g) of a catalyst preparation. The catalyst sample is usually titrated with hydrogen, and the total metal surface area is calculated from hydrogen uptake assuming one hydrogen atom per surface metal atom. The cluster studies show that this assumption is not correct for very small particles, and this observation may explain some of the anomalies reported in the catalysis literature when conventional dispersion analysis is used.

Knowledge of the number of adsorption sites on a cluster and, more importantly, the variation in this number with cluster size, can provide us with important clues as to cluster structure. We will now consider this possibility in more detail.

2.8.7   Chemical Probes of Metal Cluster Structure

Undoubtedly one of the most important properties of a cluster is its structure
– the geometrical arrangement or packing of its constituent atoms. Not only is
structure vital to meaningful theoretical treatments of cluster physics and
chemistry, but if we are to understand the relationships between cluster structure and chemical reactivity we obviously need structural information. Unfortunately, such information is not easy to obtain, chiefly because available cluster
densities are generally too low for traditional X-ray or electron absorption or
diffraction studies. The problem is compounded if cluster structure changes with
size, since single-sized samples would then be necessary for meaningful experiments. The primary tool we *can* use is mass spectrometry. As has been discussed
in other chapters of this book, the mass spectra of rare gas, alkali metal and semi
conductor clusters often show "magic numbers," cluster sizes of enhanced ion
signal intensity. This enhancement is ascribed to clusters (or perhaps ionized
clusters) of particular stability, so that either they exist in greater quantity in the
distribution of neutral clusters, or they are stable ion fragments created in the
ionization process. The interpretation of rare gas cluster magic numbers is that
the relatively short range and spherically symmetric atom-atom interaction
potential leads to icosahedral packing, which maximizes the number of atom-
atom bonds (see Vol. I, Chap. 4.6). For alkali and some other main group metals,
magic stability is conferred on clusters having filled electronic shells (Vol. I,
Chaps 4.1–4.2). For semiconductor clusters, magic numbers are usually interpreted in terms of bonding rules (Vol. I, Chap. 4.4).

No traditional magic numbers have ever been seen in the mass spectra of
transition metal clusters. The most likely reason is that the laser vaporization
cluster generation procedure results in a cluster growth process that is kinetically driven, i.e., not at equilibrium, so that relative cluster density does not
reflect cluster stability. An attempt to make particularly stable ionized cluster
fragments via multiphoton ionization likewise fails, most likely because, unlike
rare gas clusters, transition metal clusters will melt before they are hot enough to
evaporate atoms, so they will lose any geometrically determined stability they
might have had. We thus must find other ways to determine transition metal
cluster stabilities, and from the stabilities hope to deduce something about
cluster structure. One method, collision-induced dissociation of mass-selected
ionized clusters, will be discussed in the next chapter. Here, we will discuss
chemical probes that provide information about neutral metal cluster structure.
Size-dependent effects are seen that are analogous to magic numbers. We will
see that the magic is related to a cluster's surface morphology rather than to its
stability, although the two are related.

As discussed in Sect. 3, adsorbate uptake plots often show plateaus at
particular m values, i.e., a cluster might bind a given (integral) number of
adsorbate molecules over an extended range of reagent pressures in the FTR.
For example, the uptake plot for NH_3 on Fe_{13} shows *two* plateaus, one at $m = 4$

and a second (at higher NH_3 pressure) at $m = 6$ [2]. This implies that Fe_{13} has four, presumably equivalent, strong binding sites for NH_3, and two additional sites, equivalent to each other but more weakly binding than the first four. There is a substantial surface science literature, both experimental [17] and theoretical [18], that NH_3 binds to metal surfaces through a single atom in an atop or apex position. If we assume similar binding to a cluster, then we are looking for a 13-atom cluster with four apex metal atoms, and two other possible sites. This is satisfied by both fcc (face-centered cubic, shown in I) and bcc (body-centered cubic, II) packing. Each cluster has four corner atoms (shown in gray) that can serve as primary NH_3 binding sites, and two open faces that can provide the remaining two sites. On the basis of the uptake of NH_3, we cannot distinguish which of these structures might prevail for Fe_{13}.

For nickel and cobalt clusters, the NH_3 uptake data provide more definite structural information. The uptake plots for Ni_{19} and Co_{19}, for example, are quite different, with Co_{19} showing a plateau at $m = 6$ beginning at an NH_3 pressure of 0.05 Pa, while Ni_{19} saturates at $m = 12$, but not until the NH_3 pressure reaches 13 Pa. Six binding sites are consistent with fcc packing (III) for cobalt, and 12 sites imply icosahedral structure (IV) for nickel. Note that the 12 apex sites on IV are quite close together, so that it is not surprising that a high NH_3 pressure is needed to saturate this cluster.

Both Co_{55} and Ni_{55} show extensive plateaus at $m = 12$. This suggests a particularly stable (closed shell) cluster with 12 apex atoms. Again, two structures satisfy this requirement, fcc (V) and icosahedral (VI). However, for both nickel and cobalt the 71-atom cluster is also a stable species that binds 12 NH_3 molecules, and the most likely structure for this is the 55-atom icosahedron with a 16-atom cap (VII). No such "closed subshell" species exists for fcc packing

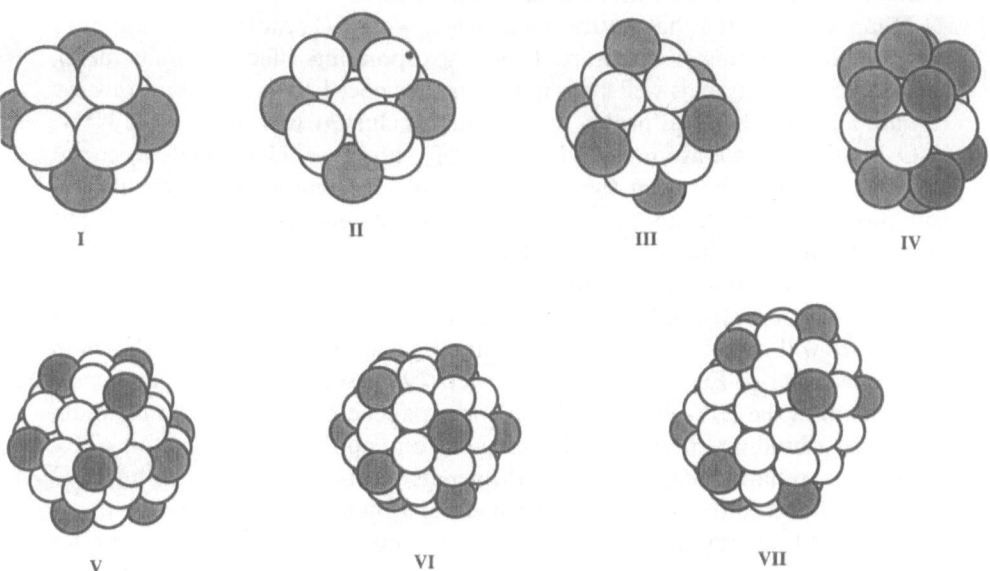

I   II   III   IV

V   VI   VII

at 71 atoms. This is strong evidence that clusters in this size region have icosahedral structure. In fact, NH_3 uptake experiments suggest that most nickel and cobalt clusters in the 50- to 160-atom size range are icosahedral in structure.

As discussed above, NH_3 binds to apex sites. H_2, on the other hand, dissociatively chemisorbs to yield H atoms that bind in bridging (between two metal atoms) or hollow (between three or four atoms) sites. These binding patterns are consistent with the fact that a cluster saturated with hydrogen atoms binds just as many NH_3 molecules as a cluster without the hydrogens, and does so without losing any of its hydrogens. Ammonia and hydrogen, in other words, bind noncompetitively. While hydrogens do not block the ammonia, they do lower the cluster-NH_3 binding energy somewhat [19]. This has the effect that, at least for hydrogenated cobalt and nickel clusters, the icosahedral structure becomes particularly evident in NH_3 saturation experiments. This is illustrated in the upper panel of Fig. 5, which shows $\bar{m}$ values for saturated coverages of NH_3 on hydrogenated cobalt clusters Co_nH_x for $n = 82$ to 155. As can be seen, $\bar{m}$ repeatedly drops to a value near 12. The sequence of n values where $\bar{m}$ has a minimum includes many of the magic numbers seen in rare gas (in other words, icosahedral) cluster mass spectra. The numbers correspond to stripping off successive caps and faces from the third closed icosahedral shell at $n = 147$. Since the 147-atom icosahedron has 12 apex atoms, it is not surprising that it binds 12 NH_3 molecules. The stripping off of the third shell uncovers increasing portions of the second shell ($n = 55$, shown in VI), which of course also has 12 apex sites, so it is not surprising that the intermediate partially closed shell clusters will likewise bind 12 ammonias. If a second-shell face has just one or two metal atoms on it, they are relatively exposed and act much like apex atoms, so they bind additional NH_3 molecules. Thus between the "magic" numbers the clusters bind more than 12 molecules.

It might be argued that saturating a cluster with NH_3 molecules could have an effect on the cluster structure. (The corresponding effect on bulk metal, surface reconstruction, is well known in surface science [20]). To test for this, we can use another chemical probe, the reaction of clusters with H_2O. Like NH_3, H_2O binds to single-atom sites. Unlike NH_3, however, H_2O binds to metal cluster so weakly that even the first H_2O molecule reacts under equilibrium conditions [10]. As discussed above in Sect. 5, the equilibrium constant K_{eq} is related to the binding energy $-\Delta H^0$ by $RT\ln(K_{eq}) = -\Delta G^0 = -\Delta H^0 + T\Delta S^0$. If we assume that the entropy ΔS^0 of adsorbing the first H_2O molecule is not a strong function of cluster size, then a plot of $RT\ln(K_{eq})$ vs cluster size will reveal any trends in the binding energy. Such a plot is shown in the lower panel of Fig. 5 for Co_nH_x clusters [21]. We see that in many cases a minimum in the NH_3 $\bar{m}$ values in the upper panel correlates with a maximum in the H_2O binding energy for clusters with one additional metal atom. This extensive correlation implies that hydrogenated cobalt clusters with an H_2O molecule bound to them also have icosahedral structure. The correlation arises from the fact that, as pointed out above, an additional metal atom on the surface of a closed icosahedral subshell will be relatively exposed and will thus have

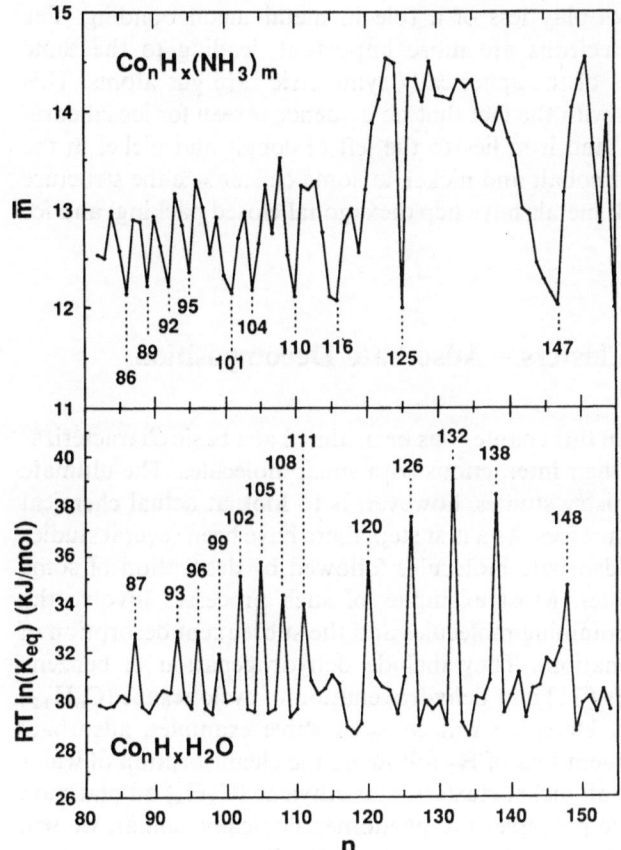

Fig. 5. The average number $\bar{m}$ of NH$_3$ molecules that saturate hydrogenated cobalt clusters (*upper panel*) and RT times the logarithm of the equilibrium constant for the binding of an H$_2$O molecule to the same clusters (*lower panel*). The n values for specific clusters having minima in $\bar{m}$ or maxima in $RT\ln(K_{eq})$ are indicated. The lower panel is adapted from [21]

fewer metal–metal bonds than atoms *in* the closed subshell. This reduced metal coordination leads to an enhancement in the atom-H$_2$O binding energy (just as it provides an additional NH$_3$ binding site), so the observed pattern is not unexpected. Since it is unlikely that the (rather weak) binding of a single H$_2$O molecule to clusters of this size would have any effect on cluster structure, we conclude that for the most part hydrogenated cobalt clusters are icosahedral whether or not they are saturated with NH$_3$ molecules. Of course, the detailed picture must be somewhat more complicated, since there is not a perfect correlation for all cluster sizes in Fig. 5.

The appearance of icosahedral structure for metal clusters is somewhat surprising, since icosahedral packing is not a bulk crystalline packing scheme. Presumably the nearly filled (and less spatially extended) d orbitals at the right side of the transition series play less of a role in metal–metal bonding. The spherically symmetric s electrons are more important, leading to the same structures seen for clusters of the spherically symmetric rare gas atoms. This interpretation is consistent with the fact that no evidence is seen for icosahedral structure for iron clusters, and iron lies to the left of cobalt and nickel in the periodic table. But even for cobalt and nickel, at some cluster size the structure must change, since the bulk metals have hcp (hexagonal closed packing) and fcc packing, respectively.

2.8.8   Chemistry on Clusters – Adsorbate Decomposition

The work described so far in this chapter has been aimed at a basic characterization of metal clusters and their interactions with small molecules. The ultimate goal of metal cluster chemistry studies, however, is to look at actual chemical transformation on cluster surfaces. As a first step, there have been several studies of the decomposition of adsorbate molecules followed by desorption of some new species from the cluster. Most examples of such processes involve the adsorption of hydrogen-containing molecules and the subsequent desorption of H_2 molecules (dehydrogenation). They include dehydrogenation of benzene (C_6H_6) on niobium clusters [22] and dehydrogenation of cyclohexane (C_6H_{12}) on platinum clusters [23]. Here, we will consider three examples: adsorbate decomposition and subsequent loss of H_2 following the chemisorption of water on iron clusters, ammonia on nickel clusters, and ethylene (C_2H_4) on platinum clusters. Although the three processes are phenomenologically similar, we will see that their detailed mechanisms appear to be quite different.

As discussed above, the reaction of iron clusters with a single H_2O molecule is an equilibrium one under the usual experimental conditions, implying a reversible process in which an intact H_2O molecule can repeatedly adsorb and desorb from a cluster. However, when a second H_2O molecule adds to an iron cluster, an irreversible loss of H_2 from the cluster is seen [10]. While the detailed mechanism of this reaction is not known, we can speculate that it probably involves the concerted loss of H_2, one H atom coming from each H_2O molecule. Even though there are many sites on the cluster surface for H atoms to bind, they apparently do not. This is because the dissociative chemisorption reaction of H_2 with iron clusters is kinetically limited (see Sect. 4), so it would be energetically difficult for the reverse reaction, the recombination of two surface-bound H atoms and the desorption of H_2, to occur. Thus we conclude that in the process of forming H_2 the H atoms never interact directly with the metal surface. Formation of strong cluster-OH bonds appears to be the driving force for the reaction.

The situation is quite different when NH_3 reacts with small nickel clusters. Figure 6 shows portions of the TOF mass spectra that illustrate the species seen for Ni_7 reacting with NH_3 for a long interaction time (upper spectrum) and

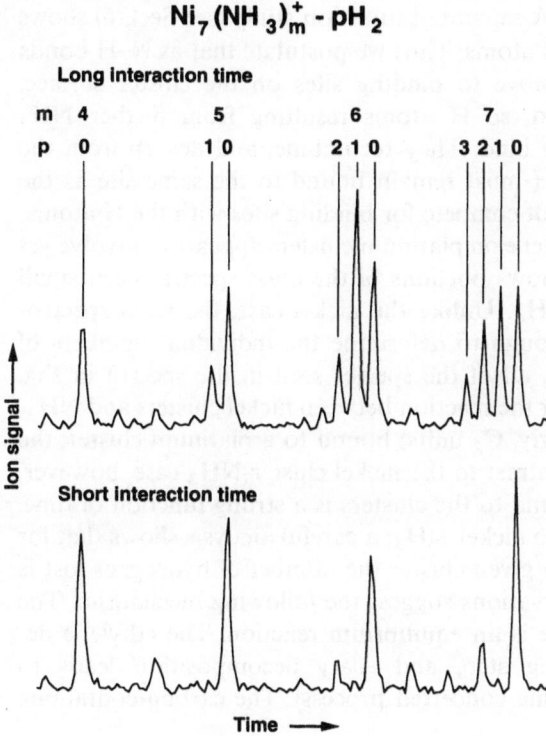

$$Ni_7(NH_3)_m^+ - pH_2$$

Long interaction time

m	4		5		6		7	
p	0		1 0		2 1 0		3 2 1 0	

Short interaction time

Ion signal ⟶

Time ⟶

Fig. 6. Portions of the mass spectra illustrating the products formed when Ni_7 reacts with NH_3 for a long interaction time (*upper spectrum*) and for a short interaction time (*lower spectrum*). Adapted from [24]

a short interaction time (lower spectrum) [24]. As can be seen, for the long interaction time there is extensive H_2 loss for the most ammoniated species $Ni_7(NH_3)_7$, while decreasing loss is seen for progressively less ammoniated species. For the short interaction time, essentially no H_2 loss is seen. (Note that the distribution of total cluster intensity is nearly independent of time. At this level of NH_3 coverage, the initial chemisorption of NH_3 is at equilibrium, as discussed in Sect. 5.) This time dependence implies that some step in the overall process that leads to H_2 loss is rate limiting, and does not occur on the short interaction timescale. Two-reagent studies, in which both NH_3 and D_2 are added to the FTR, indicate that the initial breakage of the N–H bond(s) is the rate limiting step. When the cluster-NH_3 interaction time is kept short, D_2 adds to the cluster just as if the NH_3 weren't there, indicating that intact NH_3 molecules are still present on the cluster. If N–H bonds had broken, some of the H- (or D-) bindings sites would be filled, and the cluster would not be able to accommodate as many D atoms as it does. When D_2 is added to clusters that have had a long interaction time with NH_3, a very complex spectrum results that is indicative of H/D exchange. Clearly, under these conditions, H atoms are bound to the cluster surface, so N–H bonds have broken.

The best clue to the mechanism of this reaction is afforded by the systematics of H_2 loss seen in Fig. 6. If we assume that the final nitrogen-containing species left on the cluster is NH, then a little numerical calculation will show that the maximum extent of H_2 loss from Ni_7 corresponds to leaving eight H atoms on the cluster surface in each case. A separate saturation study (see Sect. 6) shows that indeed Ni_7 can bind eight H atoms. Thus we postulate that as N–H bonds break, the resulting H atoms move to binding sites on the cluster surface. Eventually, these sites are filled, so H atoms resulting from further NH_3 decomposition have no place to bind. They recombine, and desorb from the cluster as H_2. Note that the NH must remain bound to the same site as the original NH_3, or at least must not compete for binding sites with the H atoms.

The dehydrogenation of ethylene on platinum clusters appears to involve yet another mechanism. Figure 7 shows portions of the mass spectra when small platinum clusters react with C_2H_4. Unlike the nickel case, the mass spectrometer resolution is not high enough to determine the individual numbers of H_2 molecules lost. Nevertheless, all of the species seen in the spectra in Fig. 7 show substantial H_2 loss. As for the reaction between nickel clusters and NH_3, the more ethylenes (more properly, C_2 units) bound to a platinum cluster, the greater the hydrogen loss. In contrast to the nickel cluster-NH_3 case, however, here the number of C_2 units bound to the clusters is a strong function of time. Furthermore, again in contrast to nickel-NH_3, a careful analysis shows that for a given number of C_2 units on a given cluster the number of hydrogens lost is *independent* of time. These observations suggest the following mechanism. The initial chemisorption of ethylene is an equilibrium reaction. The ethylene decomposition is the rate-limiting step, and *every* decomposition leads to H_2 desorption (probably via some concerted process). The carbon-containing

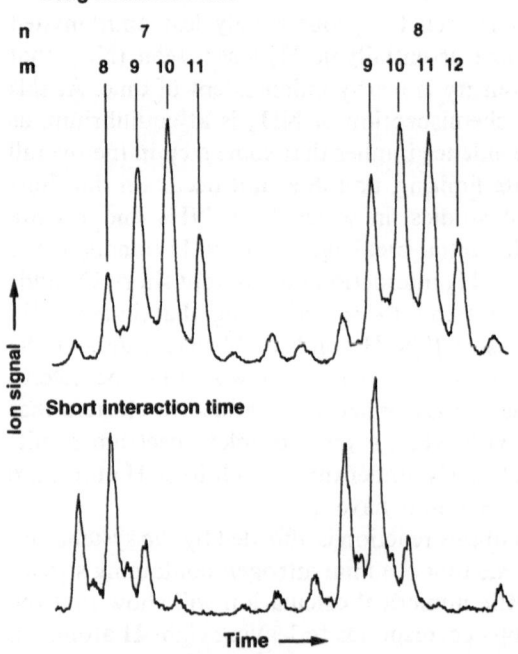

Fig. 7. Portions of the mass spectra illustrating the products of the reactions of Pt_7 and Pt_8 with C_2H_4 for a long interaction time (*upper spectrum*) and a short interaction time (*lower spectrum*)

decomposition product binds to a different site than the initial ethylene does, so the liberated ethylene-binding sites rapidly equilibrate with more ethylenes. The final composition of the cluster (i.e., the number of C_2 units) will depend on the reaction time, because the decomposition, and hence the number of new sites made available, is kinetically limited. But for a given number of C_2 units, corresponding to a given number of decompositions, the cluster will have lost the same number of hydrogens, since each decomposition is accompanied by H_2 loss.

Such studies of chemistry on clusters illustrate the powerful techniques available to us for determining detailed reaction mechanisms. The ability to change and monitor reaction time and temperature, and the ability to determine, via mass spectrometry, the number and nature of the species on the clusters, will be important in future studies of model catalytic reactions.

2.8.9   Future Prospects

It should be clear that transition metal cluster chemistry is an important field in its own right. Much has been learned about the fundamental chemical processes that occur on cluster surfaces. However, if the promise to revolutionize heterogeneous catalysis is to be met, one great challenge still faces us. If indeed we can demonstrate that a certain cluster size or range of sizes is particularly efficient at catalyzing a particular reaction, then we must be able to prepare a size-specific catalyst sample that can be used in a practical reactor. This is not an easy task. First, to achieve general size selectivity a beam of clusters must be ionized and the desired size mass-selected. Next, the ionized cluster must be neutralized and deposited onto a suitable substrate, such as silica (silicon dioxide) or alumina (aluminium oxide). This step is crucial, and there are several processes that may destroy the size selectivity. The neutralization may cause cluster fragmentation. The landing on the support surface may also cause fragmentation. Once on the support, the clusters may migrate and agglomerate to form larger clusters. The interaction with the support itself may dramatically change the chemical properties of the cluster. These are all important considerations, and several of them are treated in detail in Chap. 3.6 of this book.

While methods for the preparation of supported clusters are being researched, there still remains much gas phase work to be done. The identification of promising systems for modeling typical industrial catalytic processes still remains to be done. Also, the sorts of structural studies described here should be done for other catalytically important metals. As we expand our understanding of metal-promoted reactions at the microscopic level, we further enhance our ability to design improved catalysts, even by more conventional means. Finally, metal cluster chemistry is proving to be an exciting field for chemists. Clusters do behave in strange and mysterious ways, and our chemical intuition is constantly being challenged in our efforts to explain our observations.

2.8.10   Recent Developments

The most significant recent discovery has been the identification of the nitrogen molecule as an important probe of cluster structure. Unlike ammonia, nitrogen binds molecularly to *every* atom on a cluster's surface, and some atoms can even bind two molecules. Binding rules have been developed to relate an atom's binding ability to its metal-metal coordination. Measurement of nitrogen saturation levels has provided structural information for small nickel and cobalt clusters. A preliminary report is given in [25].

Acknowledgments. I would like to thank all members of the Metal Cluster Chemistry Group at Argonne, past and present, especially Dr. Eric K. Parks. This work was supported by the U.S. Department of Energy, Office of Basic Energy Sciences, Division of Chemical Sciences, under Contract W-31-109-Eng-38.

References

1. E.K. Parks, G.C. Nieman, L.G. Pobo, S.J. Riley: J. Chem. Phys. **86**, 1066 (1987)
2. E.K. Parks, G.C. Nieman, L.G. Pobo, S.J. Riley: J. Chem. Phys. **88**, 6260 (1988)
3. See, for example, R.P.H. Gasser: *An Introduction to Chemisorption and Catalysis by Metals.* Oxford: Clarendon Press 1987. pp. 10–16
4. S.C. Richtsmeier, E.K. Parks, K. Liu, L.G. Pobo, S.J. Riley: J. Chem. Phys. **82**, 3569 (1985)
5. S.J. Riley, E.K. Parks: In: *Proceedings of the International Symposium on the Physics and Chemistry of Small Clusters*, ed. by, P. Jena, B.K. Rao, S.N. Khanna, NATO ASI Series **B**: 158. New York, London: Plenum Press 1987. p 727
6. For a recent review, see M.F. Jarrold: In: *Advances in Gas-Phase Photochemistry and Kinetics* 2: *Biomolecular Collisions*, ed. by, M.N.R. Ashford, J.E. Baggot, London: Royal Society of Chemistry 1989. p. 337
7. D.M. Cox, K.C. Reichmann, D.J. Trevor, A. Kaldor: J. Chem. Phys. **88**, 111 (1988)
8. M.B. Knickelbein, S. Yang: J. Chem. Phys. **93**, 1533 (1990)
9. R.L. Whetten, D.M. Cox, D.J. Trevor, A. Kaldor: Phys. Rev. Lett. **54**, 1494 (1985)
10. B.H. Weiller, P.S. Bechthold, E.K. Parks, L.G. Pobo, S.J. Riley: J. Chem. Phys. **91**, 4714 (1989)
11. E.K. Parks, B.H. Weiller, P.S. Bechthold, W.F. Hoffman, G.C. Nieman, L.G. Pobo, S.J. Riley: J. Chem. Phys. **88**, 1622 (1988)
12. G.A. Somorjai: *Chemistry in Two Dimensions: Surfaces.* Ithaca: Cornell University Press 1981; F.C. Tompkins: *Chemisorption of Gases on Metals.* New York: Academic Press 1978
13. E.K. Parks, G.C. Nieman, L.G. Pobo, S.J. Riley: J. Phys. Chem. **91**, 2671 (1987)
14. E.K. Parks, K. Liu, S.C. Richtsmeier, L.G. Pobo, S.J. Riley: J. Chem. Phys. **82**, 5470 (1985)
15. H.H. Storch, N. Golumbic, R.B. Anderson: *The Fischer-Tropsch and Related Syntheses.* New York: Wiley 1951
16. D.M. Cox, P. Fayet, R. Brickman, M.Y. Hahn, A. Kaldor: Catal. Lett. **4**, 271 (1990)
17. C.W. Seabury, R.J. Rhodin, R.P. Merrill: Surf. Sci. **93**, 117 (1980); C. Klauber, M.D. Alvey, J.T. Yates, Jr.: Chem. Phys. Lett. **106**, 477 (1984)
18. C.W. Bauschlicher: J. Chem. Phys. **83**, 3129 (1985)
19. W.F. Hoffman III, E.K. Parks, S.J. Riley: J. Chem. Phys. **90**, 1526 (1989)
20. G. Ertl: In: *Metal Clusters in Catalysis* (Studies in Surface Science and Catalysis, Vol. 29). B.C. Gates, L. Guczi, H. Knötzinger (eds.). Amsterdam, New York: Elsevier 1986. p. 577

21. T.D. Klots, B.J. Winter, E.K. Parks, S.J. Riley: J. Chem. Phys. **92**, 2110 (1990)
22. R.J. St. Pierre, E.L. Chronsiter, M.A. El-Sayed: J. Phys. Chem. **91**, 5228 (1987); R.J. St. Pierre, M.A. El-Sayed: J. Phys. Chem. **91**, 763 (1987); M.R. Zakin, D.M. Cox, A. Kaldor: J. Phys. Chem. **91**, 5224 (1987)
23. D.J. Trevor, R.L. Whetten, D.M. Cox, A. Kaldor: J. Amer. Chem. Soc. **107**, 518 (1985)
24. S.J. Riley: Z. Phys. D **12**, 537 (1989)
25. E.K. Parks, L. Zhu, J. Ho, S.J. Riley: Z. Phys. D **26**, 41 (1993)

2.9 Chemistry with Cluster Ions

S.L. Anderson

2.9.1 Introduction

This section will give an overview with a few specific examples, of the wide range of chemistry experiments that have been performed with cluster ions. I will focus on chemistry of cluster ions composed of metal and semi-metal atoms, and will not attempt to cover non-covalently bound clusters (van-der-Waals, hydrogen bonded, etc.). As a further limitation of scope, I will only discuss studies which treat a broad range of cluster sizes.

Using various techniques it is possible to measure rate constants, cross sections, and thermochemical parameters such as bond, reaction, and activation energies. From this data it is frequently possible to deduce the mechanisms of cluster ion reactions, and to draw some inferences about the physical properties of the cluster ions. Indeed, given the small number of chemically interesting cluster systems for which detailed characterization from spectroscopy or theory is available, the cluster chemist is generally forced to rely on these inferences to a large degree.

2.9.1.1 Advantages to Working with Ions

In most experimental studies of neutral cluster chemistry, reaction of the whole distribution of cluster sizes occurs simultaneously, and the cluster size dependence must be extracted in the detection process (usually ionization followed by mass analysis). The charge on ionized clusters allows the reactants to be mass-selected and manipulated in other ways, before the reactions are allowed to occur. A few of the major advantages that result are:

- Mass-selecting the reactant ions eliminates the problem of different size reactant clusters producing identical mass products, which allows study of product channels in which the cluster changes size (e.g. $M_n^+ + O_2 \rightarrow M_{n-2}O^+ + M_2O$). These are difficult to examine for neutrals due to interference from reactions of smaller clusters.
- Since the reactants and products are already ionized, one doesn't have to worry about differences in ionization efficiency (and thus detection sensitivity)

for reactant and product species, allowing more accurate measurements of reaction rates.

- Since reactants and products are not ionized in the detection process, there is no problem with electron or photon induced fragmentation which can alter the apparent chemistry.
- The charge on the ionized clusters also gives the experimentalist a handle to use in controlling reactants. This allows study of reactions over a wider range of conditions than is possible with neutral clusters. For example, it is easy to vary the collision energy for an ion-molecule reaction from tens of meV to tens of eV. In addition, it is possible to store ions in radio frequency or electromagnetic traps for any desired length of time. This can be used to equilibrate the internal and translational energy of cluster ions at a well controlled temperature, and is also a powerful tool for chemical kinetics measurements.
- A final potential problem for neutral cluster studies in which the clusters are generated in a laser vaporization source then immediately reacted in a flow tube, is that there are a large number of ions produced the source, and some of the "neutral" cluster chemistry may in fact be due to reaction of ions which are subsequently neutralized by recombination in the flow tube.

Since it is by nature complicated enough to unravel cluster chemistry, a large number of research groups have turned to ion-molecule reaction techniques to help make these problems more tractable.

2.9.1.2   Differences between Neutral and Ion-Molecule Chemistry

Having listed a few of the experimental advantages to working with ionized clusters, it is reasonable to ask how their chemistry differs from that of the equivalent neutrals. A few studies have been reported where neutral and ionized clusters have been studied under similar conditions, and they suggest that the differences in reactivity are small (see below). This however is not a sufficient basis to draw any general conclusions, so it may be useful to discuss briefly what we might expect. Cursory perusal of the literature [1] shows that many ions are much more reactive than their neutral precursors. For example, Ar^+ is highly reactive while neutral argon is essentially inert. There are two reasons for the high reactivity of typical ions, both of which are less important for cluster systems.

A major difference between the chemistry of stable neutral species and the concomitant ions is that nearly all chemically stable molecules are even electron species, and most have singlet ground states with all electrons paired. To react, it is necessary to unpair electrons, which costs energy and frequently results in activation barriers even for exoergic reactions. In contrast, ions produced from these precursors are by necessity odd electron species, and like most radicals they are highly reactive. Thus the comparison of neutral and cationic argon is inappropriate. Ar^+ should be compared with the isoelectronic species – atomic chlorine, which is also reactive. The ion which is isoelectronic with argon is K^+, and it is indeed rather unreactive.

For metal and semi-metal clusters this even/odd electron difference between neutral and ionized species should become relatively unimportant as size increases. This is because the promotion energy required to unpair electrons becomes small as the discrete, well separated orbitals in the isolated atoms merge into bands. This appears to be true even for small clusters of elements such as boron, where one might expect the density of states to remain low. Since boron has 3 valence electrons, B_n^+ clusters with $n = 2, 4, 6, \ldots$ have an odd number of electrons, while odd size clusters have even numbers of electrons (and singlet ground states [2]). If electron pairing was the dominant factor controlling reactivity, we might expect to see a pervasive even/odd pattern of reactivity, however as shown below this is not the case.

The other major difference between ion and neutral chemistry is in the interaction potential governing collisions. The long range part of the potential for collisions between neutrals is only weakly attractive, while for an ion-molecule collision there are ion-dipole or ion-induced dipole forces which are stronger and operate at longer range [3]. This has two effects. Large impact parameter collisions which for neutrals would not reach the range of intermolecular separations where reaction can occur (i.e. would "miss colliding"), are captured by the long range forces and may react. At low collision energies, this results in collision cross sections more than an order of magnitude larger than the hard-sphere cross section. As cluster size increases, this effect becomes less important because the hard-sphere collision cross section increases rapidly, and soon becomes comparable to the capture cross section except at very low energies. A second effect of the attractive forces is to accelerate the reacting species into the collision, which may help drive reaction by overcoming small activation barriers. Again for cluster ions, this effect should become less important as physical size increases, since the reactants don't come so close together before chemical forces begin to dominate. Indeed we frequently observe activation barriers for cluster ion reactions which grow in as cluster size increases.

All this is not to say that there will be no differences between the chemistry of neutral and ionized clusters. What does appear to be true is that the chemistry is qualitatively similar, and differences due to charge state generally are smaller than differences due to changes in cluster size or geometric structure. Clearly as the cluster size increases, the effect of one additional or missing electron must become negligible. (N.B. For van-der-Waals clusters the convergence of ion and neutral chemical properties will be much slower, since the binding forces in that case are weak and perturbations from electrostatic forces are relatively large).

2.9.2 Experimental Methods

Most of the techniques that have been applied to cluster ion chemistry were originally developed for the study of molecular ion chemistry, and detailed descriptions can be found in recent review books [4]. This section gives a *brief* description of some major techniques, with an emphasis on the particular advantages and disadvantages as applied to clusters. I will concentrate on experiments which take full advantage of the ability to mass-select the reactant clusters.

2.9.2.1 Flow Tube Experiments

One of the simplest approaches for studying metal cluster ion chemistry is the mini-flow tube, which also allows measurements on both ions and neutrals under the same conditions. This technique is essentially identical to the flow tube reactor discussed in the previous section on chemistry of neutral metal clusters. A pulsed laser is used to vaporize a small volume of a solid target mounted in a buffer gas flow. As the metal plasma is cooled by buffer gas collisions, condensation, recombination, and autodetachment processes occur, resulting in a collection of neutral and positively and negatively charged clusters entrained in the buffer gas. To study chemistry, a reactant gas is injected into the metal-containing buffer gas flow at some point downstream. The clusters and reactant gas are mixed by turbulence, allowed to flow together for some distance (on the order of 5 cm) then co-expand into the vacuum system. For studies of ionized clusters, the mass distribution is directly sampled by a mass spectrometer. Neutral clusters require ionization (usually by excimer laser) followed by mass spectrometric sampling. Chemistry occurs in a high pressure of buffer gas, and three-body stabilization of adducts is commonly observed. Work along these lines has been reported by a number of groups including those of *Cox* [5] and *Castleman* [6].

The principle virtue of this technique is its simplicity, which allows rapid surveying of cluster size effects. There are also a number of problems. For example, the internal temperature of the reactant clusters is not well defined. It is commonly assumed that they are equilibrated at the flow tube temperature, however recent measurements by *Kolenbrander* and *Mandich* [7] suggest that this is overly optimistic. The major problem lies in the fact that reactions of the whole ensemble of neutral and charged clusters are studied together. One thus loses all the advantages (discussed above) that can result from being able to mass-select, accelerate, store, and collect ions. In addition, a variety of processes such as association, detachment, and recombination can cause products of ion reactions to appear as neutrals and vice versa. This potentially confuses the issue of what chemistry one is observing in both neutral and ion flow tube experiments.

For studies of neutral cluster chemistry there are no easy alternatives, however for ions these problems can be greatly ameliorated by mass-selecting the ions before injecting them into the reactor (the SIFT technique) [4]. To date this has not been used for flow tube studies of large metal or semi-metal cluster ions, however Jarrold and co-workers have reported mass-selected cluster ion chemistry studies in a drift tube. In this experiment [8], clusters are produced in a laser vaporization source, allowed to expand into vacuum, mass-selected, then injected through a small hole into a 2.5 cm drift cell. A uniform electric field together with a high density (~ 0.4 Torr) of neon buffer gas cause the ions to reach a drift velocity through the cell, and chemistry can be observed by adding a small fraction of neutral reactant to the buffer gas. A small fraction of the ions exit the cell through a second small aperture, and products are analyzed by mass spectrometry. This is a nice technique for kinetics measurements on size selected cluster ions, and also allows crude measurements of collision energy effects by changing the drift field and/or buffer gas pressure.

2.9.2.2 Ion Cyclotron Resonance

The ICR technique is a powerful method for studying ion reactions. In essence, ions are trapped by a combination of a strong magnetic field and weaker electric field. In a magnetic field, charged particles undergo circular motion with an orbit frequency which is dependent only on the charge/mass ratio and the field strength. It is possible to monitor the mass distribution of ions in the trap by several methods. In Fourier transform mode (FT-ICR or FTMS) ICR is capable of measurements on massive ions with resolution ranging up to $\sim 10^6$, which is useful in studying large cluster ions, particularly if isotopes confuse the mass spectrum. By applying a driving field at the cyclotron frequency for a particular mass ion, it is possible to increase its translational energy, and eject it from the ICR cell. This selective excitation can be used to mass-select the nascent size distribution of cluster ions.

A typical sequence of events used to study cluster ion chemistry goes something like this: Create an ensemble of cluster ions in the trap. (Optionally) store the cluster ions for some period of time with or without buffer gas to cool their initial internal and translational excitation. Use selective excitation to drive all masses except the cluster of interest out of the trap. Inject a pulse of neutral reactant into the trap. Monitor the conversion of reactant cluster ions to products as a function of time. (Optionally) use selective excitation to drive out all masses except a particular product ion, then either study collision induced dissociation or subsequent chemistry of the primary product ion.

The major experimental "trick" in doing cluster ion chemistry in an ICR, is getting a sufficient intensity of cluster ions into the strong magnetic trapping field. One approach which has been used extensively by *Ridge* and co-workers [9], is to flood the ICR cell with a volatile organo-metallic precursor (e.g. $Fe(CO)_5$), ionize it, then allow clustering ion-molecule reactions to grow cluster ions. Another approach is to mount a solid metal or semi-metal target in wall of the trap and hit it with a focussed laser. For some metals [10] and semi-metals [11] this results in production of cluster ions containing up to tens of atoms, however for other materials only small cluster ions are formed [12]. A more general method is to make the cluster ions outside the ICR apparatus, then inject them into the trap. As Smalley and co-workers have shown [13], injection is difficult because of the strong magnetic fields. McIver and co-workers have found that an rf-only quadrupole field can be used to efficiently guide large cluster ions into the magnetic field [14].

ICR is a good technique for measuring ion-molecule reaction kinetics and equilibria. As mentioned, it also allows one to follow a sequence of reactions by isolating products and carrying out further chemistry. By trapping ions for extended times (seconds if necessary) with buffer gases it is possible to equilibrate the internal and translational energy of the clusters at a well defined temperature. The principle disadvantage of the ICR methods is that it is not straightforward to increase the center-of-mass frame collision energy in a well defined way, and thus quantitative kinetics work tends to be done under thermal conditions.

2.9.2.3 Low Energy Ion Beam Techniques

In these methods, cluster ions are produced in one of a variety of sources, mass-selected, then formed into a beam with well defined and controllable kinetic energy. The cluster beam passes through a beam or gas cell containing a target molecule, and products are detected mass spectrometrically. Since the ion beam kinetic energy is easily controlled, it is possible to measure reactions as a function of collision energy. This typically gives clues to the reaction mechanisms, and in some cases allows measurement of energy thresholds which can be used to derive thermochemical information about both ionized and neutral cluster reactions. This information is extremely valuable, and largely unavailable from other techniques. It is also possible to measure recoil energy and angular distributions, which provide further mechanistic insight. Beam techniques tend to be most useful in getting mechanistic and thermochemical data, while true rate measurements are best done with ICR or other methods.

One of the cluster ion beam instruments at Stony Brook [15] is shown in Fig. 1. The unique feature of this machine is its storage ion source. Cluster ions produced by either fast atom bombardment (for metals) or laser ablation (for semi-metals) of a solid target are internally and translationally hot and must be cooled before use. We inject the cluster ions into a labyrinthine radio-frequency trap, where they are stored with a buffer gas for on the order of 100 ms to equilibrate them at the trap temperature. By adding a reactant gas into the buffer gas, we also can pre-react the cluster ions to produce oxides, hydrides, . . . for further chemistry studies. The cooled clusters are then mass-selected and guided through a gas cell containing a neutral reactant. Product ions and unreacted cluster ions are collected by the ion guides, mass analyzed, and counted. Other cluster ion beam instruments [16, 17] are similar, except they use laser vaporization/nozzle type cluster sources.

Top View

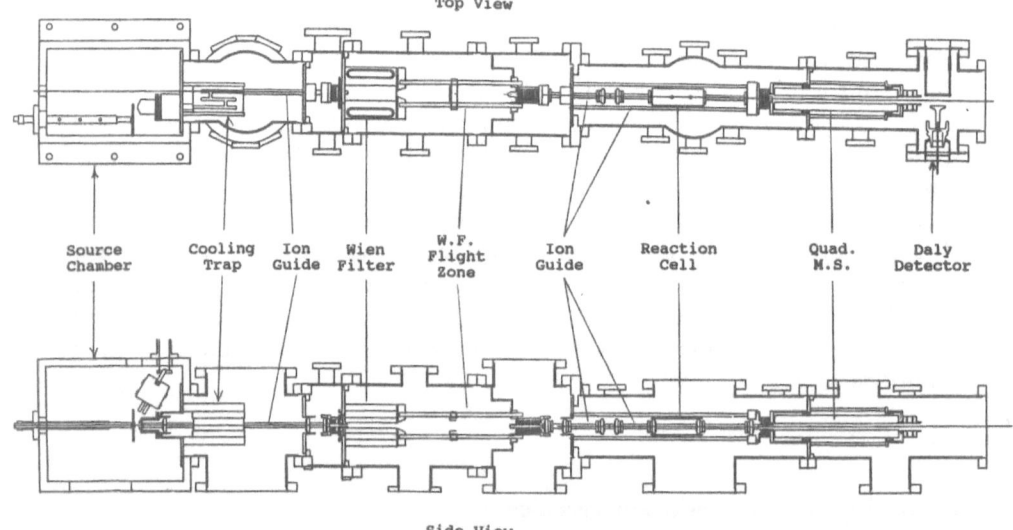

Source	Cooling	Ion	Wien	W.F.	Ion	Reaction	Quad.	Daly
Chamber	Trap	Guide	Filter	Flight	Guide	Cell	M.S.	Detector
				Zone				

Side View

Fig. 1. Stony Brook cluster ion instrument #1. Cluster ions can be made either by fast atom bombardment or laser ablation of a solid target. The nascent hot cluster ions are cooled by storing them in a maze-shaped rf trap which is filled with buffer gas (N_2). Cooled cluster ions are mass-selected by a Wien filter. Reactions are studied using guided-beam ion optics, and products are passed through a quadrupole mass filter and counted

2.9.3 Comparison of Ion and Neutral Cluster Reactivity

Despite the problems inherent to the non-mass-selected flow tube experiments, they have revealed some fascinating chemistry, and do allow direct comparison of the chemical reactivity of neutral and ionized clusters. Figure 2 shows results from one such experiment by *Zakin* et al. [5], in which they studied the dehydrogenation of benzene by niobium clusters. Benzene is injected into the Nb_x containing helium flow as described above, and they monitor the appearance of products of the form $Nb_x C_6 H_m$. The data in this figure is for complete dehydrogenation, i.e. $m = 6 \rightarrow m = 0$. The top frame shows the conversion probability observed for positively charged products emerging from the flow tube, and the bottom frame shows the equivalent results for neutral products which are ionized by 6.42 eV photons. The assumption is that the top frame represents the conversion probability for reaction of Nb_x^+ cations with benzene, while the bottom frame is for reaction of neutral Nb_x. Note the rather striking similarity – the differences in reactivity between Nb_x and Nb_x^+ are smaller than the differences due to variation in cluster size.

Ionized and neutral clusters are not always so similar. For example *Zakin* et al. [18] studied chemisorption of D_2 on neutral, and positively and negatively

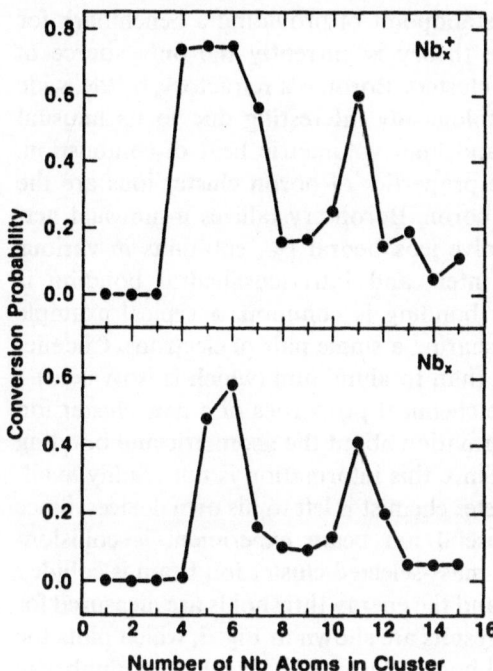

Fig. 2. Probability for complete dehydrogenation of benzene adsorbed on niobium clusters. Top frame is for positively charged clusters. Bottom frame is for neutral clusters. (from M.R. Zakin, R.O. Brickman, D.M. Cox, and A. Kaldor: J. Chem. Phys. **88**, 5943 (1988))

charged Nb clusters. In general the reactivities and saturation uptake limits were quite similar for the three charge states, however for certain size clusters there were substantial differences. This was attributed to existence of structural isomers for these anomalous species. *Elkind* et al. [13] also have compared neutral and charged cluster reactivity for niobium, and they concluded that it is strikingly similar. In their ICR treatment of the ion chemistry they were able to clearly identify structural isomerism for certain size cluster ions.

2.9.4   Boron Cluster Ions: A Case Study with Ion Beams

Cluster ion beam techniques have been used to study a variety of types of metal and semi-metal cluster ion chemistry. *Jarrold* and co-workers have focused on aluminum [19] and silicon [20] cluster ions. Work at Stony Brook has concentrated on boron [2, 21], aluminum [15, 22], and carbon [23] cluster ions. *Armentrout* and co-workers have recently extended their long-standing efforts in atomic transition metal ion chemistry [24] to transition metal cluster ions [25]. The designs and methodology for all these experiments are conceptually similar, and will be illustrated by a single example – the series of boron cluster ion studies performed by Anderson and co-workers at Stony Brook.

We chose to concentrate on boron because it is a simple (5 electron) material and thus should be a good system for theoretical studies of covalently bound clusters. This is important from the standpoint of providing a benchmark for theoretical studies, but also because theory is currently the only source of detailed structural information about clusters. Boron is a refractory, brittle, wide band-gap semiconductor. It is technologically interesting due to its unusual strength, thermoelectric properties, and high volumetric heat of combustion. More germane to understanding the properties of boron cluster ions are the bonding and chemical properties of boron. Boron crystallizes in unusual net-work structures, which typically involve icosahedral B_{12} sub-units in various 3-dimensional arrangements. Both inter- and intra-icosahedral bonding is strong and directional. Multi-center bonding is common, a typical example being triangular sets of three atoms, sharing a single pair of electrons. Chemic-ally, boron is more similar to silicon than to aluminum (which is iso-valent).

A logical first step in probing the chemical properties of a new cluster ion system, is to attempt to get some information about the geometric and bonding structure of the clusters themselves. Since this information is not readily avail-able from spectroscopic work, the cluster chemist is left to his own devices. Since ions can easily be accelerated, a useful ion beam experiment is collision-induced-dissociation (CID). Here the mass-selected cluster ion beam is collided with an inert target (typically xenon) and the energy thresholds are measured for each observed fragment ion. Typical results are shown in Fig. 3, which plots the CID cross sections for a typical small boron cluster [2]. There are a number of things that can be learned from this type of data. Note that there are only two significant fragmentation channels, and that both consist of loss of a single B atom:

$$B_n^+ + Xe \rightarrow B^+ + B_{n-1} + Xe$$
$$\rightarrow B + B_{n-1}^+ + Xe \ .$$

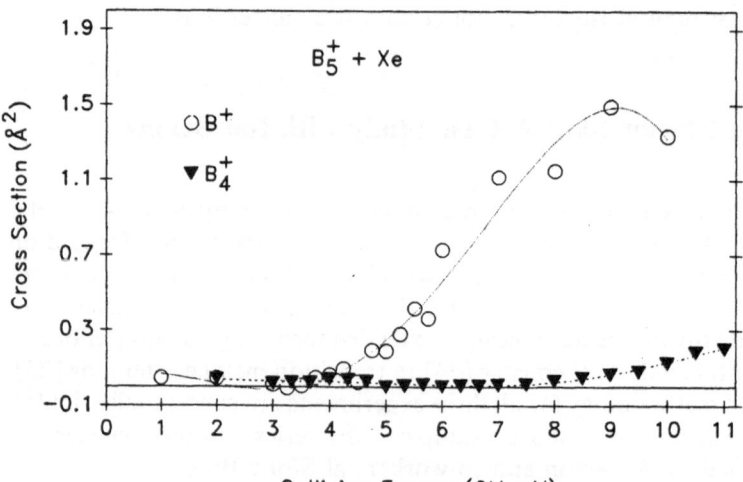

Fig. 3. CID cross sections for B_5^+ as a function of collision energy

This observation, which holds for B_{2-13}^+, suggests compact cluster structures, since one might expect that linear or other extended structures would have substantial probability for fission processes. In addition, the fragmentation threshold energies give an estimate of the cluster ion stabilities (binding energies), which range from ~ 2 to ~ 8 eV in this size range. Finally we note that the difference in thresholds for loss of B vs. B^+ gives an estimate of the difference in ionization potentials (IP) between B and B_{n-1}. This assumption leads to the prediction that the IPs for B_{2-4}^+ are greater than that of B (8.298 eV), while for the larger clusters the IPs are less than IP(B), and gradually decrease with increasing cluster size.

The cluster size dependence of the CID results are summarized in Fig. 4, which shows the cluster stabilities obtained from threshold measurements, and the CID cross sections at 10 eV collision energy. Note that the stability generally increases with increasing cluster size, but with large fluctuations. The shift from B^+ to B_{n-1}^+ as the dominant fragment ion is also evident. Finally note the high stability and low CID cross section of B_{13}^+.

Having obtained some information about how the physical properties of boron cluster ions change with size, it is interesting to examine chemical properties. We have looked at reactions with quite a few molecules, however

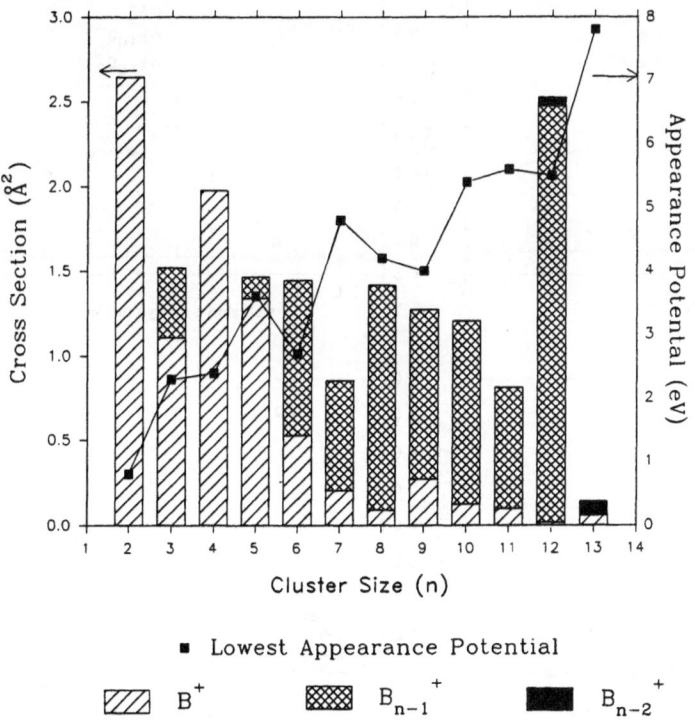

■ Lowest Appearance Potential

▨ B^+ ▨ B_{n-1}^+ ■ B_{n-2}^+

Fig. 4. Summary of CID results for B_n^+ ($n = 2$–13), showing stabilities obtained from fragmentation thresholds, and the CID cross sections at 10 eV collision energy

let's just examine two which give a very different picture of the relationship between physical and chemical behavior. Figure 5 shows the absolute cross sections for reaction of a typical small and medium size boron cluster ion with D_2 and CO_2. In the case of CO_2 only the major product channels are shown.

Consider the reaction with D_2 first. The major product channel for all size reactant clusters is D atom transfer:

$$B_n^+ + D_2 \rightarrow B_nD^+ + D \ .$$

This reaction has cluster size dependent energy thresholds, which indicate the existence of activation barriers for reaction. Measurements of the thresholds allow us to estimate lower limits on the B_n^+–D bond energy. Another interesting feature of this reaction is that the peak reactivity (i.e. peak cross section for formation of B_nD^+) has a strong even/odd alternation. Reactant cluster ions with an odd number of electrons (even number of atoms) react 2–5 times more efficiently than even-electron cluster ions. Clearly for this case, electronic structure has a profound effect on reactivity. It is important to emphasize however, that hydrogen is the *only* reactant for which we observe this kind of effect. For

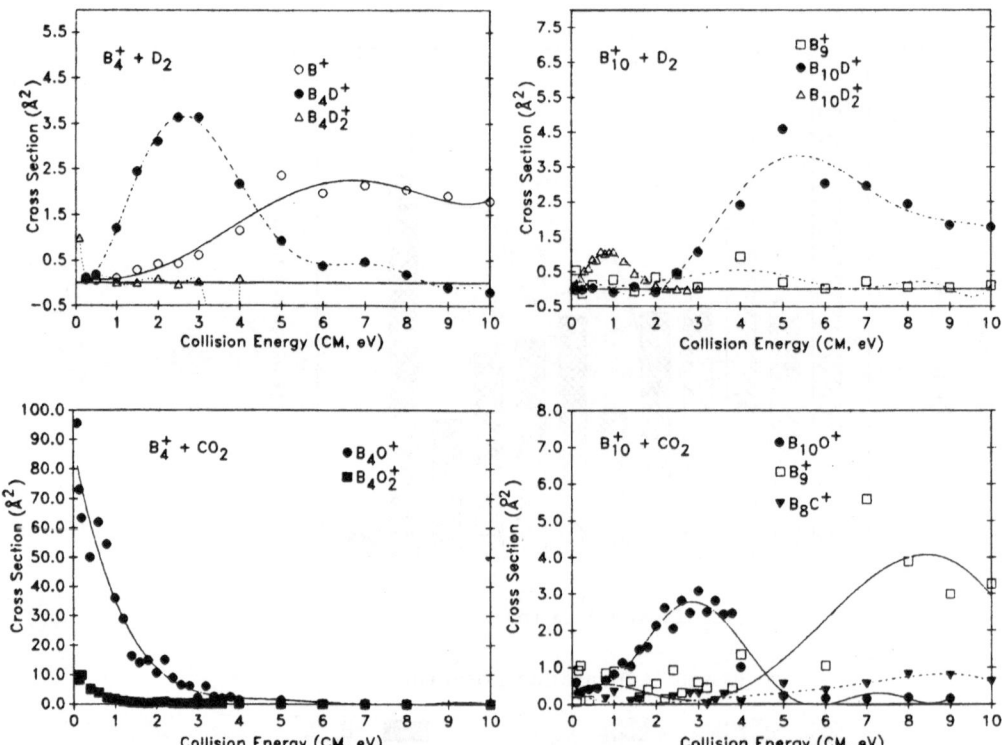

Fig. 5. Absolute cross sections for reaction of B_4^+ and B_{10}^+ with D_2 and CO_2, showing all major product ion channels

all other reactions, the dependence of reactivity on electronic structure, if any, is not simply correlated to the existence of unpaired electrons in the cluster. For boron, and probably for reaction of many other metal and semi-metal clusters, reactivity is more strongly dependent on geometric structure. Note that peak reactivity is low – only about 18% of the hard sphere collision section even for the even-sized cluster ions.

At high collision energies we observe products such as B^+ and B_{n-1}^+, which we know from our CID measurements can result from simple fragmentation. Comparison of the threshold energies shows however, that chemical reactions must be responsible for some of these products. The likely reaction is:

$$B_n^+ + D_2 \rightarrow B_n D^+ + D \rightarrow B^+ + B_{n-1} D + D \rightarrow B_{n-1}^+ + BD + D \ ,$$

i.e. decomposition of the primary $B_n D^+$ product.

A more interesting phenomenon is the observance of the $B_n D_2^+$ product, which is a metastable adduct, with lifetime just long enough to give some probability for detection. The adduct first appears for B_4^+, and becomes the dominant low energy product for clusters around B_9^+. For larger clusters such as B_{10}^+, there are activation barriers for adduct formation. The observance of this product and RRKM modeling of its lifetime as a function of energy, gives us a better estimate of the B_n^+–D bond energies.

In addition, we gain some mechanistic insight. As for most of the cluster ion reactions we have examined, the mechanism for low energy reaction of B_n^+ with hydrogen appears to involve formation and decay of an intermediate complex. In this case, D_2 is dissociatively attached to the boron cluster ion. For the larger cluster ions, the energy required to dissociate D_2 results in activation barriers to complex formation. Adducts can only be observed in reactions with the correct energetics. For many molecules, the adduct formed can decay exoergically, and its lifetime is too short to allow observation. Adducts with D_2, CO and perhaps other species can be observed because they have the right combination of strong complex binding energy, and no low energy decay pathways. As cluster size becomes large, we expect to see adducts even in exoergic reactions.

The lower half of the figure gives the cross sections for the major product channels in reaction of B_4^+ and B_{10}^+ with CO_2. The major low collision energy product for all size reactant cluster ions is production of $B_n O^+$ via the reaction:

$$B_n^+ + CO_2 \rightarrow B_n O^+ + CO \ .$$

Note the huge difference in cross section between B_4^+ and B_{10}^+. Small clusters (B_n^+, $n = 2$–$7, 9$) react with essentially 100% efficiency at low collision energies, but larger cluster ions are unreactive, and thresholds to oxidation are observed. The drop off in reactivity is due to energetics. Evidently the reaction becomes endoergic for the larger cluster ions, implying that the B_n^+–O bond energy decreases to less than 5.45 eV for $n > 9$. This decrease in reactivity with cluster ion size is something that we have observed for reaction with many neutral molecules. Our proposed explanation is that as the cluster size increases, for clusters with 3-D structures the number of bonds/atom increases. This as we have seen, makes the cluster ions more stable as they increase in size, but also leads to coordinative saturation, and reduced reactivity.

There are also a variety of other channels (not shown) which though minor, are important in that they give some insight into the mechanism. For example we observe the reaction:

$$B_n^+ + CO_2 \rightarrow B_{n-1}CO^+ + BO .$$

This is consistent with a mechanism for interaction of B_n^+ with CO_2 involving a complex of the form $[OB_nCO]^+$. The lowest energy decomposition channel for this species should be CO elimination (giving the B_nO^+ product) however BO elimination (giving $B_{n-1}CO^+$) should also be a low energy process, assuming that CO has a reasonable binding energy to the cluster ion.

At high collision energies we observe B_{n-1}^+ production. (This channel is not plotted for B_4^+ because it does not show up when plotted on a $100\,\text{Å}^2$ scale). There are two contributions to B_{n-1}^+ production:

$$B_n^+ + CO_2 \rightarrow B_nO^+ + CO \rightarrow B_{n-1}^+ + BO + CO ,$$

and simple CID

$$B_n^+ + CO_2 \rightarrow B_{n-1}^+ + B + CO_2 .$$

We observe a small amount of $B_nO_2^+$ product which was shown by pressure dependence studies to be the result of a sequential reaction:

$$B_n^+ + CO_2 \rightarrow B_nO^+ + CO$$
$$B_nO^+ + CO_2 \rightarrow B_nO_2^+ + CO .$$

In fact if we increase the scattering cell pressure to 10^{-3} Torr (20 times our normal pressure) we can drive this sequence of reactions for many steps. For example, for B_6^+ we observe $B_6O_x^+$ species containing up to seven O atoms. This raises two interesting points. The fact that we see such a long sequence of reactions means that each step in the sequence must be exoergic with no activation barriers. This is in spite of the change in composition from pure boron to boron oxide. In addition, the fact that we see very little fragmentation of the cluster during the sequence of oxidation reactions suggests that the energy deposited in the cluster-oxide product ion is less than about 1 eV, otherwise we would begin to see fragmentation. In sum, it appears that for reactant clusters for which first step of the oxidation sequence (B_nO^+ production) is exoergic, subsequent steps are also exoergic, but not very.

These experiments represent only a subset of the capabilities of beam experiments. For example, it is possible to produce beams of oxides, nitrides, ... which are products of cluster reactions, and then study their chemistry and fragmentation properties. Running in pulsed mode using guided-beam techniques [26], allows measurements of angular and recoil energy distributions without significant loss of signal. These experiments can provide further dynamical and thermochemical information, difficult to obtain by other means.

2.9.5   Chemistry Studies with ICR Methods

2.9.5.1   Tracking Sequential Chemistry

One of the most powerful features of the ICR method is the ability of follow sequences of reactions. As for any cluster reaction experiment, a distribution of cluster ions is produced, then selective excitation is used to eject all but a single mass cluster from the cell. This is allowed to react and the product distribution is monitored to give rate constants. After some time, selective excitation is again used, but now to eject all ions except one of the products of the primary reaction. One can then monitor secondary reactions of the primary product, then repeat the process to monitor further steps in a sequence of reactions. This provides a much cleaner method for probing sequential chemistry than the pressure dependent beam experiment described in the previous section. Alternatively, the product ions can be accelerated and collided with an inert target to do collision-induced-dissociation, which can provide qualitative information about the structure of product ions.

Mandich, Reents, and co-workers [27] have published a series of papers describing experiments of this type, looking at sequential reactions of silicon cluster ions with various gases. They have been greatly aided in understanding their results by quantum chemical calculations performed by *Raghavachari* [28], which shed some light on the factors which control reactivity. One example is reaction of Si_n^+ and Si_n^- ($n = 1$–7) with silane, which is of interest as a potential heterogeneous nucleation mechanism for particle growth during chemical vapor deposition of silicon.

What is observed, is that Si_n^+, $n = 1$–3, 5 undergo clustering reactions of the form:

$$Si_n^+ + SiD_4 \rightarrow Si_{n+1}D_2^+ + D_2 \quad \text{or}$$

$$\rightarrow Si_{n+1}^+ + 2D_2 \; .$$

For Si_2^+ and Si_5^+ only one step occurs, while for Si_3^+ reactant, three SiD_2 addition steps occur resulting in production of $Si_6D_6^+$. Even for reaction of Si_3^+, there are several side branches of the chain which terminate early. For positively charged reactants containing 4, 6, and 7 atoms and for all the negatively charged clusters, no reaction was observed. Clearly something inhibits reaction of the clusters or results in chain termination after only a few steps. They conclude that reaction with SiD_4 requires a divalent Si center to activate the Si–D bonds, and that for the larger clusters these are absent due to coordinative saturation.

2.9.5.2   Chemical Measurements of Cluster Ionization Potentials

ICR methods are excellent for determination of some types of thermochemical data for both neutral and charged clusters. Examples are proton affinities, bond energies, and ionization potentials. This approach has recently been applied to electron transfer measurement of neutral carbon cluster ionization potentials (IP) by *Bach* and *Eyler* [29]. The idea is very simple; mass-selected, positively charged cluster ions are allowed to collide with a reactant gas. If the carbon cluster ion is able to extract an electron from the target molecule, then its IP must be higher than that of the target, otherwise it must be lower. Repeating the experiment with a series of target species with closely spaced IPs allows the carbon cluster ionization potentials to be bracketed closely.

The experiment works as follows. A low pressure of the target molecule is admitted to the ICR vacuum system. Carbon cluster ions (6–24 atoms) are produced by laser ablation from a graphite target mounted in one wall of the ICR cell, then a series of excitation sweeps is used to eject all but a single size carbon cluster ion from the ICR. Since the laser-created cluster ions are undoubtedly hot, a pulse of argon and then SF_6 gas is injected to collisionally relax them during a 0.6 second storage period. A second series of ejection sweeps is then used to eject any cluster ions which have reacted during the cooling period. The resulting carbon cluster ions are then allowed to collide with the target molecule for periods up to 3 seconds, and the extent of electron transfer is monitored.

The resulting IPs are quite interesting. In the 6 to 24 atom size range, the IPs generally decrease with increasing size, however they observe that clusters containing 7, 11, 15, 19, and 23 atoms (i.e. $4n + 3$ atoms) have anomalously low IPs. This was tentatively attributed to the observation that the resulting cluster ion of these sizes has $4n + 2$ electrons, and is thus aromatic. IP's can also be measured by photoionization spectroscopy of the neutral, however the ICR method is much easier (once you have an FT-ICR) and avoids problems with fragmentation, multiphoton absorption, and sensitivity.

2.9.6 Chemical Identification of Isomers

As cluster size increases, it is expected that there will be an increasing number of low-lying, stable isomers for each cluster size. Particularly for semi-metal clusters where bonding is strongly directional, these different isomers may have quite different properties. So far, the most unambiguous evidence for structural isomerism comes from cluster ion chemistry studies. Let's consider the case of carbon which has been studied extensively using the ICR technique by *McElvany* et al. [11], *Parent* and *McElvany* [30], and *McElvany* [31]. In these experiments, carbon cluster ions containing up to 20 atoms are created inside the ICR cell by laser ablation from a graphite target. All ions are ejected from the trap except for the desired size cluster ion, which then is allowed to react with a low pressure of a neutral reactant. The extent of reaction is measured as a function of time, yielding rate constants for each size cluster ion.

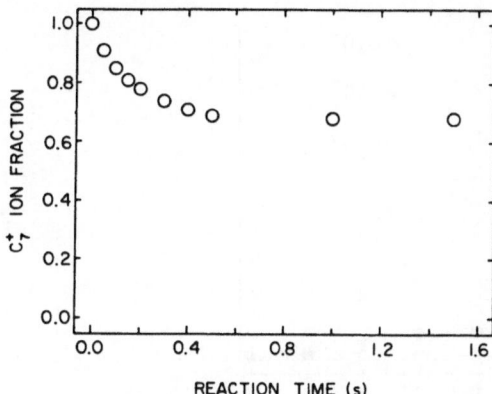

Fig. 6. C_7^+ loss from reaction with D_2, as a function of time. Note convergence to ~ 0.67 – the primary cluster ion fraction which is unreactive. (from S.W. McElvany, B.I. Dunlap, A. O'Keefe: J. Chem. Phys. **86**, 715 (1987))

They observe that for a variety of neutral molecules, reactivity decreases with increasing cluster size, and in particular there seems to be a transition occurring in the cluster size range around 7–9 atoms. Smaller cluster ions are generally reactive, while 10 atom and larger ones are much less so. This observation is consistent with theoretical work [32] which suggests (with some disagreement on details) that the most stable structures for small carbon clusters is linear, while for 10 and larger, cyclic structures are lowest in energy. Since one might expect linear structures to be more reactive, these chemical results appear to be a nice test of the structural predictions.

For the transitional clusters containing 7 to 9 atoms, something interesting is observed. For example, Fig. 6 shows the loss of C_7^+ by reaction with D_2 (yielding C_7D^+) as a function of reaction time. For smaller carbon cluster ions, a plot like this shows a simple exponential decay. For C_7^+, note that the initial ion population decays exponentially, but not to zero. Instead it asymptotically approaches a constant (unreactive) fraction of about 0.67. Similar results were obtained for reaction of C_{7-9}^+ with HCN. The clear interpretation of these results is that for these cluster sizes, the laser ablation process produces two isomers – one which has high reactivity similar to the small clusters, and one that is unreactive similar to the large clusters. This suggests that these transitional size clusters exist in both linear (reactive) and cyclic forms.

Figure 7 shows how this shows up in an ion beam experiment [23]. The cross section for the reaction: $C_7^+ + D_2 \rightarrow C_7D^+ + D$ is plotted as a function of collision energy. There appear to be two components to the cross sections – a small one that peaks at low collision energies, and a larger one that appears to have an energy threshold and peaks at about 5 eV. The low energy component indicates some probability for reaction without activation energy, which presumably is due to reaction of the linear isomer. The higher collision energy component is attributed to reaction of the cyclic isomer, which under the conditions of the ICR work (low collision energies) is unreactive. For the larger size clusters which are unreactive in the ICR, the ion beam data shows only a high energy component with substantial activation energy. Curiously, our ion beam results for some smaller clusters also show signs of two component

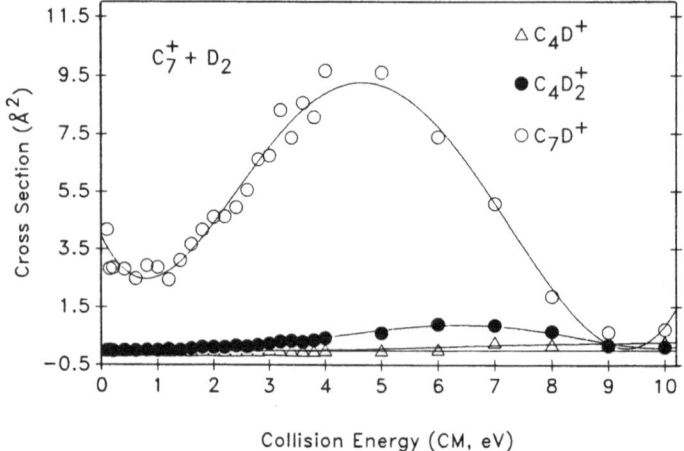

Fig. 7. Cross sections for reaction of C_7^+ with D_2 as a function of collision energy. Note the presence of both high and low energy components

behavior. This is still consistent with the ICR results, since the activation energies for reaction of the higher energy component appear to be small enough to allow reaction under the conditions used in the ICR work. These results suggest however, that even for small carbon cluster cations, laser ablation may produce both linear and cyclic isomers that are somewhat reactive at thermal energies. *McElvany* [31] has examined the properties of carbon cluster anions (C_n^-, $n = 4$–13) as well using ICR techniques and has found no evidence for structural isomerism.

2.9.7 Future Directions

In the past few years, a number of powerful methods have been developed for studying the chemistry of metal and semi-metal cluster ions. These methods also allow direct measurement of some cluster physical properties (e.g. ionization potentials), and estimation of many others (e.g. stabilities from CID, heteroatom binding energies from thresholds). This has allowed closer examination of cluster ions than has been possible for neutral clusters, and a reasonable amount has been learned about reaction mechanisms, thermochemistry, and cluster ion structures.

A real limiting factor has been the lack of theoretical or spectroscopic insight into the detailed structures of the cluster ion reactants. For clusters of simple atoms, it seems reasonable to expect that the next few years will bring computational studies which will clear up some of the many mysteries, and correct some of the mistaken interpretations which have no doubt been made. These systems are also the most promising candidates for detailed spectroscopic work. For heavy metal clusters, quantitative calculations are much more difficult, and it is not so clear that experimentalists can expect much help from theory. Even for simple systems, in the near term only a limited (but extremely valuable) set of theoretical or spectroscopic results is likely to become available. (Equilibrium properties such as geometric and electronic structures, binding energies, and the binding sites and energies for heteroatoms are obvious candidates).

Fortunately there are a number of experimental techniques which can provide some additional information on cluster ion properties and chemistry. For example, mass-selected ion dip or monitor ion spectroscopy provide a sensitive method for measuring vibrational spectra of ions. To date they have not been applied to metal or semi-metal cluster ions, however as cluster sources gain in intensity, this will no doubt be attempted. Extension of the ICR method for determination of thermochemical properties beyond measurement of IPs, should add to the database of bond energies. Both ICR and beam techniques will benefit from CID probing of cluster reaction product structures and binding energies. Beam methods will be extended to measurements of product recoil energy and angular distributions. Isomer properties can be probed by variable temperature ion sources and/or collisional or laser heating techniques.

In summary, substantial progress has been made in unraveling the reaction mechanisms and energetics for a few simple cluster ion systems. Quantitative understanding will require a great deal more work from both experimentalists and theorists. Nonetheless, it seems reasonable to predict that at least for a few simple "benchmark" systems, a full picture of the relationship between cluster physical properties and the resulting chemistry will be developed within the next few years.

2.9.8   Recent Developments

There has been considerable progress in the cluster ion chemistry field recently. For example, the ion chromatography technique has been used to sort out the isomers of carbon and silicon clusters, to study interconversion between isomers, and to examine the effects of isomerism on chemistry. This and other topics have recently been reviewed by *Parent* and *Anderson* [33]. The development of bulk fullerene synthesis/separation methods has resulted in an explosion in that area, including many ion-beam and ICR studies of chemistry. Several aspects of fullerene ion chemistry have been reviewed recently by *Schwarz* et al. [34], *McElvany* et al. [35], and *Basir* et al. [36].

References

1. Y. Ikezoe, S. Matsuoka, M. Takebe, A. Viggiano: *Gas Phase Ion-Molecule Reaction Rate Constants Through* 1986, The Mass Spectroscopy Society of Japan, Tokyo, 1987
2. L. Hanley, J.L. Whitten, S.L. Anderson: J. Phys. Chem. **92**, 5803 (1988)

3. J.O. Hirschfelder, C.F. Curtiss, R.B. Bird: *Molecular Theory of Gases and Liquids*, Wiley, New York (1964) chapter 1
4. see for example, J.H. Futrell, ed. *Gaseous Ion Chemistry and Mass Spectrometry*, Wiley, New York (1986)
5. M.R. Zakin, R.O. Brickman, D.M. Cox, A. Kaldor: J. Chem. Phys. **88**, 5943 (1988)
6. R.E. Leuchtner, A.C. Harms, A.W. Castleman, Jr.: J. Chem. Phys. **91**, 2753 (1989)
7. K.D. Kolenbrander, M.L. Mandich: J. Chem. Phys. **92**, 4759 (1990)
8. M.F. Jarrold, J.E. Bower, K.M. Creegan: J. Chem. Phys. **90**, 3615 (1989)
9. D.P. Ridge, W.K. Meckstroth: Gas Phase Inorg. Chem. 93–113 ed. by D.H. Russel, Plenum, New York (1989)
10. M.M. Ross, S.W. McElvany: J. Chem. Phys. **89** 4821 (1988)
11. S.W. McElvany, H.H. Nelson, A.P. Baronavski, C.H. Watson, J.R. Eyler: Chem. Phys. Lett. **134**, 214 (1987); S.W. McElvany, B.I. Dunlap, A. O'Keefe: J. Chem. Phys. **86**, 715 (1987); S.W. McElvany, W.R. Creasy, A. O'Keefe: J. Chem. Phys. **85**, 632 (1986)
12. see for example Y. Huang, B.S. Freiser: J. Am. Chem. Soc. **110**, 4434 (1988)
13. J.L. Elkind, F.D. Weiss, J.M. Alford, R.T. Laaksonen, R.E. Smalley: J. Chem. Phys. **88**, 5215 (1988)
14. C.B. Lebrilla, D.T. Wang, R.L. Hunter, R.T. McIver, Jr.: Anal. Chem. **62**, 878 (1990)
15. L. Hanley, S.A. Ruatta, S.L. Anderson: J. Chem. Phys. **87**, 260 (1987)
16. M.F. Jarrold, J.E. Bower, J.S. Kraus: J. Chem. Phys. **86**, 3876 (1987)
17. S.K. Loh, D.A. Hales, L. Lian, P.B. Armentrout: J. Chem. Phys. **90**, 5466 (1989)
18. M.R. Zakin, R.O. Brickman, D.M. Cox, A. Kaldor: J. Chem. Phys. **88**, 3555 (1988)
19. M.F. Jarrold, J.E. Bower, J.S. Kraus: J. Chem. Phys. **86**, 3876 (1987)
20. M.F. Jarrold, J.E. Bower: J. Am. Chem. Soc. **111**, 1979 (1989)
21. L. Hanley, S.L. Anderson: J. Chem. Phys. **89**, 2848 (1988); S.A. Ruatta, L. Hanley, S.L. Anderson: J. Chem. Phys. **91**, 226 (1989); P.A. Hintz, S.A. Ruatta, S.L. Anderson: J. Chem. Phys. **92**, 292 (1990); S.A. Ruatta, S.L. Anderson: J. Chem. Phys. **94**, 2833 (1991)
22. S.A. Ruatta, L. Hanley, S.L. Anderson: J. Chem. Phys. **89**, 273 (1988)
23. P.A. Hintz, M.B. Sowa, S.L. Anderson: J. Chem. Phys. **95**, 4719 (1991); ibid Chem. Phys. Lett. **177**, 146 (1991)
24. see for example P.B. Armentrout, J.L. Beauchamp: Acc. Chem. Res. **22**, 315 (1989)
25. S.K. Loh, L. Lian, P.B. Armentrout: J. Chem. Phys. **91**, 6148 (1989); S.K. Loh, L. Lian, P.B. Armentrout: J. Am. Chem. Soc. **111**, 3167 (1989)
26. S. Scherbarth, D. Gerlich: J. Chem. Phys. **90**, 1610 (1989)
27. M.L. Mandich, W.D. Reents, Jr.: J. Chem. Phys. **90**, 3121 (1989); W.D. Reents, Jr., A.M. Mujsce, V.E. Bondybey, M.L. Mandich: J. Chem. Phys. **86**, 5568 (1987); M.L. Mandich, W.D. Reents, Jr., V.E. Bondybey: J. Phys. Chem. **90**, 2315 (1986)
28. K. Raghavachari: J. Chem. Phys. **92**, 452 (1990), and references therein
29. S.B. Bach, J.R. Eyler: J. Chem. Phys. **92**, 358 (1990)
30. D.C. Parent, S.W. McElvany: J. Am. Chem. Soc. **111**, 2393 (1988)
31. S.W. McElvany: J. Chem. Phys. **89**, 2063 (1988); S.W. McElvany: Int. J. Mass Spectrom. Ion Proc. (in press)
32. see for example K. Raghavachari, J.S. Binkley: J. Chem. Phys. **87**, 2191 (1987); A.K. Ray: J. Phys. B **20**, 5233 (1987) D.H. Magers, R.J. Harrison, R.J. Bartlett: J. Chem. Phys. **84**, 3284 (1986)
33. D.C. Parent, S.L. Anderson: Chem. Rev. **92**, 1541 (1992)
34. H. Schwarz, T. Weiske, D.K. Böhme, J. Hrušák: In *Buckminsterfullerenes*, W.E. Billups and M.A. Ciufolini, eds. (VCH Pubs., New York, 1993), Ch. 10
35. S.W. McElvany, M.M. Ross, J.H. Callahan: Acc. Chem. Res. **25**, 162 (1992)
36 Y. Basir, Z. Wan, J.F. Christian, S.L. Anderson: Int. J. Mass Spectrom. Ion Process. (in press)

3 Embedded, Supported, and Compressed Clusters

3.1 Optical Properties of Silver Clusters in Dielectric Matrices

K.-P. Charlé and *W. Schulze*

3.1.1 Introduction

For practical applications such as e.g. heterogeneous catalysis and photography one has to accumulate a sufficient amount of metal clusters on a support or in a matrix. Especially for matrices there is the additional advantage of increased stability, because they inhibit the coagulation of the clusters. Dilute colloids of metal clusters embedded in a glass matrix, i.e. coloured glasses can be stable for hundreds of years. The basic needs for practical applications, accumulation and stability, are also advantageous for the investigation of physical properties of clusters, because they help to overcome intensity problems. On the other hand one has to pay for this gain, because the properties of clusters are affected by interactions with the environment in general. Therefore the investigation of clusters embedded in a macroscopic environment brings about the additional problem of how to recognize the influence of the environment and how to separate the intrinsic contribution to the properties under consideration.

3.1.2 Optical Absorption of Colloids

The optical properties of metal clusters can be studied conveniently by embedding them in a dielectric host, which is transparent in the frequency range of interest (VIS/UV here) and measuring the absorption of such a colloid or suspension. According to classical optics the absorption spectrum of the colloid becomes independent of the size of the clusters as quantified by their largest linear dimension D, if the size is much smaller than the wavelength λ of the incident light, i.e. if the non-retarded limit

$$D/\lambda \ll 1 \tag{1}$$

applies. This is certainly the case for $D < 100$ Å and $\lambda > 3000$ Å. This classical prediction is, however, not plausible, because one expects that the electromagnetic response of a cluster depends on its size in some way. Consequently one expects that the absorption of a colloid depends on the size of the embedded clusters in the non-retarded limit as well. The classical absorption spectrum serves as a reference in the sense that any observed change, which can be correlated with a change of the cluster size, is called "size effect".

In general there is no simple relation between the experimentally accessible size effect in the absorption spectrum of a colloid, which usually contains an ensemble of clusters with different sizes and shapes, and the size dependence of the electromagnetic response of a single cluster. To understand the limitations of this method as well as the favourable conditions, the knowledge of a few basic facts of the classical theory of colloidal absorption will be helpful.

The most important condition is that the colloids are so dilute that the electromagnetic interaction among the clusters via their induced dipole moments can be neglected. This condition is fulfilled, if the filling factor f (ratio of metal volume to colloid volume) is not larger than 10^{-2}. For dilute colloids with

$$f < 10^{-2} , \tag{2}$$

the total adsorption is just the sum over all single cluster contributions, which are proportional to the volumes v of the particles. Since the absorption coefficient of the colloid is by definition the absorption per unit volume of the colloid, this absorption coefficient is simply proportional to the filling factor f and the frequency dependence is that of a single embedded cluster averaged over all spatial orientations (taken for granted, that the clusters are randomly oriented).

If all the clusters have the same shape, the classical absorption coefficient of a dilute colloid in the non-retarded limit is given by [1]

$$\alpha(\omega) = f \cdot \sqrt{\varepsilon_h} \cdot \frac{\omega}{c} \cdot \sum_m C_m \operatorname{Im} \left\{ \frac{\varepsilon - \varepsilon_h}{\varepsilon_h + n_m(\varepsilon - \varepsilon_h)} \right\}$$

$$= f \cdot \varepsilon_h^{3/2} \cdot \frac{\omega}{c} \sum_m \frac{C_m}{n_m^2} \operatorname{Im} \left\{ - [(n_m^{-1} - 1)\varepsilon_h + \varepsilon]^{-1} \right\} , \tag{3}$$

with

ε_h = dielectric constant of the matrix,

$\varepsilon(\omega) = \varepsilon_1(\omega) + i\varepsilon_2(\omega)$ = dielectric function of the bulk metal

and $\operatorname{Im} \{ \ \}$ denoting the imaginary part of the term in the curly brackets.

n_m is the depolarization factor associated with the mth surface plasmon mode and the weight C_m is a relative measure of the dipole moment of the corresponding surface plasmon. The total number of modes as well as their depolarization factors and weights depend on the shape only. One can easily imagine that this shape dependence of the absorption spectrum will complicate both the detection and the analysis of the size effect, if several surface plasmons contribute. Therefore spherical clusters would be most suitable, because for spheres there is only one surface plasmon with depolarization factor 1/3 and weight 1. In this case the absorption coefficient has the simple form

$$\alpha(\omega) = f \varepsilon_h^{3/2} \frac{\omega}{c} \cdot 9 \left(- \operatorname{Im} \left\{ \frac{1}{2\varepsilon_h + \varepsilon} \right\} \right) \tag{4}$$

and the frequency of the surface plasmon is determined by the well known resonance condition

$$2\varepsilon_h + \varepsilon_1(\omega) = 0 \tag{5}$$

for spheres. However, since it is not possible to control the shape of clusters during growth, one has to check a posteriori, whether the spherical approximation is applicable or not.

Though it is frequently observed (electron microscopy) in agreement with theoretical expectations that large clusters ($D > 20$ Å) have polyhedral shapes with flat facets and more or less sharp edges and corners, the spherical approximation usually works surprisingly well. As an example of this experimental finding the absorption spectra of dilute colloids ($f < 10^{-2}$) of large Cu, Ag and Au clusters ($D \sim 100$ Å) embedded in solid Ar are shown in Fig. 1. The clusters are grown by the gas aggregation technique (GAT, see Chap. III) and subsequently condensed together with the matrix gas on a cooled transparent target ($T \sim 10$ K) [2].

The measured absorption (solid lines) is satisfactorily described by the calculated absorption (dashed lines) for spherical particles according to the classical formula (4), using recent experimental bulk dielectric functions (Cu, Au [3]; Ag [4]) and $\varepsilon_h = (1,29)^2$ for the Ar matrix [2]. The absolute value of the calculated absorption coefficient has been adjusted by fitting the maximum (Ag, Au) or the shoulder (Cu) to facilitate the comparison. The overall satisfactory agreement for all three metals with the classical formula indicates furthermore that for very large clusters with $D \sim 100$ Å size effects are negligible, i.e. they are at most of the order of magnitude of the experimental error and the uncertainty

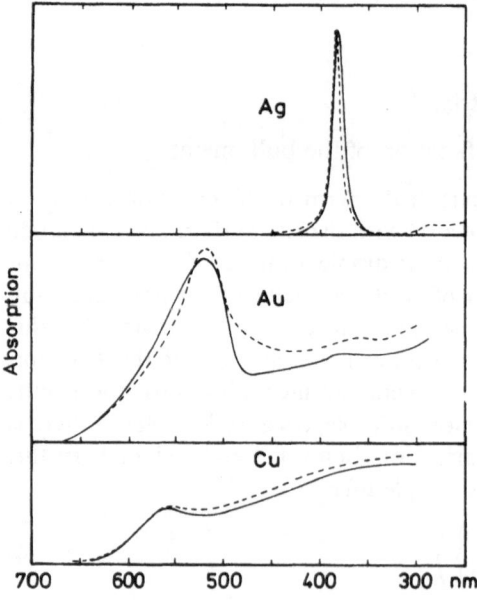

Fig. 1. *Solid lines*: Measured absorption spectra of dilute colloids of large Ag, Au and Cu Clusters (mean diameter $D = 100$ Å) embedded in solid Ar. *Dashed lines*: Calculated absorption spectra for spherical shape

associated with the scatter of the input data and, not to forget, the spherical approximation. For example, the position of the surface plasmon of Ag ($\hbar\omega = 3.25$) is in excellent agreement with the calculated position ($\hbar\omega = 3.23$ eV) as obtained from the resonance condition (5) for spherical particles. In view of the fact that even the factor 2 in this relation is only approximate (spherical approximation), the agreement is remarkable. Likewise, the FWHM (full width at half maximum) of this absorption band, which depends on the magnitude of the imaginary part $\varepsilon_2(\omega)$ and the derivative $d\varepsilon_1/d\omega$ of the real part of the bulk dielectric function near the resonance frequency, is very well reproduced ($\Gamma_{exp} = 0.1$ eV, $\Gamma_{calc} = 0.07$ eV).

Since the measured spectra of Fig. 1 agree with the prediction of classical optics, they serve as a reference for a study of the size effect (in Ar matrices). They demonstrate still another point of practical importance. It is readily seen that not all metals are equally suitable for a quantitative investigation of the size dependence of optical properties, because systematic changes of a narrow resonance band like the absorption of Ag clusters are much easier identified and more accurately to measure than changes of a smooth and almost structureless spectrum like that of Cu, which is essentially due to the strong interband absorption of bulk Cu above 2 eV. A sufficiently narrow and prominent surface plasmon absorption band is also found for Na and K, which thus would be as good as or even better candidates than Ag, since they are better suited for theory (jellium!). However, because of the high reactivity and the low melting point of the alkali metals a reliable determination of the cluster size by electron micros-copy is extremly difficult. Since electron microscopy is the only working method for large clusters, a quantitative investigation of size effects in large clusters appears not feasible at the time being. Correspondingly, up to now a sufficient amount of experimental information is available for Ag only. This holds espe-cially for the influence of the embedding matrix.

3.1.3   Size Effect and the Influence of the Matrix

3.1.3.1   The Size Effect in Ar

As cluster growth is a stochastic phenomenon, one always has to deal with size distributions. The absorption of a dilute colloid is therefore the sum over appropriately weighted contributions from clusters of different sizes and the measured size effect is an averaged one, from which the underlying size depend-ence of the absorption of a single cluster has to be recovered. For this reason a direct determination of the size distribution, which allows for reliable estimates of the mean size and the width, is necessary for quantitative studies. Of course, narrow size distributions are highly desirable, as they facilitate the analysis of experimental results considerably. Since the determination of the distribution by electron microscopy becomes more and more unreliable for mean diameters below about 20 Å (corresponding to Ag clusters with about 250 atoms), a quantitative investigation is essentially confined to the range

$$20 \text{ Å} < \bar{D} < 100 \text{ Å} \ .$$

By GAT one obtains narrow, almost normal distributions with a dispersion σ/D of about 0.2 in this range.

The final result of the direct size determination is a histrogram of a distribution, which represents the concentration fraction of clusters with a certain diameter in the colloid. This distribution is commonly referred to as size distribution. The relevant distribution for the absorption of a colloid, however, is associated with the filling factor fraction, which up to a constant is equal to the concentration fraction multiplied by the third power of the diameter for essentially spherical clusters. The corresponding filling factor distribution can thus be calculated from the experimentally determined size distribution (or the corresponding histogram) in a very simple way. The mean diameter D of this distribution will be used throughout in the following, which is a little bit larger than that obtained from the size distribution.

Figure 2 shows absorption spectra of dilute Ag/Ar colloids prepared by GAT and subsequent embedding of the clusters in solid Ar. The mean diameter D of the Ag clusters decreases from bottom to top. The spectrum at the bottom for $D = 100$ Å is the reference spectrum already discussed and shown in Fig. 1.

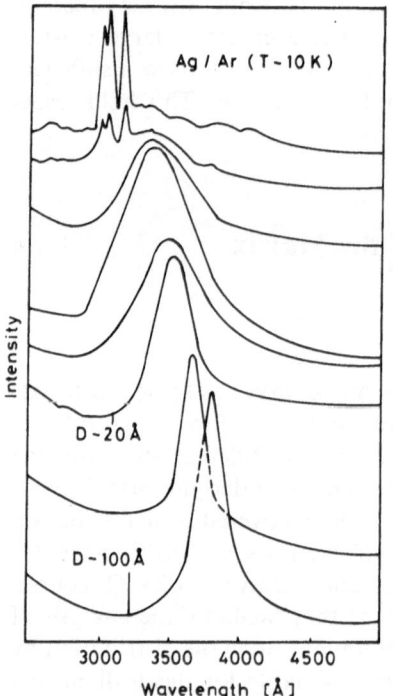

Fig. 2. Absorption spectra for Ag/Ar colloids. The mean diameter D of the Ag clusters decreases from bottom to top

A size effect is clearly visible, as the surface plasmon absorption band shifts towards higher energies ("blue shift") and broadens with decreasing diameter. The spectrum at the experimental limit for direct size determination ($D = 20$ Å) is also shown and the next three spectra demonstrate that the surface plasmon just continues to shift and to broaden for $D < 20$ Å in the same manner as before. The third spectrum in this sequence is presumably due to clusters with some ten atoms (see below and Fig. 3). In the next top spectrum the broad absorption band has almost disappeared and the spectrum at the top, which has been included for completeness, is essentially due to atoms, dimers and multimers.

A quantitative analysis (17 spectra, least square fit) of the size effect in the range 20 Å $< D <$ 100 Å reveals that both the peak position and the FWMH Γ of the surface plasmon of spherical Ag clusters embedded in solid Ar depend linearly on the reciprocal mean diameter. The exact result is [5]

$$\hbar\omega = 3.21 + 0.58 \cdot 1/D \ ,$$

$$\Gamma = 0.04 + 0.59 \cdot 1/D \ ,$$

with

$$[\hbar\omega] = [\Gamma] = \text{eV}; [D] = \text{nm} \ .$$

Note, that the intercepts of these linear regression lines agree within statistical

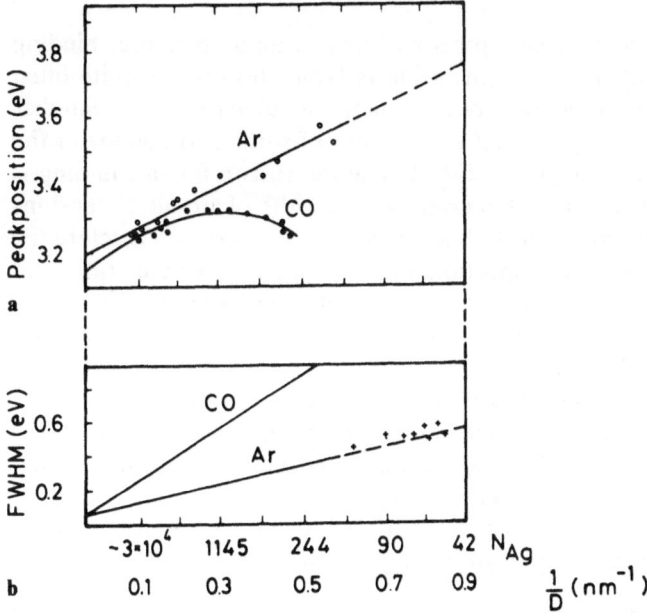

Fig. 3a, b. The position of the maximum **a** and the FWHM of the surface plasmon absorption band **b** in dependence of the reciprocal mean diameter of the particles for Ag/Ar and Ag/CO colloids. The number of atoms per article is also indicated. For the *dashed part* of the curves see text

error bounds with the reference values according to (4) and (5), and that there are already size effects for very large clusters with $D = 10$ nm containing some 10^4 atoms. As discussed above they are too small, however, to be clearly recognized in a single measurement. To explore the size effect in the range $D < 20$ Å, it is noticed that the relation between the peak shift

$$\Delta\hbar\omega = \hbar\omega - \hbar\omega_0 \tag{6a}$$

and the broadening

$$\Delta\Gamma = \Gamma - \Gamma_0 \tag{6b}$$

is independent of the cluster size for Ag/Ar in the range $D > 20$ Å, where $\Delta\Gamma = \Delta\hbar\omega$ within statistical error. It turns out that this relation holds true in the range $D < 20$ Å as well. As it stands this result proves only that both the broadening and the peak shift depend on the size in the same way. However, a few size distributions of Ag clusters with $D < 20$ Å have been determined ex-situ by (time-of-flight) mass spectroscopy. This additional information strongly suggests that the size effect remains linear for clusters consisting of some ten atoms. The corresponding extrapolation of the above experimental results for $D \geq 20$ Å to smaller sizes is shown in Fig. 3 as dashed lines.

3.1.3.2   The Influence of the Matrix

Finite size corrections to intensive properties (e.g. vapour pressure, binding energy per atom, etc.) of small systems such as large clusters are quite often found to be proportional to the reciprocal diameter or cubic root of the number of particles, i.e., they are proportional to the ratio of surface to volume of the system. From a phenomenological point of view the reason for this finding is that physical surfaces are actually thin layers with a width of atomic dimensions and specific surface properties, which differ from bulk properties. Therefore the leading correction term will be proportional to the surface to volume ratio and the constant of proportionality depends on the difference between the macroscopic surface properties and bulk properties. This general consideration applies also to the observed size effect in Ag/Ar colloids, which is not described by classical optics, because the physical nature of surfaces and interfaces is completely ignored. Adding the fact well known from surface physics that the properties of surfaces can be influenced especially by chemisorbing adsorbates, one expects also an influence of the embedding matrix on the size effect. To be specific, one expects that the position and the width of the surface plasmon absorption band of spherical Ag clusters are given by

$$\hbar\omega = \hbar\omega_0 + a \cdot 1/D + \ldots \tag{7a}$$

and

$$\Gamma = \Gamma_0 + b \cdot 1/D + \ldots, \tag{7b}$$

where eventually terms of higher order in $1/D$ have to be included as indicated. The classical reference values $\hbar\omega_0$ and Γ_0 as calculated from (4), (5) depend on the bulk dielectric function of Ag and the dielectric constant ε_h of the matrix only, whereas the coefficients a and b as well as possible higher order terms can additionally depend on the chemical properties of the matrix substance.

An experimental proof of the influence of the matrix is provided by studies of the size effect in solid Xe, C_2H_4 [5] and glass [6, 7], which all have about the same dielectric constant ($\varepsilon_h \cong 2.25$, $\hbar\omega_0 \cong 3.05$ eV). In Xe and C_2H_4 the peak shift is virtually the same as observed in Ar (see above), whereas for glass a very weak blue shift, i.e., $a \approx 0$ has been observed in one case [6] and a red shift with $a \cong -0.2$ eV nm [7] in the other case. The broadening in Xe is the same as in Ar. In C_2H_4 the coefficient b in (7b) is about twice as large as in Xe and in glass it is three times larger [6, 7]. The red shift found in [7] is possibly due to the preparation method of the colloid, which is not free from objections. Even if this result is disregarded, the remaining three examples still clearly demonstrate that the influence of the matrix on size effects can be quite drastic.

For a systematic investigation of this influence the inert gases Ne, Ar, Kr, Xe and N_2 have been studied as well as the weakly interacting gases O_2, C_2H_4 and finally the strongly interacting CO [5]. Except for Ne, where a minor quantitative change is found, the size effect in all inert matrices is the same so that Ar can be taken as representative for this group. In the more reactive matrices O_2 and C_2H_4 only the broadening of the absorption band is influenced and larger than in Ar, whereas the peak shift is the same as in inert matrices. This indicates that the broadening is more sensitive to interactions with the environment than the shift. As can be seen in Fig. 3, where the size effect in the inert Ar matrix is shown for comparison, the influence of the CO matrix is dramatic. The broadening is three times larger than in Ar like in a glass matrix and the shift is no longer linear in $1/D$. A quadratic term has to be included in Eq. (7a) to fit the size dependence of the peak position.

3.1.3.3   The Intrinsic Size Effect

Since the shift of the peak position is the same for inert and weakly interacting matrices it is likely to be an intrinsic property of large Ag clusters, which is due to an extra contribution from the surface. The theory of the electromagnetic response of plane metal surfaces [8, 9] has been adapted to a spherical geometry in [10]. The result of interest in the present context is the modified resonance condition

$$\varepsilon_1 + 2\varepsilon_h + \frac{2}{R} \, \mathrm{Re} \left\{ (\varepsilon - \varepsilon_h) d \left(\omega, \frac{1}{R} \right) \right\} = 0 \tag{8}$$

for a sphere with radius R. The complex quantity $d = d_\perp - d_\parallel$, which has the dimension of a length, is the difference between the surface response functions for electric fields normal and parallel to the surface. These surface response functions account essentially for the general non-locality of the electromagnetic response of conduction electrons, which cannot be neglected in strongly in-homogeneous regions such as met around surfaces. It is this non-locality, which renders the familiar classical boundary conditions of electromagnetic theory invalid and enforces a detailed examination of the surface region [8, 9]. As $\varepsilon_2(\omega)$ for Ag is much smaller than $(\varepsilon_1(\omega) - \varepsilon_h)$ in the frequency region of interest here $(3.0 \text{ eV} \leq \hbar\omega \leq 3.6 \text{ eV})$, the size dependent correction term in (8) depends essentially on $\text{Re}\{(d_\perp - d_\parallel)\} = \delta$, which can thus be determined by inserting into Eq. (8) the measured resonance frequencies for various cluster diameters and matrices. The result of this analysis is within experimental error independent of both the matrix (with Ne included) and the cluster size at least in the range $D \geq$ 20 Å. In the above quoted accessible frequency range it can be represented as

$$\delta(\omega) = 0.87 \left(1 - \frac{\omega}{\omega_p}\right)^{-1/2} \text{Å}$$

with ω_p being the actual frequency of the volume plasmon of Ag ($\hbar\omega_p = 3.78$ eV [4]). Since a more detailed consideration [11] shows that this experimental result compares well with the theory of electromagnetic surface response, the observed size dependence of the position of the surface plasmon absorption band in inert and weakly interacting matrices, namely the blue shift, appears to be an intrinsic property of large Ag clusters.

In summary, an extensive experimental investigation of the size dependence of the surface plasmon absorption of large Ag clusters embedded in various substances shows that the observed size dependence originates from the specific optical properties of metal surfaces. In this way the possible influence of the environment, which increases with decreasing cluster size, is readily understood.

References

1. R. Fuchs: Phys. Rev. **B 11**, 1732 (1975)
2. H. Abe, W. Schulze, B. Tesche: Chem. Phys. **47**, 95 (1980)
3. P.B. Johnson, R.W. Christy: Phys. Rev. **B 6**, 4370 (1972)
4. G.B. Irani, T. Huen, F. Wooten: J. Opt. Soc. Am. **61**, 128 (1971)
5. K.-P. Charlé, F. Frank, W. Schulze: Ber. Bunsenges. Phys. Chem. **88**, 350 (1984)
6. L. Genzel, T.P. Martin, U. Kreibig: Z. Phys. B – Condensed Matter **21**, 339 (1975)
7. J.D. Garnière, R. Rechsteiner, M.A. Schmithard: Solid State Commun. **16**, 113 (1975)
8. P. Feibelman: Prog. Surf. Sci. **12**, 287 (1982)
9. F. Forstmann, R.R. Gerhards: *Springer Tracts in Modern Physics*; Vol. 109 (Springer 1986) especially Chpt. 5
10. P. Apell, Å. Lyungbert: Solid State Commun. **44**, 1367 (1982)
11. K.-P. Charlé, W. Schulze, B. Winter: Z. Phys. D – Atoms, Molecules and Clusters **12**, 471 (1989)

3.2 Aerosols, Large Clusters in Gas Suspensions

H. Burtscher and *H.C. Siegmann*

3.2.1 Introduction

Small particles in gas suspension (in air aerosols) have played an important role in the development of modern physics. We recall the discovery of the Brownian motion, the study of light scattering which explains the spectacular phenomena when the sun is rising or setting, and the famous determination of the elementary charge by Millikan. However, this paper will deal with much smaller particles in the submicron range with diameters between 0.1 and 0.001 µm. These particles are practically invisible as the scattering cross section for light becomes very small. Yet, particles of this size range are omnipresent in the air in which we live. Normally, their concentration ranges between 10^4–10^6 particles/cm^3. Depending on the chemical composition, the particles can be toxic or lead to a series of undesired chemical reactions. However, without these particles, there would also be no cloud formation and hence no rain. Today still very little is known about these particles, specifically their surface properties. The reason for this gap in our knowledge is the lack of experimental methods to investigate particularly the surface properties. This has to be done in situ, as the particles, when precipitated in a filter, will react with the substrate and the other particles in the filter and thus change their surface chemical composition so that not much insight can be gained from the chemical analysis of precipitates regarding the initial state of the surface when the particle is still suspended in its original environment. Therefore, it is essential to develop experimental methods with which these submicron particles can be studied in their natural gaseous environment and with which the change of the surface chemistry can be observed when spurious amounts of other gases are admitted or when light is incident inducing photochemical reactions.

In the following first some tools, frequently used in aerosol science will be introduced. Then some features of photoelectric charging, which is one important method to achieve the objective mentioned above, will be presented.

3.2.2 Some Tools of Aerosol Science

If one looks at particles on a substrate, one usually is stuck with particles of various sizes. However, in the case of particles in gas suspension, one has the possibility to select a quite narrow size range of the particles by means of a differential mobility analyzer (DMA) [1] shown in Fig. 1. An electric field is generated between the inner rod and the outer tube by applying an electrical voltage. Gas which is free of particles ("sheath air") flows through the cylindrical condensor in a laminar flow. The carrier gas with electrically charged particles enters through an outer ring, and the particles acquire a drift velocity $v = b \cdot E$ in the electric field E according to their electrical mobility b. As b depends on the friction in the gas, it is related to the aerodynamic radius R of the particles. The drift velocity v is perpendicular to the carrier gas flow and only particles with a specific velocity v will enter through the slit in the inner electrode and can thus be separated.

Another remarkable instrument of aerosol technology is the diffusion battery [2], where particles can be size-selected by their diffusion constant. For example the penetration P of a screen type diffusion battery [3] is given by

$$P = \exp(-cD^{2/3}) \, ,$$

where D is the diffusion constant of the particles. The constant c depends on the flow rate and the geometry of the battery. For $R \leq 100\,\mathrm{nm}$, the free molecule relation $D \propto R^{-2}$ is a good approximation. If the penetration rises from P to P',

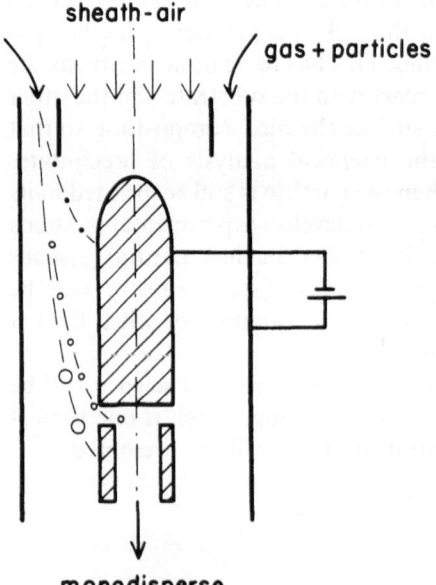

sheath-air

gas + particles

monodisperse
particles

Fig. 1. Differential mobility analyzer (DMA) to select a narrow size range of charged particles

the radius increases by $\Delta R = R_0 [(\ln P'/\ln P)^{-3/4} - 1]$. If the initial penetration is small, say $P = 10^{-2}$, even very small increase in the size of the particles by $\Delta R = 0.1$ nm will lead to a measurable change of P. This means that the growth of a particle of, e.g. $R = 10$ nm, by adsorption of 1 monolayer of molecules can still be detected [4].

A further very useful instrument is the condensation nucleus counter with which less than 1 particle/cm³ can readily be detected. The particle is brought into a gas with a supersaturated vapor for which it acts as a condensation nucleus. After sufficient condensation of vapor, the particle grows to an optical size and can be detected by light scattering [5].

3.2.3 Diffusion Charging of Particles

As shown in 3.2.2., particles can be size selected by measuring the electric mobility as soon as they carry an electric charge. Electrical charging of the particles can be achieved by generating ions in the gas in which the particles are suspended and letting the ions diffuse onto the particles. The ions are generated by a radioactive source [6], or by an electrical discharge [7], as in the electrical filters that are widely used to clean exhaust gases from industrial processes. This charging of the particles by diffusion depends on the size of the particles, but not on their chemical nature or their surface properties. In the case of particles with $R \gg \lambda$, where R is the radius and λ the mean free path in the gas, the size dependence of diffusion charging can be easily derived by considering a spherical particle of radius R, and a large sphere of radius r concentric to the particle. The current of charged gas molecules through the sphere with radius r is given by

$$4\pi r^2 \left(D\frac{dn}{dr} + bn(r)\frac{dU}{dr} \right) = \text{const.} , \tag{1}$$

where n is the number of charged gas molecules per cm³, D the diffusion constant and b the electrical mobility. U is the potential energy between the diffusing gas molecule and the particle. For initially uncharged particles, $U = 0$, and (1) can easily be integrated yielding $n(r) = -\text{const}/4\pi rD + n(\infty)$ and $i = 4\pi r^2 \, dn/dr = 4\pi RD(n(\infty) - n(R))$, where $n(\infty)$ is the concentration of charged gas molecules at infinite distance from the particle and $n(R)$ the concentration of charged gas molecules at the particle surface. This can of course be applied also to uncharged gas molecules diffusing to the particle surface; $n(R)$ depends then on the sticking probability of these gas molecules at the particle surface. In the case of charged gas molecules, the sticking probability is unity and $n(R) = 0$. The rate of diffusion charging is then given by $i = 4\pi RDn(\infty)$. It becomes smaller as the particles radius R decreases. This explains why very small particles cannot be efficiently removed from industrial exhaust in electrofilters. These particles are then found in great densities in the atmosphere and often have lifetimes of month or even years particularly if they are hydrophobic, that is if they cannot act as condensation nuclei for water. It should be noted that a correction to the above formula becomes necessary if the particle radius R is much smaller than the mean free path of the molecules. There is then a ballistic fall of the gas molecule close to the particle just as in vacuum and the charging rate decreases even faster with R for $R < 100$ nm as was shown by *Hoppel* [8].

3.2.4 Photoelectric Charging of Particles

This can now be contrasted with photoelectric charging of the particles, where the gas containing the particles is irradiated with ultraviolet light. The energy hv of the photons has to be below the ionization threshold of the gas molecules, but above the photoelectric work function of the particle. In this way, the gas molecules remain electrically neutral, but the particle can absorb a photon and emit a photoelectron. This process obviously depends on the chemical composition of the particle and particularly on the state of its surface, as the photoelectric work function is a very sensitive function of the nature and amount of adsorbates at the surface. However, whether or not the particle will remain with a positive charge depends on whether or not the photoelectron diffuses back to it. Hence photoelectric charging depends also on diffusion charging, but in an opposite sense. It is large, when the backdiffusion probability of the photoelectron is small. This shows that precisely those particles, that cannot be charged efficiently by diffusion charging, will be charged with great efficiency by photoelectric charging and vice versa. It turns out that in air at ambient pressure and temperature, the limiting size range for photoelectric charging is just around a radius of $R \approx 1$ μm. This arises, because $R/\lambda_e \approx 1$ for $R = 1$ μm where λ_e is the free path of electrons. It should also be noted that the photoelectron attaches itself to an oxygen molecule after being thermalized by collisions with air molecules and forms a negative oxygen ion in $\sim 10^{-5}$ s with much smaller electrical mobility and diffusion constant. The fact that the natural limit of photoelectric charging is around a particle size of 1 μm means that exactly the particle size range that can penetrate through the human respiratory tract far into the deepest part of the lung [9] may be measured selectively. Hence photoelectric charging should become important for surveying the air quality just for this reason alone. There are however more important reasons why it has very great potential for improving and surveying the quality of the air which has deteriorated to a deplorable extent in recent years due to various factors including excessive use of the automobile.

It is then expected that photoelectric charging of particles in gas suspension has the following 3 important characteristics:

1) It is effective only for particles below a certain radius R_0. R_0 depends on the diffusion of the electrons; it is different in carrier gases where negative ions are

formed such as air as opposed to gases in which the photoelectrons remain free such as He or clean N_2.

2) It depends sensitively on the energy hv of the light and on the photoelectric work function of the particles. Of particular interest are electropositive adsorbates, that is adsorbates that donate part of their electrons to the particle when adsorbed at the surface. In this case an electric dipole layer is formed at the surface with the positive pole at the particle gas interface. This means that the photoelectric work function Φ is lowered when the molecule is adsorbed. If now hv is chosen such that it is below the work function Φ_0 of the "naked" particle surface, but above the work function Φ of the surface with adsorbates, only those particles can be photoelectrically charged that have adsorbates on their surface. This leads to an incredibly sensitive yet very simple detection scheme for electropositive adsorbates. For instance in the interesting case of polycyclic aromatic hydrocarbons adsorbed on carbon particles, detection sensitivities of 10^{-9} g/m^3 have been realized with a small commercial low pressure mercury lamp as light source.

3) For weak light intensities, photoelectric charging should be proportional to the photoelectric yield Y of the particles, to the density of the photon current j_{hv}, and to the part F of the total surface of the particles that has a work function $\Phi \leq hv$. The number of photoelectrically charged particles can simply be measured by letting the gas carrying the particles flow through a filter in which all particles, charged and uncharged, are caught, and by measuring the current i that flows to ground potential. We have then for weak light intensity:

$$i = Y j_{hv} F \ . \tag{2}$$

If, on the other hand, the intensity of the light is strong, the particles acquire a limiting charge $P_{max}(hv)$ that is reached when a subsequent further photo-electron can no longer escape from the Coulomb potential of the previously charged particle. The photoelectric work function of the particle is given by

$$\Phi(R) = \Phi_\infty + \frac{e^2(p+1)}{4\pi\varepsilon_0 R} - \frac{5}{8}\frac{e^2}{4\pi\varepsilon_0 R} \ , \tag{3}$$

where Φ_∞ is the work function of a flat surface and p the number of positive charges which are already on the particle. This was shown by *Wood* [10] and others; the last term is the image potential. From the condition $hv \leq \Phi(R)$ for the emission of a photoelectron from a particle, we obtain for the limiting charge obtainable in photoelectric charging:

$$P_{max}(hv) = \frac{hv - \Phi_\infty}{e^2} 4\pi\varepsilon_0 + \frac{5}{8} \ . \tag{4}$$

We see that much larger charges are expected in photoelectric charging as compared to diffusion charging where the limiting charge is given by the thermal energy of the diffusing ion. With sufficiently large photon energies hv as avail-

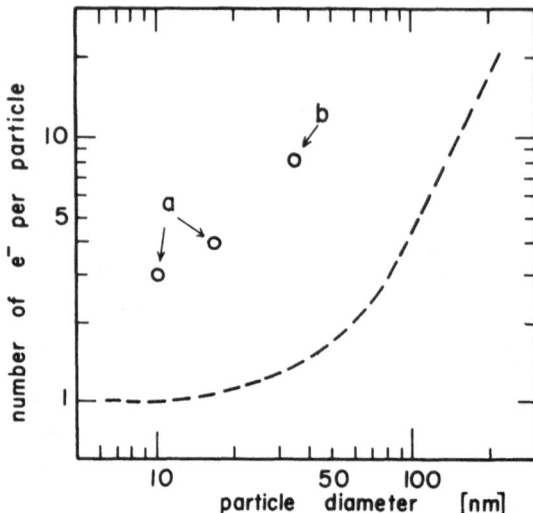

Fig. 2. Number p of elementary charges on a particle vs. particle diameter as obtained in field charging after Dennis [12] (*dashed line*), and as obtained by **a** photoelectric charging of Ag-particles with a low pressure mercury discharge (4.9 eV) [11], and **b** photoelectric charging of C-particles with an excimer-laser of 6.4 eV photon energy, from [21]

able in modern synchrotron radiation sources, the charge on the particle could be large enough to let it explode by the Coulomb repulsion of the charges.

Multiple charging of particles has been observed in experiments by *Jung* et al. [11] for silver particles using a 4.9 eV Hg-low pressure lamp. The maximum number of charges on the particles is found to agree with the prediction of Eq. (4).

Figure 2 compares the number of elementary charges p that can be put on a particle with radius R by field charging to the one obtainable in photoelectric charging. In field charging the electrons migrate along electric field lines and acquire a much higher energy than in diffusion charging. Therefore they have a higher probability to overcome the repulsive Coulomb potential for multiple charging. Field charging is the most efficient of the commonly used charging mechanisms [12]. It is obvious that much higher charging is achieved by photoemission. The limit charge is given by the photon energy, the threshold $\Phi(R)$ and the particle radius R.

3.2.5   The Photoelectric Yield of Small Metal Particles

For basic investigations of photoemission from particles, experiments with Ag- and Au-particles suspended in very clean He gas seem most appropriate. Figure 3 shows the apparatus used to measure the absolute photoelectric yield Y from the metal particles vs. particle diameter [13]. Before each measurement, the apparatus is evacuated to ultra high vacuum (10^{-9} mbar) and baked. Then very clean He gas as obtained by evaporating liquid He is admitted. The He gas is at atmospheric pressure and temperature, and a flow of 5 1/min through the apparatus is maintained. A silver wire of 99.999% purity is heated in the flow of

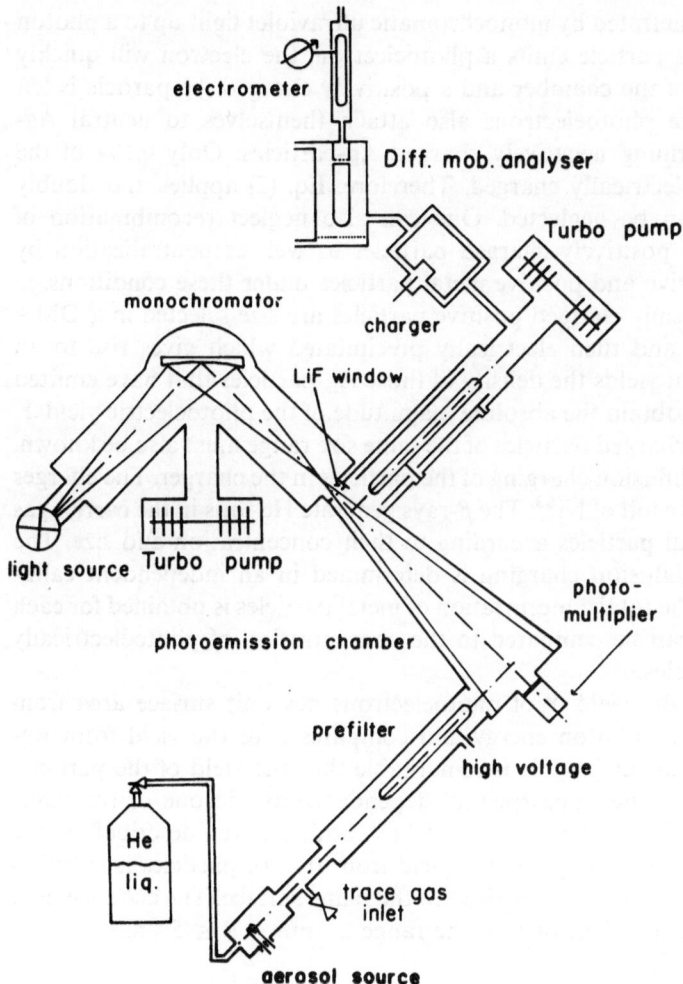

electrometer

Diff. mob. analyser

Turbo pump

monochromator

charger

LiF window

light source   Turbo pump

photo-
multiplier

photoemission chamber

prefilter

high voltage

He
liq.

trace gas
inlet

aerosol source

Fig. 3. Apparatus to measure the photoelectric yield of metal particles suspended in He-gas as a function of the radius of the particles

He. The Ag-atoms evaporating from the hot Ag-wire are cooled by collisions with the He-gas and condense by homogeneous nucleation into particles ranging from 2–10 nm. Electron microscopy on precipitated particles reveals that the particles obtained by this method have the form of spheres. The smaller particles from 2–5 nm are single crystals, but in the larger ones more than one crystalline orientation may exist.

There is then a possibility to admit traces of a gas like O_2 into the He. The trace gas has about 10 s to adsorb on the particle surface. Simultaneously, the particles charged in the process of production are removed from the flow in the electric field of the prefilter. The neutral particles enter the photoemission chamber which is penetrated by monochromatic ultraviolet light up to a photon energy of 11 eV. If a particle emits a photoelectron, the electron will quickly diffuse to the walls of the chamber and a positively charged Ag-particle is left behind. Some of the photoelectrons also attach themselves to neutral Ag-particles thereby forming negatively charged Ag-particles. Only 0.1% of the particles are photoelectrically charged. Therefore, Eq. (2) applies and doubly charged particles can be neglected. One can also neglect recombination of photoelectrons with positively charged particles as well as neutralization by coagulation of negative and positive metal particles under these conditions.

The photoelectrically charged positive particles are size selected in a DMA as shown in Fig. 1 and then electrically precipitated which gives rise to an electric current which yields the density of those Ag-particles that have emitted a photoelectron. To obtain the absolute magnitude of the photoelectric yield Y, the density of the uncharged particles of the same size range must also be known. This is achieved by diffusion charging of the particles in the charger. The charger contains a radioactive foil of Ni^{63}. The β-rays generate He-ions in the carrier gas that charge the metal particles according to their concentration and size. The percentage of this diffusion charging is determined in an independent calibration. In this way, the total concentration of metal particles is obtained for each range of sizes and can be compared to the concentration of photoelectrically charged metal particles.

Figure 4 shows the yield Y of photoelectrons per unit surface area from a plane Ag-surface vs. photon energy and compares it to the yield from Ag-particles with $R = 3.8$ nm [14]. It is remarkable that the yield of the particles near threshold shows the same spectral dependence as the one of the plane surface, hence apart from a kink at $hv = 4.9$ eV, both are well described by the law of Fowler Nordheim [15]. Yet the yield from the Ag-particles is substantially larger by a factor of 200, compared to the plane surface. The enhancement turned out to be independent of R in the range 2.7 nm $\leq R \leq$ 5.4 nm.

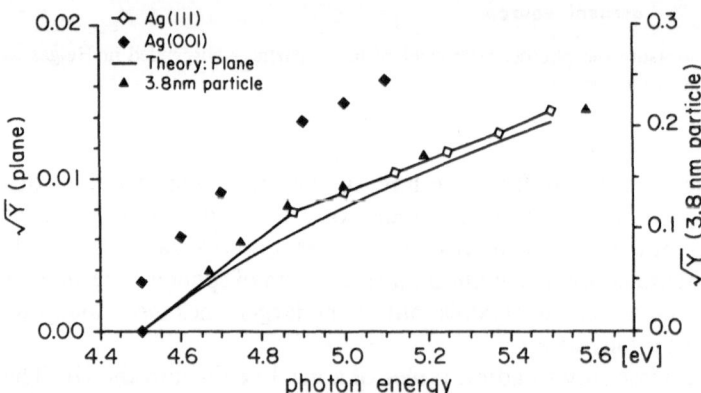

Fig. 4. Photoelectric yield (emitted electrons/incident photon) of a plane Ag-surface compared to the yield from Ag particles with a radius of 3.8 nm. The left abscisse is for a plane surface, the right one for the particles. Note the much larger photoyield of the latter

Figure 5 shows that the photoelectric threshold $\Phi(R)$ which is obtained from extrapolating $Y \to 0$ is in perfect agreement with electrostatics [10].

Figure 6 finally shows the photoelectric yield of Ag-particles with $R = 5$ nm up to a photon energy of 11 eV. It is remarkable that a photoelectric yield close to unity can be reached at photon energies of 11 eV. The full curve is considerably different from that of a plane surface [16] but may be derived by a simple

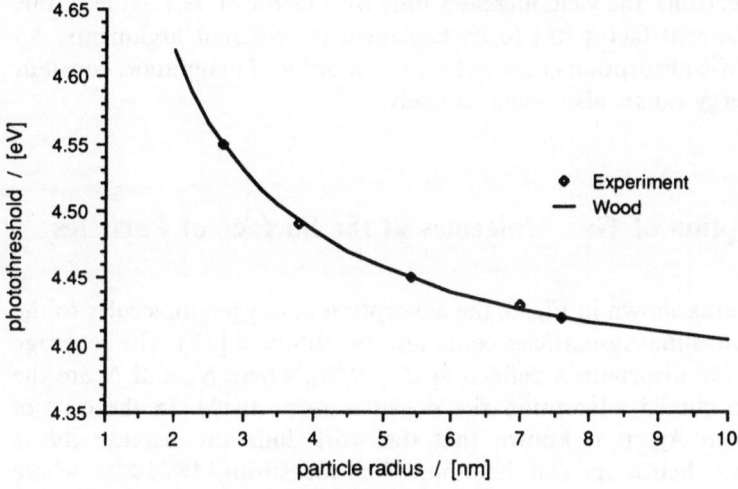

Fig. 5. Photoelectric threshold Φ vs. radius of Ag-particles. Full curve is obtained from simple electrostatics according to Wood [10]

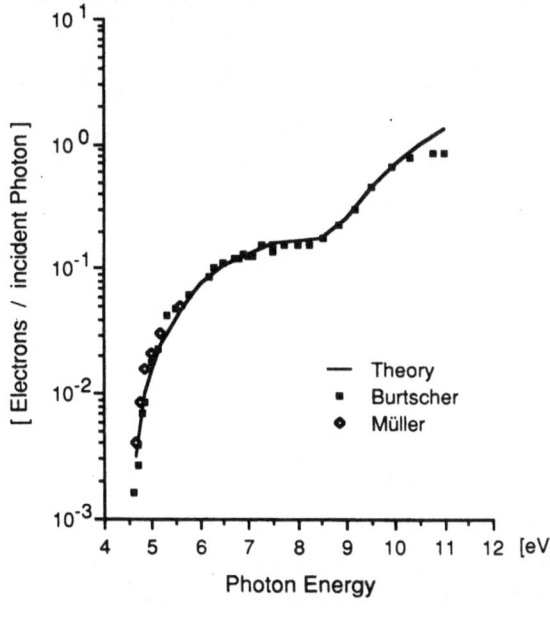

Fig. 6. Photoelectric quantum yield of Ag-particles with $R = 5$ nm. The full curve is obtained from a three step model assuming an escape function adapted to the spherical shape of the particles and a large mean free path of the photoexcited electrons [17]. The theoretical curve is further multiplied by a constant factor to achieve best agreement with the data. The experimental data is taken from *Burtscher* et al. [16], and *Müller* [13]

application of the familiar 3-step model of photoemission, yet accounting for the spherical shape of the particles [17]. It explains the anomalous shape of the yield curve compared to a plane Ag-surface. However, these geometrical arguments can not explain the enhancement of the yield in the small particles. The absolute value of the yield obtained from the simple theory is ∼ 40 times lower than the observed yield over the entire range of photon energies. When the escape probability is set to unity which is equivalent to an infinite mean free path of the photoexcited electrons, the yield increases only by a factor of ≈ 7. At least this residual enhancement factor has to be explained by different arguments. An increase of the Mie absorption cross section by an order of magnitude, constant over a wide energy range, also seems unlikely.

3.2.6   Adsorption of Gas Molecules at the Surface of Particles

With the apparatus shown in Fig. 3, the adsorption of oxygen molecules to the surface of free ultrafine Ag-particles could also be observed [18]. The coverage of a surface by an adsorbate is defined as $\theta = N/N_0$ where N_0 and N are the total and the occupied adsorption site densities respectively. In the case of O_2-adsorption on Ag, it is known that the work function increase $\Delta\Phi$ is proportional to θ, hence one can determine θ by measuring $\Delta\Phi/\Delta\Phi_{max}$, where $\Delta\Phi_{max}$ is the increase of the work function when all adsorption sites are occupied. Figure 7 shows $\Delta\Phi/\Delta\Phi_{max}$ as function of the O_2 concentration. From θ, one can in turn evaluate the sticking parameter α, as the partial pressure of O_2 in the He gas is determined with a mass spectrometer. In contrast to the findings

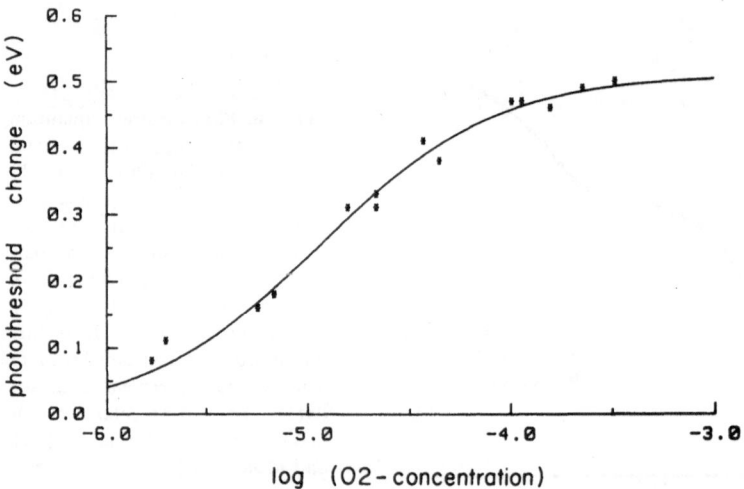

Fig. 7. Oxygen adsorption on small silver particles: threshold change $\Delta\Phi/\Delta\Phi_{max}$ of the particles due to oxygen adsorption on the particle surface as function of O_2 concentration in the gas [18].

on macroscopic surfaces, it was found that $\alpha = \alpha_0(1 - \theta)^2$. This indicates simple dissociative adsorption with no physisorbed precursor state. Additionally, α_0 is smaller by 2 orders of magnitude with small particles compared to the flat Ag-surface. It is clear that such measurements are crucial to the understanding of catalytic phenomena on finely dispersed catalysts, and it is definitely necessary to do more measurements of this type on different materials.

Other experiments concerning the adsorption of molecules onto fine particles were performed by *Burtscher* and *Schmidt-Ott* [4] and by *Niessner* [19].

The experimental set-up is shown in Fig. 8. Fine C-particles are generated by electric sparks between C electrodes in Ar-gas (impurities < 1 ppm) [20]. The particles are of spherical shape with a radius $R \cong 50$ nm. They are carried by the flow of the Ar-carrier gas into a neutralization chamber, in which a radioactive source produces enough ionized Ar^+ and electrons to establish a well defined charge distribution on the particles; with the temperature and particle sizes present, doubly charged particles do not occur. Hence the following DMA, compare Fig. 1, selects one size of negatively charged C-particles. The stream of the Ar-gas containing these particles now passes through a section in which an elevated partial pressure of perylene is maintained by appropriately heating

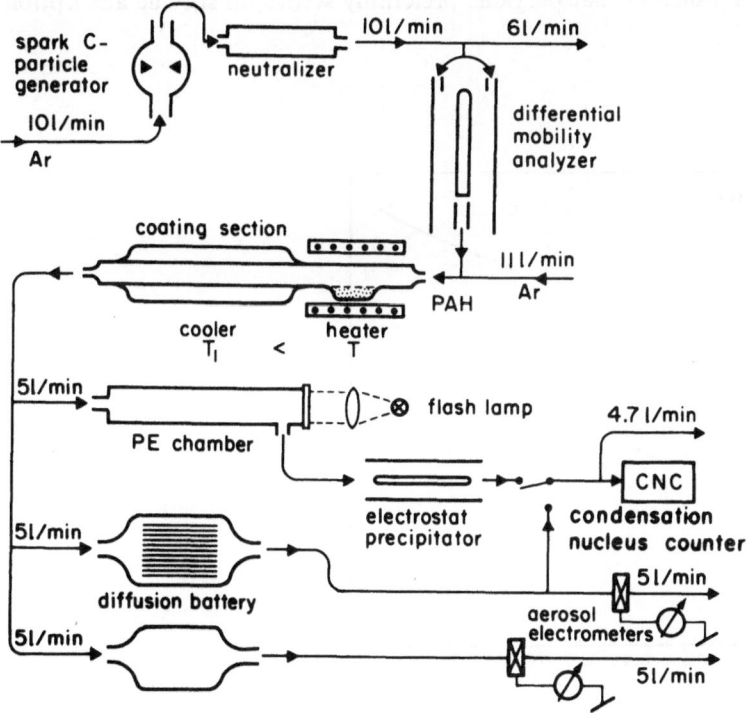

Fig. 8. Experimental set-up for observing the adsorption and condensation of perylene on C-particles, from [4]

a perylene container. In the subsequent cooler section, the molecules condense onto the C-particles. The stream containing the particles can now enter a photo-ionization chamber in which the negatively charged particles may be neutralized by emission of a photoelectron. After removal of all particles that have not emitted a photoelectron, the number of neutral particles, that is the number of the photoelectrically active particles, is determined in a condensation nucleus counter. The stream of gas molecules can also be directed through a diffusion battery, and the transmission of this battery can simply be measured by register-ing the current of negatively charged particles. If the particles have grown in size by adsorption of perylene, the transmission of the diffusion battery changes and the physical growth can be determined. The penetration of the diffusion battery is directly related to the growth ΔR of the aerodynamic radius of the particles in the coating section. Finally, the flow of the charged and coated particles can also be directly measured to insure and check the constancy of the concentration of the C-particles before they enter the diffusion battery. Figure 9 shows results obtained with this setup in the range of the monolayer adsorption of perylene on C-particles with $R = 6.7$ nm [21]. Plotted is the photoelectric activity ε with a pulsed mercury high pressure arc vs. the increase of the aerodynamic radius ΔR. First, it is noted that adsorption of perylene obviously increases ε. Second, one observes 2 flat sections in ε vs. ΔR, which indicates 2 different adsorp-tion sites. This indicates that perylene preferably settles on specific adsorption sites.

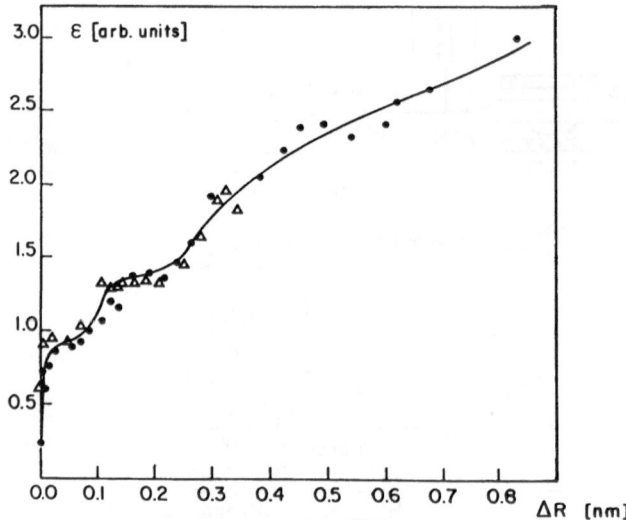

Fig. 9. Probability ε of photoelectric removal of a previously acquired negative charge of a C-particle of $R = 6.7$ nm vs. the increase ΔR of the aerodynamic radius of the particle by adsorption of perylene, from [10]

3.2.7   Applications of Photoemission of Electrons from Particles in Gas Suspension

Several interesting applications of photoemission on gas suspended particles have been identified. They include the observation of submicron particle properties in the plume of volcanoes [22], and the investigation of particle formation in combustion processes [23, 24]. These applications make use of the fact that photoelectric charging can be achieved without electrodes, hence can be done in aggressive environments, and is sensitive to the condensation or chemical reactivity of matter on the particle surface as the hot gases containing the particles mix with air and/or become cooler. Other applications such as monitoring the air quality make use of the fact that photoelectric charging of particles is selective to respirable particle size ranges and also responds specifically to particles coated with polycyclic aromatic hydrocarbons that contain highly carcinogenic species such as benzo(a)pyrene [25].

As an example it will be shown how photoelectric charging of particles can be used to control the combustion of oil in a stove and may serve to achieve maximum efficiency of the combustion at a minimal rate of pollution of the environment with soot particles [24]. The exhaust gases are irradiated with ultraviolet light from a low pressure mercury discharge. Photoelectrons are emitted from the particles and a positive charge of the aerosol is detected. The experiments with adsorption of perylene on C-particles [4] described above lead to the hypothesis that polycyclic aromatic hydrocarbons (PAH) with three or more rings are present at the surface of exhaust particles if the combustion process proceeds with lack of oxygen, and causes the emission of electrons. Hence the presence of PAH in the exhaust gas can readily be detected by photoelectric charging and indicates that the combustion was incomplete e.g. due to lack of oxygen. A feedback mechanism can then regulate the air inlet in such a way that the PAH-signal just disappears. It should be noted that excess of air is also unwanted as it simply cools the furnace.

However, other substances may also condense on the exhaust particles and quench photoemission. For example films of water or acids have been found to be very effective in quenching photoelectric charging of particles. Hence the photoemission chamber has to be maintained at an elevated temperature to prevent the condensation of acids and water. A detailed description of an appropriate device to achieve photoelectric charging of particles generated in combustion is given in [23].

Figure 10 shows results obtained with a small domestic oil stove. The photoelectric signal (aerosol photoemission APE) shows a steep increase close to the maximum value of the efficiency η of the burner. Comparing the APE-signal with the oxygen signal reveals that the changes in the concentration of oxygen are much smaller. Therefore, the measurement of the oxygen concentration for combustion regulation must be very precise, in fact to $\approx 0.1\%$, to achieve reliable combustion regulation. However, even a very rough measure-

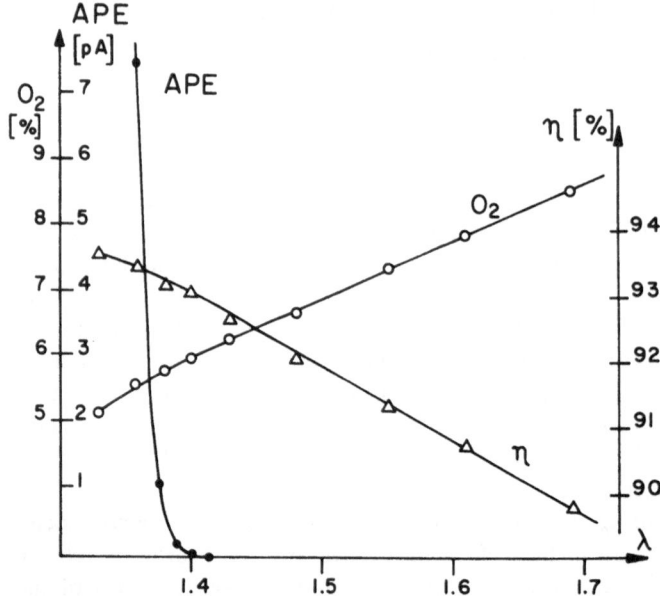

Fig. 10. The positive current from photoelectrically charged exhaust particles (APE), the efficiency η of the combustion, and the oxygen concentration in the exhaust gases of an oil furnace is plotted against the air/fuel ratio λ

ment of photoelectric charging provides an accurate criterium to find the optimum value of the air/fuel ratio λ.

The sharp increase of the photoelectric signal occurs when soot and CO are produced in the combustion. The soot production is commonly measured by the Bacharach number which is a logarithmic optical scale for the amount of soot. The relation between log (photoelectric signal), Bacharach number and log (CO-concentration) was found to be linear over more than 3 decades. It should be noted that relatively small changes in the air supply already cause a large increase in the amount of soot and CO, and also the photoelectric charge of the aerosol.

The PAH concentrations on filter samples of the exhaust has also been determined by gas chromatography [26]. A linear relation between the photo-emission signal and the chemically determined total PAH-concentration has been found. This is in agreement with the hypothesis that the lowering of the work function is actually due to condensation of PAHs or due to formation of PAHs on the surface of the exhaust particles, which are in fact submicroscopic soot particles. The linear relation between PAH-concentration and photo-electric signal is predicted by Eq. (2).

Practically all particles occurring in large quantities in the ambient air in cities, near traffic areas, and in cigarette smoke are produced in incomplete combustion and contain PAHs. An important source are Diesel engines. As the PAHs contain extremely carcinogenic species such as benzo(a)pyrene and as the photoelectrically detected particles are small enough to penetrate into the human lung, these particles may present a hazard to human health besides being responsible for unwanted photochemical reactions in the atmosphere. The photoelectric aerosol sensor is a simple tool to detect the particles and can help to reduce their concentrations by properly adjusting the various combustion processes in which they are generated.

3.2.8   X-ray Absorption of Particles

So far photoelectron emission upon irradiation of the particles with light of photon energy of a few electron volts, i.e. close to the threshold, has been regarded. Then electrons from valence and conduction bands are emitted, the information obtained mainly concerns the particle surface.

If photon energies in the X-ray range are used, core electrons may also be emitted. Due to the carrier gas it is not possible to measure the electron energy, but the integral yield which in the case of X-rays mainly depends on the absorption of the radiation.

X-ray absorption yields information on composition and chemical binding by measuring the "absorption edges" [27]. The absorption edges are related to the binding energy of inner shell electrons (core levels), which depend on the element, but are shifted by chemical binding. Typical binding energies are in the order of some keV.

Additional information can be obtained by measuring the "fine structure", i.e. oscillations near absorption edges. This technique (EXAFS) yields information on the neighbourhood of atoms (atomic distances etc.). These oscillations occur within about 50 eV from the edges [28].

For the analysis of solids X-ray extinction is measured directly. For particles a direct measurement is not practicable, as the product of main absorption and particle mass concentrations typically is less than 10^{-5}. However, the charge the particles may acquire by direct photoelectron emission or secondary electron emission (e.g. by Auger-processes) can be detected with high sensitivity.

As the carrier gas is also ionized by the radiation, the produced ions have to be precipitated before they can attach to particles. This is achieved by electrostatic precipitation, making use of the much higher electrical mobility of the ions compared to that of charged particles. Figure 11 shows the result of a first experiment on the L2 and L3 absorption edges of tin [29]. Synchrotron radiation at Hazylab in Hamburg has been used as X-ray source. The results obtained from the particles correspond well with those from a tin foil. This preliminary result clearly demonstrates the feasibility of in-situ X-ray absorption measurements from small particles.

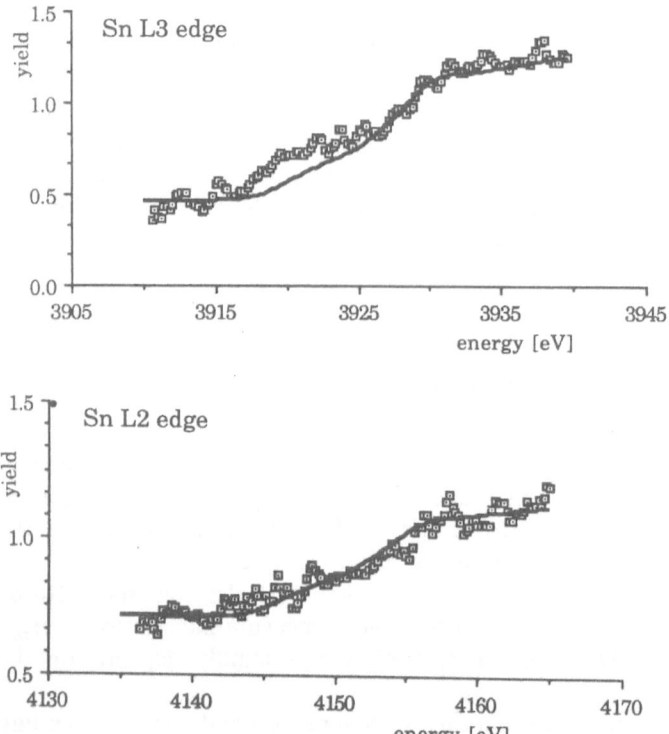

Fig. 11. L_2 and L_3 absorption edges for tin particles and a tin foil (*solid line*)

3.2.9   Conclusions

In conclusion it should be noted that photoemission experiments on particles suspended in gases are only at the beginning. The study of photoemission properties of submicroscopic particles is a promising way to obtain information on surface chemistry in the natural gas environment of the particles and on nucleation or growth of particles even in hostile environments such as combustions and the plumes of volcanoes.

The experiments clearly show that photoemission from particles can easily be observed. The general characteristics of photoelectric charging, namely its dependence on the particle radius R, and on the presence of adsorbates that effect the work function agree with expectations. However, some observations as the large enhancement of the photoelectric yield with Ag-particles are not understood at present.

3.2.10   Recent Developments

The work on combustion particles has been continued by investigating ad- and desorption of volatile material on the particle core using a thermodesorption technique [30, 31]. These experiments show a considerable dependence of sorption properties on the particle size. Two interesting new techniques for aerosol analysis are low pressure impaction, which allows to determine mass and density of particles of only few nanometers in diameter [32] and particle imaging by the scanning tunneling microscope [33]. Characterization of salt particles by photoelectric charging was studied and applied to volcanic emissions [34]. A new branch of rapidly increasing importance is production of new materials via aerosol routes [35].

Acknowledgments. The authors would like to acknowledge helpful discussions with many colleagues, especially Andreas Schmidt-Ott.

References

1. G.P. Reischl: Aerosol Sci. Technol. **14**, 5–24 (1991)
2. W.C. Hinds: Aerosol Technology, J. Wiley, New York, 1982, p. 148 ff
3. H.G. Scheibel, J. Porstendörfer: J. Aerosol Sci. **15**, 673–682 (1984)
4. H. Burtscher, A. Schmidt-Ott: J. Aerosol Sci. **17**, 699–703 (1986)
5. J.K. Agarwal, G.J. Sem: J. Aerosol Sci. **11**, 343–358 (1980)
6. M. Adachi, Y. Kousaka, K. Okuyama: J. Aerosol Sci. **16**, 109–124 (1985)
7. W.C. Hinds: Aerosol Technology, J. Wiley, N.Y. 1982, p. 299 ff
8. W.A. Hoppel: In *Electrical Processes in Atmospheres*, ed. by H. Dolezalek, R. Reiter, Steinkopf, Darmstadt, 60–69 (1977)
9. J. Heyder, J. Gebhart, G. Rudolf, C.F. Schilter, W. Stahlhofen: J. Aerosol Sci. **17**, 811–826 (1986)
10. D.M. Wood: Phys. Rev. Lett. **46**, 749 (1981)
11. Th. Jung, H. Burtscher, A. Schmidt-Ott: J. Aerosol Sci. **19**, 485–490 (1988)
12. R. Dennis: *Handbook on Aerosols*, National Technical Information Service, US Dept. of Commerce (1976); W.C. Hinds: Aerosol Technology, J. Wiley, New York, ·1982, p. 292 ff
13. U. Müller: Diss. Nr. 8544 ETH Zürich (1988)
14. U. Müller, A. Schmidt-Ott, H. Burtscher: Z. Phys. B **73**, 103–106 (1988)
15. M. Cardona, L. Ley: In *Topics in Applied Physics*: Photoemission in Solids I, ed. by M. Cardona et al., Springer, Berlin, Vol. 26, (1978)
16. H. Burtscher, A. Schmidt-Ott, H.C. Siegmann: Z. Phys. B **56**, 197–199 (1984)
17. U. Müller, H. Burtscher, A. Schmidt-Ott: Phys. Rev. B **38**, 7814–7816 (1988)
18. U. Müller, A. Schmidt-Ott, H. Burtscher: Phys. Rev. Lett. **58**, 1684–1686 (1987)
19. R. Niessner, P. Wilbring: Anal. Chem. **61**, 708–714 (1988)
20. S. Schwyn, E. Garwin, A. Schmidt-Ott: J. Aerosol Sci. **19**, 639–642 (1988)
21. A. Schmidt-Ott: *Experimente an kleinen Teilchen in Gassuspension*, Habilitationsschrift ETH Zürich, 1988
22. M. Ammann, H. Burtscher: Bull. Volcanology, **52**, 577–583 (1990); M. Ammann, L. Scherrer, W. Müller, H. Burtscher, H.C. Siegmann: Geophys. Res. Lett. **19**, 1387–1390 (1992)
23. H. Burtscher: J. Aerosol Sci. **23**, 549–595 (1992)
24. H. Burtscher, A. Schmidt-Ott, H.C. Siegmann: Aerosol Sci. Technology **8**, 125–132 (1988)

25. U. Heinrich: *Assessment of Inhalation Hazards, Integration and Extrapolation Using Diverse Data*, ed. by U. Mohr et al., Springer Verlag, 301–313 (1989)
26. S.R. McDow, W. Giger, H. Burtscher, A. Schmidt-Ott, H.C. Siegmann, Atm. Environment **24**, 2911–2916 (1990)
27. E.P. Bertim: *Principles and Practice of X-ray Spectrometric Analysis*, Plenum Press, New York, 1975
28. *EXAFS and Near Edge Structure*, Series in Chemical Physics **27**, ed. by A. Bianconi, L. Incoccio, S. Stipcidi, Springer, 1983
29. U. Müller, A. Schmidt-Ott, H.C. Siegmann, S. Krummacher, W. Niemann: Jahresbericht Hazylab, Hamburg (1988) 123
30. D. Steiner: Diss. Nr. 10191 ETH Zürich (1993)
31. D. Steiner, H. Burtscher: Water, Air, and Soil Pollution, **68**, 149–157 (1993)
32. N.P. Rao, J. Fernandez de la Mora, P.H. McMurry: J. Aerosol Sci. **23**, 11–26 (1992)
33. B. Schleicher, Th. Jung, H. Burtscher: J. Coll. Int. Sci. **161**, 271–277 (1993)
34. M. Ammann, R. Hauert, H. Burtscher, H.C. Siegmann: J. Geophys. Res. **98B**, 551–556 (1993)
35. A. Gurav, T. Kodas, T. Pluym, Y. Xiong: Aerosol Sci. Technol. **19**, 411–453 (1993)

3.3 Metal Clusters in a Liquid Environment. Photographic Development

J. Belloni, J. Amblard, J.L. Marignier, and *M. Mostafavi*

3.3.1 Introduction

For a long time, the ultra-divided state of matter has raised basic questions in various fields, viz. nucleation–growth dynamics [1], colloid science [2], radiochemistry at tracer scale [3], electrocrystallization [4], heterogeneous catalysis [5], and photography [6]. They all converge on the problem of the extent to which the laws valid at a macroscopic scale still hold when only a few atoms constituting a cluster are involved. Although clusters of atoms are made of the same element, their physical and chemical properties may differ markedly from those of the bulk, due to size quantization effects. Actually, each cluster has to be studied as a new species.

The phenomena which at first suggested size-dependent properties of metal clusters occurred in the presence of surrounding environments [2–7]. However, available quantitative data concerning surrounded clusters are still scarcer at present than data concerning isolated clusters which were extensively studied by beam techniques [8–10] during the last decade. This is in part due to the obvious difficulty of finding experimental methods appropriate for identifying and studying the properties of such small objects when these objects are immersed in a medium with which they interact strongly. However, interest has recently been stimulated in size-quantization effects on metal clusters in liquid environments due to the possibility of their investigation by pulse techniques [11–13], and because of their catalytic efficiency in the photochemical or voltaic conversion and storage of solar energy [14, 15].

The free diffusion of metal atoms and clusters in liquids results in their coalescence to larger sizes, and restricts the lifetime of the smallest aggregates. Another feature of the smallest aggregates is their propensity to undergo oxidation by the surrounding liquid [7] more easily than does the bulk metal. Both processes, coalescence and corrosion, contribute to the transient character of the small clusters, so that the latter can be observed only by time-resolved detection, coupled with an intense, short laser or electron pulse able to produce them at an appreciable concentration.

The scope of this Section is: 1) to describe how metal aggregates of low nuclearity may be prepared in polar liquids (e.g. through radiation chemistry techniques) and their properties studied; 2) what is the mechanism of their growth and the change of their properties with size (investigated by pulse techniques); and 3) to what extent they may be stabilized against either coalescence or corrosion.

3.3.2   Synthesis

3.3.2.1   Transient Aggregates Under Pulse Techniques

The short lifetime of aggregates consisting of a few metal atoms, due to coalescence or to corrosion, usually makes their observation possible only through pulse techniques. As in other methods (chemical, electrochemical), the precursors are metal ions M^+ which may be reduced at room temperature into metal atoms by the fast transfer of electrons issued from electron donors induced within a short pulse. The best conditions for studying quantitatively the kinetics of the further processes are provided by operation in homogeneous media, i.e. in solutions of the metal ions.

The electrons required for the reduction are produced by the absorption of either a laser pulse (pulsed laser photolysis), or an electron accelerator or X-ray pulse (pulse radiolysis). As a matter of fact, the energy of radiation such as X-rays, γ photons or accelerated particles (e^- or ion beams) exceeds by far the bond energies of any molecule, so that the cross section of the interaction with the solution is small (the penetration throughout the medium is deep), and not specific (the solvent and the solute absorb the energy proportionally to their electron abundance) [16]. A consequence of this is that the solvent molecules (more abundant) undergo most of the ionization and excitation events, yielding e.g. in aqueous solutions H_2O^*, H_2O^+, e^-, and that the solute molecules are almost exclusively transformed indirectly by reactions with the radiolytic species issued from the solvent [17]:

$$H_2O \longrightarrow \!\!\!\!\!\! \text{\Large\char"2D29\char"2D29\char"2D29} \longrightarrow e^-_{aq}, H_3O^+, H^{\cdot}, OH^{\cdot}, H_2, H_2O_2 \ . \tag{1}$$

The solvated electron is the strongest reducing agent [18] (E^0 $[H_2O/e^-_{aq}] \approx -2.7 \ V/NHE$ [19]), which reacts with metal ions of any valency at almost diffusion controlled rates [19, 20]:

$$M^+ + e^-_{aq} \rightarrow M_1 \ . \tag{2}$$

Insofar as the $OH^{\cdot}$ radicals could cause reverse oxidation, they must be scavenged by previously added solutes (either formate ions or alcohols such as isopropanol). These compounds remove $OH^{\cdot}$ (also $H^{\cdot}$) radicals and replace them by secondary radicals with strong reducing properties, $RR'\dot{C}OH$ in the case of secondary alcohols:

$$^{\cdot}OH \ (or \ ^{\cdot}H) + RR'CHOH \rightarrow H_2O \ (or \ H_2) + RR'\dot{C}OH \ , \tag{3}$$

$$M^+ + RR'\dot{C}OH \rightarrow M_1 + H^+ + RR'CO \ . \tag{4}$$

The first reduction step of bivalent metal ions may also be followed by the disproportionation of the transient monovalent states, which leads to zero valency for half of them. It is noteworthy that the reducing species ($RR'\dot{C}OH$ or e_{aq}^-) are generated homogeneously, thus yielding, after independent reactions with the ions, a population of isolated atoms also homogeneously distributed throughout the bulk.

Soon after the pulse, the population of atoms is created. As a general feature [21–28], they readily associate with excess ions:

$$M_1 + M^+ \rightarrow M_2^+ \ , \tag{5}$$

where M^+ symbolizes the dominant form of the excess metal ions. In fact, the precursor ions may also be in the form of a negatively charged complex. Reaction (5) is quite fast [19, 27], e.g. $k = 5.9 \times 10^9 \, l \, mol^{-1} \, s^{-1}$ when M is Ag [21]. Then the atoms complexed with ions coalesce, thus leading to charged aggregates. Electrostatic repulsion is doubtless a factor of stabilization of the final colloid against further coalescence.

After the pulse, the evolution of the process in solution is observed by time-resolved techniques [16], currently spectrophotometry or conductimetry. The assignment of the optical absorption bands to the transients results from the correlation between their formation and the decay of their precursors, and from material balance. The rate constants are derived from these kinetics at known concentrations. A lot of information has been collected [12, 13, 19, 27] on various reduced states, isolated metal atoms and small clusters, either neutral or charged. The pulse radiolysis method may be used as well for non aqueous solutions where the radiation effects are homologous [22–24] (Table 1), as for microheterogeneous systems [29] such as micelles, vesicles, emulsions, or for ion exchange membranes [30, 31] (Fig. 1).

In a similar way, photons emitted by a pulsed laser in the visible and most commonly in the UV range may excite [29, 32] in a transparent medium certain anions A^- which yield electrons by photodetachment:

$$A^- \xrightarrow{h\nu} (A^-)^* \rightarrow A^\cdot + e^- \ , \tag{6}$$

$$e^- + M^+ \rightarrow M_1 \ , \tag{7}$$

Table 1. Optical absorption maxima (λ nm) of transient silver atom or aggregates in various solvents

Solv.	H_2O	NH_3		EDA	HF
		23 °C	− 50 °C		
	[21]	[22]	[23]	[23]	[24]
Ag_1	360	435	450	440	
Ag_2^+	310	390	380	345	300
Ag_4^{2+}	275[25]				

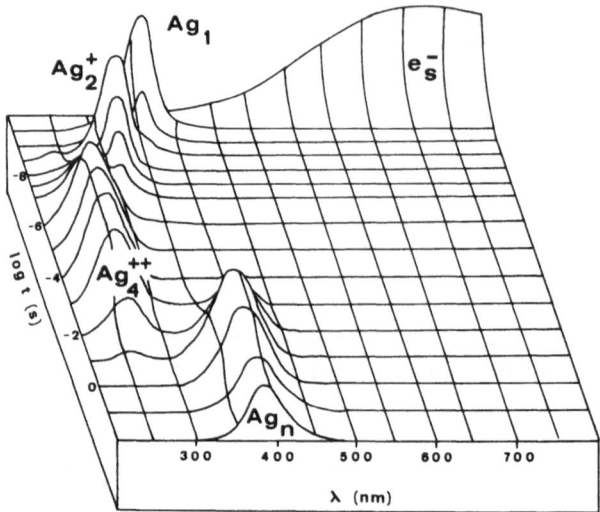

Fig. 1. Optical absorption spectra of transient species formed during the pulse radiolysis of a per-fluorinated ion exchange membrane swollen with a 10^{-1} mol l^{-1} aqueous–alcoholic Ag_2SO_4 solution [31]. Early steps (< 1 ns) were inferred from experiments at lower Ag^+ concentration. The assignments were given from data in free solution [33]. Note that the absorbance of the silver aggregates at 400 nm decreases in the time range 0.1–10 s. This decrease results from the slow corrosion of the nascent silver aggregates ($n < 8$) [31]

after which the coalescence starts as above. The mechanism of coalescence is a succession of steps; many of them were extensively studied in the case of Ag [21–25, 31, 33, 34]. Similar features have been found also for Tl [26, 28], In, Cu, Hg, and Pb [27], or Au [35]. Both processes of coalescence and association slow down when the number of atoms per nucleus increases (and the nuclei concentration decreases), producing finally a metastable colloid, generally charged. The formation of the colloid of Pt is clearly accelerated at increasing ionic strength, which screens electrostatic repulsions between charged aggregates, and conversely slowed down by addition of a surfactant [36].

3.3.2.2 Stable Aggregates

It must be kept in mind that the strength of the metal–metal bond favors the coalescence of the clusters inasmuch as they are free to diffuse and encounter each other. Different methods are used to slow down or inhibit the coalescence in order to stabilize the clusters at the smallest size. However, the final properties strongly depend on the kinetics of growth, as well as on the age of the preparation.

Under steady-state conditions of irradiation by γ photons or X-rays, the same species as in the pulsed regime are transitorily produced, but only stable states may now be observed: either stable low valency ions, or the final stable aggregates, often of a few nanometers (although much smaller under certain conditions). Again, the advantages of the high energy radiation are the generation, at room temperature, of strong reducing agents directly from the solvent (i.e. without additives), and the production of atoms originally isolated and homogeneously dispersed. Due to the penetration properties of the radiation, these atoms may be produced even in the cavities of microheterogeneous systems. The initial homogeneity of the reactants ensures a narrow size distribution of the final nanoaggregates, and better control of the process reproducibility.

The synthesis of stable low nuclearity aggregates must overcome two difficulties: 1) the coalescence must be slowed down sufficiently, but 2) when aggregates arise from isolated atoms and contain only a few atoms, they are liable to be corroded by the medium, due to their lower electrochemical potential (see 3.3.4.2). For subcolloidal solutions, the coalescence is easily controlled by surfactant molecules, as in the chemical synthesis of colloids [2]. In the radiolytical procedure, the surfactants (polyvinyl alcohol (PVA), sodium dodecyl sulfate (SDS) [37], sodium polyvinyl sulfate (PVS) [38], carbowax [14], etc.) are selected either for their inert character or for their positive behaviour under irradiation. For example, PVA also acts as an oxidizing radical scavenger [39]. Its efficiency is related to the small fraction of non hydrolyzed acetate groups which it contains and which strongly complex the precursor ions (as shown for instance in the case of nickel by the lowering of the reaction rate with e_{aq}^- [38, 39]), thus stabilizing the metal colloid [40].

The radiolytic species are also able to initiate polymerization of monomers: this has been used to generate at the same time Pt clusters and their stabilizing surfactant (polyacrylamide, PAM, or poly N methylacrylamide, PNMAM) from a solution containing both metal ions and monomers [41]. Similarly, metal clusters of Pt, Pd, Cu, Ag, embedded in PVA further crosslinked under γ irradiation have been obtained [39].

The corrosion effect is less severe for noble metal aggregates (Pt, Ir, Ag, . . .). Under certain conditions, the final yield is lower than that of the reducing radiolytic species, while molecular hydrogen arising from the corrosion by protons is evolved. This has been shown for nascent Cu in liquid ammonia, thus suggesting a shift of the redox potential towards negative values [7]. In the case of non-noble metals, the corrosion would be fast enough to even inhibit completely any synthesis of the aggregates. Therefore, it is necessary to add a base to the medium to scavenge the protons, and the initial concentration must be higher than for noble metals, so as to favor fast coalescence [39, 45, 46]. Additionally, the presence of a surfactant prevents flocculation of the metal hydroxide at high pH values. Table 2 gives a list of metals, noble and non-noble, for which subcolloidal solutions of nanoaggregates have been obtained through irradiation of aqueous solutions.

Metal nanoaggregates may also be generated inside hydrophilic domains of microheterogeneous systems [29], such as vesicles, micelles, polymeric materials

Table 2. Optical absorption properties of stable metal nanoaggregates obtained by radiolysis of aqueous solutions. All of them show a "brown" band corresponding to a UV–VIS scattering decreasing with λ. In addition to this, some colloids exhibit a specific surface plasmon band whose λ_{max} is given below. The symbol (∗) indicates that the earliest steps of the colloid formation have been studied by pulse radiolysis

M_n	Optic. abs.	Refs	M_n	Optic. abs.	Refs	M_n	Optic. abs.	Refs
Mo	Brown band	[39]	Pd	Brown band	[39]	*Hg	+ 500 nm	[27, 39]
Ru	"	[39]	*Pt	"	[39]	*In	+ 270 nm	[27]
Os	"	[39]	*Cu	+ 570 nm	[27, 39]	*Tl	+ 300 nm	[26, 28]
Co	"	[39]	*Ag	+ 380 nm	[21, 33]	Sn	Brown band	[39]
Rh	"	[39]	*Au	+ 520 nm	[35]	*Pb	+ 300 nm	[44]
Ir	"	[39, 42]	Zn	Brown band	[39]	Sb	Brown band	[39]
Ni	"	[39, 40]	*Cd	"	[43]	Bi	+ 280 nm	[27, 39]

and ion exchange membranes [30], or upon various supports (carbon, silica, metal oxides such as Al_2O_3, SnO_2, bulk metal electrodes [46]), or upon colloidal particles such as TiO_2 [47]. For this purpose, the support is immersed in the solution to be irradiated in the absence of any surfactant.

3.3.3   Physical Properties

3.3.3.1   Size, Structure

As shown below (3.3.3.3), size determination of unstable aggregates composed of a few atoms is made indirectly from the coalescence rate and the time elapsed since the atoms were formed.

For stable aggregates of higher nuclearity, the size may be observed by high resolution electron microscopy (HREM): very small Ir aggregates (≤ 1 nm diameter, i.e. less than 37 atoms) [42] or Pt (≈ 1 nm, i.e. 35 atoms) [48] were thus obtained in irradiated aqueous solutions with PVA as the surfactant. The size of other metal clusters, including non-noble metals which cannot be stabilized below a minimum size, generally lies between 2 and 5 nm. Some of them, viz. Ag, Pd or Ni, often exhibit pentagonal shapes and typical fivefold symmetries [39] suggesting a decahedral (Dh) or an icosahedral (Ih) structure. However, electron diffraction patterns indicate that, as soon as the clusters are big enough to be observed, the structure is fcc as in the bulk metal. Actually these clusters are fcc multi-twinned particles originating from a perfect Dh or Ih nucleus [4].

Mixed solutions of two metal ions with quite different properties, irradiated at room temperature, yield ultra-divided alloys exhibiting superlattice reflections which indicate a perfectly-ordered arrangement [45]. Depending on the initial composition, Cu_3Pd, CuPd, Cu_3Au, CuAu, Ni_3Pt, etc. . . . were thus found [39].

Although this has not yet been confirmed by direct observations, it is likely that such a perfect alloying of both kinds of metal atoms, M and M', stems from the growth kinetics in the presence of excess precursor ions M^+ and M'^+. Both kinds of ions exhibit the same reactivity toward e_{aq}^- and moreover, comparable thermodynamical properties (see 3.3.4). Aggregation involves homo- and cross-ed associations of atoms and ions (reaction 5), followed by further electron transfers. Both processes progressively build a nanocrystal whose composition reflects the probability of finding the two kinds of ions in the radiolytic solution and the probability of reduction by e_{aq}^- (or radicals) of the growing nucleus [23].

Sizes in the 1–100 nm range may also be determined in situ by small angle X-ray scattering (SAXS). The SAXS method is well adapted for subcolloidal solutions exhibiting homogeneous distributions of spherical-shaped metal particles. The growth inhibition brought about by PVA has been demonstrated in this way for Pd nanoaggregates [39, 49].

3.3.3.2 Magnetic Susceptibility

Clusters of Ni or Co exhibit ferromagnetic properties and their solutions constitute ferrofluids. When they are included in polymers, their support can be displaced by a magnet. They also may be separated from the solution: however, due to their very small size, a strong magnetic field must then be applied for a long time [40]. The magnetic susceptibility also enables one to determine e.g. the amount of Ni in Ni–Ru alloys embedded as clusters in PVA [30].

3.3.3.3 Optical Properties

The interaction of light with most metal clusters results in a broad, non-specific absorption band of increasing intensity toward the UV. Table 2 gives a list of stable metal nanoaggregates. For some of them an additional surface plasmon band shows more specificity.

At the very beginning of the aggregation process, the absorption spectra of short lived atoms and low nuclearity clusters are observed mostly by time-resolved spectroscopy. The spectrum of the isolated atom M_1 differs from that of the complexed form M_2^+ [21, 27], as shown above for silver (Fig. 1). It is also known that the position of the maximum depends both on the nature of the solvent and on temperature [23] (Table 1). This would support the conclusion of a partial charge transfer to the solvent (CTTS), as proposed by *Kevan* [50], who studied Ag_1 by spin echo in frozen polar matrices, and by *Walker* et al. [32], who photoionized Ag_2^+ or Tl_2^+ in water at 300 nm by a double flash technique.

The dimerization of M_2^+, yielding M_4^{2+} (or M_3^+), has been observed for Ag, Cu, Hg, In and Bi [27] with a maximum in the UV. The assignment results from correlation of the second-order decay of M_2^+ ($k_{dim} = 2 \times 10^8$ l mol^{-1} s^{-1} for

Ag_2^+ [51]) with increase of the UV band, which also makes possible a determination of the respective extinction coefficients [51]. Then the dimer M_2, complexed with one (M_3^+) or two (M_4^{2+}) ions M^+, decays through further coalescence.

The increase of the absorption band due to the clusters of higher nuclearity (possessing more or less charge) is somewhat delayed. For Ag_n, which has been studied most extensively, the increase at 380 nm corresponds to an absorption band that is not specific for the aggregation number beyond $n = 13$. Below this value, the extinction coefficient per atom increases with n [51]. Assuming a coalescence rate roughly size-independent, thus keeping the same value as k_{dim} measured for the earliest step (reaction 5), the predominant size at a given time may be calculated from the initial atom concentration [34, 51].

As shown in Fig. 1, the aggregation mechanism of Ag atoms inside the hydrophilic cavities of a polymeric membrane such as NAFION is very similar to that in a free solution. However, the rate of diffusion between hydrophilic cavities is lowered by a factor of 5×10^4, so that the coalescence is slowed down by the same factor as soon as the stage of Ag_4^{2+} dimerization is reached. Moreover, the diffusion of clusters, e.g. larger than $n \approx 5$ under given swelling conditions, through the channels linking the cavities becomes sterically hindered [31].

A somewhat different example of ultraslow aggregation dynamics is offered by silver sulfate solutions containing unprotonated polyacrylate ions which exhibit both surfactant and ligand properties [52]. Again the ultraslow aggregation allows long-lived clusters of a few atoms to be produced by γ irradiation, and to be observed spectrophotometrically. With increasing dose, new bands at 300 and 335 nm are observed which are due to precursors with $n \leq 4$. In addition, a broad component at 400–600 nm, depending on the polyacrylic acid (PAA) concentration, is assigned to a strong interaction between PAA and Ag_n. In contrast with the preceding system, the clusters are stable toward corrosion by oxygen, even over a few months. However, the strong ligand effect does not protect the smallest oligomers, either from corrosion by H_3O^+ [53], or from a growth catalyzed by the couple Cu^{2+}/Cu^+.

The role of the surfactant appears in the above case of primary importance. This is also true for silver perchlorate solutions containing sodium polyphosphate (SPP) as surfactant. The main feature of these solutions is a slow and progressive evolution of the spectrum for each supplementary dose, with new bands at 275 nm, followed by 300 nm, then 325 nm, and finally shifted to 380 nm. The 275 nm band was assigned to $(Ag_4^{2+})_{SPP}$; however, the species does not react with oxygen, which suggests a much weaker reducing character than for other even more condensed silver species. The reactivity of these species toward oxygen, Cu^{2+}, CCl_4, and nitrobenzene confirms the low nuclearity of the clusters, in spite of their surprising stability. Combined γ irradiation (to produce a stable solution of $(Ag_4^{2+})_{SPP}$) and pulse radiolysis (to observe the products of its reduction) enabled the authors [25] to assign the 300 nm band to a trimer species $(Ag_{3+x}^{x+})_{SPP}$, and the 330 nm band to clusters of higher nuclearity.

3.3.4 Chemical Properties

Some data given above have already indicated that the reactivity of metal aggregates depends on their nuclearity as well as on the environment (pH, surfactant, solvent, etc.). The time range over which this reactivity may be studied depends on the lifetime of the aggregates. Thus the existence of a cluster with given n starts with the coalescence of its precursors and ends with its own various reactions, including coalescence.

3.3.4.1 Stable Aggregates

The stability of nanoaggregates first refers to their possible synthesis in the medium and means that the aggregates no longer react with the surrounding species.

Nevertheless, stable aggregates (mostly of less noble and non-noble metals: Cu_n, Ni_n, Co_n, Zn_n, etc.) are more easily oxidized, for instance when brought into contact with oxygen, than the corresponding bulk metals.

Advantage may be taken of this reactivity of small clusters toward oxidants to solve, generally by spectrophotometry, the problem of measuring the total amount of reduced metal atoms surrounded by excess precursor ions in a given sample, and hence to calibrate the optical absorption spectra. The oxidant chosen depends on the metal and on the other molecules present: it must be stronger for nanoaggregates of Ag [33] or Tl [26] than for Ni [39, 40] aggregates of comparable size. Reactivity differences enable one to analyze Ni even when alloyed with a more noble metal in a bimetallic cluster [39, 40]. Electron transfer from clusters of Ni [39, 40] or Cu [39, 54] to ions of more noble metals (such as Ag^+, Pd^{2+}, or $AuCl_4^-$) yields quantitatively colloidal solutions of these metals, thus providing another analytical method.

Since the electron transfer requires that the binding energy of the electron, measured by the standard redox potential E^0 [M_n^+/M_n], be higher for the acceptor than for the donor, the redox potential of clusters may be estimated relatively to that of a series of reference redox systems. For instance, for $(Ni)_{PVA}$ aggregates of a few nanometers [40], $-0.45 \text{ V} < E^0[Ni_n^+/Ni_n] < -0.33 \text{ V}$.

The strong reactivity toward O_2 or nitrobenzene of silver oligomers stabilized by SPP [25], which contrasts with the inert character of nanoaggregates and a fortiori of the bulk metal, is obviously also a consequence of the size quantization effects.

3.3.4.2 Short-Lived Aggregates

Unless surfactants introducing strong ligand interaction are added, coalescence shortens the lifetime of the smallest aggregates. The reactivity of an isolated atom or of a small cluster must therefore be studied by pulse radiolysis in the presence of a reactant at the proper concentration to compete efficiently with spontaneous coalescence.

Most of the available data concern electron transfer (i.e. redox) processes. The observed occurrence of a given reaction means that the electrons do transfer from one species to the other, and hence provides information on the relative values of their standard redox potentials. In the absence of any reaction, no conclusion can be drawn, since the process may be inefficient due to a possible activation energy barrier. The method of indirect determination of redox potentials through pulse radiolysis (or photolysis) has been extensively used for short-lived radicals [16].

It must be pointed out that a charged cluster M_{n+x}^{x+} is not totally reduced; from a strict point of view, it contains n reduced atoms only. Depending on the reactant S, the cluster may behave either as an electron donor (thus being oxidized to $M_{n+x}^{(x+1)+}$), or as an electron acceptor (thus being reduced to $M_{n+x}^{(x-1)+}$):

$$M_{n+x}^{x+} + s \longrightarrow M_{n+x}^{(x+1)+} + s^- , \tag{8}$$

$$\phantom{M_{n+x}^{x+} + s} \longrightarrow M_{n+x}^{(x-1)+} + s^+ . \tag{8'}$$

Upper and lower limits of the redox potential (E^0 [S/S$^-$] and E^0 [S$^+$/S] respectively) can be obtained by reference to those of the solutes S with which the aggregate does react.

Monomers

A series of observed reactions of isolated and complexed silver monomers (Ag_1 and Ag_2^+) with different electron acceptors is shown in Table 3 [33]. The

Table 3. Reaction rate constants (in $l\ mol^{-1}\ s^{-1}$) of Ag_1, Ag_2^+ and Ag_2^+ (am) with various substances, as determined by Tausch-Treml et al. [33]

Substance	Ag_1	Ag_2^+	Ag_2^+ (am)
Fe^{3+}	1.2×10^9	3×10^8	
Cu^{2+}	6.5×10^8	$< 10^5$	
Ni^{2+}	$< 5 \times 10^6$	$< 10^5$	
Cr^{3+}	$< 3 \times 10^6$	$< 10^5$	
O_2	5.0×10^9	4.6×10^8	7×10^9
H_2O_2	3.5×10^9	8.0×10^6	1×10^9
$CHBr_3$	3.0×10^9	5.0×10^8	2×10^9
$CHCl_3$	1.1×10^9	$< 2 \times 10^5$	2×10^8
CCl_4	1.1×10^9	1.5×10^7	1×10^8
$ClCH_2COOH$	1.5×10^5	$< 1 \times 10^6$	3×10^6
CH_3NO_2	2.3×10^9	1.1×10^8	1.5×10^8
$C_6H_5NO_2$	2.8×10^9	3.0×10^8	9×10^8

data allowed the authors to conclude that the potentials E^0 [Ag$^+$/Ag$_1$] or E^0 [2Ag$^+$/Ag$_2^+$] were lower than the lowest of the referred systems (E^0(CH$_3$NO$_2$/CH$_3$NO$_2^-$] = $-$ 1.0 V). On the other hand, the isolated Ag$^+$ ion cannot be reduced by the strong electron donor (CH$_3$)$_2$ĊOH. This suggests a strong negative value for E^0 [Ag$^+$/Ag$_1$] [55].

In addition, a simple thermodynamical estimation, based on the potential of the bulk silver electrode and on the sublimation energy of the bulk metal Ag$_\infty$ (ΔG_{sub} = 2.6 eV), and assuming the hydration energy of Ag$_1$ to be negligible, yields [33]

$$E^0[Ag^+/Ag_1] = E^0[Ag^+/Ag_\infty] - \Delta G_{sub}/e = 0.799 - 2.60 = -1.8 \text{ V .} \quad (9)$$

The high negative value thus obtained for E^0[Ag$^+$/Ag$_1$] means that the isolated silver atom behaves as a strong electron donor. This value is also compatible with the recently observed electron transfer [34] from the still stronger donor Ni$^+$ to Ag$^+$ (E^0[Ni^{2+}/Ni$^+$] $\approx -$ 1.9 V [39]), and actually no reaction was observed between Ni^{2+} and Ag$_1$ (Table 3). In a similar manner, the potentials have been calculated for the monomers of Cu or Tl (Table 4), and found to be much lower than for the bulk metals [26, 56].

Provided the potential E^0[M^{2+}/M$^+$] is known, the value E^0[M$^+$/M$_1$] may also be derived in the case of a divalent metal such as nickel (Table 4) [39]:

$$E^0[M^+/M_1] = 2\,E^0[M^{2+}/M_\infty] - E^0[M^{2+}/M^+] - \Delta G_{sub}/e \ .$$

Using critically evaluated data from the literature in the above equation, it has been shown that E^0[M$^+$/M$_1$] is lower than -2.73, -0.60, and $+0.28$ V for Co, Zn and Cd respectively [39].

Oligomers

The redox potential of silver oligomers of nuclearity greater than 1, involved in reactions where they behave as electron acceptors, has been obtained by pulse radiolysis of silver sulfate solutions containing the precursor of an electron donor [34, 51]. The initial reaction with e$_{aq}^-$ and reducing radicals yields silver atoms and the electron donor. When the donor is Cu$^+$, of potential E^0[Cu^{2+}/Cu$^+$] = 0.16 V, the electron transfer does not occur, either with Ag$^+$ whose potential is very negative, or with Ag$_n^+$, as long as the coalescence of silver atoms has not proceeded far enough for the potential E^0[Ag$_n^+$/Ag$_n$] to exceed

Table 4. Electrochemical potentials (in V/NHE) of metal electrodes (Handbook or [39]) and monomers in aqueous solutions

Silver [56]		Thallium [26]		Copper [56]		Nickel [40]	
Couple	E^0	Couple	E^0	Couple	E^0	Couple	E^0
Ag$^+$/Ag$_\infty$	+0.799	Tl$^+$/Tl$_\infty$	$-$0.336	Cu$^+$/Cu$_\infty$	+0.521	Ni/Ni$_\infty$	+1.40
Ag$^+$/Ag$_1$	$-$1.8	Tl$^+$/Tl$_1$	$-$1.9	Cu$^+$/Cu$_1$	$-$2.7	Ni/Ni$_1$	$-$2.6

$E^0[Cu^{2+}/Cu^+]$. The threshold is reached at a time which, accounting for the dimerization rate constant, corresponds to the coalescence stage $n = 11$. This means that $E^0[Ag_{11}^+/Ag_{11}] \approx 0.16$ V, i.e. a potential still lower than that of the bulk electrode [51].

Another monoelectronic donor, the reduced form of sulfonato propyl viologen (SPV), of lower potential ($E^0[SPV/SPV^-] = -0.41$ V), has been used to gain access to the redox potential of silver clusters of very low nuclearity, like those which govern the developability of the latent image in photography (see 3.3.5). The SPV$^-$ species plays the role of a photographic developer. Moreover, SPV$^-$ offers the important advantage that it is detectable by its intense, specific optical spectrum and thus provides more quantitative information about the electron transfer mechanism. As with Cu$^+$, a time delay is observed before which no transfer occurs from SPV$^-$, either to any of the successive silver clusters, or obviously to isolated Ag$^+$. The first size for which electron transfer starts is $n = 4$, hence the first electron acceptor is Ag$_5^+$ [34].

After the time delay, the SPV$^-$ decay, which corresponds stoichiometrically to the new silver atoms produced by transfer, obeys pseudo first-order kinetics for a much longer time than complete reaction with all Ag$_5^+$ species would require. From the detailed analysis of the results [34], it has been concluded that the kinetics do correspond to a catalytic transfer of successive electrons (k_t) to the same silver cluster acting as a growth center and successively accreting silver cations (k_a):

$$Ag_5^+ + SPV^- \xrightarrow{k_t} Ag_5 + SPV , \tag{10}$$

$$Ag_5 + Ag^+ \xrightarrow{k_a} Ag_6^+ , \tag{11}$$

$$Ag_6^+ + SPV^- \xrightarrow{k_t} Ag_6 + SPV , \tag{12}$$

$$\cdot \cdot \cdot \cdot \cdot \cdot \cdot \cdot \cdot \cdot \cdot \cdot \cdot \cdot \cdot \cdot \cdot \cdot \cdot \cdot \cdot \cdot \cdot$$

$$Ag_{n-1} + Ag^+ \xrightarrow{k_a} Ag_n^+ , \tag{13}$$

$$Ag_n^+ + SPV^- \xrightarrow{k_t} Ag_n + SPV . \tag{14}$$

The result is an autocatalytic growth of the silver cluster. The transfer is faster than the coalescence [34], but when SPV$^-$ is strongly depleted, the coalescence again becomes more competitive, and dimerization occurs. The concentration of the clusters now twice as large is divided by two, and the catalytic decay of SPV$^-$ continues with a rate half that of the preceding one (Fig. 2). No change is observed when both SPV and Ag$^+$ concentrations are increased by a factor of 3 (curves 1 and 5 of Fig. 2), confirming that the direct transfer to isolated Ag$^+$ ions is unlikely and also indicating that the complexation of Ag$_n$ with Ag$^+$ is fast and does not control the cluster growth rate. Thus, observation of the SPV$^-$ decay provides a detailed description of the autocatalytic growth mechanism of a cluster and, in addition, yields the redox potential value $E^0[Ag_5^+/Ag_5] \approx -0.40$ V, of Ag$_5^+$ [34].

The recently observed role of cupric ions, Cu^{2+}, in the catalytic coalescence of oligomers stabilized by PAA [53] or SPP [25] ions has been tentatively

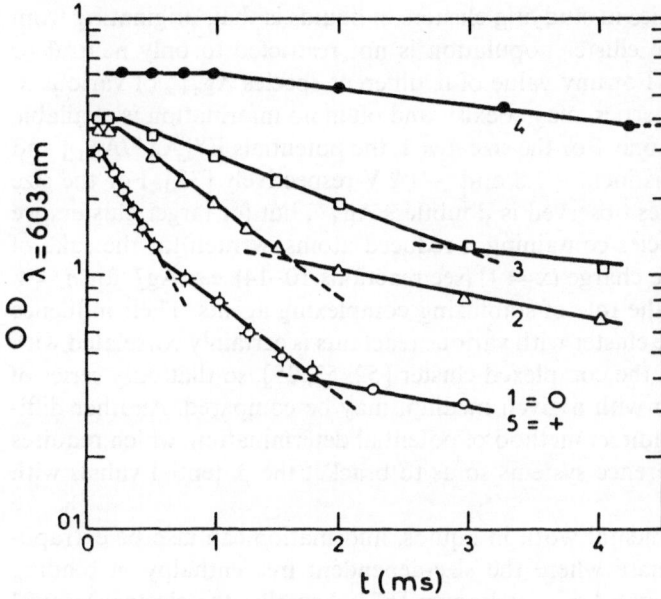

Fig. 2. Logarithmic plot of SPV⁻ decay at $\lambda = 603$ nm with the same dose per pulse and the same $(CH_3)_2CHOH$ concentration $= 0.2$ mol l⁻¹, for various $[Ag^+]/[SPV]$ ratios. The total amount of reduced species per pulse is 5.2×10^{-5} mol l⁻¹. 1: $[Ag^+] = 1.2 \times 10^{-3}$ and $[SPV] = 3 \times 10^{-4}$ mol l⁻¹, respectively; 2: 8×10^{-4} and 3×10^{-4} mol l⁻¹; 3: 6 × 10⁻⁴ and 3×10^{-4} mol l⁻¹; 4: 2×10^{-4} and 8×10^{-4} mol l⁻¹; 5: 3.6×10^{-3} and 9×10^{-4} mol l⁻¹

explained by a possible relay transfer of the electrons from the small clusters (of more negative potential) to Cu^{2+} and the subsequent transfer from Cu^+ to big clusters (of higher potential). Direct electron transfer from small to big clusters had also been suggested as a consequence of the size-dependent potential and recognized as an alternative coalescence process which could account for the narrow final size distribution [23].

3.3.4.3 Electrochemical and Ionization potentials

Still more marked is the influence of a surfactant exhibiting strong ligation properties. The reactivity of low nuclearity oligomers, stabilized by SPP or PAA, seems to indicate higher stability toward oxygen $(E^0[O_2/O_2^-] \approx -0.33$ V) of $(Ag_4^{2+})_{SPP}$ $(n = 2)$, as compared to clusters with n ≥ 3–7. This implies, unless an activation energy barrier exists, that the redox potential of the donor, $(Ag_4^{2+})_{SPP}$, is more positive than -0.33 V and, moreover, is higher than the potentials of 3–7 clusters. These results constitute the first indication of an oscillating dependence of the potential for aggregates in a liquid environment, as *Henglein* concluded [25]. Similar odd–even oscillations of the ionization potential of bare clusters have been observed in the gas phase [8] (see Chap. 4).

One of the difficulties in studying clusters in liquids is that, originating from unreduced cations, the cluster population is not restricted to only neutral or monocharged species. For any value of n, different species Ag_{n+x}^{x+} of various x, hence of different properties, may coexist, and often no information is available on the predominating one. For the size $n = 1$, the potentials $E^0[Ag^+/Ag_1]$ and $E^0[2Ag^+/Ag_2^+]$ are distinct, -1.8 and -1.2 V respectively [33]. For the size $n = 2$, the major species observed is doubtless Ag_4^{2+}, but for larger clusters we refer to the major species containing n reduced atoms, written for the sake of simplicity with a single charge $(x = 1)$ (see reactions 10–14), e.g. Ag_5^+ for $n = 4$. A second difficulty is the role of stabilizing/complexing agents. Their influence on reaction rates of the cluster with various reactants is certainly correlated with the actual potential of the complexed cluster [52, 53, 25], so that only series of clusters in equilibrium with a given medium may be compared. Another difficulty stems from the indirect method of potential determination, which requires series of adequate reference systems so as to bracket the potential values with greater accuracy.

Besides the experimental work in liquids, information can also be extrapolated from the gas phase where the size-dependent free enthalpy of binding a silver atom to a cluster Ag_{n-1} is known [57]. Actually, the electrochemical potential of the bulk metal relates to the equilibrium between the electrode and isolated aquated ions $E^0[Ag^+/Ag_\infty]$, and thus differs from the monomer potential $(-1.8$ V) by the sublimation energy (3.3.4.2). Accounting now for the free enthalpy of binding one atom to the monomer or to the dimer [57], calculation of $E_n^0[(Ag^+ + Ag_{n-1})/Ag_n]$ yields $E_2^0 = -0.1$ V and $E_3^0 = -0.8$ V, revealing a marked oscillation, at least over this limited range. Strictly, these values can be compared with the metal-electrode potential, i.e. $E^0[Ag^+/Ag_\infty] = 0.799$ V, when $n \to \infty$.

The potential $E_n^0[Ag_n^+/Ag_n]$, named ionization redox potential [58], as obtained by electron transfer reactions, differs from the electrochemical one (except for $n = 1$ where they coincide) by the size-dependent desorption energy of Ag^+ [58]. Figure 3 shows the variation of $E^0[Ag_n^+/Ag_n]$ with n as the consequence of size-quantization effects in liquids.

It may be directly compared to the size-dependent ionization potential in the gas phase, provided the Fermi level of the hydrogen electrode $(-4.5$ eV) is accounted for, to fix the relative scales of energies required for extracting one electron from a cluster either in solution or in the gas phase [34]. However, the available data concerning silver are still scarce. By analogy with other metals, e.g. copper [59], the general trend of the IP variation with n in the gas phase is expected to follow the law $IP_n - IP_\infty = -87.05 (n V)^{1/3}$ (dotted line of Fig. 3), where V is the volume of one atom expressed in nm^{-3} [8].

Moreover, superimposed oscillations due to odd–even variations of the binding energy between the cluster atoms have been observed in the case of copper [8, 59]. No oscillations have been observed at yet experimentally for the potential of Ag_n in the liquid phase, although they are expected, as shown by the

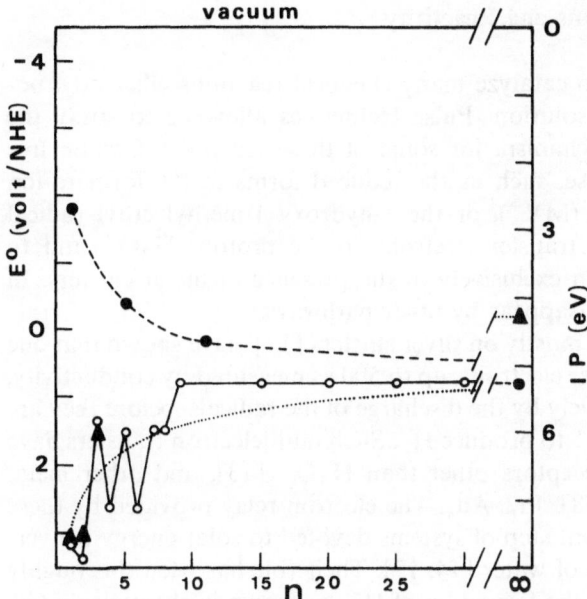

Fig. 3. Size-dependence of the redox potential $E^0[Ag_n^+/Ag_n]$ relative to NHE in the condensed phase (● and *dashed line*), and of the ionization potential IP of Ag_n in the gas phase (▲ from [59, 60] and *dotted line* computed from the drop model [8]). The variation of IP from Ag_4 to Ag_∞ in the gas phase is probably similar to that of Cu_n which presents a rough decrease of IP with a few oscillations (○ and *solid line*) at low n values (from [59])

thermodynamical calculations above [57] and by the unusual stability toward O_2 of $(Ag_4^{2+})_{SPP}$ [25].

The redox curve of Fig. 3 is confirmed by other experiments where the silver clusters also behave as electron donors toward the proton H_3O^+. Actually, when the coalescence of silver atoms included in polymeric membranes containing acid ions at pH 1 is observed [31], the slow decay at long time of the total amount of atoms (Fig. 1) must be attributed to the corrosion of the smallest clusters. The phenomenon is enhanced at low dose when the coalescence, still slower, cannot protect the aggregates from their reaction with H_3O^+. Assuming, as a first approximation, that the corrosion rate constant value is independent of n, it is concluded that all clusters with $n < 8 \pm 2$ may react (hence their redox potential is negative relative to NHE), and that the rate constant is $k_{corr} \approx 0.8\,l\,mol^{-1}\,s^{-1}$ [31]. This quite low value implies an activated process, possibly slowed down by hydrogen evolution. Direct observation of the corrosion was permitted by the severe rate decrease of the competing coalescence in the membrane. Reciprocally, in free H_2O (NH_3 or EDA), with highly-concentrated H_3O^+ (NH_4^+ or RNH_3^+) added, the failure to observe such a corrosion had been attributed to the efficient competition of coalescence [23].

3.3.4.4   Catalytic Mechanisms and Reactivity

Metal clusters are known to catalyze many chemical reactions efficiently, notably electron transfers in solution. Pulse techniques allow us to study the elementary steps of the mechanism for some of these reactions. Certain free radicals produced by a pulse, such as the reduced forms of the formate ion (CO_2^-), of methyl viologen (MV^+), or the 1-hydroxy 1-methyl ethyl radical $((CH_3)_2\dot{C}OH)$, are able to transfer electrons to the protons H_3O^+ and to produce molecular hydrogen exclusively in the presence of metal clusters. In their absence, the radicals disappear by other pathways.

Extensive investigations, mostly on silver clusters [12], have shown that one cluster may store many excess electrons, up to 500 as measured by conductivity, which are provided successively by the discharge of the radicals, before they are transferred pairwise to H_3O^+ to produce H_2. Such multielectron transfers have also been observed with acceptors other than H_3O^+ [13], and other metal clusters, e.g. Ir_n, Cd_n, Tl_n [58], Pt_n, Au_n. The electron relay provided by these small particles is an important step of systems devoted to solar energy conversion through photocleavage of water [14, 15]. Their role has been thoroughly examined, mostly regarding the formation of H_2, as shown in the review [13]. Through pulse radiolysis, the catalytic mechanism of the superoxide anion O_2^- dismutation by Pt clusters has also been investigated [47].

A systematic study of the catalytic efficiency of clusters [14, 15, 41], measured by the quantum yield of H_2 in photochemical systems, has shown that the protection against coalescence ensured by surfactants does not prevent the reactants from approaching the clusters. This preserves the catalytic efficiency.

Metal nanoaggregates grafted upon transparent SnO_2 electrodes for photoelectro-chemical cells improve greatly electron transfers at the interface. The overpotential of these modified, yet transparent electrodes, is almost suppressed when they are grafted with the equivalent of a few monolayers of Pt or Ir nanoaggregates [61].

More generally, mono- and mostly multi-metallic clusters are of great interest in supported catalysis wherein their electronic structure is thought to be an important parameter. For bimetallic clusters, this structure depends on the regularity of the arrangement of both kinds of atoms. Thus, perfectly ordered alloys such as those synthesized by radiolytic means [46] should exhibit enhanced catalytic efficiency.

The size-dependent redox potentials of the metal clusters in a liquid environment of course play a major role when they relay electrons. This was emphasized above in the case of the autocatalytic growth of silver clusters in the presence of an electron donor. As a fundamental rule, the occurrence of electron relay implies that the potential value of the cluster is intermediate between that of the donor and of the acceptor [23]. The efficiency of the catalyst thus appears to be a consequence not only of its high specific area, which increases the transfer rates at the interface, but more basically of an adequate size-dependent potential which governs whether the transfer be thermodynamically allowed or not. The optimum potential (and hence size) of the cluster depends on the potentials of both reactants.

This interpretation of catalytic efficiency gives an additional reason for acquiring systematic data about the size-dependent potentials of metal clusters.

3.3.5 Photographic Development

Photographic development is a special example of catalysis: the autocatalytic growth of a silver cluster fed by a developer acting as an electron donor [6].

The gelatin emulsion contains silver bromide microcrystals. During the short exposure, electrons are photodetached by light from Br^- anions and are transferred to Ag^+, thus yielding, in the bright parts of the image, a few silver atoms per AgBr microcrystal. These atoms coalesce and the distribution of the AgBr crystals containing the silver clusters (or specks) constitutes the imprint of light, the so-called latent image, yet too weak to be seen by our eyes.

Development consists basically of a chemically selective amplification, delayed relative to the exposure, which involves immersion of the exposed emulsion into an aqueous solution of the electron donor. It substitutes for the initial spatial distribution of microcrystals containing clusters of variable nuclearity a new distribution of microcrystals: either unchanged, or totally reduced into silver grains, depending on the size of the initial speck.

The undeveloped AgBr crystals, still photosensitive, are then eliminated by dissolution (the fixation step). The contrast between the areas without or with small silver grains scattering the light in the bright parts of the image then makes it visible as a negative picture. A second photograph of the negative provides the positive picture.

The possibility of development for the silver specks of the latent image depends on their size, which must be larger than a critical size n_c [6]. This critical size depends in turn on the developer strength, i.e. its redox potential. The development mechanism has been tentatively interpreted by numerous theories [6], either founded on the phase model (comparable to a supersaturation of silver atoms in the AgBr crystal) or on an atomistic model (referring to the size-dependent IP of the specks as known from the gas phase – see Fig. 3). However both models encounter serious difficulties, the first one in predicting a correct critical size, the second one in accounting for an increasing facility for the electron transfer from the developer to the speck [62].

Conversely, the experiments on the autocatalytic growth of silver clusters in the presence of an electron donor (3.3.4) showed quite similar features as those well known in photographic development [34]. Therefore, the same interpretation given there in a free solution may now be extrapolated to the selective amplification involved in the photographic development process.

The interpretation is founded 1) on the fact that the electron transfer is the rate-determining step, responsible for the initiation of the autocatalytic growth, and 2) on the general positive size-dependence of the aggregate ionization (or redox) potential with n (Fig. 3). Due to the difference with respect to the fixed value of the developer potential, such a positive variation induces a thermodynamical threshold which readily explains the existence of a minimum critical size for the developable speck, and why the critical size depends on the developer strength. The threshold implies no transfer below the critical size, and *a fortiori* no direct transfer to unexposed AgBr crystals.

However, the environments are not strictly identical: silver sulfate solution for the simulation experiments [34], interface between an AgBr crystal and the developer solution in photography. Therefore, the absolute values of the cluster potentials in AgBr may be systematically shifted relative to the curve of Fig. 3, thus modifying the critical size for a given developer. Nevertheless, the interpretation still holds.

Let us order in a diagram (Fig. 4) the AgBr microcrystals of the latent image more or less exposed to light according to the number of silver atoms they

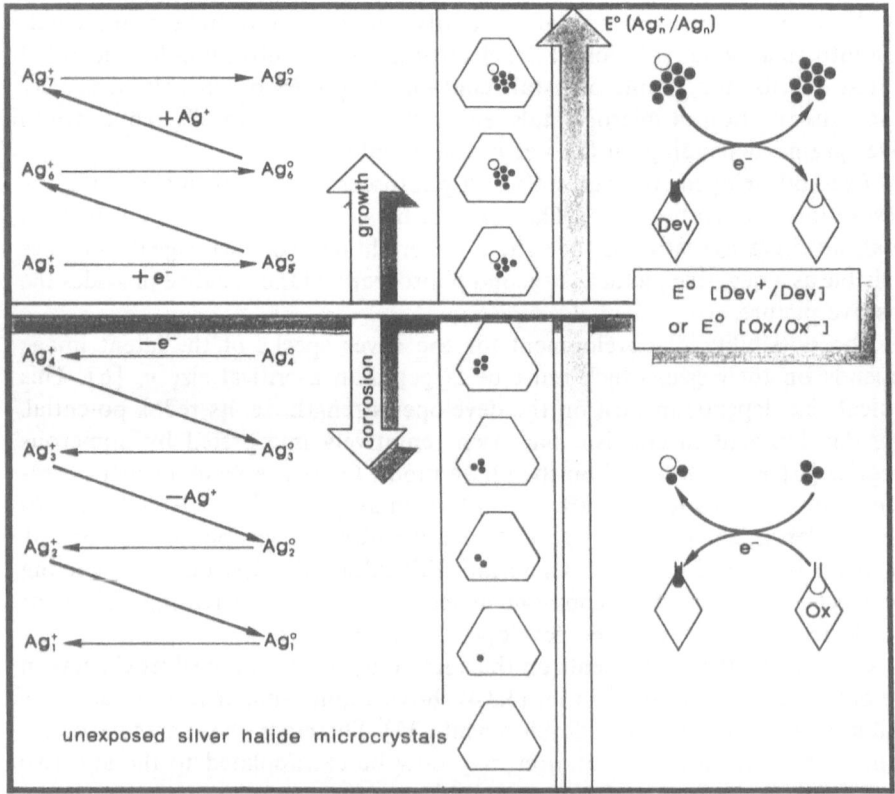

Fig. 4. Interpretation of the photographic development based on a size-dependent redox potential of the silver specks (see text)

contain, and consider that the redox potential of the silver specks, once brought into contact with the solvent of the developer solution, increases with n. The possible transfer of the electron from the developer to the speck, supposed to be (as in solution) the rate determining step, depends on the respective values of both potentials, that of the developer and that of the speck. Below the speck size for which the potential difference is zero, the transfer is not thermodynamically allowed. Conversely, beyond this critical size the transfer may occur. Then the aggregates is larger by one atom, and after a silver ion accretion it is ready again to accept another electron, thus starting the autocatalytic growth which does not stop up to the complete reduction of the crystal. Actually, photographic development mostly uses multielectron donors. In such a case, it is the most negative redox potential among the various monoelectronic steps which fixes the critical number of atoms in a developable speck. It cannot be ruled out that, when more data are available, the size-dependence will show oscillations such as those observed in the gas phase, instead of being monotonic. The same explanation of development would still hold, except that the critical size would be that above which the potential of any speck exceeds that of the developer.

Another somewhat puzzling phenomenon is that small specks, if not readily developed, may undergo easy oxidation which may cause accidental regression of the latent image (Fig. 4). The phenomenon has long been known and tentatively attributed to reverse oxidation. However, since the potential of the specks was thought to be very positive, as that of bulk silver, and hence higher than that of most oxidants, it is clear that the explanation was unacceptable [3]. The size-dependence of the redox potentials for Ag_n now answers this objection.

3.3.6   Conclusion

Ultrafine metal clusters may be synthesized in homogeneous or heterogeneous liquid systems, in a particularly easy, reproducible way, by means of radiolytic reduction of precursor ions. When performed in presence of proper additives, the method may be extended to the formation of other ultra-divided chemical compounds, e.g. oxides, sulfides, chlorides, or even molecular metal clusters [63].

The size-dependent properties of small metal clusters in a liquid environment are now clearly demonstrated and are being actively investigated. In spite of the transient character of the smallest clusters, huge differences as compared to bulk metal properties are observed, mostly by pulse techniques, and the general trends of variation with n are known. They allow us to understand certain features of the behavior of transient metal clusters: stability, corrosion, catalytic properties, and nucleation/growth dynamics, which are of great interest in different fields of fundamental and applied research.

3.3.7 Recent Developments

Still smaller stable solvated clusters have been synthesized and their growth dynamics studied: the main results concern Ag or Cu oligomers of a few atoms ligated by polyacrylate anions [64–68], other larger Au-coated Ag particles [65]. The catalytic reactivity of bulk electrodes has been enhanced through grafting of PtRu and NiRu [69]. The redox potentials of other transient Ag_n and Cu_n clusters have been evaluated [64–68, 70], and compared with the corresponding IPs for the same bare clusters [71]: the difference obeys rather well the Born's model in $n^{-1/3}$ [70, 72].

An evaluation of the electrochemical potential of clusters [65] concludes to some parity effects as for the IPs in the gas phase. A review compares the binding energy, the structure or the optical properties of the solvated atoms and clusters [70]. Results concerning the IPs of $Hg_n(NH_3)_x$ ($n = 1$ or 2, $x = 0$ to 6), are now bridging the gap between the gas phase and the condensed phase [73].

References

1. A.A. Chernov: *Modern Crystallography III: Crystal Growth Springer Series in Solid-State Sciences.* Springer-Verlag, Berlin (1984)
2. H.B. Weiser: *Colloid Chemistry,* John Wiley & Son, New York (1939)
3. M. Haïssinsky: *La chimie nucléaire et ses applications,* Masson, Paris (1957)
4. G. Maurin: In *Growth and Properties of Metal Clusters. Applications to catalysis and the photographic Processes,* J. Bourdon Ed., Elsevier, Amsterdam, 101 (1980)
5. G.A. Somorjaï: Adv. Catal. **26**, 1 (1977)
6. E. Moisar, F. Granzer: Photogr. Sci. Eng. **26**, 1 (1982)
7. M.O. Delcourt, J. Belloni: Radiochem. Radioanal. Letters **13**, 329 (1973)
8. M.D. Morse: Chem. Rev. **86**, 1049 (1986), and references therein
9. M.M. Kappes: Chem. Rev. **88**, 369 (1988), and references therein
10. E. Schumacher: Chimia **42**, 357 (1988), and references therein
11. J.H. Baxendale, E.M. Fielden, J.P. Keene: In *Pulse Radiolysis,* Acad. Press, London, 207 (1965)
12. A. Henglein: Top. Current Chem. **143**, 113 (1988), and references therein
13. A. Henglein: Chem. Rev. **89**, 1861 (1989), and references therein
14. M. Graetzel: Acc. Chem. Res. **14**, 376 (1981); M. Graetzel In *"Frontiers of Science"*, A. Scott Ed., Basil Blackwell Publishers, London, 1990, p. 83–97, and references therein
15. J.M. Lehn, J.P. Sauvage, R. Ziessel: Nouv. J. Chim. **4**, 623 (1980)
16. *"The Study of Fast Processes and Transient Species by Electron Pulse Radiolysis"*, J.H. Baxendale, F. Busi (Eds.), NATO ASIS Series C – Mathematical and Physical Sciences **86** D. Reidel, Dordrecht (1982). *"Radiation Chemistry. Principles and Applications"*, Farhataziz & Rodgers M.A.J. Eds., VCH Publishers, Weinheim (1987)
17. J. Belloni, J.L. Marignier: Radiat. Phys. Chem. **34**, 157 (1989)
18. E.J. Hart, M. Anbar: *The hydrated electron,* John Wiley & Son, New York (1970)
19. G.V. Buxton, R.M. Sellers: Coord. Chem. Rev. **22**, 195 (1977), and reference therein
20. G.V. Buxton, C.L. Greenstock, W.P. Helman, A.B. Ross: J. Phys. Chem. Ref. Data **17**, 513 (1988)
21. J. Von Pukies, W. Roebke, A. Henglein: Ber. Bunsenges. Phys. Chem. **72**, 842 (1968)
22. Farhataziz, P. Cordier, L.M. Perkey: Radiat. Res. **68**, 23 (1976)

310   J. Belloni et al.

23. J. Belloni, M.O. Delcourt, J. Amblard, J.L. Marignier: In *"6th Tihany Symp. on Radiation Chemistry 1986"*, P. Hedwig, L. Nyikos, R. Schiller Eds., Akadémial Kiadó, Budapest, 89 (1987)
24. D. Martin-Rovet, B. Hickel: Private Communication (1984)
25. P. Mulvaney, A. Henglein: J. Phys. Chem. **94**, 4182 (1990)
26. J. Butler, A. Henglein: Radiat. Phys. Chem. **15**, 603 (1980)
27. B.G. Ershov, N.L. Sukhov: Radiat. Phys. Chem. **36**, 93 (1990)
28. B. Cercek, M. Ebert, A.T. Swallow: J. Chem. Soc. A 612 (1966)
29. J.H. Fendler: Chem. Rev. **87**, 877 (1987)
30. O. Platzer: Thèse Orsay (1989); O. Platzer, J. Amblard, J.L. Marignier, J. Belloni: Atochem European Pat. Appl. No 88, 402371, 4 (26th sep. 1988)
31. O. Platzer, J. Amblard, J.L. Marignier, J. Belloni: J. Phys. Chem. **96**, 2334 (1992); J. Amblard, O. Platzer, J. Ridard, J. Belloni: J. Phys. Chem. **96**, 2341 (1992)
32. N. Basco, S.K. Vidyarthy, D.C. Walker: Can. J. Chem. **51**, 2497 (1973)
33. R. Tausch-Treml, A. Henglein, J. Lilie: Ber. Bunsenges: Phys. Chem. **82**, 1335 (1978)
34. M. Mostafavi, J.L. Marignier, J. Amblard, J. Belloni: J. Radiat. Phys. Chem. **34**, 605 (1989); Z. Phys. D-Atoms, Molecules & Clusters **12**, 31 (1989)
35. J. Westerhausen, A. Henglein, J. Lilie: Ber. Bunsenges. Phys. Chem. **85**, 182 (1981); S. Mosseri, A. Henglein, E. Janata: J. Phys. Chem. **93**, 6791 (1989)
36. M.O. Delcourt, J. Belloni, J.L. Marignier, C. Mory, C. Colliex: Radiat. Phys. Chem. **23**, 485 (1984)
37. A. Henglein: J. Phys. Chem. **83**, 2209 (1979)
38. C.D. Jonah, M.S. Matheson, D. Meisel: J. Phys. Chem. **81**, 1805 (1977)
39. J.L. Marignier: Thèse ès Sciences Orsay (1987)
40. J.L. Marignier, J. Belloni: J. Chim. Phys. **85**, 21 (1988)
41. R. Rafaeloff, Y. Haruvy, J. Binenboym, G. Baruch, L.A. Rajbenbach: J. Molec. Catal. **22**, 219 (1983), and references therein
42. J. Belloni, M.O. Delcourt, C. Leclere: Nouv. J. Chim. **6**, 507 (1982)
43. A. Henglein, J. Lilie: J. Phys. Chem. **85**, 1246 (1981)
44. M. Breitenkamp, A. Henglein, J. Lilie: Ber. Bunsenges. Phys. Chem. **80**, 973 (1976)
45. J.L. Marignier, J. Belloni, M.O. Delcourt, J.P. Chevalier: Nature **317**, 344 (1985)
46. J. Belloni, J.L. Marignier, M.O. Delcourt, M. Minana: US Patent No 4, 629, 709, dec. 16th (1986) & US Patent No 4, 745, 094, may 17th (1988)
47. J. Belloni, A. Lecheheb: Radiat. Phys. Chem. **29**, 89 (1987)
48. M.O. Delcourt, N. Keghouche, J. Belloni: Nouv. J. Chim. **7**, 131 (1983)
49. J.L. Marignier, J. Belloni, J. Amblard, T. Zemb: unpublished results
50. L. Kevan: J. Phys. Chem. **85**, 1628 (1981)
51. A. Henglein, R. Tausch-Treml: J. Coll. Interf. Sc. **80**, 84 (1981)
52. M. Mostafavi, N. Keghouche, M. O. Delcourt, J. Belloni: Chem. Phys. Letters **167**, 193 (1990)
53. M. Mostafavi, N. Keghouche, M.O. Delcourt: Chem. Phys. Letters **169**, 81 (1990)
54. J. Khatouri, M. Mostafavi, J. Belloni, J. Amblard: Chem. Phys. Letters **191**, 351 (1992)
55. A. Henglein: Ber. Bunsenges. Phys. Chem. **81**, 55 (1977)
56. A. Henglein, T. Proske: Ber. Bunsenges. Phys. Chem. **82**, 471 (1978)
57. K. Hilpert, K.A. Gingerich: Ber. Bunsenges. Phys. Chem. **72**, 842 (1968); Ber. Bunsenges. Phys. Chem., **84**, 739 (1980)
58. A. Henglein: Ber. Bunsenges. Phys. Chem. (in press)
59. L.S. Zheng, C.M. Karner, P.J. Brucat, S.H. Yang, C.L. Pettiette, M.J. Craycraft, R.E. Smalley: J. Chem. Phys., **85**, 1681 (1986)
60. P.Y. Cheng, M.A. Duncan: Chem. Phys. Letters, **152**, 341 (1988)
61. J. Bruneaux, H. Cachet, M. Froment, J. Amblard, J. Belloni, M. Mostafavi: Electrochim. Acta **32**, 1533 (1987)
    J. Bruneaux, H. Cachet, M. Froment, J. Amblard, M. Mostafavi: J. Electroanal. Chem. **269**, 375 (1989)
62. J. Belloni-Cofler, J. Amblard, M. Mostafavi & J.L. Marignier: La Recherche **21**, 48 (1990); Endeavour, **15**, 2 (1991); J. Belloni, M. Mostafavi, J.L. Marignier, J. Amblard: Imaging Sci. **35**, 68 (1991)

63. M.M. Bettahar, M.O. Delcourt: Radiat. Phys. Chem. **32**, 779 (1988); H. Remita, R. Derai, M.O. Delcourt: Radiat. Phys. Chem. **37**, 221 (1991)
64. N. Keghouche, M. Mostafavi, M.O. Delcourt, G. Picq: J. Chim. Phys. **90**, 777 (1993); M. Mostafavi, M.O. Delcourt, G. Picq: Radiat. Phys. Chem. **41**, 453 (1993)
65. A. Henglein: J. Phys. Chem. **97**, 5457 (1993), and references therein; P. Mulvaney, M. Giersig, A. Henglein: J. Phys. Chem. **97**, 7061 (1993)
66. J. Khatouri, M. Mostafavi, J. Amblard, J. Belloni: Z. Phys. D – Atoms, Molecules & Clusters **26**, S 82 (1993)
67. M. Mostafavi, M.O. Delcourt, J. Belloni: J. Imaging Science Techn. (1994) (in press)
68. S. Remita, J.M. Orts, J.M. Feliu, M. Mostafavi, M.O. Delcourt: Chem. Phys. Letters (1993) (in press)
69. J. Amblard, J. Belloni, O. Platzer: J. Chim. Phys. **88**, 835 (1991)
70. J. Belloni, J. Khatouri, M. Mostafavi, J. Amblard: *Ultrafast reaction dynamics and solvent effects*: Am. Inst. Phys., P.J. Rossky & Y. Gauduel Eds. 527 (1993)
71. C. Jackschath, I. Rabin, W. Schulze: Z. Phys. D – Atoms, Molecules & Clusters **22**, 517 (1992); G. Alameddin, J. Hunter, D. Cameron, M.M. Kappes: Chem. Phys. Letters **192**, 122 (1992); M.B. Knickelbein: Chem. Phys. Letters **192**, 129 (1992)
72. V. Russier, C. Mijoule, J. Langlet: *Proc. Int. Symp. Phys. Chem. Finite Systems: from Clusters to Crystals*: P. Jena, S.N. Khanna, B.K. Rao, Eds. Klewer Acad., Boston (1992)
73. C. Dedonder-Lardeux, C. Jouvet, S. Martrenchard, D. Solgadi, A. Tramer: Z. Phys. D – Atoms, Molecules & Clusters, **20**, 73 (1991)

3.4 Larger Semiconductor Clusters ("Quantum Dots")

L. Brus

3.4.1 Introduction

In this chapter, I outline the simple theory of size effects that occur for semiconductor crystallites in the diameter range of 2–100 nm. In this range continuous electronic band structure develops from discrete, molecule-like states; that is, the Schrodinger equation eigenfunctions are size dependent. There are also entirely independent electromagnetic size effects related to classical optical resonances, and to the evolution from molecule-like electric dipole scattering to bulk-like polariton scattering, in Maxwell's equations. Electromagnetic polariton phenomena in larger crystallites have interesting, size dependent consequences for both photon absorption and fluorescence. If the crystallite is an excised fragment of the bulk lattice, then all these phenomena can be realistically modelled using known experimental parameters. This chapter is written in the style of a tutorial, assuming a thorough knowledge of quantum mechanics and electromagnetic theory.

Semiconductors are a special class of materials with regard to cluster size effects. The fact that the chemical bonding is quite strong and directional has two important consequences: First, smaller clusters with a few tens of atoms apparently adopt unique, molecule-like structures in order to minimize broken surface bonds. The bulk like unit cell is not present. (In this chapter these smaller clusters are not considered.) Second, even in larger clusters exhibiting the bulk structural unit cell, significant purely quantum electronic size effects occur because electron delocalization near the conduction and valence band edges is commonly quite strong. This effect is the three dimensional analog of the extensively studied quantum electronic structure of thin layer semiconductor superlattices.

The lowest excited, neutral electronic state of a bulk crystalline semiconductor is well described by the Wannier exciton Hamiltonian

$$\hat{H} = \frac{-\hbar^2}{2m_e} \nabla_e^2 - \frac{\hbar^2}{2m_h} \nabla_h^2 - \frac{e^2}{\varepsilon |\bar{r}_e - \bar{r}_h|} \tag{1}$$

Here m_e and m_h are effective masses, and ε is the dielectric coefficient. The lowest eigenstate is hydrogenic in structure, internally correlated, and weakly bound (i.e. a few meV). In order to model electron–hole physics in small crystallites, the physical basis and limits of this equation with additional boundary conditions must be understood. I discuss the electrostatics of electron–hole interaction first, and then treat the applicability of bulk effective masses to modelling of electron–hole motion in small crystallites. Both effects are then combined, and finally coupling of crystallite excited electronic states to the electromagnetic field is considered.

3.4.2   Elementary Theory

3.4.2.1   Electrostatics

The fact that the Coulomb interaction in Eq. (1) is significantly screened ($\varepsilon = 5$–15) necessarily implies that the electron and hole individually have large dielectric solvation energies in the screening medium [1, 2]. At high frequency in a semiconductor, the direct Coulomb interaction is screened by valence electron polarization. As a consequence, the electron–hole Coulomb term in a neutral crystallite must incorporate electrostatic polarization at the crystallite surface, and the ionization potential of a crystallite will have an important electrostatic component. The (position dependent) interaction potential between two charges in a sphere (radius R) of dielectric coefficient ε_1 in a medium of coefficient ε_2 is [1]

$$V(r_e, r_h, R) = \frac{-e^2}{\varepsilon_1 |r_e - r_h|} + P(r_e, R) + P(r_h, R) - P_M(r_e, r_h, R) , \qquad (2)$$

where

$$P(r, R) \equiv \sum_{n=1}^{\infty} \alpha_n \left[\frac{r}{R}\right]^{2n} \frac{e^2}{2R} \quad \text{and} \quad P_M(r_e, r_h, R) \equiv \sum_{n=1}^{\infty} \alpha_n e^2 \frac{r_e^n r_h^n}{R^{2n+1}} P_n(\cos \theta) ,$$

$\alpha_n = [(\varepsilon_1/\varepsilon_2 - 1)(n + 1)]/[\varepsilon_1(\varepsilon_1 n/\varepsilon_2 + n + 1)]$. P describes a charge's interaction with its induced surface polarization, and yields a relatively weak radial force pulling the charge to the sphere center, the point of the greatest dielectric stabilization. P_M describes the interaction of one charge with the surface polarization of the other. This term, as well as the direct Coulomb term, represent attractive forces between electron and hole. Through P_M, the net electron–hole attraction increases as ε_2 decreases with respect to ε_1. The net Coulomb interaction is modestly less shielded in the crystallite than in the bulk because part of the field fringes outside the crystallite into a region of lower ε. These various electrostatic terms can be used in a Schrodinger's equation for electron–hole motion in a model semiconductor sphere [3]. For the lowest few neutral excited states, the numerical effect of including the surface polarization terms is relatively minor as relatively little of the field fringes out of the sphere.

Electrostatic effects become numerically larger if the crystallite is ionized, leaving a charged sphere with one mobile hole inside, and with electric field lines exiting the sphere and terminating on an electron at infinity. The electrostatic energy $V(r_h)$ is

$$V(r_h) = \frac{(\varepsilon_1/\varepsilon_2 - 1)e^2}{2\varepsilon_1 R} + P(r_h, R) . \tag{3}$$

$V(r_h)$ is the classical (position dependent) energy to charge a dielectric sphere, leaving an opposite charge at infinity.

3.4.2.2 Molecular Orbitals and the Quantum Size Effect

In this Sect. I discuss single carrier dynamical motion. In bulk crystals, a scalar effective mass can be used to represent carrier localization energy in Eq. (1) because the electron and hole of interest exist near band extrema – at the bottom of the conduction band and the top of the valence band respectively. These two regions predominantly contribute to the lowest excited state of the crystallite. In general, to localize a wavepacket in a spatial region of order R, a distribution of wave vector K states of width ca. π/R must be superimposed. If the band energy $E(K)$ at an extremum is expanded in a Taylor's series, then the wavepacket energy shift is $\frac{1}{2}\partial^2 E/\partial K^2 (\Delta K)^2$. As long as the superposition involves only this term the carrier can be considered to be a pseudoparticle with an effective mass inversely proportional to the band curvature. This approximation is typically valid for shifts of 0.1–0.4 eV.

Further insight is gained by viewing the crystalline as a three dimensional polymer of a repeating unit cell [4]. Tight binding single electron eigenspectra exhibit systematic aspects of the evolution of band structure from molecule-like discrete states. For example, Fig. 1 shows $E(K)$ for an infinite linear chain of identical atoms, with electron exchange between nearest neighbors only. The figure also shows the discrete eigenvalues for chains of 11 and 13 atoms, respectively [4]. The finite chain eigenvalues are those infinite chain $E(K)$ values corresponding to wavefunctions with nodes on the chain ends. Near the extrema, the eigenspectrum is a particle-in-a-box, effective mass eigenspectrum. The lowest level is a standing wave with one half wavelength inside the box.

In general, as long as the unit cell and bonding in the finite polymer are the same as in the bulk, effective mass like models can be used, over a limited energetic range, for the size dependence of the carrier localization energies. This implies that the carrier dynamics can be modelled by a Wannier type Hamiltonian (1) with appropriate boundary conditions.

The effective mass structure of discrete one electron levels (e.g., molecular orbitals) for a "spherical" crystallite of a zinc-blende, direct band gap compound semiconductor (InP, GaAs, or CdSe) is exemplified in Fig. 2 [5]. The valence band structure is complex because the valence band is three fold electronically

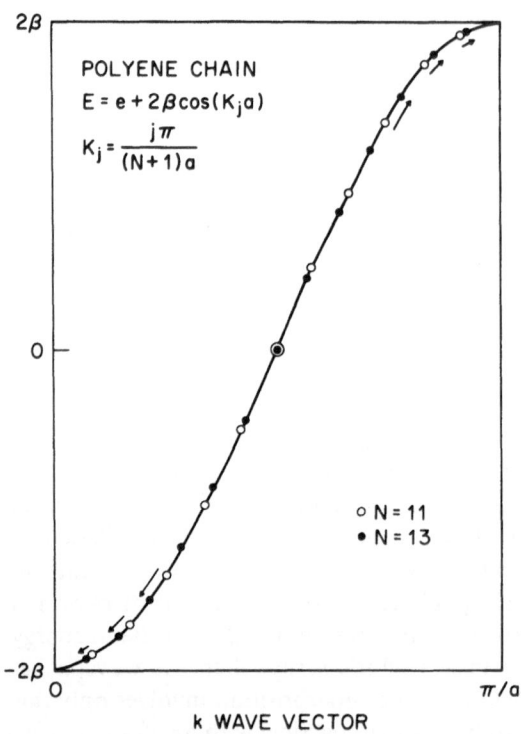

POLYENE CHAIN

$$E = e + 2\beta\cos(K_j a)$$

$$K_j = \frac{j\pi}{(N+1)a}$$

o N = 11
• N = 13

k WAVE VECTOR

Fig. 1. Single electron eigenvalues for linear chains of $N = 11$ and 13 identical atoms, shown on top of the band for an infinite chain. Adapted from [4]

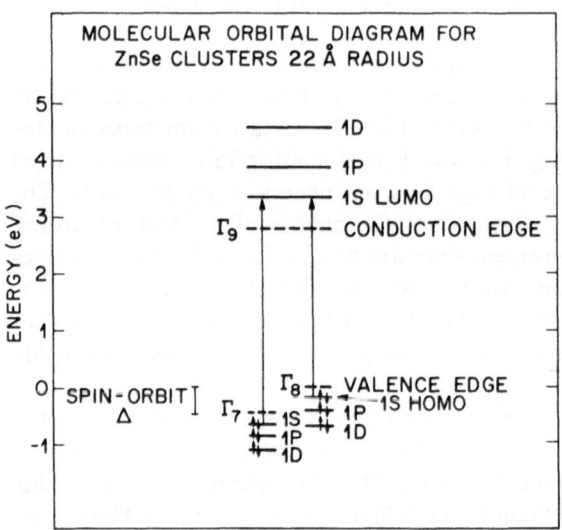

MOLECULAR ORBITAL DIAGRAM FOR
ZnSe CLUSTERS 22 Å RADIUS

Fig. 2. Molecular orbitals for a 44 Å diameter ZnSe crystallite. The lowest two allowed transitions are shown with *vertical lines*. Adapted from [5]

degenerate at its extremum. In a certain sense, the hole has internal degrees of freedom not present in the electron. The "effective mass" is a tensor that has both S and D components in the spherical point group. This eigenvalue problem for zinc-blende lattices has only recently been addressed in full detail [6].

3.4.2.3 Excited Electronic States and Ionization

In order to describe the lower excited states as a function of radius R, Eq. (1) can be modified to incorporate the complete electrostatic interaction (Eq. (2)), along with the boundary condition that the wavefunction is zero at the surface. As we have discussed above, this model effectively includes both molecular orbital effects and electron–hole correlation through the modified Coulomb interaction. The kinetic localization energies vary as R^{-2} and the electrostatic terms approximately as R^{-1} [3, 7]. Thus the quantum localization energies dominate in the limit of small R. The wavefunction is a simple product of $1S$ particle-in-a-sphere functions for electron and hole, with energy [3, 8]

$$E(R) = \frac{\hbar\pi^2}{2R^2}\left[\frac{1}{m_e} + \frac{1}{m_h}\right] - \frac{1.8e^2}{\varepsilon_1 R} + \text{polarization terms} . \tag{4}$$

In this extreme small particle limit, there is no electron–hole correlation. Fig. 3 shows the expected variation with size for representative semiconductors. In direct band gap semiconductors, this lowest excited state is electric dipole allowed with a radiative lifetime on the order of a nanosecond. The oscillator strength is size independent [3].

As size increases, the Coulomb term and P_M induce electron–hole correlation [9–15]. A true hydrogenic exciton forms when R is about twice the bulk exciton Bohr radius (about 60 Å in CdS for example). One consequence is that the electric dipole absorption cross section per crystallite begins to increase in proportion to crystallite volume. The radiative lifetime (in the absence of thermal dephasing) is predicted to decrease from nanoseconds to picoseconds.

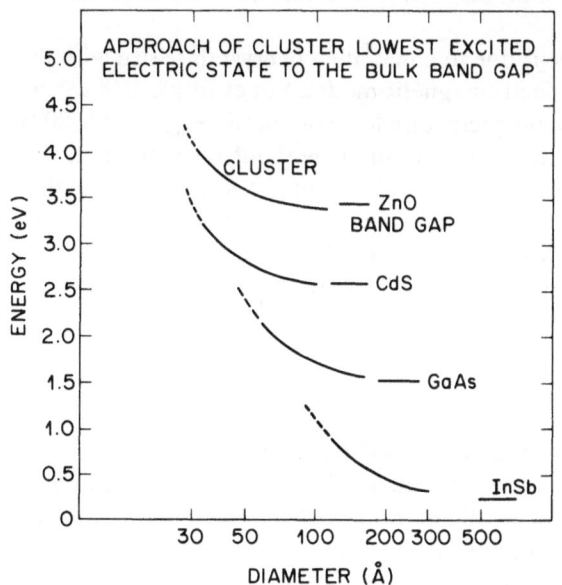

Fig. 3. Predicted size dependence adapted from [3]

As we shall see, polariton and local electromagnetic field effects related to Mie scattering theory also become important as electron–hole correlation develops.

The size dependence of the ionization potential can be modelled with a Hamiltonian that includes the hole kinetic energy and $V(r_h)$. Both terms favor the hole being in the center of the crystallite, in a near $1S$ orbital. With this eigenfunction, the ionization potential is [1, 2]

$$I_p(R) = I_p(\infty) + \frac{\hbar^2\pi^2}{2m_h R^2} + \frac{(\varepsilon_1/\varepsilon_2 - 1)e^2}{2\varepsilon_1 R} + \text{smaller terms} . \tag{5}$$

The electrostatic term dominates for most materials at intermediate sizes.

3.4.2.4 Electromagnetic Phenomena

In this section I discuss how the eigenstates of semiconductor crystallites couple to the electromagnetic field. First, consider a bulk semiconductor characterized by a wavelength (λ) dependent, complex dielectric coefficient $\varepsilon' + i\varepsilon''$. The Lorenz–Mie theory describes classical electromagnetic scattering and absorption by spheres of such bulk materials [16]. The theory gives these cross sections as infinite sums of electric and magnetic multiple moments multiplied by the corresponding powers of (R/λ). A bulk semiconductor has a continuous, non-zero ε'' at energies above the band gap, corresponding to wavelengths were optical absorption occurs. Generally speaking, if $R \gg \lambda$ scattering dominates absorption, in analogy to the fact that reflection dominates absorption above the band gap on flat surfaces. If $R \ll \lambda$, however, then the electric dipole term is dominant and absorption dominates scattering, as is also the case in molecular spectroscopy.

If $R \approx \lambda$, then resonances occur due to a matching of crystallite size with the internal wavelengths of specific electromagnetic modes. For example, a Si sphere of 112 nm diameter has a sharp, magnetic dipole resonance in σ_{abs} near 4880 Å [17]. (The bulk absorption spectrum is continuous without resonances in this region.) For this size, the internal field intensity is calculated to be 150 times that inside a macroscopic Si sample; as a consequence, a size resonance in the lattice spontaneous Raman scattering also occurs. This effect is purely classical.

If $R \ll \lambda$ then Rayleigh scattering is weak, and the electric dipole term dominates σ_{abs}:

$$\sigma_{abs} = \frac{8\pi^2 R^3}{\lambda} \text{Im}\left[\frac{\varepsilon - 1}{\varepsilon + 2}\right] . \tag{6}$$

The absorption spectral shape should be independent of R in this small crystallite limit. This result of Lorenz–Mie theory, using the experimental bulk dielectric coefficient, fails in the 1–10 nm range when the quantum effects discussed above become significant and the optical spectrum develops discrete structure. Lorenz–Mie theory also fails in that it predicts σ_{abs} increases as the crystallite volume. Actually, in the strong confinement limit when there is no electron–hole correlation, quantum oscillator strengths of discrete transitions are independent of size [3].

In this small particle limit, when the wavefunction shows strong confinement with only weak electron–hole correlation, the crystallite shows weak coupling to the electromagnetic filed. This limit is characteristic of molecular spectroscopy in general.

However, a modified Mie theory, using a nonlocal dielectric constant, is valid for larger crystallites of diameter several times the exciton Bohr radius [18, 19]. In this range hydrogenic, internally correlated excitons are postulated to exhibit center of mass confinement in the crystallite. If ω_T is the frequency of the lowest bulk exciton (a few meV below the band gap) then a nonlocal dielectric constant can be postulated

$$\varepsilon(\omega, K) = \varepsilon_0 + \frac{\omega_p^2}{\omega_T^2 - \omega^2 + DK^2 - i\gamma\omega} . \tag{7}$$

ω_p is a measure of band to band oscillator strength. The electronic polarization induced by the external field is not uniform, but has a node on the sphere surface. The K^2 term leads to a predicted shift in absorption to higher energy with decreasing size, although ambiguities remain as to the exact boundary conditions. Even in a large sphere where quantum effects are negligible, the optical resonance does not occur at ω_T but at a slightly higher frequency ω_F given by $\mathrm{Re}[\varepsilon(\omega_F)] = -2\varepsilon_2$. This electromagnetic shift is about 20 cm^{-1} for the lowest exciton in CuCl crystallites, for example [20]. This shifted mode is the electric dipole, Frohlich surface polariton of the sphere. This effect is an electromagnetic shape resonance, in that the upfield shift depends upon shape but is independent of size. This is a different phenomenon than the size resonance earlier described for large Si spheres. These phenomena represent strong interaction between the crystallite and the radiation field, and have no analogy in molecular spectroscopy.

The Frohlich surface polariton is predicted to have a remarkable effect on crystallite luminescence [20]. If the shift of a exciton center-of-mass quantized level, to higher energy then ω_T, comes into resonance with $\omega_s = \omega_F - \omega_T$, then a significant increase in the luminescence rate is predicted. This effect appears to occur in the luminescence of CuCl spheres of $R = 75$ Å; in this system the exciton Bohr radius is only 7 Å.

If the exciton state is intense and has an especially narrow linewidth on the order of 1 meV, then the crystallite optical response is predicted to be nonlinear, even bistable, due to positive feedback between local field effects and saturation of the exciton resonance. The resonance saturates because of a finite electron–hole lifetime. In this case F, the ratio of internal to external field intensities, changes dramatically across the Frohlich resonance (at low intensity) as shown in Fig. 4. Above the resonance, near the Frohlich dipolar polariton, F is large. In this region, small changes in the absorption have a large effect on F. The absorption changes as the resonance saturates, which in turn changes F,

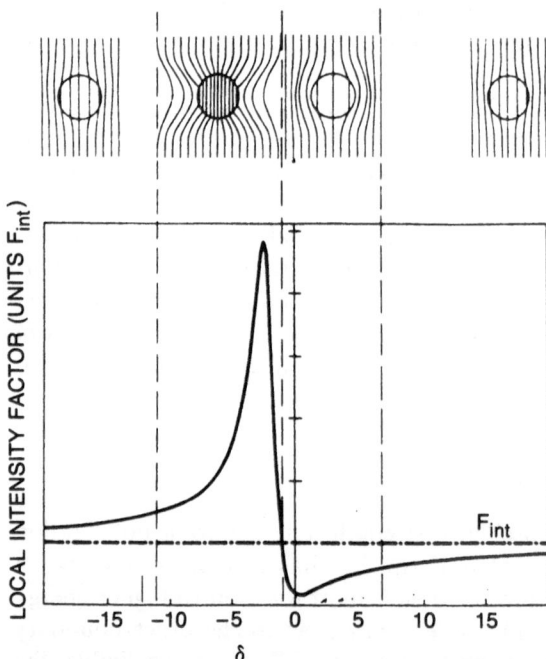

Fig. 4. Dispersion of the local intensity ratio F in the neighborhood of an isolated absorption resonance. Adapted from [21]. δ is a normalized detuning of the frequency from ω_T

which may further saturate the resonance, etc., leading to a nonlinear dependence of absorption on field intensity.

3.4.3 Summary

The dynamical motion of an electron–hole pair in a small crystallite, interacting with each other and with the electromagnetic field, is a rich physical problem that naturally lends itself to modelling, and draws upon ideas from various branches of physics. This chapter summarizes those common aspects that should apply to all direct band gap isotropic semiconductors. There are several reviews that cover experimental work [22–24]. Many of the ideas discussed here are sparsely verified predictions, and it will be interesting to see how real crystallites behave.

Acknowledgement. I am deeply grateful to many collaborators whose names are given in the references. I also thank S.L. McCall for his constructive comments on this chapter. The supportive and stimulating atmosphere inside AT&T Bell Laboratories made this work possible.

References

1. L.E. Brus: J. Chem. Phys. **79**, 5566 (1983)
2. G. Makov, A. Nitzan, L. Brus: J. Chem. Phys. **88**, 5076 (1988)
3. L.E. Brus: J. Chem. Phys. **80**, 4403 (1984)
4. L.E. Brus: New J. Chem. (France) **11**, 23 (1987)
5. N. Chestnoy, R. Hull, L. E. Brus: J. Chem. Phys. **85**, 2237 (1986)
6. J. B. Xia: Phys. Rev. **B 40**, 8500 (1989)
7. Al.L. Efros, A.L. Afros: Sov. Phys. Semicond. **16**, 1209 (1982)
8. Y. Kayanuma: Solid State Comm. **59**, 405 (1986)
9. V. Mohan, J.B. Anderson: Chem. Phys. Lett. **156**, 520 (1989)
10. S.V. Nair, S. Sinha, K.C. Rustagi: Phys. Rev. **B 35**, 4098 (1987)
11. Y.Z. Hu, M. Lindberg, S.W. Koch: Phys. Rev. **B 42**, 1713 (1990)
12. T. Takagahara: Phys. Rev. **B 36**, 9293 (1987); T. Takagahara: Surface Science 196, 590 (1987)
13. Y. Kayanuma: Phys. Rev. **B 38**, 9797 (1988)
14. E. Hanamura: Solid State Comm. **62**, 465 (1987); E. Hanamura: Phys. Rev. **B 37**, 1273 (1988)
15. T. Takagahara: Phys. Rev. **B 39**, 10206 (1989)
16. M. Kerker: *The Scattering of Light and Other Electromagnetic Radiation* (Academic, New York, 1969)
17. D.V. Murphy, S.R.J. Brueck: Optics Letters **8**, 494 (1983)
18. R. Ruppin: J. Op. Soc. Am. **71**, 755 (1981)
19. R. Ruppin, J. Phys. Chem. Solids **50**, 877 (1989)
20. A.I. Ekimov, A.A. Onushchenko, M.E. Raikh, Al.A. Efros: Sov. Phys. JEPT **63**, 1054 (1986)
21. S. Schmitt-Rink, D.A.B. Miller, D.S. Chemla: Phys. Rev. **B 35**, 8113 (1987)
22. A. Henglein: *Topics in Current Chemistry Vol.* **143** (Springer-Verlag, Berlin, 1988), Chapt. 4
23. M.L. Steigerwald, L. Brus: Annu. Rev. Mater. Sci. **19**, 471 (1989)
24. M.L. Steigerwald, L. Brus: Accounts of Chemical Research **23**, 183 (1990)

3.5 Electromagnetic Excitations of Large Clusters

U. Kreibig and *M. Quinten*

3.5.1 Introduction

The ruby color of ancient church windows, which is due to copper clusters/particles, makes, as an example, obvious that optical ones belong to the most spectacular physical cluster effects. Historically they were the first to be discovered, dating back to roman times, at least. Locating them on a size scale, optical cluster effects are observed from the molecular cluster region of some ten atoms [1] per cluster up to extremely large particles consisting of, say, 10^8 atoms each. (Hence, the term "cluster" will be used throughout the following as synonymous to "small particle".) This size region is larger than for any other known cluster size effect. For good metals it is the most prominent by far. This stems from the fact that the total oscillator strength of the "conduction" electrons, spread in the bulk from the plasma frequency to zero frequency, is compressed into narrow, extremely strong spectral bands. This is demonstrated schematically in Fig. 1.

There are two different directions on the size scale of approaching the cluster region: the molecular approach and the solid state approach. Each one is based on extrapolation of established theoretical concepts and experimental methods towards this particular intermediate region of matter. According to the purpose

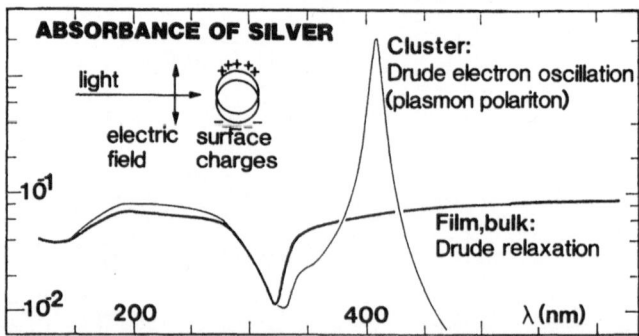

Fig. 1. Different absorbances of films and clusters of silver. Interband transitions determine the absorption below wavelength $\lambda = 325$ nm

of this chapter the approach from small cluster sizes will not be treated here; it is the topic of other chapters of this book.

The solid state approach presumes the distinction between a surface region (or in case of some embedding substance, an interface region) and a region of inner atoms, and peculiar cluster properties may then be due to the surface and/or the size-limited inner cluster region.

For most of the optical cluster effects, however, there is no clear distinction between surface and size effects. E.g. the potential box model yields discrete energy levels for "free" electrons by the *Dirichlet* boundary condition at the surface, thus determining the stationary electron wave functions throughout the whole particle.

The spectacular colors of various metal and semiconductor clusters are created by the cluster surface/cluster–matrix interface, the electric charging of which gives rise to electrical multipole moments (Fig. 2). They can be induced by external electric fields either by beams of fast electrons [2] or by electromagnetic fields. This was directly confirmed by HEELS (High Energy Electron Loss Spectroscopy) images [3]. The spectral response of such excitations, on the other hand is determined by the polarizability of the cluster as a whole.

The inner electromagnetic field of the excited cluster can show resonances at special frequencies if phase relations between this field and the external field are appropriately fulfilled by a negative real part ε_1 of the dielectric function of the particle material. In the limit of small clusters, approximately these frequencies are determined by the condition

$$\varepsilon_1(\omega_{s,l}) = -((l+1)/l)\varepsilon_m, \quad l = 1, 2, 3, \ldots \tag{1}$$

ε_m is the dielectric function of the medium surrounding the cluster. $l = 1, 2, 3, \ldots$ hold for the different multipoles (dipole, quadrupole, octupole, ...).

FAR FIELDS OF DIFFERENT SPHERICAL MODES

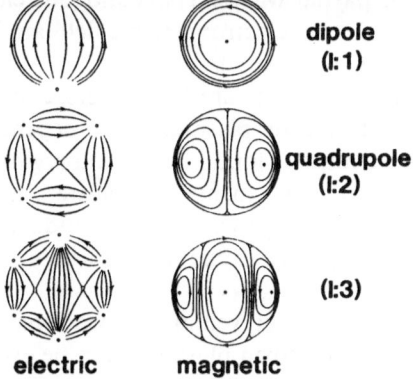

dipole
(l: 1)

quadrupole
(l: 2)

(l: 3)

electric   magnetic

Fig. 2. Symmetries of spherical modes: electric and magnetic scattering fields due to the "electrical" partial waves (the clusters are located at the centers). After [4]

In larger clusters, both, peak positions and excitation probabilities exhibit strong dependences on cluster size.

This classical effect described by *Mie* [4] and *Debye* [5] on the basis of Maxwells equations was later reinterpreted [6, 2] as being due to quantized elementary collective excitations which are well established in solid state physics [7, 8]. According to different contributions to the polarization of the matter, atomic vibrations ("cluster phonons") and electronic excitations ("cluster plasmons" of free and bound electrons) are to be distinguished. It may be pointed out that this description includes quantum mechanical cluster size effects due to the size limitation of the "electronic" and of the "ionic" particle. They are, however, hidden in the complex dynamical polarizability (i.e. the dielectric function of the particle material) which, in the frame of *Maxwells* theory, is introduced phenomenologically and deserves extra interpretation. Hence, two kinds of effects in the optical properties of clusters may be distinguished:

– the size dependences of the electrodynamic excitation ("extrinsic cluster size effects")

– the changes of the particle polarizability due to size limitations ("intrinsic cluster size effects").

The dynamic polarizability $\alpha(\omega)$ (or the dielectric function $\varepsilon(\omega)$) of the particle material represents an average taken over the contributions of all electronic and atomic constituents of the cluster i.e. over volume and surface. In this sense, the average values of α or ε ascribed to some cluster are, throughout the following, those, which would yield the experimentally observed spectral response, when introduced in *Mie*'s theory.

An arbitrary disturbance of the surface charge of a spherical cluster, induced by external *electric* fields can be represented by a multipole expansion, i.e. the sum of contributions of the, in principle, infinite number of multipoles of spherical geometry. As will be shown later, the multipole expansion corresponds to the introduction of cluster surface plasmon or phonon modes of different order. This mode concept is the physical interpretation of the formal expansion into orthogonal spherical functions or partial waves. The analogous electrical excitations by the *magnetic* part of an external electromagnetic field are infinite series of orthogonal eddy current modes (see below).

In ionic materials the cluster surface phonons are located close to typical optical lattice frequencies, i.e. 50 meV. The cluster plasmons are – depending on the electron properties (density, effective mass) – found in spectral regions from the IR (in semiconductors) to the UV (metals). In metallic clusters the ionic excitations are suppressed by electronic shielding, while in semiconductors and semimetals both types of excitations can exist and even interfere with each other [9].

The famous gold and copper ruby and silver yellow are produced by the conduction electron plasmons, yet, bound electron plasmons can be excited as well, supposed the coupling within the respective electron system is strong.

Theoretical models assume these elementary excitations [7, 8] to give quantum mechanical eigenstates of the cluster. In realistic materials, however, their lifetimes are restricted to few oscillations due to their complex interactions with electronic and ionic dissipation subsystems. Even in the metals with least damping (Na, K, Al, Ag), the number of oscillations is restricted to less, say 10^2.

The relation of cluster plasmons to coherently excited single electron excitations in coupled electronic systems has recently been pointed out in discussions regarding the strong optical excitations observed in molecular Na and K clusters [1]. Vice versa, the plasmon decay may be, in part, described as loss of coherency in a collective of single electron excitations, by individual damping processes.

As a consequence of this complex coupling of excitation and decay channels, it has not yet been possible to give a comprehensive quantum mechanical description of the elementary excitations in clusters of realistic material. Theoretical results are restricted to "free electron" or "jellium" clusters [10, 11, 12, 16].

In the case of optical excitation, the coupling of the plasmon to the external electromagnetic field induces additional radiation damping. These effects are called "polariton" effects, polaritons being microscopic elementary excitations of matter which are coupled to macroscopic electromagnetic fields. They additionally reduce the plasmon lifetimes [2]. Experiments prove that, both, the intrinsic damping and the radiation effects depend on the cluster size. The latter are what we call "extrinsic" cluster size effects. They are observed in clusters which are large enough that retardation of the electromagnetic fields is important. Hence, for small clusters, often no discrimination is made between plasmons and plasmon polaritons.

Quantitative descriptions of lifetime effects are restricted to the phenomenological model, where radiation damping is described by the electrodynamic Mie theory, while size effects of the cluster polarization are included in the imaginary part of the dynamical polarizability of the cluster material. *v. Fragstein* and his coworkers [13] were the first to evaluate and interpret these size dependent optical material properties from experiments of extinction and refraction.

As a first step it proved to be useful to start with the polarizability of the appropriate bulk and to introduce more or less rough corrections due to the limited size of the cluster, which usually are named "size effects". It is these effects that reflect the transition of the extended solid to the molecule.

In the following it will be demonstrated that a multitude of cluster effects can be treated in this semi-phenomenological way, presumed the limits of applicability are carefully regarded. A number of different intrinsic size effects influencing the optical properties of clusters were listed up, recently [14]. Some of them will be discussed in Sect. 3.5.4.

Though the total energy per atom is lower in clusters than for the isolated atom, it is still higher than in the bulk material, thus rendering the cluster state thermodynamically unstable. Neutral clusters are therefore produced in the free space only with short life-times, limited by their times of free flight in the apparatus and/or the times until fragmentation or coalescence processes take place. More extended lifetimes of free clusters may be achieved in future by trapping them as cluster ions in electromagnetic cages. Up to now, however, they are obtained only by stabilizing the neutral clusters on a supporting *substrate* or in an embedding *matrix*. The influences of the stabilizing materials are included in *Maxwells* boundary conditions by their dielectric function ε_m. This gives rise to small changes of positions and widths of the optical bands. It has become obvious [14, 15] that, due to the "spill-out" effect of the conduction electrons and to physi- and chemisorptive reactions in the cluster–matrix interface, the matrix effects on the electronic and optic cluster properties may be much stronger than these influences of the electrodynamic boundary conditions.

Some of the interface effects exhibit size dependences towards smaller cluster sizes. Yet, a detailed investigation of the physics and chemistry on cluster interfaces is still at the beginning and existing optical investigations [14, 15] do not give a clear view, at present. (Note added in proof: this topic was investigated recently in more detail [59, 60, 61].) Throughout the following, we therefore take matrices into account only via ε_m. We, also, disregard the more complicated case of supporting substrates, where ε_m differs for the two half-spaces.

In the frame of *Maxwells* theory, *Mie*'s theory gives a complete description of the extinction spectra for arbitrary cluster sizes and electromagnetic wavelengths. The according real part of the polarization was complemented by *Gans* and *Happel* [17] shortly afterwards.

Common optical experiments do not allow to investigate single clusters because of their small cross sections. To obtain quantities accessible to optical experiments, fictitious many-cluster samples are constructed in the *Mie* theory, consisting of N single, non-interacting *Mie* particles. For usual optical measurements, this N has to amount typically to $> 10^{10}$. "Non-interacting" means, that the clusters are well separated with low volume concentrations below, say 10^{-6} (see below).

At higher packing densities, interactions between separated clusters become important via the electromagnetic scattering fields. This electromagnetic coupling is due to induced multipolar excitations and, hence, resembles the *v.d. Waals* interaction with the difference, however, that only those multipolar moments are important which are excited in the clusters by the incident light wave. These interaction effects will be discussed later. They cause the optical cluster spectra to change drastically; the according mechanical coupling forces which are induced, as well, still await their final identification [62].

As an alternative way of describing the optical properties of n-cluster systems with more or less dense packing, "effective medium" theories have been developed, almost simultaneously with *Mie*'s theory [20–22]. They were extended to a very interesting general theoretical concept by *Bergman* [23]. The existing effective medium theories are limited to the quasistatic approximation, (Fig. 6), and, hence, typical size dependent cluster effects which show up in the extended *Mie* theory for larger particles, are not comprised. The present survey being devoted mainly to larger clusters which are exactly described by the alternative *Mie* concept and its extensions, the effective medium concept is disregarded here.

Mie's theory is restricted to polariton excitations i.e. the excitations by the electromagnetic field and, hence, may not be misapplied to quantitative discussions of free plasmons or phonons in clusters which can be excited, e.g. by fast electrons and analyzed by high electron energy loss spectroscopy [7]. This holds vice versa: polariton excitations should not be quantitatively interpreted by plasmon theories. The reason is, that beyond the size limits for the quasistatic region, both, position and shape of the response bands differ for the different excitation mechanisms. Rather than discussing these effects, we restrict ourselves here to present Fig. 3 which shows the measured peak maxima of Ag clusters for both kinds of excitation [2]. Both peaks coincide for clusters as small as 10 nm. While the plasmon peak exhibits no measurable dispersion when the cluster size is increased, the polariton peak is drastically shifted to lower frequencies, the difference achieving 1 eV for 100 nm particles. This effect may be illustrated as follows: In case of fast electron impact the excitation time is restricted to about one plasmon oscillation period, the plasmon thus being due to free oscillation of the *Drude* electrons in the cluster. If however, the electrons are excited by electromagnetic waves, dispersion matching to the external radiation field is required, and then re-radiation is possible, leading to the elastic *Mie*-scattering. Since the latter increases strongly with particle size beyond the quasistatic region, both, peak shift and broadening occur due to the increasing additional radiation damping. This effect was as well observed for higher order multipoles which are suppressed in the case of polariton excitation, compared to the free plasmon. This can also be extracted from Fig. 3: while, at 80 nm particle size the measured plasmon frequency is mainly due to the quadrupolar plasmon (dipolar

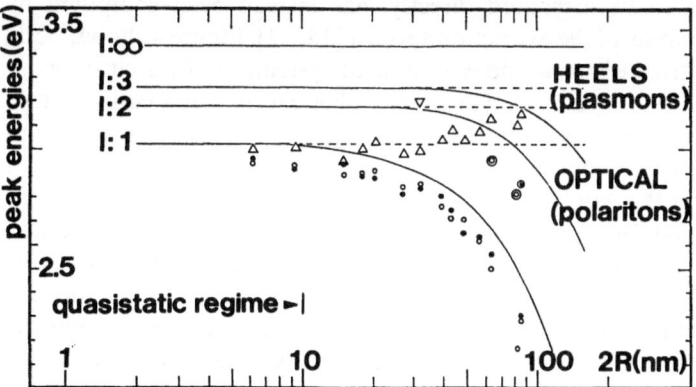

Fig. 3. Plasmon and plasmon polariton mode energies of silver clusters (after [2]). *Dashed lines*: Calculated HEELS-spectra. *Solid lines*: Optical spectra. The HEELS (high electron energy loss spectroscopy) plasmons show no dispersion, in contrast to the optical extinction plasmon polariton peaks. *Points and circles*: optical experiments. *Triangles*: HEELS experiments

and quadrupolar contributions were not resolved experimentally, so the observed maximum seems to shift continuously from the dipole to the quadrupole position), the according plasmon polariton, merely yielding a tiny dip superimposed to the dipolar band [2], is shifted towards lower frequencies. For experimental results concerning band widths, we refer to the original literature.

3.5.2   Extinction of Radiation by Single Particles

Stimulated by early thorough experiments and supported by earlier mathematical investigations concerning acoustic resonances *Mie* [4] was the first to solve the problem of the complex elastic (i.e. $\Delta\omega = 0$) scattering of electromagnetic waves by spherical particles, of arbitrary chemical composition and size, on the basis of *Maxwells* theory. Similar results were published shortly afterwards by *Debye* [5] in his investigation of pressure of light onto clusters. It is the exact solution of the problem and thus has remained unchanged until today [24, 25]. Only few extensions of this theory were performed, later. These are:
- Assumption of different particle shapes as ellipsoids [26], cylinders [28] and cubes [29] and particles surrounded by shells [27].
- Inclusion of volume longitudinal plasmons [32, 33].
- Introduction of size effects in the dielectric functions.
- Consideration of electromagnetic cluster interactions in many cluster systems.
While *Mie* had restricted himself to extinction (i.e. absorption and scattering) spectra, *Gans* and *Happel* [17] completed the theory by calculating, under certain limits, yet, the according real part of the polarizability. They expressed their results in terms of the effective refractive index of a diluted system of many clusters, which can be measured directly by interferometry [13]. This quantity being, yet, difficult to be measured directly, has instead been computed by *Kramers Kronig* analysis of the extinction spectra [18, 19]. Figure 4 shows, both, extinction and relative refractive index of diluted systems of silver clusters.

The *Mie* and *Gans-Happel* formulae are the spherical geometry equivalent to the *Fresnel* and the *Murmann* formulae of plane geometry. (Less general solutions have been presented, for other cluster geometries, like ellipsoids [26, 63] or cylinders [64].) After the ABC (additional electrodynamic boundary conditions) had been formulated by *Sautter* and *Forstmann* [30] and by *Melnik* and *Harrison* [31] for a plane interface, they were included into *Mie*'s theory by *Clanget* [32] and *Ruppin* [33]. The physical meaning of the ABC (normal component of the current density j_{norm} continuous at the surface) is to complete the divergency free rotation fields of *Mie* by rotation free divergency fields which give rise to the excitation of (longitudinal) volume plasmons. This extension of *Mie*'s theory will be included in the following, though it induces only minor optical effects (Fig. 5) which have not yet been identified experimentally.

It should be pointed out, that the distinction between surface and volume excitations looses its meaning if cluster sizes are close to the *Fermi*-wavelength

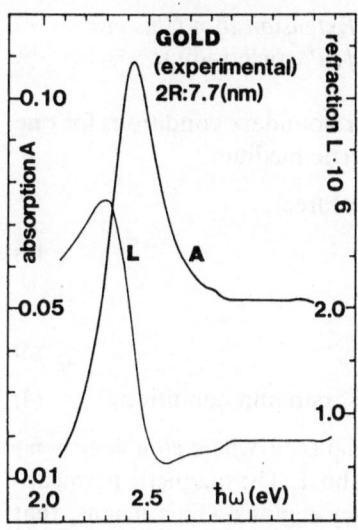

Fig. 4. Real part L and imaginary part A of the linear response of spherical gold clusters embedded in a glass matrix. A is the absorption constant (which, at the given particle size, coincides with the extinction constant). L is the relative refractive index (see text). A was measured and L was evaluated by Kramers Kronig analysis (see [18])

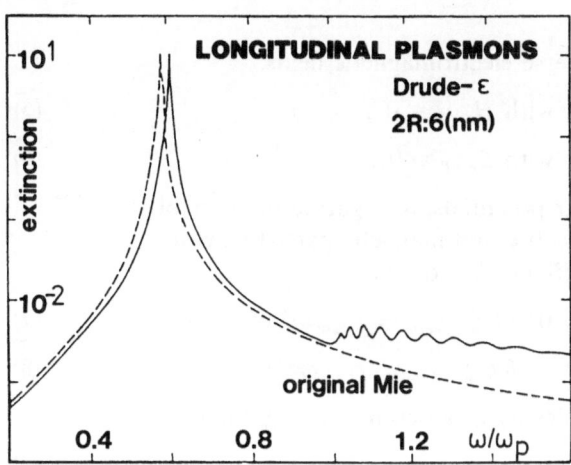

Fig. 5. Effect of the excitation of longitudinal (volume-) plasmons on the *Mie* spectrum of clusters with Drude electrons (after [33])

λ_F of the involved electrons. In the case of metals such clusters consist of some twenty atoms. The reason is, that the thickness of the electronic surface layer roughly amounts to λ_F.

In the following, an outline of the *Mie* theory for *one* particle including longitudinal excitations is given by listing up the mathematical steps. (To obtain experimentally accessible quantities, mesoscopic samples of typically 10^{10} particles are required. *Many* of the *Mie* clusters are therefore thought to be uniformly spread in a macroscopic volume with volume concentrations and sample thicknesses low enough to keep particle interactions and multiple scattering events rare.)

Calculation of the Extinction of One Cluster and Extension to n Clusters:
  1. Introduction of spherical coordinates r, θ, ϕ. Particle radius: $r = R$.
  2. Plane, monochromatic incident wave.
  3. Solution of *Maxwells* equations with proper boundary conditions for one arbitrary spherical particle, embedded in a dielectric medium

Electric fields: $E = E(\text{divergency free}) + E(\text{curl free})$,

Magnetic fields: $H = H(\text{divergency free})$.   (2)

• Boundary conditions at $r = R$:

$$E_\theta^{\text{incident}} = E_\theta^{\text{interior}}; \qquad E_\phi^{\text{incident}} = E_\phi^{\text{interior}} , \tag{3}$$

Current density: $j_{\text{total}}^{\text{normal}}$ continuous (Sautter–Forstmann condition) .   (4)

The linear response function of the particle material is $\varepsilon\mu$, where $\varepsilon(\omega) = \varepsilon_1 + i\varepsilon_2$ is averaged over the cluster volume as described above. The magnetic permeability μ is set 1 for the investigated high frequency regions. (This means, that magnetic response of the material which may be effective at lower frequencies, is excluded throughout the following.) The dielectric function of the embedding material is ε_m.

• Separation of the transverse electromagnetic fields:

"electric partial waves"   with $H_{r=R} \equiv 0$,   (5)

"magnetic partial waves"   with $E_{r=R} \equiv 0$.   (6)

Introduction of three scalar potentials, to separate the variables:
Π_E, Π_M, Γ (indices E, M: electric and magnetic partial waves)
They are solutions of the *Helmholtz*-equations:

$$\Delta\Pi_{E,M} + K_{\text{transverse}}^2 \cdot \Pi_{E,M} = 0, \quad K_{\text{transverse}}^2 = \varepsilon_{\text{trans}}\omega^2/c^2 , \tag{7}$$

$$\Delta\Gamma + K_{\text{longitudinal}}^2 \cdot \Gamma = 0, \quad K_{\text{longitudinal}}^2 = \varepsilon_{\text{long}}\omega^2/c^2 . \tag{8}$$

• Separation of the variables by product-ansatz. E.g. for Π_E:

$$\Pi_E = F_1(r) F_2(\theta) F_3(\phi) \tag{9}$$

Solutions:

Variable	Differential equation	Solution functions
r	Bessel	Cylinder-functions (Bessel, Neumann); Index *l*.
θ	Spherical harmonics	Legendre polynomials; Indices *l, m*.
ϕ	Harmonic oscillation	$\frac{\cos}{\sin}$ -functions; Index *m*.

• Back-transformation of the potentials into fields of different polar and azimuthal order *l, m*. These are divided into 3×3 groups, i.e. each of the three waves consist of three different contributions:

incident wave   electrical partial waves
waves in the particle   magnetic partial waves
scattering waves   longitudinal waves

4. Computation of intensities I from all field amplitudes and evaluation of the following experimentally accessible quantities for systems of n identical particles, packed loosely together in a macroscopic volume with volume concentration C. (This volume may be filled with some matrix material):
– the absorption constant A [mm^{-1}], (due to production of heat in the particle)
– the scattering constant S [mm^{-1}], which can be computed angle resolved $S(\theta, \phi)$ and angle integrated
– the extinction constant E including, both, absorption and scattering losses
– the relative change of refractive index of the sample

$$L = [n(\text{sample}) - n(\text{matrix})]/n(\text{matrix}); \quad n: \text{refractive index} \tag{10}$$

(This latter quantity follows from the *Gans-Happel* extension. (17)).

The quantities A, S and E are defined by the *Lambert–Beer* law. Their explicit formulae are omitted here. They can be found in the original literature (e.g. [4, 24, 25, 17]). Instead, a computer program is contained in the Appendix, which is – as we hope – self explaining by the extensive comments. The following discussion is based upon this program, and approximative formulae are added where they facilitate the understanding of the physical background.

3.5.3   Discussion of Optical Properties of Isolated Clusters

(A) In the *quasistatic case* ($R \ll \lambda$), Fig. 6, the clusters are small enough that the instantaneous phase of the wave is constant over the cluster dimensions. Then, particles of ellipsoidal and, in particular of spherical shape are homogeneously polarized in the $l = 1$ (dipolar) mode and higher l modes do not contribute essentially. The inner electric field of such particles (dielectric function

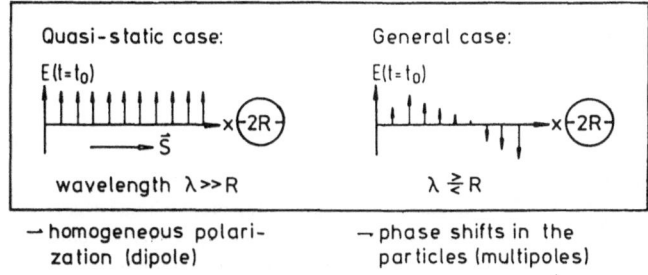

Fig. 6. Definition of the quasistatic region. S: Poynting vector; E: electric field; $2R$: cluster diameter

$\varepsilon = \varepsilon_1 + i\varepsilon_2$; depolarization parameter N) embedded in some matrix (dielectric function $\varepsilon_m = \varepsilon_{m1} + i\varepsilon_{m2}$) is

$$E^{interior}/E^{incident} = \varepsilon_m/(\varepsilon_m + N(\varepsilon - \varepsilon_m)) \ . \tag{11}$$

Special cases are the absorbing cluster ($\varepsilon_2 \neq 0$) in the dielectric matrix ($\varepsilon_{m2} = 0$), and the dielectric bubble ($\varepsilon_2 = 0$) in an absorbing medium ($\varepsilon_{m2} \neq 0$). Resonances occur at frequencies ω_s, where the following condition, derived from Eq. (11), holds:

$$[\varepsilon_1(\omega_s) - \varepsilon_{m1}(\omega_s)(N-1)/N]^2 + [\varepsilon_2(\omega_s) - \varepsilon_{m2}(\omega_s)(N-1)/N]^2 =$$

$$= minimum. \tag{12}$$

Spherical clusters ($N = 1/3$) with small ε_2 or small $(d\varepsilon_2/d\omega)$ in the vicinity of ω_s and $\varepsilon_{m2} = 0$ give

$$\varepsilon_1(\omega_s) = -2\varepsilon_{m1} \ . \tag{13}$$

Specializing further to Drude electrons, i.e. free electrons characterized by the relaxation frequency ω_D and the plasma frequency ω_P we obtain

$$\omega_s = \omega_P/\sqrt{(1 + 2\varepsilon_{m1})} \ . \tag{14}$$

The appropriate resonance frequencies ω_s of Eq. (13) follow directly from the spectrum of $\varepsilon_1(\omega)$, an example being shown in Fig. 7. At $\omega = \omega_s$,

$$Re \{E^{interior}/E^{incident}\} = 0 \tag{15}$$

and

$$Im \{E^{interior}/E^{incident}\} = maximum \ . \tag{16}$$

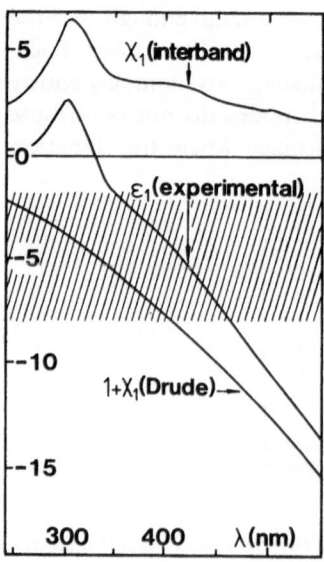

Fig. 7. Real part of the dielectric function of bulk silver $\varepsilon_1(\omega)$ and its decomposition into interband and Drude contributions, Eq. (18). χ are susceptibilities. According to Eq. (1) the resonance frequencies $\omega_{s,l}$ follow from ε_1 within the *hatched* region when ε_m is varied between 1 and 4

In this approximation scattering S is zero. The extinction $E(\omega)$ contains merely dipolar absorption losses and the extinction is given for the absorbing spherical particle in a dielectric matrix, by the famous equation

$$E(\omega) = 6C\varepsilon_{m1}^{3/2}(\omega/c)\frac{\varepsilon_2(\omega)}{(\varepsilon_1(\omega) + 2\varepsilon_{m1})^2 + \varepsilon_2^2} .$$ (17)

(C: volume concentration; c: vacuum velocity of light)

This formula has now and then been misused for larger particles. In fact, it holds for e.g. Ag and Au particles in the resonance region only if $2R \lesssim 10$ nm.

It should be stressed that Eq. (17) does not comprise explicitly the cluster size, i.e. identical optical spectra should be obtained for clusters of all sizes within the quasistatic regime. Yet, already *Mie* mentioned that this independence is only formal. Below some critical size experimental spectra have to reflect the transition from the solid to the molecular state. In the frame of Eq. (17) these ensuing intrinsic cluster size effects are included in size dependent dielectric functions $\varepsilon(\omega, R)$. The particular structure of Eq. (17) gives ease for the evaluation of intrinsic size effects which may not to be despised: *all* size dependences of $E(\omega)$ observed experimentally on small clusters can be directly attributed to the latter. Some of these effects will be discussed in the following.

In the case of non-isometric clusters (e.g. for ellipsoidal particles: different depolarization parameters N_i for the 3 axes) the appropriate extinction constant E is a tensorial quantity with spectra for the three principal axes, differing, in general, in peak positions as well as in peak widths and peak heights [34, 55].

The spectral structures of the *Mie* extinction can be varied additionally by choosing embedding media with different ε_m. As an example, the peak position of Ag-clusters can be shifted up to 0.5 eV by different dielectric embedding matrices.

In fact, from Eq. (13) it follows that only the ratio $\varepsilon_1/\varepsilon_{m1}$ obeys a defined maximum condition: $\varepsilon_1(\omega_s)/\varepsilon_{m1} = -2$. (This does not hold beyond the limits of the quasistatic case).

Influences of $\varepsilon(\omega)$ on the E-spectra are shown schematically in Fig. 8. The graphs a) and b) demonstrate that peak positions and widths are essentially determined by $\varepsilon_1(\omega)$. Flat ε_1-spectra, i.e. small $d\varepsilon_1/d\omega$ yield broad bands. Hence, the band widths are not indicative only to damping mechanisms and energy dissipation! The graphs c) and d) show that, in principle, multipeak structures can be produced by proper ε_1-spectra (c), which, however, may be damped away when ε_2 is sufficiently large (d). So, band widths are also influenced by all damping mechanisms entering ε_2. In the special case of a *Drude* material, the plasmon polariton band width equals the relaxation frequency enclosed in the *Drude*-ε. Both, low $d\varepsilon_1/d\omega$ and high ε_2 in the vicinity of resonance frequencies ω_s render the cluster polariton bands in most materials to be smeared out past recognition, and only few materials exhibit sharp and selective bands. These are among others the alkalis, some noble metals and aluminum. Narrow phonon polariton bands are found e.g. in MgO.

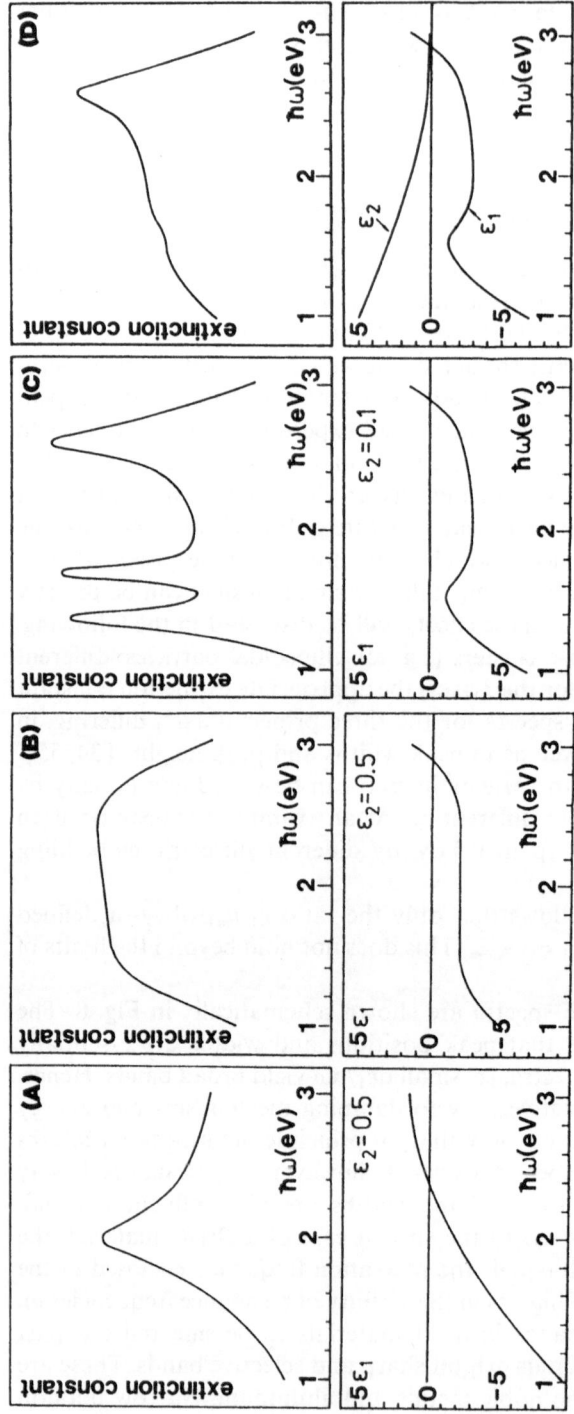

Fig. 8. Schematic *Mie* extinction spectra demonstrating the influences of the dielectric material functions $\varepsilon_1(\omega)$ and $\varepsilon_2(\omega)$ on the shape of resonance bands $(\varepsilon_m = 1)$

It should be pointed out that no assumptions whatsoever were made about the origin of the particular ε-spectra: these can be ionic contributions as well as electronic ones. For instance, one extinction peak of some semiconductor may be due to optical phonons while an other may be due to the conduction electrons and a third may appear in the interband transition region, always provided, that Eq. (12) or (13) is fulfilled. That is, extra identification of the origin is required for each observed *Mie* peak.

Interestingly, in the case of plasmon polariton excitations, all relaxation effects contained in ε and described in the theory of conductivity in the one-electron picture, as e.g. electron lattice defect scattering, contribute likewise to the plasmon band widths. The controversy between single electron relaxation and collective excitation is obvious here, again. A detailed description of the plasmon polariton decay is still missing.

In spectral regions of interband transitions, transitions including various electronic bands may be simultaneously effective and then, the unique assignment of some peak in the spectrum may not be possible. For example, in Cu, Ag, Au, the *Drude* electron excitations are hybridized with interband transitions, and hence the cluster plasmon polariton shifts from about $\hbar\omega_s \sim 10$ eV of the free *Drude* metal down to the visible region of the spectrum. This is demonstrated in Fig. 7 where the susceptibility terms of *Drude* and interband transitions contributing to $\varepsilon_1(\omega)$ are separated. The underlying physical effect is that the phase of the conduction electron response is shifted by the background polarization which is due to interband transitions from the valence (d-) band to the s-band at the Fermi level.

Electrons of high lying valence bands (i.e. bands with weak localization) for their part can as well exhibit extinction peaks due to collective excitations, yet usually, these excitations are strongly damped by high ε_2.

(B) *Beyond* the cluster size limits of *the quasistatic region* (i.e. $R \ll \lambda$ to $R < \lambda$), the extrinsic size effects arising mainly from retardation become important, increasing with the particle size. Mainly, there are four size effects which follow from *Mie*'s theory:

1. The position and shape of the polariton bands change with particle size – in contrast to Eq. (17). Usually the peak shifts towards lower frequencies and the width is increased with increasing size.
2. Beside the dipolar polariton ($l = 1$) higher order multipoles ($l = 2, 3, \ldots$) are excited. This is because of the phase modulation over the particle dimension as shown in Fig. 6.
3. Scattering contributes essentially to the extinction beside the consumptive absorption. Regarding the particles as spherical micro-antennae which act as selective receivers, they turn – in course with increasing R/λ – into also emitting antennae, their radiation resistance lessening with increased R.
4. Excitations due to the magnetic partial waves (Eq. (6)) contribute to E. Since they do not follow a resonance behaviour similar to the one of the electrical partial waves (in the optical spectral region, magnetic susceptibilities amount

to $\mu = \mu_m = 1$), their spectra, are weakly structured. However, in the case of Ag-particles they clearly reflect the onset of interband transitions. (Fig. 4 of [2]). It should be pointed out that they are the response of the electronic system onto the magnetic part of the incident wave (magnetic material reactions are excluded for the optical region as mentioned before). They are interpreted to be eddy currents of the respective electrons.

The resulting extinction spectra $E(\omega)$, an example of which is given in Fig. 9, are thus composed of a multitude of different excitations which are listed up in Table 1. The extinction E and the scattering S, both, angle resolved and angle integrated can be measured separately by conventional optical methods. Recently [57] it was shown that also the absorption A can be determined by using photothermal methods. In principal, it is necessary in calculations to sum over an infinite number of different multipolar contributions, i.e. $l = 1, 2, \ldots, \infty$. This is a consequence of the mathematical method of expanding the fields into orthogonal functions. Physically, however, even for larger clusters (yet, with $R < \lambda$), only few multipoles really contribute to the loss spectra to a measurable amount. This is demonstrated in Fig. 10 where peak heights of the various modes of silver clusters are compiled as depending on cluster size. At low cluster sizes, i.e. the quasistatic region, there is only the dipolar absorption (its height decreases towards lower sizes again due to the intrinsic cluster effects, described in Sect. 3.5.4). Towards larger particles the dipolar scattering surpasses the absorption, then the quadrupolar absorption increases, which, on its part, is surpassed by the quadrupolar scattering, then the octupolar excitations follow and so on. However, none of these contributions are able to exceed the dipolar scattering which for all larger particles is the most prominent contribution (the reason is that for $R < \lambda$ the dipole antenna couples to the external electromagnetic radiation field best). It should be kept in mind that Fig. 10 gives no direct

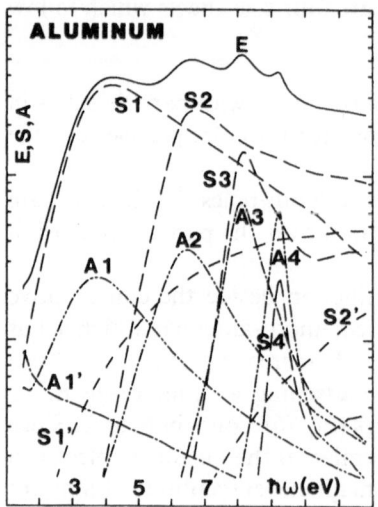

Fig. 9. Decomposition of the total *Mie* extinction spectrum E of aluminum particles ($2R = 100$ nm) into absorption A and scattering S contributions of different multipolar order $l = 1, 2, 3, 4$. Primes: "magnetic" partial waves (see also Table 1)

Table 1. Hierarchy of the *Mie* extinction contributions (after [57]). A_l, S_l: contributions of the lth mode absorption and scattering; E_R, H_R: see Eqs. (5) and (6)

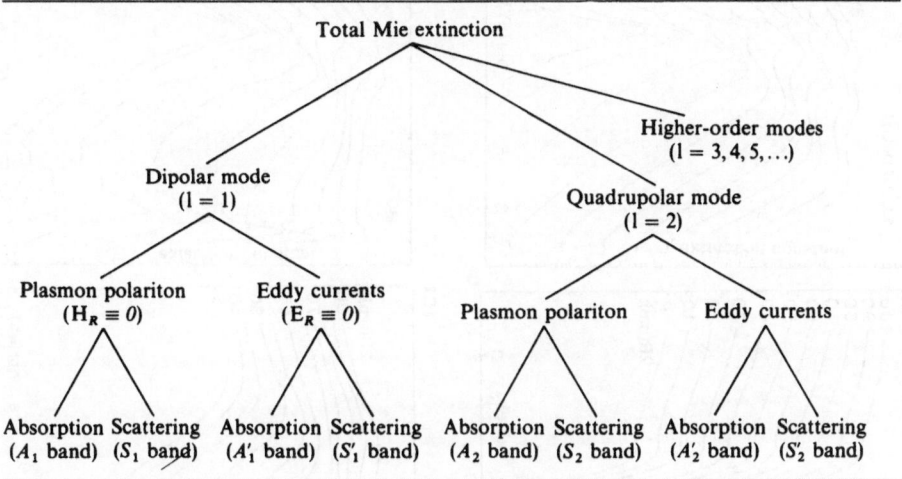

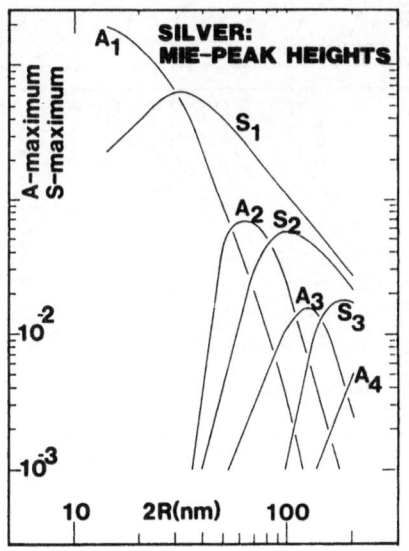

Fig. 10. Silver particles: calculated peak maxima of the *Mie* bands of different multipolar modes. A: absorption; S: scattering; $l = 1, 2, 3, 4$. Below $2R = 15$ nm the A1-curve decreases again because of intrinsic size effects

information about oscillator strengths, yet, since these are proportional to the product of height and band width, the latter also vary strongly with R.

Figure 11 gives a compilation of *Mie* extinction spectra for spherical particles of various materials and various sizes, which may be useful for practical

Fig. 11. Calculated *Mie* extinction spectra of clusters of various materials (metal, semiconductor, insulator). Parameter is the size $2R$ ($\varepsilon_m = 1$). Intrinsic cluster size effects are disregarded

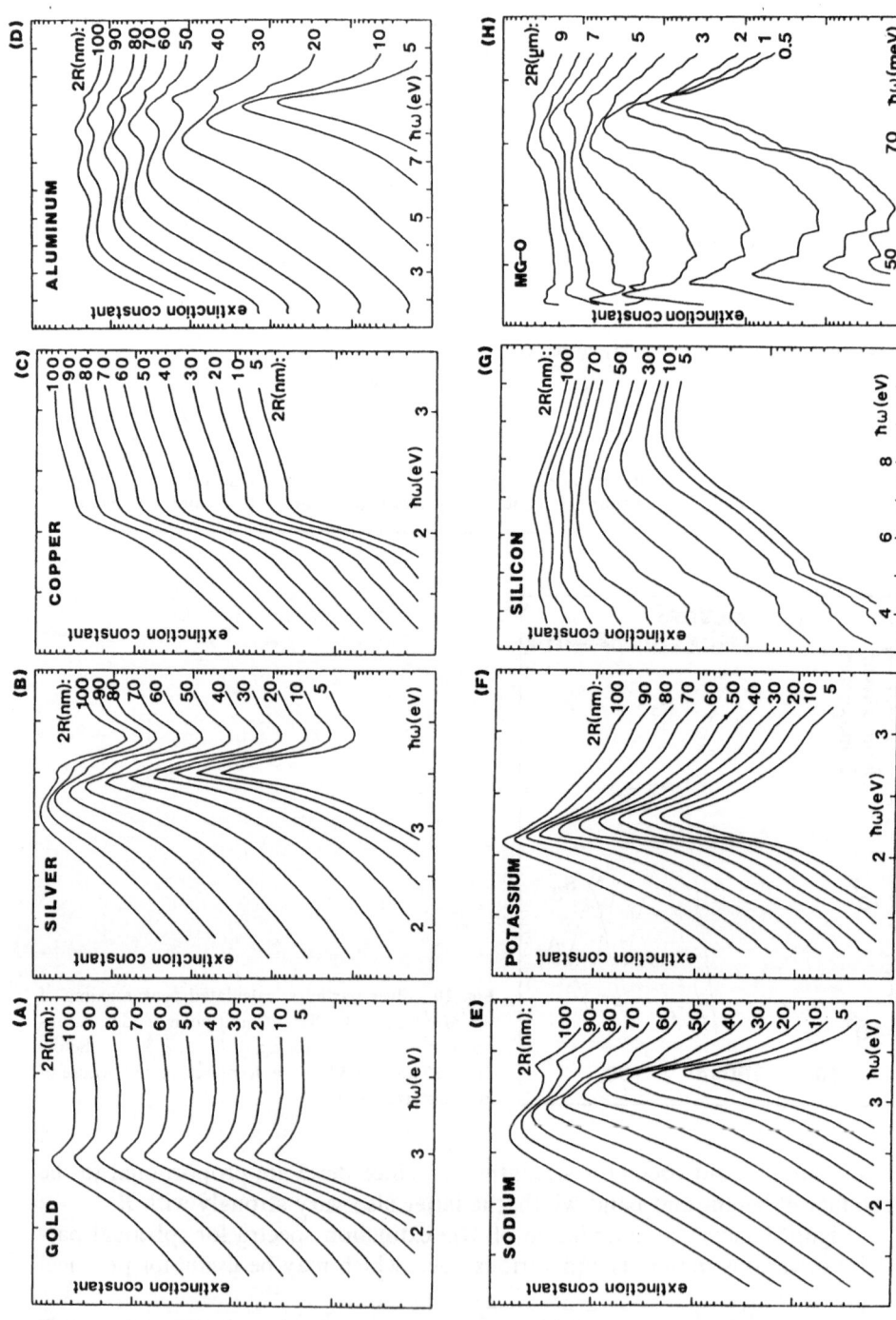

purposes and, also, demonstrates the ample variety of spectral structures to be observed.

While Fig. 11a–f are devoted to metallic clusters, 11g and h give examples for semiconductor and for dielectric particles. The extinction peaks of a) to g) are due to plasmon–polariton like excitations, while h) shows the response for an ionic system, i.e. cluster phonon polaritons. Interestingly, the spectra b), d), e) and f) due to "good" free electron metals, differ markedly, thus pointing to the extreme sensitivity of *Mie* spectra on details of the electronic structure.

3.5.4 Intrinsic Optical Cluster Size Effects

As pointed out above, the cluster state is characterized by its particular optical properties, deviating, both, from molecular and from bulk solid ones. Some of such cluster size/cluster surface/cluster interface effects are listed in Fig. 12. In the optical properties it is the intrinsic cluster size effects that reflect the transition between molecule and solid [36]. The transition takes place in the quasistatic region. These effects are frequently expressed directly by the resulting

SURFACE:

- BOUNDARY FOR ELECTRON WAVES
- DISCRETE EIGENSTATES
- STANDING WAVES
- SHELL MODEL

- SURFACE RESONANCES
- SURFACE STATES
- EVANESCENT WAVES

- REDUCED SCREENING
- INELASTIC SURFACES SCATTERING
- INCREASED ELECTRON – PHONON – COUPLING
- SURFACE INDUCED POLARIZABILITIES
- MULTIPOLAR ELEMENTARY EXCITATIONS
- SURFACE PLASMONS / POLARITONS
- SURFACE PHONONS / POLARITONS
- SURFACE MELTING

VOLUME:

- ATOMIC STRUCTURES
- STRUCTURE FLUCTUATIONS
- ATOMIC DISTANCES
- SUPPRESSION OF EXCITONS,
  MAGNETIC ORDER ETC.
- DISCRETE ATOMIC VIBRATION
  MODES

INTERFACE:
(COATINGS, MATRIX)

- ADSORPTION
- CHEMICAL BINDING
- EFFECTS ON EVANESCENT WAVES
- ELECTRICAL DOUBLE LAYERS
- STABILIZATION OF SPECIAL
  STRUCTURES
- SELECTIVE GROWTH

Fig. 12. Cluster-size/surface effects

changes of the *Mie* bands, and the discussions of red/blue shifts and of band broadening are still topical [14, 15]. Sixteen of such effects were compiled recently [14]. For a detailed discussion we refer to this paper. (Yet, the continuous approach from the behaviour of the size limited solid towards the bulk which occurs in larger clusters cannot be treated this way, since, beyond the quasistatic regime intrinsic and extrinsic size effects are both contained simultaneously in the shape of the *Mie* spectra.)

One can carry the discussion alternatively in terms of the average polarization α or the dielectric function ε of the cluster material, contained in the *Mie* formalism. The latter way is advantageous in the respect that it is not restricted to the quasistatic limit. Besides, different polarization and relaxation contributions usually are additive in α and ε and thus can be treated separately. Their real and imaginary parts are related, obeying Kramers–Kronig relations [18]. (Sum rules for oscillator strengths of the *Mie* bands, derived therefrom, have been tested successfully [18].) For example, for metals holds the additivity:

$$\varepsilon(\omega, R) = 1 + \chi^{\text{lattice}} + \chi^{\text{Drude}} + \chi^{\text{interband}} , \tag{18}$$

where χ^j are susceptibilities.

In terms of the solid state approximation, χ^{lattice} encloses optical phonon spectra, phonon lifetimes and excitation probabilities. The parameters entering χ^{Drude} are density, effective mass and relaxation frequency of the conduction electrons, and $\chi^{\text{interband}}$ is determined by the joint density of states (JDOS) of the contributing electronic bands and corresponding transition matrix elements. In principle, all of these quantities may develop size dependences.

Some of these dependences may be weak, as e.g. of the conduction electron density, others may be strong, as e.g. of the *Drude* relaxation frequency. Close to the molecular state, some of these terms/quantities loose their meaning. For example, the phonons being normal modes of the bulk, are replaced by atomic vibrations/rotations. The picture of an electronic band structure changes into the shell model [16] (which is not limited to the outermost s- or conduction electrons, necessarily) because the linear wavevectors as variables of the band-structure are no longer useful quantities, and matrix elements have to be redefined due to the lack of the wavevector conservation law in clusters.

In free electron, jellium and quantum dot models, some of these effects have been theoretically investigated [10, 33, 11, 12, and Lit. in 14]. Due to the cluster surface boundary conditions for electron waves, the electron energies are a sequence of discrete values. As a consequence, electronic excitations between these levels yield discrete spectra, and this is reflected in the appropriate dielectric functions (Fig. 13a, b). The effect of the discrete levels on the *Mie* band (Fig. 13c) is occasionally called *Landau* damping, in analogy to the magnetic Landau level systems. It should be kept in mind that this discrete level picture holds under the assumption of stationary eigenstates of electrons with formally infinite lifetimes, and, hence, may not be applied to excitations in realistic clusters. As an ultimate limit, discrete level effects vanish in the single cluster if lifetimes of excited states equal the level spacings close to the *Fermi* level. Regarding also that there are

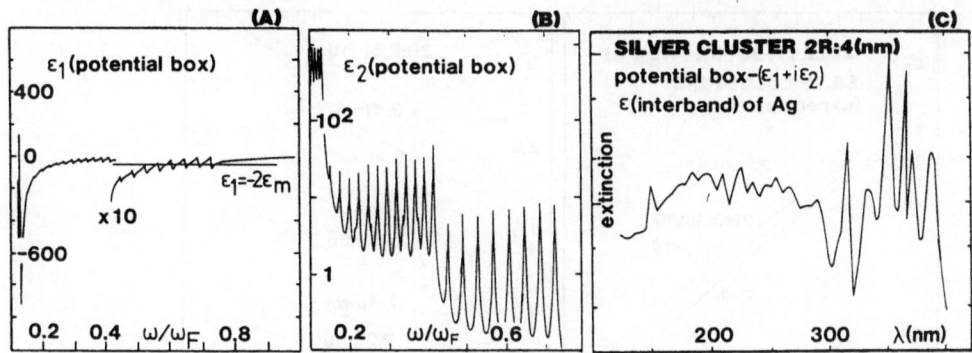

Fig. 13. Cubic potential box model applied to silver clusters (after [11]). a), b): dielectric function ε_1 and ε_2. c): resulting *Mie* extinction with the potential box – ε used instead of the Drude part in the $\varepsilon(\omega)$ of silver. Interband contributions are kept unchanged

varying cluster sizes in realistic macroscopic *n*-cluster systems, the discrete energy level structure of the single cluster will be hidden and smeared out in experimental excitation spectra of the *n*-cluster sample. It has been shown [10, 11] that then the remaining cluster size effect is mainly given by a $1/R$-law in ε_2 (see Fig. 14):

$$\varepsilon_2(\omega, R) = \varepsilon_2^{\text{bulk}}(\omega) + A \cdot (1/R) \ . \tag{19}$$

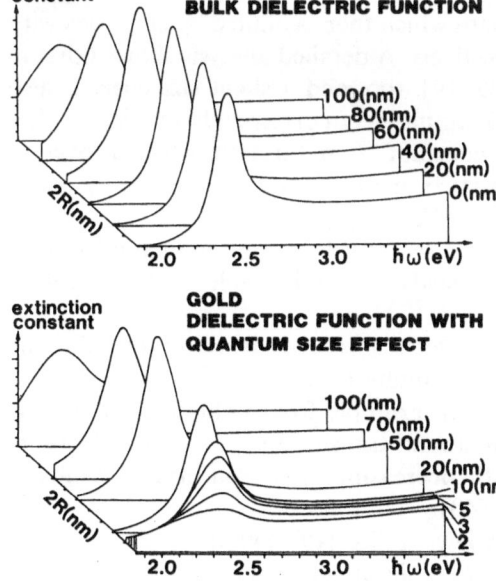

Fig. 14. Calculated influence of the size dependence of ε (**Eq. (19)**) on the *Mie* spectra of gold particles of various sizes: Upper graph without, lower graph with quantum size corrections of ε. At larger sizes the spectra coincide. Toward small clusters the band is damped down (A of Eq. (19) = 1)

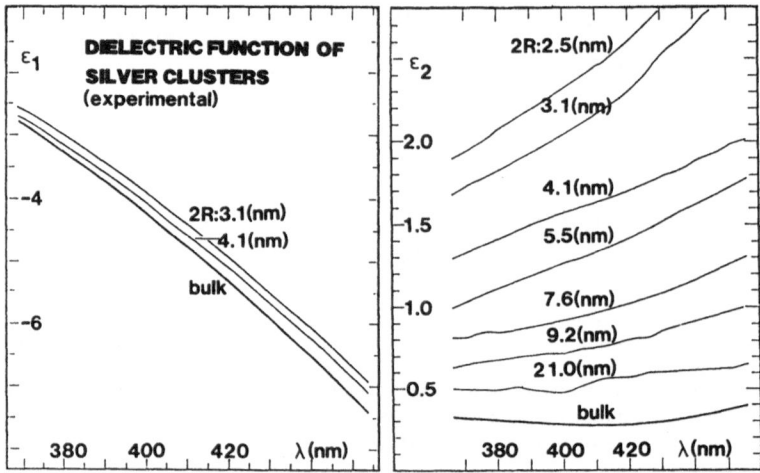

Fig. 15. Experimental spectra of the dielectric function $\varepsilon_1(\omega, R) + \varepsilon_2(\omega, R)$ of silver clusters embedded in glass (after [18]). As shown in the original Ref. [18] the size dependence follows the assumption of a size dependent Drude relaxation frequency, which, roughly, gives Eq. [19]

In fact, this dependence has been verified for various clusters containing conduction electrons and Fig. 15 shows an example. Obviously, this size dependence is perceptible up to particles of more than 10^6 atoms and, hence, proves to be one of the most prominent intrinsic cluster size effects.

An alternative explanation has been given, holding in the solid state regime of larger clusters. Assuming a quasicontinuous conduction electron energy band to be present, the classic one-electron theory of electrical conduction suggests the model of an electron mean free path which then is limited by collisions with the particle surface or interface, respectively. A detailed analysis showed that in this model $\varepsilon_2(\omega)$ follows Eq. (19) [39, 19], too, and a slight size effect is also identified for the ε_1-spectra which quantitatively corresponds to Fig. 15. The discussion of precise values of the constant A in Eq. (19), which proves to depend, both, on the particle and matrix material and on the details of the interface scattering [19, 59, 60, 61, 14, 15], still persists.

It has been calculated [39, 56] that a radially decaying conduction electron density (as e.g. the spill-out effect) also yields the $1/R$-dependence of Eq. (19) yet with $A = A(\omega)$.

Beside these "conduction" electron effects, size dependences of the interband transitions have been identified experimentally though for very small clusters, only. We, again, show, for simplicity an example from noble metals: in gold clusters of less, say 10^3 atoms, the interband transition edge was found to shift and to be broadened (Fig. 16) with decreasing size. Concerning results of semiconducting clusters we refer to [40].

For practical purposes, the $1/R$-law of Eq. (19) comprises the strongest intrinsic optical size effect. So, it is enclosed in the *Mie* computer program given

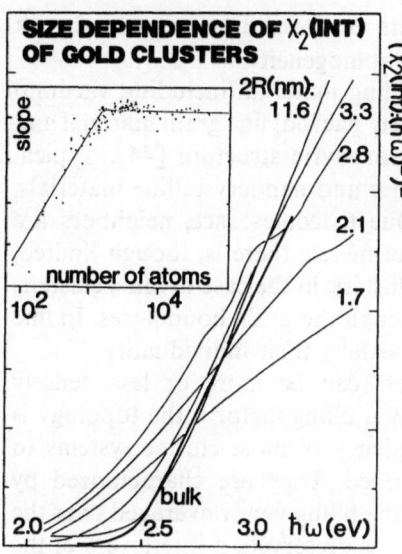

Fig. 16. Cluster size dependence of the interband transition edge in gold clusters evaluated from measured extinction spectra. (Clusters were embedded in glass.) χ_2(int): imaginary part of $\chi^{interband}$ of Eq. [18]. (After [36]). The inset shows the slope of the linear parts of experimental edge spectra as depending on the cluster size. (Assuming unchanged density of the cluster material, the cluster sizes are converted into the according number of atoms per cluster)

in the Appendix. It should be mentioned, however, that there are cases, where this relation has not been verified, e.g. the Au_{55}-clusters, which were stabilized in a metalorganic compound [41]. Experiments [42] proved that the optical extinction of these clusters does not show any hints of a plasmon polariton at all.

Only a few experiments have been performed concerning nonlinear response of clusters onto electromagnetic waves. There are size effects, but they are small in the case of noble metals [43].

3.5.5 Optical Properties of Cluster Matter

The preceding sections were devoted to the optical properties of individual clusters, and the step to macroscopic samples, which are accessible to conventional optical experiments was performed by simply multiplying the single cluster extinction by the number of clusters in a many cluster system. Such systems are a particular example of what is called "cluster matter", with the special properties of
- missing interaction effects among the clusters which would destroy the assumed additivity of the *Mie* extinctions,
- missing multiple scattering effects which, as well, would destroy this additivity.

"Cluster matter" designates a peculiar class of inhomogeneous materials which are of great practical and technical importance. They are distinguished from homogeneous matter by larger building units (i.e. the clusters) and lower structural correlation. Preferentially, the units are formed and arranged with statistical spatial disorder and/or chemical inhomogeneity.

Depending on the existence of an embedding material (including vacuum) there are two different kinds: the (usually densely packed) fine grain material and the inclusion-matrix material or matter of porphyric structure [44]. Typical representatives of the former are nanoceramics and nanocrystalline materials, the clusters of which are closely packed. Due to coalescence, neighbors are separated by grain boundaries. In the case of metals there is, though limited, electronic conduction between neighboring clusters; in the case of ionic clusters, merely atomic vibrations are transported through the grain boundaries. In fine grain materials, the clusters thus have lost partially their individuality.

In inclusion-matrix materials the clusters can be more or less densely packed, the mean packing being described by a filling factor if the topology is statistically random. In fact, due to the tendency of most cluster systems to coagulate, larger aggregates are usually formed. They are characterized by a local filling factor which markedly exceeds the filling factor averaged over the total sample. The individuality of the clusters is preserved, if interlayers of the matrix material separate neighboring clusters in the aggregates. For metallic clusters in dielectric matrices this means that the electrons are confined to the individual clusters. Such aggregates then form building blocks of the inhomogeneous matter. This kind of cluster aggregates, "the coagulation aggregates", will be treated in the following.

Two alternative ways of description have been developed. The first, the effective medium approach [45], is based upon some statistical assumptions on the topography of the clusters distributed in the matrix. An effective *macroscopic* dielectric function is introduced, replacing the inhomogeneous by a fictitious homogeneous material with identical macroscopic optical properties. This implies that there is no internal scattering – one characterizing property of inhomogeneous matter –, and, hence, these models are restricted to the quasi-static region.

The second way is to describe the optical properties of typical examples of the existing cluster aggregates in detail and then to compose the macroscopic sample by multiplying their optical responses by their partial filling factors. Such a spectral analysis of cluster matter has recently been performed quantitatively [46]. The effect of aggregation on the optical response is a drastic one: not only intensities are changed but the structure of the spectra is strongly influenced as is shown by the experimental examples of Fig. 17. Obviously, by increasing aggregation the spectrum is altered continuously from the narrow, selective single cluster band to broad extinction features which resemble the optical absorption of a plane film of the same material. Recently, it has been found [58], that for sufficiently dense packing the *Mie* scattering of the single clusters changes into the regular reflection, due to interference effects ("*Oseen* transition").

The underlying physical model is the following: A coagulation aggregate is assumed to contain a limited number n of identical spherical clusters (the case of

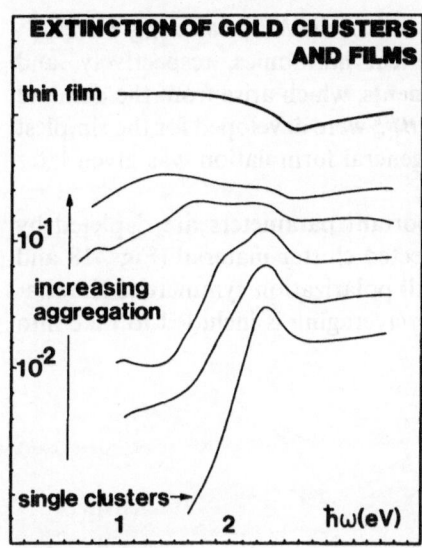

Fig. 17. Measured extinction spectra of gold cluster matter of various mesoscopic structures. Parameter: mean size of the cluster aggregates. Single cluster size: $2R = 10$ nm. The uppermost curve is due to a thin evaporated gold film. For sake of clearness different spectra are separated by shifting them along the ordinate for arbitrary amounts (i.e. constant factors due to the logarithmic scale of the ordinate)

different cluster sizes in small aggregates was also investigated, [47]) with equal center-to-center distances D between all next neighbors. All n-aggregates of similar shape are then replaced by one "typical" n-aggregate of regular topography (linear chain, triangle, icosahedron etc.). The optical response of these aggregates is computed from the "Extended Mie Theory" (EMT). Finally it is averaged over all aggregate orientations to describe more closely realistic samples.

As done before with Mie's theory, we have to omit a comprehensive theoretical description here which is to be found in the original literature [48–51] and, instead, give a qualitative brief survey of how the Mie theory has to be extended for this purpose. In the EMT, the plane wave incident on one cluster, is superimposed by the near-field Mie scattering fields of all other likewise excited clusters of the aggregate. The EMT focuses on to the electric fields and, hence, magnetic coupling which influences the eddy currents is neglected. The resulting electric excitation field may – according to the retardation phases of the waves – become larger or smaller than the incident field. In principle – though not for numerical computations – the extension of Mie's theory is straightforward: a fourth potential describing the scattering fields impinging a certain cluster has to be added to the potentials Π_E, and Π_M of Eq. (7).

This potential is of the form

$$\Pi_{\text{interaction}} = \sum_{j \neq i}^{n} \Pi_{\text{scattering}}(j) = \frac{1}{K^2 r_i} \sum_{l=1}^{\infty} \sum_{m=-l}^{+l} \Psi_l(Kr)_i \, Y_{l,m}(\theta_i, \phi_i)$$

$$\times \sum_{j \neq i}^{n} \sum_{q=l}^{\infty} \sum_{p=-q}^{+q} [A_{l,m}^{q,p} a_{q,p}(j) + B_{l,m}^{q,p} \cdot b_{q,p}(j)] \ . \tag{20}$$

$a_{q,p}(j)$ and $b_{q,p}(j)$ are the expansion coefficients of the *Mie* scattering waves, Ψ_l and Y_{lm} are cylinder functions and spherical harmonics, respectively, and $A_{l,m}^{q,p}$ are $B_{l,m}^{q,p}$ are the interaction matrix elements, which arise from the demand for self-consistency of the problem. $A_{l,m}^{q,p}$ and $B_{l,m}^{q,p}$ were developed for the simplest case of a particle pair by Trinks [48]; the general formulation was given later [49–51].

In the following, the effects of the important parameters are depicted by some informative graphs for arbitrarily selected cluster material (Figs. 18 and 19). The graphs contain the excitations of all polarization symmetries simultaneously, since three-dimensional orientation averaging is included to take into

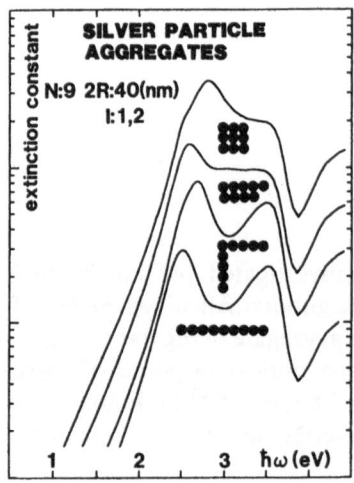

Fig. 18. The Extended *Mie* Theory (EMT) including cluster interaction effects: extinction spectra of silver cluster aggregates with $N = 9$ for varying aggregate shapes. Single cluster size $2R = 40$ nm. Computations included $l = 1$ and $l = 2$. For sake of clearness different spectra are separated by shifting them along the ordinate for arbitrary amounts (i.e. constant factors due to the logarithmic scales of the ordinate)

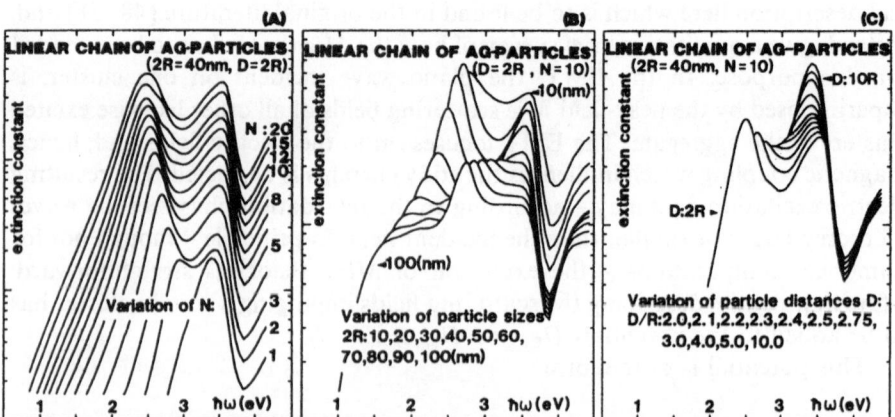

Fig. 19. The Extended *Mie* Theory (EMT): extinction spectra of linear aggregates of silver clusters. Graphs A–C: variation of different parameters. Parameters are: **A** number N of clusters; **B** single cluster size; **C** next neighbor distance. For sake of clearness different spectra are separated by shifting them along the ordinate for arbitrary amounts (i.e. constant factors due to the logarithmic scales of the ordinate)

account the arbitrary orientations of aggregates in realistic samples. It should be mentioned, that in the case of two-dimensional arrangements of aggregates on a substrate the averaging has to be altered. As a consequence, strong effects due to the angle of incidence and of the polarization state of the incident light can be observed in such samples [52].

Even for given, constant n, the spectra strongly differ for different topographic structures (Fig. 18). The typical feature of the theoretical spectra is the appearance of a multitude of different bands. They are due to different normal modes of excitation of the aggregate as a whole. Usually there are two prominent peaks. In the case of the linear chain they can easily be attributed to the two polarization states of electric field $E \parallel$ chain axis (lower frequency) and $E \perp$ chain axis (higher frequency). This attribution is more difficult for complex aggregate structures. It should be mentioned that this distinct multiple peak structure is not given by the simple effective medium theories, as e.g. the one of *Maxwell-Garnett* [20]. They can be obtained by including supplementary direct cluster–cluster interactions [21].

In general, aggregate samples exhibit strong scattering which, in the frame of the EMT, can be computed numerically. The relative contributions of absorption and scattering are not only determined by the size of the individual clusters but also by the size and shape of the aggregate as a whole and are different for different polarizations.

Thus, a theoretical description of cluster matter should include the complete *Mie* theory for various multipoles and also the retardation of the scattered waves, since it is the retardation that generates scattering.

Though the different *l*-modes are orthogonal in the single particle, they are not in the aggregate. Therefore, each *l*-mode of one particle interacts with each *l*-mode of all other particles, and for larger clusters/aggregates the dipole–dipole, dipole–quadrupole, quadrupole–dipole, quadrupole–quadrupole, etc. interactions have all to be taken into account. It is obvious that the maximum treatable value of *l* and, hence, the maximum sizes of, both, particles and aggregates are strongly limited by computer capacities. Figure 19 affords an insight into the properties of the results, for the special case of linear chains:

Figure 19a demonstrates that the typical two-peak structure mentioned above is increasingly split and shifted towards low frequencies when the length of the chain grows. As has been shown recently, there is a crossover of this behaviour to constant splitting for long chains, if the computation is restricted to the dipolar excitations, only [53].

The additionally occuring smaller peaks are due to the dipolar opposite phase excitations and the quadrupolar excitations, of different polarization symmetry, where the terms "dipolar" and "quadrupolar" refer to the single particle excitations.

The extinction spectra of more complex two- or three-dimensional aggregates differ from the linear chain. In realistic samples with various aggregate structures most of the calculated spectral features are therefore smeared out in measured spectra by superimposition effects. Yet, the two principal peaks have repeatedly been detected.

Figure 19b demonstrates the cluster size dependence for the example of the linear chain of $N = 10$. Obviously, the peak splitting as well as widths, positions and heights of the peaks depend strongly on size. The small peaks occuring for larger R are again due to dipolar opposite phase and quadrupolar excitations. Thus it is proven that quasistatic approximations may be applied only for, both, small individual particles and small aggregates. However, the splitting decreases rapidly, when the next-neighbor distance D is increased (Fig. 19c). At $D > 8R$ the interaction effects have almost vanished.

The maximum size of coagulation aggregates, simulated by now with EMT, amounts to about 80 particles including $l = 1, 2$. Larger aggregates are interesting to investigate e.g. the fractal structures produced by DLA (diffusion limited aggregation) [54].

It should be remembered that these aggregates are all assumed to be nonconductive due to the insulating matrix interlayers between neighboring clusters even if their thickness $d = D - 2R$ (D: next neighbor center-to-center distance) was, formally, extrapolated to zero in the preceding graphs. Electrical percolation effects are restricted to samples of aggregates where neighboring particles are connected by coalescence. Such samples are particular examples of the above mentioned fine grain materials. They are out of the scope of the present paper since the individuality of the particles is lost by the coalescence, and new, larger, particles are produced which – from their irregular shapes – would not be described by Mie-like theories. The transition of given aggregates from coagulation to coalescence has recently been observed experimentally [53] and percolation pre-states were discovered. These experiments showed that this transition is connected with marked changes of the optical extinction spectra. Hence, the latter prove to be a reliable tool for investigation of electric percolation in cluster matter. It awaits, yet, a detailed theoretical description.

3.5.6 Appendix: Computer Program of the Mie Theory

```
c         ========================= PROGRAM MQMIE ========================
c
c         EXTINCTION AND TOTAL SCATTERING OF COMPACT SPHERES AND SPHERES
c         WITH ONE LAYER, EMBEDDED IN A HOMOGENEOUS NONABSORBING MEDIUM
c
c         by      Dr. M. Quinten
c                 1. Physikalisches Institut
c                 RWTH Aachen
c                 D - 52056 Aachen, Germany
c                 phone   ++49 (241) 807167
c                 Fax     ++49 (241) 8888-331
c                 e-mail quinten@acds10.physik.rwth-aachen.de
c
c         The program source language is FORTRAN 77 with MS-extensions
c         ================================================================
$NOTRUNCATE                                     ! only for MS-FORTRAN
$NOTSTRICT                                       ! only for MS-FORTRAN

          INCLUDE 'FGRAPH.FI'                    ! only for MS-FORTRAN

          IMPLICIT REAL*8 (A-H,O-Z)

          PARAMETER (MULT=1000)

          CHARACTER*72 COREFILE,SHELLFILE,FILENAME,CHA*1

          LOGICAL LOGO1,LOGO2,PARTIC,YES
c         partic = .true.   means       : compact spheres
c         yes = .true.      means       : with cluster size effect
c         logo1, logo2 = .true. means : equidistant data

          COMPLEX*16 AMIE(MULT),BMIE(MULT),DKS,DKC,X,XC,Y,YC

7         FORMAT(F10.3)
70        FORMAT(E16.8)
71        FORMAT(F10.3,E16.8)
72        FORMAT(A)
73        FORMAT(A\)


c         **** Initialization *****

1000      partic = .false.
          yes = .false.
          logo1 = .false.
          logo2 = .false.
          pi = dacos(-1D0)
          call clearscreen(0)                    ! only for MS-FORTRAN

c         **** Data Input ****

          WRITE(*,73)'  compact (c) or layered sphere (l)
          READ(*,72)CHA

          IF (CHA.EQ.'c'.OR.CHA.EQ.'C') THEN
           PARTIC=.TRUE.
           WRITE(*,73)'  filename of dielectric constant
           READ(*,72)COREFILE
           SHELLFILE=COREFILE
1001       WRITE(*,73)'  dielectric constant of host medium EM
           READ(*,*,ERR=1001)EM
1002       WRITE(*,73)'  particle radius R [nm]
           READ(*,*,ERR=1002)R
           RC=R
```

```
          ELSE
          PARTIC=.FALSE.
          WRITE(*,73)' filename of core dielectric constant      '
          READ(*,72)COREFILE
          WRITE(*,73)' filename of shell dielectric constant      '
          READ(*,72)SHELLFILE
1003      WRITE(*,73)' dielectric constant of host medium EM      '
          READ(*,*,ERR=1003)EM
1004      WRITE(*,73)' core particle radius RC [nm]      '
          READ(*,*,ERR=1004)RC
1005      WRITE(*,73)' particle radius R [nm]      '
          READ(*,*,ERR=1005)R
          ENDIF
1012      WRITE(*,73)' maximum multipolar moment LMAX      '
          READ(*,*,ERR=1012)LMAX
          WRITE(*,73)' with cluster size effect          (y/n)      '
          READ(*,72)CHA
          IF (CHA.EQ.'y'.OR.CHA.EQ.'y') YES=.TRUE.
          WRITE(*,73)' Filename for the result      '
          READ(*,72)FILENAME
          call clearscreen(0)                    ! only for MS-FORTRAN

          IF (RC.EQ.0D0.OR.RC.EQ.R) PARTIC=.TRUE.
          EM=DSQRT(EM)
          X0=2D0*PI*R
          XC0=2D0*PI*RC
          OPEN(UNIT=10,FILE=FILENAME)

          OPEN(UNIT=20,FILE=COREFILE)
          READ(20,*,ERR=1009)W_ANF
          READ(20,*)W_END
          READ(20,*)NWEL,NSPALT
          READ(20,*)OMPC2,QC,VFC,AAC
          IF (NWEL.GT.1) DW = (W_END - W_ANF)/(NWEL-1)
          IF (W_ANF.NE.W_END) LOGO1 = .TRUE.

c         The files containing the dielectric constant of the ( core )
c         particle  ( here : unit 20, corefile ) and of the shell ma-
c         terial ( here : unit 30, shellfile ) contain the following
c         data :
c
c         first wavelength (nm) W_ANF
c         last wavelength (nm) W_END
c         number of rows NWEL   number of y-data columns NSPALT ( here : 2 )
c         omp2  q  vf  a
c         (first wavelength)  first y1-data  first y2-data
c              ...              ...        ...
c              ...              ...        ...
c         (last wavelength)   last  y1-data  last  y2-data
c
c         In the case that W_ANF = W_END, the wavelength column is given.
c         In the case that W_ANF < W_END, the wavelength column is mis-
c         sing and the data are assumed to be equidistant.
c
c         The third row contains parameters for the cluster size effect :
c         omp2 : the squared volume plasmon frequency normalized to 1e30
c         q    : the intrinsic damping constant normalized to 1e15
c         vf   : the Fermi velocity normalized to 1e6
c         a    : the cluster size parameter

          WDC=QC+AAC*VFC/R
          WDC2=WDC*WDC
          QC2=QC*QC
```

```
        IF (PARTIC) GOTO 20

        OPEN(UNIT=30,FILE=SHELLFILE)
        READ(30,*,ERR=1010)W_ANF2
        READ(30,*)W_END2
        READ(30,*)NWEL2,NSPALT2
        READ(30,*)OMPS2,QS,VFS,AAS
        IF (NWEL2.GT.1) DW2 = (W_END2 - W_ANF2)/(NWEL2-1)
        IF (W_ANF2.NE.W_END2) LOGO2 = .TRUE.

        WDS=QS+AAS*VFS/(R-RC)
        WDS2=WDS*WDS
        QS2=QS*QS


20      WL2=0D0
        IWL2=0
        IWL=0

C       **** WAVELENGTH LOOP ****

30      IWL= IWL+1
        IF (LOGO1) THEN
         WL = W_ANF + DW*(IWL-1)
         READ(20,*,ERR=1200)EC1,EC2
        ELSE
         READ(20,*,ERR=1200)WL,EC1,EC2
        ENDIF

        IF (PARTIC) GOTO 50
        IF (WL-WL2) 30,50,40

40      IWL2=IWL2+1
        IF (LOGO2) THEN
         WL2 = W_ANF2 + DW2*(IWL2-1)
         READ(30,*,ERR=1200)ES1,ES2
        ELSE
         READ(30,*,ERR=1200)WL2,ES1,ES2
        ENDIF

        IF (WL-WL2) 30,50,40

C       === computation of cluster size effect ===
50       IF (YES) THEN
         OM=1883.651555D0/WL
         OM2=OM*OM
         EC1=EC1+OMPC2/(OM2+QC2)-OMPC2/(OM2+WDC2)
         EC2=EC2+OMPC2/OM*(WDC/(OM2+WDC2)-QC/(OM2+QC2))
         IF (.NOT.PARTIC) THEN
         ES1=ES1+OMPS2/(OM2+QS2)-OMPS2/(OM2+WDS2)
         ES2=ES2+OMPS2/OM*(WDS/(OM2+WDS2)-QS/(OM2+QS2))
         ENDIF
        ENDIF

        IF (PARTIC) THEN                          ! compact sphere

        EC0=DSQRT(EC1*EC1+EC2*EC2)/2D0
        DKC=DCMPLX( DSQRT(EC1/2D0+EC0), DSQRT(-EC1/2D0+EC0) )
        X=X0/WL*EM
        Y=X0/WL*DKC
        CALL MIE_CO(X,Y,LMAX,AMIE,BMIE)

        ELSE                                      ! layered sphere
```

```
          EC0=DSQRT(EC1*EC1+EC2*EC2)/2D0
          ES0=DSQRT(ES1*ES1+ES2*ES2)/2D0
          DKC=DCMPLX(DSQRT(EC1/2D0+EC0),DSQRT(-EC1/2D0+EC0))
          DKS=DCMPLX(DSQRT(ES1/2D0+ES0),DSQRT(-ES1/2D0+ES0))
          X=X0/WL*EM
          Y=X0/WL*DKS
          XC=XC0/WL*DKC
          YC=XC0/WL*DKS
          CALL SHELL_CO(X,Y,XC,YC,LMAX,AMIE,BMIE)

       ENDIF

C     === computation of extinction and scattering efficiencies ===

       QEXT = 0D0
       QSCA = 0D0

       DO 102 L=1,LMAX

       QEXT = QEXT - (2*L+1)*DREAL(AMIE(L)+BMIE(L))

       QSCA = QSCA + (2*L+1)*( AMIE(L)*DCONJG(AMIE(L)) +
     #                         BMIE(L)*DCONJG(BMIE(L)) )

102    CONTINUE

       QEXT = 2D0*QEXT/X/X
       QSCA = 2D0*QSCA/X/X

c      *** Data output ***

       WRITE(10,*) WL,QEXT,QSCA

       GOTO 30                              ! END OF WAVELENGTH LOOP

1200   CLOSE(20)
       CLOSE(30)
       CLOSE(10)
1008   write(*,*)
       write(*,*)
       write(*,73)'   Further computation   (y/n)
       read(*,72) cha
       if ( cha.eq.'y'.or.cha.eq.'Y') goto 1000
       STOP

1009   write(*,72)'  !!!  WARNING  !!! '
       write(*,72)'  The Core-File '//corefile
       write(*,72)'  could not be found ! '
       pause
       close(10,status='DELETE')
       close(20,status='DELETE')
       close(30)
       goto 1008
1010   write(*,72)'  !!!  WARNING  !!! '
       write(*,72)'  The Shell-File '//shellfile
       write(*,72)'  could not be found ! '
       pause
       close(10,status='DELETE')
       close(30,status='DELETE')
       close(20)
       goto 1008

       END
```

```
c       S U B R O U T I N E S   A N D   F U N C T I O N S


        SUBROUTINE MIE_CO(X,Y,L,A,B)
c       =================================================================
c       Calculation of the Scattering Coefficients for compact
c       spheres

        REAL*8 FAKT
        COMPLEX*16 A(1),B(1),X,Y,DK
        COMPLEX*16 JX0,JX,JY0,JY,HX0,HX,JH
        COMPLEX*16 SPHBES1,SPHHAN1

c       ------------------------------------------------------------

        DK=Y/X
        JX0=SPHBES1(X,0)/SPHBES1(X,1)
        JY0=SPHBES1(Y,0)/SPHBES1(Y,1)
        HX0=SPHHAN1(X,0)/SPHHAN1(X,1)
        JH=SPHBES1(X,0)/SPHHAN1(X,0)

        DO 1 K=1,L

          FAKT=2*K+1
          JX=JX0-K/X
          JY=JY0-K/Y
          HX=HX0-K/X
          JH=JH*HX0/JX0

          A(K)=-JH*(DK*JX-JY)/(DK*HX-JY)
          B(K)=-JH*(JX-DK*JY)/(HX-DK*JY)

          JX0=X/(FAKT-X*JX0)
          JY0=Y/(FAKT-Y*JY0)
          HX0=X/(FAKT-X*HX0)

1       CONTINUE

        RETURN
        END



        SUBROUTINE SHELL_CO(X,Y,U,V,L,A,B)
c       ==============================================================
c       Calculation of the Scattering Coefficients for a sphe-
c       re with one layer

        REAL*8 FAKT
        COMPLEX*16 A(1),B(1),X,Y,U,V,S,T,AH,BH
        COMPLEX*16 JX0,JX,JY0,JY,HX0,HX,JV0,JV,JU0,JU
        COMPLEX*16 NV0,NV,NY0,NY,JH,JN,JNY,DK1,DK2
        COMPLEX*16 SPHBES1,SPHBES2,SPHHAN1

c       ------------------------------------------------------------

        DK1=Y/X
        DK2=U/V
        JX0=SPHBES1(X,0)/SPHBES1(X,1)
        JY0=SPHBES1(Y,0)/SPHBES1(Y,1)
        NY0=SPHBES2(Y,0)/SPHBES2(Y,1)
```

```
HX0=SPHHAN1(X,0)/SPHHAN1(X,1)
JV0=SPHBES1(V,0)/SPHBES1(V,1)
JU0=SPHBES1(U,0)/SPHBES1(U,1)
NV0=SPHBES2(V,0)/SPHBES2(V,1)
JN=SPHBES1(V,0)/SPHBES2(V,0)
JNY=SPHBES1(Y,0)/SPHBES2(Y,0)
JH=SPHBES1(X,0)/SPHHAN1(X,0)


DO 1 K=1,L

  FAKT=2*K+1
  JX=JX0-K/X
  JY=JY0-K/Y
  NY=NY0-K/Y
  HX=HX0-K/X
  JV=JV0-K/V
  JU=JU0-K/U
  NV=NV0-K/V
  JN=JN*NV0/JV0
  JNY=JNY*NY0/JY0
  JH=JH*HX0/JX0

  S=-JN*(DK2*JV-JU)/(DK2*NV-JU)
  T=-JN*(JV-DK2*JU)/(NV-DK2*JU)

  AH=(JY*JNY+S*NY)/(JNY+S)
  BH=(JY*JNY+T*NY)/(JNY+T)

  A(K)=-JH*(DK1*JX-AH)/(DK1*HX-AH)
  B(K)=-JH*(JX-DK1*BH)/(HX-DK1*BH)

  JX0=X/(FAKT-X*JX0)
  JY0=Y/(FAKT-Y*JY0)
  NY0=Y/(FAKT-Y*NY0)
  HX0=X/(FAKT-X*HX0)
  JV0=V/(FAKT-V*JV0)
  JU0=U/(FAKT-U*JU0)
  NV0=V/(FAKT-V*NV0)

1      CONTINUE

RETURN
END


       complex*16 function sphbes1(x,1)
c      -------------------------------------------------------
c
c
c      calculation of the spherical bessel function j(x,1)
c      with a series expansion for arguments with abs(x) lo-
c      wer than 10 and by recurrence relations otherwise
c
c      parameters : x = free variable, optionally complex
c                   1 = multipolar order
c
c      this function has to be defined in the main program
c                     as   COMPLEX*16 sphbes1
c      -------------------------------------------------------
       implicit real*8 (a-h,o-z)
       complex*16 x,x2,y,fx,sum,j0,j1,j2,infinity
       data infinity/(1D308,1D308)/,eps/1D-20/

       xr=dreal(x)
```

```
      xi=dimag(x)

      IF (CDABS(X).LT.1D1) THEN              ! SERIES EXPANSION

      if (xi.eq.0D0) then                    ! real argument

      yr=1D0
      fxr=1D0
      sumr=1D0
      xr2=xr*xr/2D0
      do 10 k=1,1,-1
10    fxr = fxr*xr/(2*k+1)
      k1=2*1+1
      k=0
20    k=k+1
      yr = -xr2/(2*k+k1)/k*yr
      sumr = sumr + yr
      if (dabs(yr)-eps) 30,30,20

30    sphbes1 = fxr*sumr

      else                                   ! complex argument

      y=1D0
      fx=1D0
      sum=1D0
      x2=x*x/2D0
      do 11 k=1,1,-1
11    fx = fx*x/(2*k+1)
      k1=2*1+1
      k=0
21    k=k+1
      y = -x2/(2*k+k1)/k*y
      sum = sum + y
      if (cdabs(y)-eps) 31,31,21

31    sphbes1 = fx*sum

      endif

      ELSE                                   ! RECURSION

      if (xi.eq.0D0) then                    ! real argument

      j0 = dsin(xr)/xr
      j1 = j0/xr - dcos(xr)/xr
      do 40 11=2,1
       j2 = (2*11-1)/xr*j1 - j0
       j0=j1
       j1=j2
40    continue

      else                                   ! complex argument

      if (xi.gt.709D0) then
      pause' imaginary part of x too large, function returns 1D308'
      j2 = infinity
      j1 = infinity
      j0 = infinity
      else
      j0=cdsin(x)/x
      j1=j0/x-cdcos(x)/x
      do 50 11=2,1
       j2=(2*11-1)/x*j1-j0
```

```
          j0=j1
          j1=j2
50        continue
          endif

          endif

          sphbes1 = j2
          if (l.eq.1) sphbes1 = j1
          if (l.eq.0) sphbes1 = j0

          ENDIF

          end


          complex*16 function sphbes2(x,l)
c         ------------------------------------------------------
c
c         calculation of the spherical neumann function y(x,l)
c         with a series expansion for arguments with abs(x) lo-
c         wer than 10 and by recurrence relations otherwise
c
c         parameters : x = free variable, optionally complex
c                      l = multipolar order
c
c         this function has to be defined in the main program
c                        as   COMPLEX*16 sphbes2
c         ------------------------------------------------------
          implicit real*8 (a-h,o-z)
          complex*16 x,x2,y,fx,sum,n0,n1,n2,infinity
          data infinity/(1D308,1D308)/,eps/1D-20/

          xr=dreal(x)
          xi=dimag(x)

          IF (CDABS(X).LT.1D1) THEN                    ! SERIES EXPANSION

          if (xi.eq.0D0) then                          ! real argument

          yr=1D0
          sumr=1D0
          xr2=xr*xr/2D0
          fxr=-1D0/xr
          do 10 k=l,1,-1
10        fxr = fxr/xr*(2*k-1)
          kl=2*l+1
          k=0
20        k=k+1
          yr = xr2/(kl-2*k)/k*yr
          sumr = sumr + yr
          if (dabs(yr)-eps) 30,30,20

30        sphbes2 = fxr*sumr

          else                                         ! complex argument

          y=1D0
          sum=1D0
          x2=x*x/2D0
          fx=-1D0/x
          do 11 k=l,1,-1
11        fx= fx/x*(2*k-1)
          kl=2*l+1
```

```
          k=0
21        k=k+1
          y = x2/(kl-2*k)/k*y
          sum = sum + y
          if (cdabs(y)-eps) 31,31,21

31        sphbes2 = fx*sum

          endif

          ELSE                                        ! RECURSION

          if (xi.eq.0D0) then                         ! real argument

          n0 = -dcos(xr)/xr
          n1 = n0/xr - dsin(xr)/xr
          do 40 ll=2,1
            n2 = (2*ll-1)/xr*n1 - n0
            n0 = n1
            n1 = n2
40        continue

          else                                        ! complex argument

          if (xi.gt.709D0) then
          pause' imaginary part of x too large, function returns 1D308'
          n2 = infinity
          n1 = infinity
          n0 = infinity
          else
          n0=-cdcos(x)/x
          n1=n0/x-cdsin(x)/x
          do 50 ll=2,1
            n2 = (2*ll-1)/x*n1 - n0
            n0=n1
            n1=n2
50        continue
          endif

          endif

          sphbes2 = n2
          if (l.eq.1) sphbes2 = n1
          if (l.eq.0) sphbes2 = n0

          ENDIF

          end


          complex*16 function sphhan1(x,l)
c         ------------------------------------------------------------
c
c         calculation of spherical hankel function of first kind by
c                   sphbes1 + i*sphbes2
c
c         parameters :  x    = free variable, optionally complex
c                       l    = multipolar order
c
c         this function has to be defined in the main program as
c                   COMPLEX*16 sphhan1
c         ------------------------------------------------------------
          complex*16 x,one,sphbes1,sphbes2
          data one/(0D0,1D0)/
```

```
      sphhan1 = sphbes1(x,l)+one*sphbes2(x,l)

      end


      complex*16 function sphhan2(x,l)
c     -------------------------------------------------------------
c
c     calculation of spherical hankel functions of second kind by
c               sphbes1 -i*sphbes2
c
c     parameters :  x    = free variable, optionally complex
c                   l    = multipolar order
c
c     this function has to be defined in the main program as
c                   COMPLEX*16 sphhan2
c     -------------------------------------------------------------
      complex*16 x,one,sphbes1,sphbes2
      data one/(0D0,1D0)/

      sphhan2 = sphbes1(x,l)-one*sphbes2(x,l)

      end
```

References

1. W.A. deHeer, K. Selby, V. Kresin, J. Masui, M. Vollmer, A. Chatelain, W.D. Knight: Phys. Rev. Lett. **59**, 1805 (1987); K. Selby, M. Vollmer, J. Masui, V. Kresin, W.A. deHeer, W.D. Knight: Phys. Rev. **D 12**, 477 (1989); phys. Rev. **B 40**, 5417 (1989); C. Brechignac: Section 4.1 of Vol. I of this book.
2. U. Kreibig, P. Zacharias: Z. Physik **231**, 128 (1970)
3. P.E. Batson: Surface Sci. **156**, 720 (1985)
4. G. Mie: Ann. d. Physik **25**, 377 (1908)
5. P. Debye: Ann. d. Physik **30**, 57 (1909)
6. e.g. M. Brack: Phys. Rev. **B 39**, 3533 (1989)
7. H. Raether: *Excitation of Plasmons.* Springer Tracts in Mod. Phys. **38** (1980)
8. e.g. J.M. Ziman: *Electrons and Phonons*, Clarendon (1960); H.P. Baltes: *Proceedings of the ISSPIC 1*, Lausanne (1976)
9. L. Genzel: Festkörperprobleme **14**, 183 (1974)
10. A. Kawabata, R. Kubo: J. Phys. Soc. Japan **21**, 1765 (1966); W.P. Halperin: Rev. Mod. Phys. **58**, 540 (1986)
11. L. Genzel, P.T. Martin, U. Kreibig: Z. Physik **B 21**, 339 (1975)
12. W. Ekardt: Phys. Rev. **B 29**, 1558 (1984); Phys. Rev. **B 31**, 6360 (1985); D.E. Beck: Solid State Commun. **49**, 381 (1984)
13. C.v. Fragstein, H. Roemer: Z. Physik **151**, 54 (1958); C.v. Fragstein, F.J. Schoenes: Z. Physik **198**, 477 (1967)
14. U. Kreibig, L. Genzel: Surface Sci. **156**, 678 (1985) and refs. therein
15. W. Schulze: Section 3.1 of this book
16. W.de Heer, W.D. Knight, M.Y. Chou, M.L. Cohen: Solid State Physics **40**, 93 (1987)
17. R. Gans, H. Happel: Ann. d. Physik **29**, 277 (1909)
18. U. Kreibig: Z. Physik **234**, 307 (1970)
19. U. Kreibig: J. Phys. F: Metal Phys. **4**, 999 (1974)
20. J.C. Maxwell Garnett: Phil. Trans. R. Soc. London **203**, 385 (1904); **205**, 237 (1906); D.A.G. Bruggemann: Ann. d. Physik **24**, 636 (1935)
21. B.U. Felderhof, R. Jones: Z. Physik **B 62**, 43 (1985)
22. C.G. Granqvist: Z. Physik **B 30**, 29 (1978)
23. D. Bergman: *Bulk Physical Properties of Composite Media*, Edition Eyrolles (1985); W. Theiß Thesis RWTH Aachen (1989)
24. M. Born: *Optik*, Springer, Berlin (1965) 274 ff
25. C.F. Bohren, D.R. Huffman: *Absorption and Scattering of Light by Small Particles*, Wiley (1983) and refs. therein
26. R. Gans: Ann. d. Physik **37**, 881 (1912)
27. A. Güttler: Ann. d. Physik **6/11**, 65 (1952); M. Kerker: *The Scattering of Light*, Academic Press (1969) 189 ff
28. C.v.d. Hulst: *Light Scattering by Small Particles*, Wiley (1957)
29. R. Fuchs: Phys. Rev. **B 11**, 1732 (1975)
30. F. Sauter: Z. Physik **203**, 488 (1967); F. Forstmann Z. Physik **203**, 495 (1967)
31. A.R. Melnyk, M.J. Harrison: Phys. Rev. Lett. **21**, 85 (1968); Phys. Rev. **B 2**, 835 (1970)
32. R. Clanget: Optik **35**, 180 (1972)
33. R. Ruppin: Phys. Rev. Lett. **3**, 1434 (1973); Phys. Rev. **B 11**, 2871 (1975)
34. E. Rohloff: Z. Physik **132**, 643 (1952)
35. Optical material constants from Landolt-Börnstein; For MgO: P. Grosse (1988) unpublished
36. U. Kreibig: Sol. State Commun. **28**, 767 (1978); *Growth and Props. of Metal Clusters* ed. J. Bourdon Elsevier (1980) p. 371
37. R. Ruppin, H. Yatom: Phys. Stat. Solidi **B 74**, 647 (1976) K. Clemenger: Phys. Rev. **B 32**, 1359 (1985)
38. U. Kreibig, C.v. Fragstein: Z. Physik **224**, 307 (1969)

39. P. Apell, D.R. Penn: Phys. Rev. Lett. **50**, 1316 (1983)
40. L.E. Brus: Section 3.3 of this book
41. G. Schmid: Chemie in unserer Zeit **22**, 85 (1988)
42. K. Fauth, U. Kreibig, G. Schmid: Z. Physik **D 12**, 515 (1989)
43. F. Hache, D. Ricard, C. Flytzanis, U. Kreibig: Appl. Phys. **A 47**, 347 (1988)
44. U. Kreibig: Z. Physik **D 3**, 239 (1986); Z. Physik **D 12**, 505 (1989); U. Kreibig, M. Quinten, D. Schoenauer: Physica **A 157**, 244 (1989)
45. G. Niklasson, C.G. Granqvist, O. Hunderi: Appl. Optics **20**, 26 (1981)
46. M. Quinten, D. Schoenauer, U. Kreibig: Z. Physik **D 12**, 521 (1989)
47. D. Schoenauer: Diploma work Saarbruecken (1985)
48. W. Trinks: Ann. d. Physik **22**, 561 (1935)
49. D. Langbein: In *Springer Tracts in Mod. Phys.* 72 (1974)
50. J.M. Gerardy, M. Ausloos: Phys. Rev. **B 22**, 4950 (1980); **B 25**, 4204 (1982); **B 27**, 6446 (1983)
51. M. Quinten, U. Kreibig, D. Schoenauer, L. Genzel: Surface Sci. **156**, 741 (1985); M. Quinten, U. Kreibig: Surface Sci. **172**, 557 (1986); M. Quinten, U. Kreibig: Appl. Optics **32**, 6173 (1993)
52. U. Kreibig, A. Althoff, H. Pressmann: Surface Sci. **106**, 308 (1981)
53. D. Schoenauer, M. Quinten, U. Kreibig: Z. Physik **D 12**, 527 (1989)
54. T. Vicsek: *Fractal Growth Phenomena*, World Scientific (1989)
55. K.J. Berg, A. Berger, H. Hofmeister: Z. Physik **D20**, 309 and 313 (1991)
56. S.W. Koch: Phys. Blätter **46**, 167 (1990)
57. U. Kreibig, B. Schmitz, H.D. Breuer: Phys. Rev. **B 36**, 5027 (1987)
58. B. Dusemund, A. Hoffmann, T. Salzmann, U. Kreibig, G. Schmid: Z. Physik **D20**, 305 (1991)
59. B.N.J. Persson: Surface Sci. **281**, 153 (1993)
60. A. Liebsch: Phys. Rev. **B48**, 11 317 (1993); J. Tiggesbäumker, L. Kötter, K-H. Meiwes-Broer, A. Liebsch: Phys. Rev. **A48**, R1749 (1993)
61. H. Hövel, S. Fritz, A. Hilger, U. Kreibig, M. Vollmer: Phys. Rev. **B48**, 18 178 (1993)
62. H. Eckstein, U. Kreibig: Z. Physik **D26**, 239 (1993)
63. S. Asano, G. Yamamoto: Appl. Optics **14**, 29 (1975)
64. W.V. Ignatowsky: Ann. Physik **18**, 495 (1905)

3.6 Supported Clusters

S.B. DiCenzo and *G.K. Wertheim*

3.6.1 Introduction

The fundamental science motivating the study of clusters involves the development of band structure with increasing cluster size and the insulator–metal transition that takes place as the number of atoms in a cluster grows large enough to produce a partially filled band. Studies of these fundamental phenomena will help in understanding processes of technological interest, such as the origin of the catalytic properties of supported clusters and the silver-halide-based photographic process. Our primary goal, then, is to understand the evolution of electronic structure, from that of the isolated atom to that of the bulk solid. In this scheme the first ionization potentials of the isolated atom and small, molecular cluster must go over into the work functions of the metallic cluster and, finally, the bulk metal, while the molecular orbitals of the small cluster transmute into the band structure of the macroscopic solid. In this chapter we will describe the progress that has been made towards this goal in studies of supported clusters, focusing on the key role played by photoelectron spectroscopy in elucidating the electronic structure.

Free clusters, uncomplicated by the interaction with the substrate, would seem to be the preferred system for the fundamental questions posed above. However, the early photoemission work on supported clusters [1] started long before means had been devised to prepare mass-selected, free clusters. Moreover, despite the proliferation of cluster beam machines, supported clusters will continue to be studied for various reasons. The greater number of clusters that can be held in the target area offers a significant advantage over in-flight studies, and the influence of the substrate can be minimized by using weakly interacting substrates. Even polydisperse, supported clusters, which can be produced with more modest equipment than mass-selected clusters, may enjoy continued attention despite the disadvantage of a broader size distribution. Finally, since most applications involve supported clusters, and since the catalytic properties, depend on the influence of the substrate, it is necessary that the clusters be studied in this state. We emphasize that the role of the substrate interaction remains one of the most important but least understood aspects of catalytic clusters.

Our discussion of the electronic structure of clusters will focus on results of photoelectron spectroscopy, which has emerged as the preeminent technique for obtaining information about electronic structure. Angle-resolved ultra-violet photoelectron spectroscopy, the most powerful method, directly maps the dispersion of the electronic bands, but this technique requires oriented, single-crystal samples and it is not readily applied to small clusters. Angle-integrated photoemission from the valence bands of randomly-oriented microcrystals or clusters yields a view of the entire occupied density of states (DOS) and provides less detailed, but nevertheless valuable, information about the band structure. The best representation of the occupied DOS is obtained with high energy X-rays rather than ultra-violet or synchrotron radiation, because the effects of direct transitions into the well defined bands of the low-lying empty DOS are avoided. As a result X-ray photoelectron spectroscopy (XPS) has been a mainstay in the study of supported clusters.

XPS has the advantage of also providing core-electron spectra. In fact, because the photoelectric cross section is larger for the more tightly bound electrons and because the core levels are sharp and usually occur at an energy where the substrate spectrum is featureless, the core spectra are sometimes studied to the exclusion of the valence spectra. The amplitude of the core-electron signal provides information about the density of cluster material on the substrate, but the most important information resides in the binding energy shift. (In principle the line shape also contains useful information, but that turns out to be problematical for polydisperse clusters, and perhaps even for all supported clusters.) The core-electron binding-energy shift, ΔE_B, reflects changes in the outer electronic structure. *Kai Siegbahn*, emphasizing the core electrons' sensitivity to chemical changes in the valence shell, developed core-electron XPS into a useful technique [2], which he called ESCA (Electron Spectroscopy for Chemical Analysis). The valence shell changes in a small cluster will generally be more subtle than the sort of phenomena suggested by the phrase "chemical changes;" however, as observed for atoms at the surface of a solid, even such subtle changes produce core-level shifts that are easily measured and interpreted.

3.6.2   Preparation of Supported Clusters

For catalytic studies, clusters are generally prepared by chemical means, e.g., by adding a metal salt solution to a powdered support and reducing the salt at high temperature. Because photoelectrons typically have a mean-free-path of 5–25 Å in solids, such samples are not suitable for photoemission studies, since only the few clusters exposed on the surface are visible. For photoemission spectroscopy, clusters must be prepared on well defined, clean, smooth surfaces.

3.6.2.1   Mass-selected Supported Clusters

Although mass-selected clusters are clearly desirable for experiment, their collection presents a number of problems. The cluster beam machines based on the time-of-flight principle are not well adapted for collection of monodisperse clusters, because all clusters follow the same flight path. Dispersive instruments, which constitute a tiny minority, require only a slit to select the mass. In either case the major question is whether the integrity of the clusters can be maintained during the collision with the surface. It is essential that the cluster kinetic energy per atom be small compared to the cohesive energy, else the cluster may fragment on impact. Kinetic energies in excess of 1 eV per cluster atom may well allow the low-coordinated, and hence weakly-bound, atoms on the cluster surface to be shaken off during the collision. These atoms may diffuse back to their cluster of origin, but have a significant probability of reaching other clusters already on the surface, destroying the monodisperse character of the ensemble of clusters. If the clusters are to be studied by photoemission, or by any other electron spectroscopy, the samples must be prepared in ultra-high vacuum, and not exposed to the atmosphere while being transferred to the spectroscopy chamber. Finally, the coverage that can be achieved without the coalescence of sequentially deposited clusters is relatively small. Certainly no more than 10% of the area can be covered by clusters, making for relatively weak photoemission signals. Nevertheless, a number of experiments of this type have already been carried out [3, 4].

3.6.2.2   Forming Clusters by Aggregation

The usual method of sample preparation is to deposit atoms from a Knudsen cell, i.e., a thermal effusion oven, onto a clean surface in ultra-high vacuum. This allows the atoms to diffuse on the surface and to aggregate to form clusters. The choice of substrate is critical, since well-formed clusters will obtain only if the cohesive interaction between the deposited atoms is much greater than that between the atoms and the substrate. Moreover, the density of nucleation sites must be high enough so that many small clusters are formed when the requisite density of atoms is deposited. Amorphous carbon has been found to meet these requirements admirably.

Amorphous carbon may be prepared by evaporating carbon from an arc in vacuum onto a room temperature substrate, or by sputtering a graphite surface with energetic rare gas ions; either method is compatible with the ultra-high vacuum conditions required for photoemission experiments. The nucleation site density on an amorphous carbon surface is typically on the order of 3×10^{12} cm^{-2} [5]. For coverages of 10^{14}–10^{15} atoms/cm^2, or ~ 0.1 to ~ 1 monolayer (a reasonable coverage for photoemission measurements), this implies clusters of about 30 to 300 atoms each, on average. The nucleation site density is generally taken from the literature, because it is not practical to measure it for each surface prepared. This, of course, introduces some uncertainty into the quoted cluster size. In order to avoid this source of error, data are often presented as a function of coverage rather than average cluster size.

Carbon has other properties that make it highly advantageous as a support for photoemission experiments. Most important is the fact that the photoelectric cross section of the C $2sp$ valence electrons is much smaller at X-ray energies than that of noble or transition metal d-electrons, so that the response from a small number of cluster atoms can be obtained in the presence of the much larger number of substrate atoms. That there is only one core-electron level in C, the K shell at 285 eV, is also advantageous in minimizing the interference with the core-electron response from the clusters. Amorphous carbon consequently remains the substrate of choice, even when diffusive aggregation is not required, e.g., when mass-selected clusters from a beam source are collected.

3.6.3 Cluster Growth

3.6.3.1 Stochastic Model for Aggregation

It is generally agreed that clusters grow by diffusion-limited aggregation when atoms are randomly deposited on a smooth substrate. Well formed clusters obtain when the cohesive interaction between the adatoms is much greater than their interaction with the substrate. The nature of the nucleation sites has not been established, even for the widely-used amorphous carbon substrates. It is likely that clusters nucleate at dangling carbon bonds, since on cleaved graphite clusters grow only at steps. This observation also argues against the suggestion that dimers, formed during the early stages of deposition, constitute the nucleation centers. Another argument against the dimer hypothesis is that, on amorphous carbon, the cluster density does not depend strongly on the *rate* of deposition, in contrast to the situation on the other surfaces, such as cleaved crystals of insulating compounds.

If clusters grow by diffusion to randomly distributed, fixed nucleation sites, then the cluster size distribution can be estimated theoretically. One plausible starting point is the assumption that the number of atoms in a cluster should be proportional to the area associated with the nucleation site in a Wigner-Seitz construction (known to mathematicians as a Voronoi tessellation). This predicts a cluster size distribution whose width is $\sim 42\%$ of its mean. Although this is in good agreement with measured cluster sizes, the starting assumption is fundamentally flawed. In Monte Carlo simulations of diffusion to randomly placed, fixed nucleation sites, the number of atoms reaching a site has a square-root, rather than a linear, dependence on the area of the Wigner-Seitz cell [6]. However, if the model is made more realistic by recognizing that the clusters grow as they capture diffusing atoms, then the square-root dependence persists only while the clusters are small compared to their separation. Thereafter the dependence on cell area becomes stronger, becoming approximately linear when the clusters cover about 10% of the total surface area [6]. This, of course, corresponds to the physical situation whenever size distributions of supported clusters have been measured by electron microscopy, which explains the above-noted agreement with experiment. However, note that a linear relationship between cluster volume and cell area cannot be assumed in experiments on the smaller supported clusters (i.e., at low coverages), where the size distribution is in fact predicted to be comparatively narrow.

3.6.4   Band Structure of Clusters

To date only a start has been made toward the fundamental goal of understanding the transition from atomic energy levels to the band structure of solids, perhaps the most fundamental goal in the study of clusters. Most of the effort in the study of supported clusters has been devoted to comparing the band structure of polydisperse metallic clusters to that of the bulk solid. Limited experience with mass-selected supported clusters has so far shown little gain in resolution [3, 4].

3.6.4.1. Photoemission Studies – UPS and XPS

Figure 1 shows some early UPS work on the transition metal Pd [7]. Most of the essential features subsequently found for all metal clusters, of both noble and transition metals are apparent. The broad feature just above zero binding energy is the $4d$ band of Pd. With decreasing cluster size the band becomes narrower and the centroid shifts, while the cut-off that indicates the Fermi level becomes less sharp and moves toward larger binding energy. Data for Ag clusters [8] are shown in Fig. 2. In this noble metal the $4d$ band lies 4 eV below E_F, but the shifting and the narrowing of the d band are just as well defined as in Pd.

3.6.4.2   Bandwidth vs. Cluster Size

A qualitative understanding of the difference between the band structures of clusters and that of the bulk metal rests on our understanding of the behavior of surface atoms on bulk metals. At the surface the dominant change is band-narrowing due to the increased localization. Since clusters are strictly finite, the analogy to the surface of a bulk solid may seem weak, but we may draw confidence from the awareness that it is often useful to characterize localization only by the number of nearest neighbors (the coordination number) of an atom.

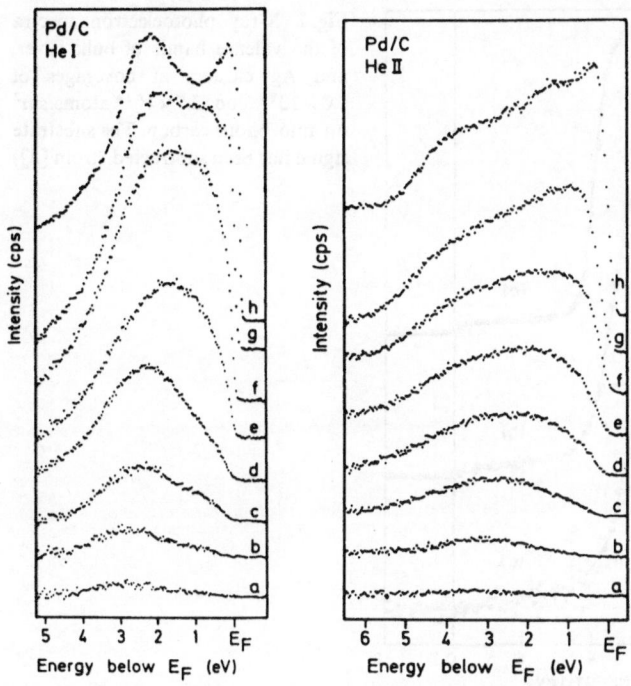

Fig. 1. He I and He II photoelectron spectra of the valence bands of Pd clusters at coverages of (a) 0.1, (b) 0.3, (c) 0.7, (d) 2.0, (e) 3.3, (f) 6.8, and (g) 9.8×10^{15} atoms/cm^2, and (h) polycrystalline bulk Pd. The signal from the carbon support has been subtracted. (from [7])

For example, at the (1 1 1) surface of an fcc crystal, such as a noble metal, each atom has 9 nearest neighbors, compared to 12 for atoms in the bulk. The bandwidth, to a first approximation, is expected to vary as the square root of the coordination number, and so the d-band of a noble metal should be reduced in width by 13% at the surface; such a reduction has indeed been observed [9]. Densities of states for the individual atomic layers of a Ni(1 0 0) slab, obtained from a band structure calculation [10], clearly show the narrowing of the bands at the surface, and confirm that even the second layer is more like the bulk than the surface. The d-band in Fig. 2c is about 20% narrower than in bulk Ag. If we follow the simple approximation above, this implies an average coordination number of 7.7. Assuming the Ag clusters are approximately spherical sub-units of the fcc structure, a coordination number of 7.7 corresponds to a 55-atom cluster [1], not too far off from the average cluster size of 80 atoms inferred from the coverage and the usual cluster density of 3×10^{12}/cm^2. Comparable narrowing is found in the d-band of Pd and of other metal clusters.

Beyond the narrowing, the other salient observation is the loss of detailed structure coupled with the tendency to assume a symmetrical shape with relatively long tails. This is especially obvious at the top of the band, where the sharp cut-off rapidly becomes smeared out. This may well be an effect of the

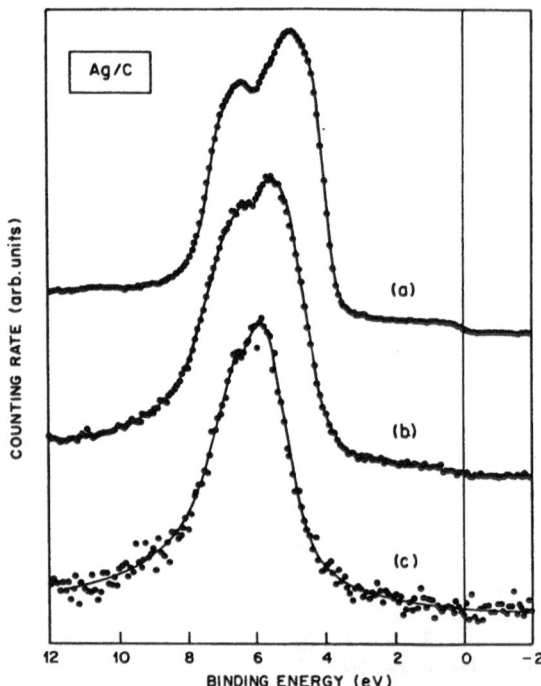

Fig. 2. X-ray photoelectron spectra of the valence bands of bulk silver, and Ag clusters at coverages of 1.0×10^{15} and 2.5×10^{14} atoms/sm^2 on amorphous carbon. The substrate signal has been subtracted. (from [8])

relatively broad distribution of cluster sizes, and not a property of the individual clusters. The contact with the substrate may also contribute to the broadening. Given that quite sharp d-band structure has been found in free mass-selected Cu clusters of comparable size [11], there is little point in considering these broadened structures in greater detail.

3.6.5 Core Electron Spectra

3.6.5.1 Systematics of Shifts in Metal Clusters

We have seen that there is a similar narrowing of the electron bands both in clusters and at the surface of bulk metals. This narrowing also affects the core-level binding energies in well-understood ways. The surface-atom core-level shifts have been successfully predicted for most metals [12]; they differ not only in magnitude but also in sign. One consequently expects that the band-narrowing in clusters will produce core-level binding-energy shifts with the same sense as the surface shifts. However, there is a surprising consistency in the core-electron binding-energy shifts that have been measured for supported clusters of simple, noble, and transition metals, in fact, of all metals yet studied. Without exception the binding energy is larger in clusters than in the bulk metal, and

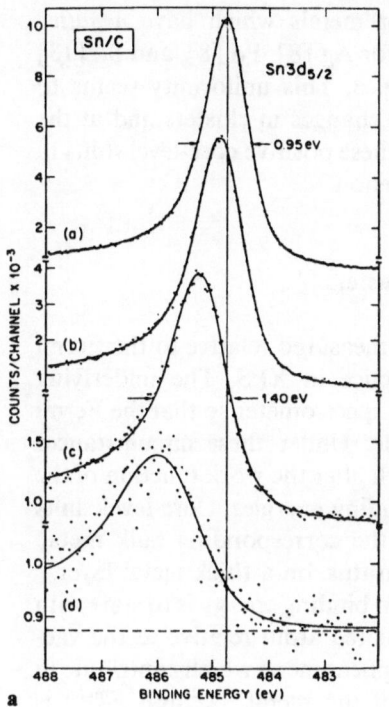

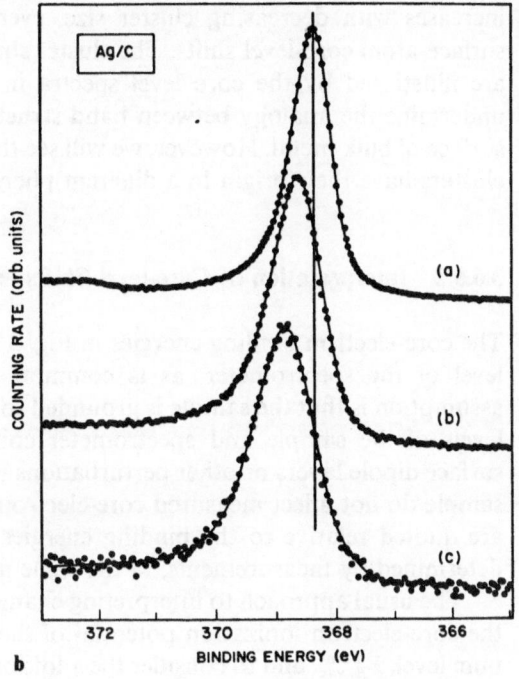

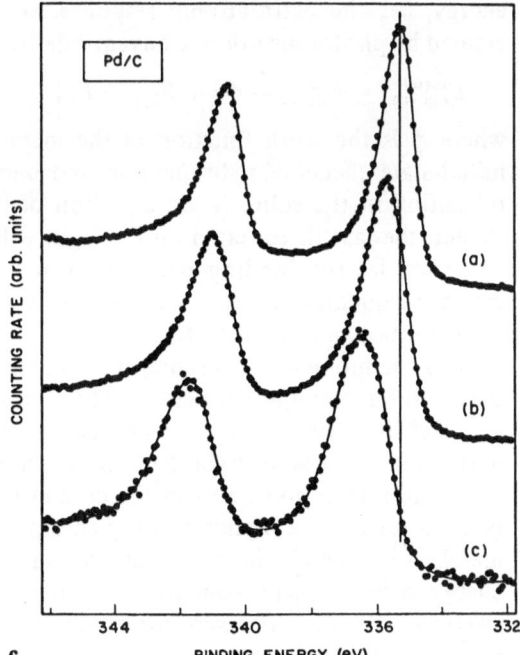

Fig. 3. Core electron spectra of tin, silver and palladium clusters, showing the Sn $4d$, Ag $3d_{5/2}$ and Pd $3d$ states. The bulk metal is labeled (*a*); the cluster size decreases from (*b*) to (*c*) and (*d*). (from [8] and [13])

increases with decreasing cluster size, even for metals which have *negative* surface-atom core-level shifts. The cluster shifts for Ag [8], Pd [8], and Sn [13] are illustrated by the core level spectra in Fig. 3. This uniformity seems to undermine the analogy between band structure changes in clusters and at the surface of bulk metal. However, we will see that these positive core-level shifts in clusters have their origin in a different phenomenon.

3.6.5.2   Interpretation of Core-level Shifts in Clusters

The core-electron binding energies in Fig. 3 are measured relative to the Fermi level of the spectrometer, as is common practice in XPS. The underlying assumption is that the sample is grounded to the spectrometer so that the Fermi levels of the sample and spectrometer coincide. Under these circumstances surface dipole layers or other perturbations which alter the work function of the sample do not affect measured core-electron binding energies. Core-level shifts are quoted relative to the binding energies in the corresponding bulk metal, determined by measurements, in the same apparatus, on a thick metal layer.

The usual approach to interpreting changes in binding energy is to start with the core-electron ionization potential of the isolated atom relative to the vacuum level, $E_{B, vac}^{atom}$, and to consider the additional phenomena which contribute in the metal. There are: (1) the work function of the metal, because E_B^{metal} is measured from the Fermi level, (2) the redistribution of charge among the outer orbitals attendant upon band formation, and (3) the final state relaxation energy, i.e., the extra-atomic response to the (positively charged) core hole created by photoionization. Conventionally, this is summarized by the equation:

$$E_{B, Fermi}^{metal} = E_{B, vac}^{atom} - e\phi + E_{hyb} - E_{rlx} \; , \tag{1}$$

where ϕ is the work function of the metal, and the subscripts *hyb* refers to initial-state effects of hybridization and band formation and *rlx* to final-state relaxation in the solid. In an equation of the same form, written for a finite cluster, the last three terms on the right will, in general, have different values. The core-electron binding-energy shift in a cluster (relative to the bulk metal) can consequently be discussed in terms of the changes in work function, hybridization, and final state screening.

For the moment we restrict the discussion to clusters sufficiently large that they exhibit metallic properties. The hybridization term will be different in a cluster, because of the band-narrowing we have discussed above. The band-narrowing changes the hybridization, which may result in a stronger or a weaker Coulomb interaction between the core electron and the valence electrons and hence a lower or a higher binding energy, as observed at the surfaces of bulk metals. In a metal, the final-state relaxation term, which lowers the binding energy, arises largely from the screening of the core hole by the conduction electrons. Because the screening will be inhibited as the conduction band separates into discrete molecular orbitals in a small cluster, one expects that the contribution of this term in small clusters will tend to increase the core level binding energy relative to the bulk. So far the outlook for core level shifts seems straightforward enough: for fairly large clusters, we expect shifts of the same sign and of a magnitude related to the surface atom core level shifts of the bulk metals, whereas for smaller clusters, the relaxation term, at least, should lead to higher binding energy. This expectation is at variance with experiment where, as illustrated by the examples in Fig. 3, all shifts are positive, suggesting some other mechanism, common to all supported metal clusters, is at work. The source of these positive shifts is a simple Coulomb effect, which arises when the clusters are not adequately grounded to the substrate.

It is generally believed that the work required to remove an electron from an isolated metal sphere is larger than that for the semi-infinite metal by $3e^2/8R$; that is, the work function of a spherical metallic cluster should be greater than that of the bulk metal by that amount [14, 15]. This term results from a simple classical calculation of the Coulomb interaction between the electron and its image charge. This formulation fits the ionization potentials measured in small, free clusters quite well, even when applied to small, non-metallic clusters. It has recently been argued, however, that the correct value for large metallic clusters is $e^2/2R$ [16]. For very small and marginally metallic clusters there are corrections that reduce the shift, accounting for the apparent agreement of the older value (with coefficient $\frac{3}{8}$) with experimental data. The formulation with coefficient $\frac{1}{2}$ is consistent with simple energy considerations, because the extra energy lost by the emerging electron is then equal to the electrostatic energy, $e^2/2R$, of the charged cluster which results. This also makes it clear that the $e^2/2R$ term should disappear if the cluster were grounded.

Core-electron photoemission leaves a cluster with a unit positive charge, which, in a free spherical metallic cluster will appear as a surface charge, resulting in a Coulomb energy of $e^2/2R$ [17, 18]. Whether this is relevant to the core-level shift mentioned above depends on whether or not a cluster is so effectively grounded that it is neutralized within the lifetime of the core hole, typically 10^{-15} s. The positive shifts suggest that they are not. When they are not neutralized, Eq. (1) must be modified by adding a Coulomb term:

$$E_B^{cluster} = E_{B,\,vac}^{atom} - e\phi + E_{hyb} - E_{rlx} + E_{coul} \tag{2}$$

and the binding energy shift in a cluster, measured relative to the bulk metal, becomes

$$\Delta E_B^{cluster} = - e\Delta\phi + \Delta E_{hyb} - \Delta E_{rlx} + E_{coul} \,, \tag{3}$$

where Δ denotes changes produced by forming the cluster.

Eventually the cluster will be neutralized by charge transfer from the substrate, but until then the substrate responds by the formation of an image charge, which reduces the final state Coulomb energy by about one-half. Even this reduced Coulomb energy amounts to a shift of $+ 0.7$ eV for a cluster 10 Å in diameter, whereas typical surface atom core level shifts, which are indicative of expected band-structure-induced shifts for clusters, are ± 0.4 eV or less. This final-state charge mechanism [17] correctly predicts a positive contribution to the binding energy shift for all metallic clusters; a shift that depends only on the cluster's size and shape, and not on the details of the electronic structure.

The essential independent proof that this mechanism is responsible for the cluster shifts is provided by Fermi edge spectra [8, 13]. If the clusters were well grounded the measured Fermi edge would coincide with that of the spectrometer. The data for Ag clusters in Fig. 4 show that the Fermi energy cut-off shifts to higher binding energy with decreasing cluster size. This can only be a final-state effect, since in the initial state E_F must be the same for the clusters and for the spectrometer. The Coulomb shift results from the delay in the neutralization of the photoionized cluster, due to carbon's low density of states at E_F. Confirmation comes from the observation that clusters on a metallic substrate do not show these large positive shifts [13, 17].

Given a measured core-electron binding-energy shift for a cluster, there is no *a priori* way to partition the shift among the four contributing mechanisms of Eq. (3). The magnitude of the hybridization term will depend on the nature of the metal, being much smaller in a simple or noble metal than in a transition metal,

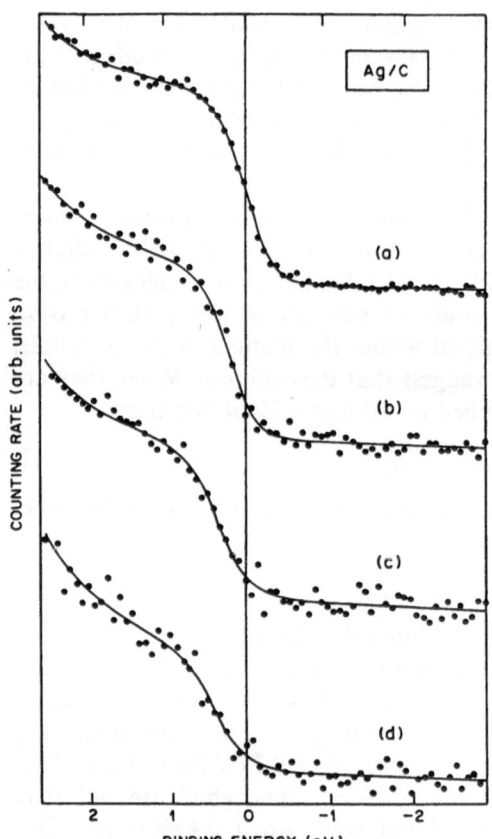

Fig. 4. Shift of the Fermi edge of small supported Ag clusters relative to that of the amorphous carbon substrate. Bulk metal, (a), and decreasing cluster size, (b)–(d). (from [8])

where it may amount to a few eV. The magnitude of the relaxation term in the bulk metal, which may also be a few eV, depends on the density of states at the Fermi energy. Even the Coulomb term, which is on the order of 1 eV for a cluster with a 10 Å *diameter*, is not easy to calculate precisely, as it depends on the shape of the cluster and the contact with the substrate.

The most direct experimental approach is to determine the Coulomb energy independently, by measuring the shift of the Fermi level, recalling that *all* photoelectrons experience the same Coulomb shift. Figure 5 shows the results of experiments on the noble metals Ag and Au which have negative surface-atom core-level shifts. In Ag, the small difference between the Fermi-level and core-level shifts has the sign and magnitude expected for a band structure contribution. In Au, the larger deviation at small cluster size is in accord with the

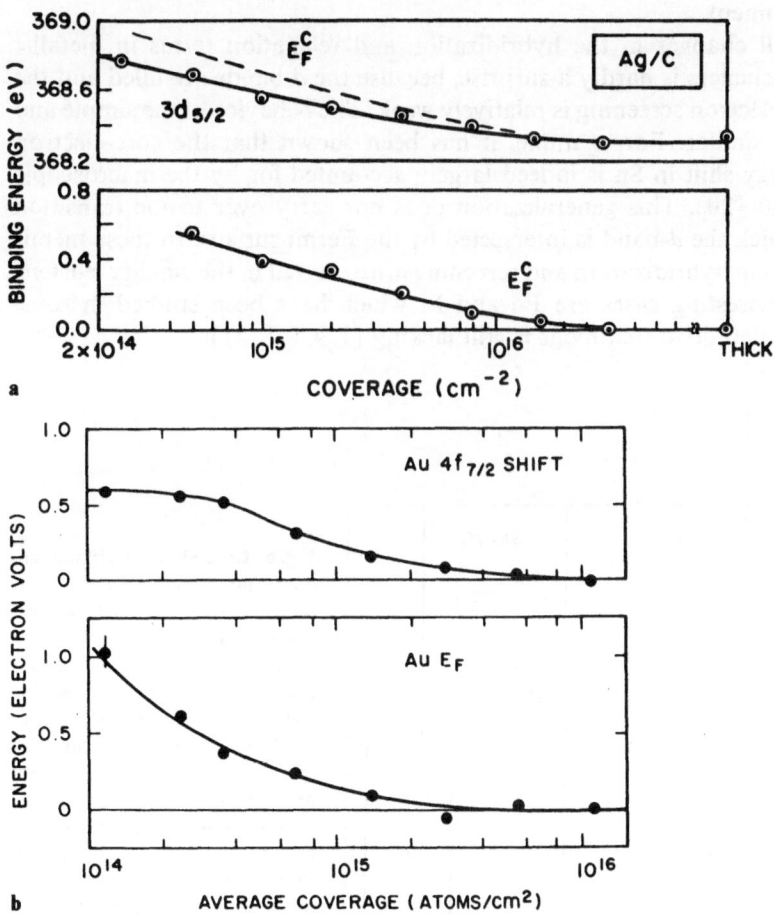

Fig. 5. Comparison of Fermi edge and core level shifts of supported Ag and Au clusters. (from [8] and [17])

− 0.4 eV surface-atom core-level shift, but many indicate the loss of metallic character (see below).

Further confirmation is found in Fig. 6, where the shift is shown to vary as the metal coverage to the − $\frac{1}{3}$ power [19], as expected for the 1/R dependence of the Coulomb shift, if the average cluster volume is proportional to the coverage. The strong deviation at large coverage in Fig. 6 is due to the coalescence of the clusters into a contiguous metallic layer. This means that the interesting physics of the core-electron shift must be sought in deviations from the coulomb term at small coverage, a difficult prospect since the signal from the Fermi edge is so weak as to be easily lost in the noise. One of the interesting observations is that much of the narrowing of the valence band takes place in a regime of cluster size where the core-electron binding-energy shift is dominated by the macroscopic coulomb effect. It follows that the bandwidth due to long-range order has only a weak effect on core-electron binding energies, which depend mainly on the local environment.

The small changes in the hybridization and relaxation terms in metallic noble metal clusters is hardly a surprise, because the d-bands are filled and the conduction electron screening is relatively weak. The behavior of the simple and sp metals is similar. For example, it has been shown that the core-electron binding energy shift in Sn is indeed largely accounted for by the macroscopic coulomb shift [20]. This generalization does not carry over to the transition metals in which the d-band is intersected by the Fermi surface. In these metals large changes in hybridization and screening are expected in the smaller clusters. The most interesting cases are Pd and Pt which have been studied in some detail, but a definitive treatment is still lacking [1, 4, 7, 8, 21].

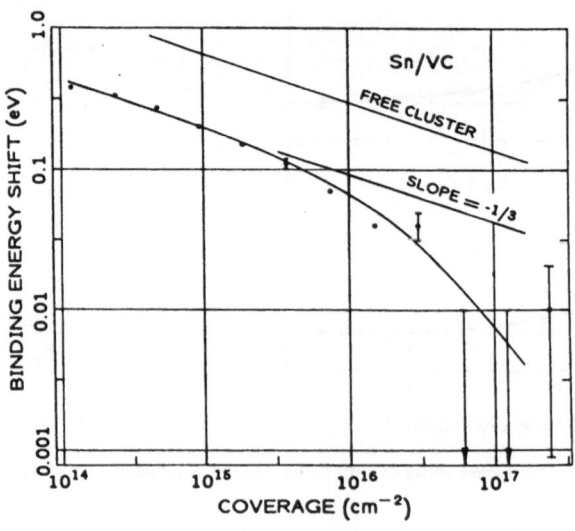

Fig. 6. Core-electron binding energy shift of tin clusters on vitreous carbon plotted against coverage. For coverages less than 3×10^{15} cm^{-2} the data exhibit a slope of − $\frac{1}{3}$, indicating that the shift is inversely proportional to the linear dimension of the clusters. At higher coverage the shift drops rapidly below this line indicating that the clusters have begun to coalesce. The coulomb shift for free clusters of the same diameter is also indicated. (from [19])

3.6.5.3   Born-Haber Cycle Approach to Core-electron Binding Energies

The above discussion of the core-electron binding energies suffers from the fact that neither E_{hyb} nor E_{rlx} in Eq. (1) is computable without a major theoretical effort, i.e., a band structure calculation for the cluster. There is, however, a complementary formalism that relates the core-electron binding energies in metal and atom to thermodynamic variables, without reference to the electronic structure [22]. This method uses the Born–Haber cycle for the bulk metal illustrated in Fig. 7, whih is readily extended to provide a treatment of the binding-energy shifts in clusters. The objective of the cycle is to express the energy required to move a core electron in the metal to the Fermi level, i.e., the core-electron binding energy $E_{B, Fermi}^{metal}$, in terms of other known quantities. The cycle starts at the lower left hand corner, with the removal of one atom from the bulk metal, at the cost of an energy E_{coh}^Z, the cohesive energy of the metals with atomic number Z. The desired core electron is then excited from the resulting free atom to the vacuum level, requiring an energy $E_{B, vac}^{atom}$. The atom now has unit positive charge. Next we apply the "equivalent-cores" approximation, which states that an atom of atomic number Z with a core hole is chemically equivalent to an atom with atomic number $Z + 1$, missing its outermost valence electron. This atom is now neutralized using the electron previously left at the vacuum level, yielding an energy equal to the first ionization potential of the $Z + 1$ atom, I^{Z+1}. This resulting neutral $Z + 1$ atom is next condensed into a piece of metal made up of $Z + 1$ atoms, yielding the cohesive energy E_{coh}^{Z+1}. In the final step the core-ionized atom is allowed to diffuse from the $Z + 1$ metal back into the original piece of metal from which it was removed in the first step. This yields the so-called implantation energy E_{impl} for a $Z + 1$ impurity into the Z metal. Using the equivalent-cores approximation in the reverse direction, the

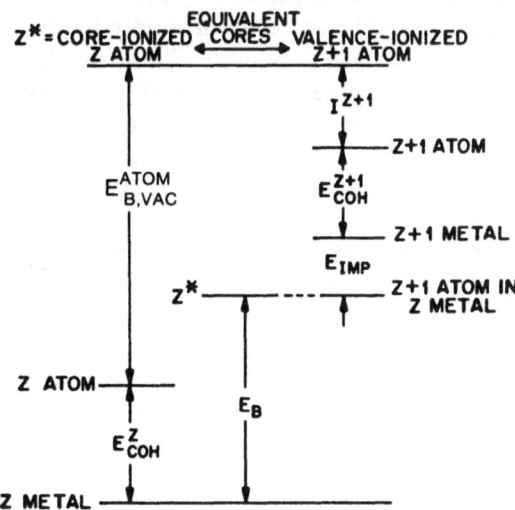

Fig. 7. Born–Haber cycle relating the core-electron binding energy in a metal to that of the free atom and various thermodynamic quantities

final result of these steps is a fully screened Z atom with a core hole in its normal environment, a state which is produced directly by exciting the core electron of an atom in the metal to the Fermi level. The energy required to traverse the cycle is consequently equal to the core-electron binding energy in the metal, $E_{B, Fermi}^{metal}$. Summing the energies around the cycle, we obtain

$$E_{B, Fermi}^{metal} = E_{B, vac}^{atom} + E_{coh}^{Z} - E_{coh}^{Z+1} - I^{Z+1} + E_{impl} . \qquad (4)$$

The cohesive energies and first ionization potentials of metals are well known, and the implantation energy can be estimated using the empirical tabulations of *Miedema* and coworkers [23]. Direct tests of this formalism, in cases where the atomic core-electron binding energy is known, have confirmed the validity of this analysis.

This formalism is readily extended to finite clusters [24], by paralleling the calculation of the core-electron binding energy shift of the atoms in the first atomic layer of a metal by *Johansson* and *Martensson* [22]. In this simple formulation, the altered properties of the first layer atoms are represented simply by a reduction of the cohesive energies, assuming them to be proportional to the coordination number, N. The surface-atom core-level binding energy shift is then obtained by taking the difference of two equations with the form of Eq. (2), one for the bulk metal and the other for the surface layer, with its reduced cohesive energies. The purely atomic terms cancel, leaving

$$\Delta E_{B}^{surf} = \delta(E_{coh}^{Z+1} - E_{coh}^{Z}) , \qquad (5)$$

where δ is the fractional reduction of the cohesive energies at the surface, $1 - N_{surf}/N_{bulk}$. The difference in implantation energy from surface to bulk has been omitted, because it is only a small correction. The sign of the core-level shift is then determined entirely by the relative size of the two cohesive energies. For Au, the cohesive energy of Hg, the $Z + 1$ metal, is much smaller than that of Au itself, predicting a negative shift, in agreement with experiment [12].

The average core-level shift in clusters should also be given simply by the fractional reduction of the average cohesive energy in the cluster. Consequently, the sign of the core-level shift in clusters of a given metal should be the same as the shift at the bulk surface. This conclusion is, necessarily, the same as that obtained from arguments based on changes in the electronic structure, but with the advantage that the shift can be trivially computed, being just the difference between two cohesive energies.

The Born–Haber analysis produces the same discrepancy with experiment as did our earlier analysis, and with the same source, that is, the Coulomb energy of the charged cluster. This happens because no charge is withdrawn from the system in the *B–H* cycle. Rather, the core electron, which is ultimately placed at the Fermi level, is recycled to serve as the screening charge. In experiments on bulk samples, the screening charge is understood to come from ground. The proper use of results from a Born–Haber cycle for clusters is to compare the calculated shift with the difference between the core-level shift and the Fermi level shift. The extant data for supported Au and Ag clusters show that this difference has the expected negative sign for the smaller clusters, but becomes too small to measure for larger clusters where the measured shifts are dominated by the coulomb energy. For Au, where the surface-atom core-level shift is − 0.4 eV for atoms with coordination number 9, the cluster data do show a comparable shift for 100 atom clusters. For Ag, where the surface atom shift is − 0.08 eV, the relative shifts are correspondingly smaller, see Fig. 5.

The B–H treatment provides a number of significant advances. It establishes a connection between cohesive energies and core-electron binding energies, which relate the seemingly remote electronic measurement to a familiar thermo-dynamic property. It provides a way of assessing the combined effects of changes in hybridization and relaxation on the core level shift. And finally, it confirms that there is a close connection between surface shifts and cluster shifts, putting a significant constraint on the interpretation of the latter.

3.6.6   Other Types of Measurement

3.6.6.1   Photoemission from Rare Gas Matrix

A method of monitoring the evolution of metallic properties of clusters that remain neutral throughout the experimental process was beautifully demon-strated [25] for Pd clusters grown in a 30–60 Å layer of Xe condensed on a Nb substrate. Here the Xe $5p$ binding energy provided a measure of the clusters' work function, which apparently reached the bulk value at about 20 atoms per cluster, and of the clusters' metallic screening response (to the positive hole in the adsorbed, and photoionized, Xe atom), which reached the bulk value at about 50 atoms. Use of the Xe photoemission allows these quantities to be measured without removing an electron from a cluster, which might be a signifi-cant perturbation in a 10- or 20-atom cluster. At the same time, the valence band spectrum of the Pd clusters is obtained, allowing one to make correlations between, e.g., bandwidth and the degree of metallic screening.

3.6.6.2   Auger Spectroscopy

Among the other techniques used to characterize the electronic properties of supported metal clusters is Auger Electron Spectroscopy (AES), a technique used mainly for chemical analysis of surfaces. In AES a core hole decays by capturing an electron from a less tightly bound level, and the excess energy is carried off by another electron, whose kinetic energy is measured. The core hole that forms the initial state of this process is identical to the final state studied by XPS. The peaks measured in AES are relatively broad because of the two-hole final state, but the shifts are larger than XPS shifts. Auger spectroscopy can make an important contribution to the study of clusters by helping to distinguish between initial state and final state effects [26].

This application of AES depends on the fact that band structure effects enter linearly in the two hole state while coulombic effects appear to the second power. We first write an expression, analogous to Eq. (3), for the change in the Auger kinetic from metal to cluster:

$$\Delta K^{\text{cluster}} = e\Delta\phi - \Delta E_{\text{hyb}} + 3\Delta E_{\text{relax}} - 3E_{\text{coul}} \; . \tag{6}$$

In the derivation of this expression it is assumed that *all the levels involved in the Auger process are core levels*. The applicability of this approach is severely constrained by this proviso, which has unfortunately been ignored in some work.

The Auger parameter, α, is defined by

$$\alpha = \tfrac{1}{2}(\Delta E_{\text{B}}^{\text{cluster}} + \Delta K^{\text{cluster}}) = \Delta E_{\text{rlx}} - E_{\text{coul}} \; . \tag{7}$$

Note that the initial state terms has been eliminated, leaving only the two final state terms. Once these have been determined, the initial state terms can be obtained using Eq. (3).

For Sn clusters it has been shown by this method that the observed shifts are largely of final state origin [20]. This is to be expected, since the sp conduction band provides little opportunity for significant rehybridization.

3.6.6.3 Absorption Edges

Absorption edge spectra can add significant information in favorable cases. For example, in Ni, Pd, and Pt clusters the so-called white line of core p-shell excitations depends on the existence of unoccupied d-band character just above the Fermi level. With suitable calibration one should be able to measure the redistribution of charge between the s- and d-bands with cluster size. Opposite effects are expected in Ni and Pd which have comparable unoccupied d-character in the metallic state, but respectively have $3d^8 4s^2$ and $4d^{10}$ configurations in the atomic state. In Ni one would consequently expect the d-character to decrease in small clusters, while in Pd it should increase.

3.6.6.4 Extended X-ray Absorption Fine Structure

Extended X-ray absorption fine structure (EXAFS) data can provide precise measurements of interatomic distances. A number of attempts have been made to use this technique to look for anomalies in the interatomic spacing in small clusters. Sizable contractions have been reported in some [27] but not all [28] experiments. In molecular dynamics simulations of Cu clusters the bond lengths were found to be essentially independent of cluster size [29]. The authors of that work suggest that there may be a fundamental difficulty in EXAFS determination of bond lengths at surfaces, because the standard formalism is not applicable to surface atoms with anharmonic vibrations.

3.6.7   Metal-Insulator Transition

3.6.7.1   Kubo's Model

A size-dependent metal-insulator transition is a safe prediction for any cluster made up of atoms of an element that is a metal in bulk form. A cluster of a few atoms cannot be classified as a metal, even when the valence electrons are delocalized over the entire cluster, as they are in clusters of the alkali metals. Such systems may be highly polarizable, but not metallic. Metallic properties presuppose the existence of a partially occupied band with level spacing sufficiently small near the Fermi level so that a small external potential can create electron-hole pairs, allowing a flow of current. In a cluster a simple indicator of conductivity is the ability of the outer electrons to screen a core hole. In a non-metallic cluster, screening is effected by the polarization of electrons in filled valence bond states.

Kubo [30] pointed out that metallic properties will be obtained in clusters when the states near the Fermi level have spacing no greater that kT, allowing charge to hop by thermal activation. The cluster size where this is realized is easy to estimate. For a metal that has one outer s-electron in a band 5 eV wide, metallic conductivity would be realized at room temperature in a cluster containing more than 200 atoms. For a d-band transition metal, 40 atoms will suffices, because of the larger number of d states. This range of cluster sizes is readily accessible in cluster beams

The insulator–metal transition in clusters is not first-order like those found in some transition metal oxides. Rather, it occurs gradually with increasing temperature and increasing cluster size. It should be noted that the group-II elements, with two outer s electrons, provide a special situation in which metallic behavior requires the overlap of the s- and p-band states in addition to *Kubo*'s criterion. Experiments with free Hg clusters have shown behavior related to this phenomenon [31, 32]. The case of Be has also been treated theoretically [33].

3.6.7.2   Experimental Observation

A clear indication of a metal-insulator transition was found in studies of mass-selected Au clusters deposited on amorphous carbon [3]. Recall that, as long as the electronic structure of the cluster resembles that of the bulk metal, the binding energy shift measured in XPS will be simply the Coulomb energy, or $\sim e^2/2R$, with a reduction due to the effects of the substrate screening. In the final state, one unit of negative charge occupies the (metallic) screening state, and the complementary unit of positive charge is distributed over the cluster surface. For clusters so small as to be non-metallic, the shift will not follow this $1/R$ dependence. In Fig. 8 the Au $4f$ core-electron binding energy plotted as a function of average coordination number shows a break from the Coulomb energy shift (dashed line) for a coordination close to 9. This corresponds to a cluster containing circa 100 atoms, in reasonable agreement with the estimate given above. Comparison with earlier work on polydisperse Au clusters grown on amorphous carbon also shows evidence for this change. As seen in Fig. 5, the core-electron binding energy shift becomes smaller than the Fermi level shift for coverages smaller than 3×10^{14} cm^{-2}, yielding an average cluster size of 100 atoms, if the nucleation site density is 3×10^{12} cm^{-2}.

While a deviation from the $1/R$-dependence of the core level shift indicates a cessation of metallic screening, and while a loss of metallic character causes a cessation of metallic screening, it is possible for metallic screening to fail in a cluster, even though the cluster is fully metallic. This will happen whenever the Coulomb energy exceeds the energy reduction due to metallic screening, that is, whenever the energy required to put a charge of $+e$ on the cluster's surface is greater than the difference between the metallic screening energy and the energy associated with some alternative response to the core hole, such as polarization. For example, the diameter of a 100 atom Au cluster is ~ 14 Å, making the coulomb energy of the isolated cluster ~ 1 eV. Although this is probably large enough to suppress the conduction- electron screening, the substrate response should reduce this by about one-half, making it somewhat more likely, although not certain, that it is *Kubo's* criterion, and not the Coulomb energy, that is responsible for the observed insulator-metal transition.

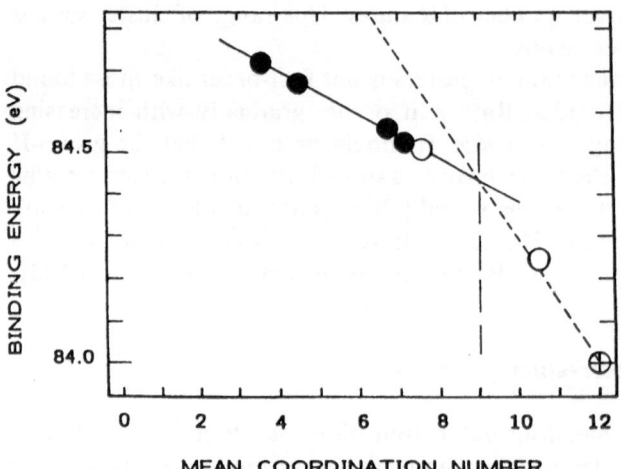

Fig. 8. Core-electron binding energies of the Au $4f$ electrons in Au clusters supported on amorphous carbon plotted against the mean coordination number. Size-selected clusters with 5, 7, 27, and 33 atoms shown as *filled circles*; Polydisperse clusters as *open circles*. (from [3])

3.6.8 Prospects for the Future

Figure 9 gives an inkling of what may be hoped for in future studies of mass-selected clusters [11]. These photoemission spectra were obtained with intense laser radiation from monodisperse Cu clusters in flight. The real accomplishment of this work is to exhibit part of the structure of the *d*-band of few-atom clusters. In the future the new synchrotron sources should make it possible to probe the complete electronic structure of free clusters. This work also provides a clear demonstration of the importance of the coulomb energy. In this experiment, in which the clusters were *negatively* charged in the initial state and neutral in the final state, the coulomb energy serves to *decrease* the electron binding energies with decreasing cluster size. As expected, the gross features of

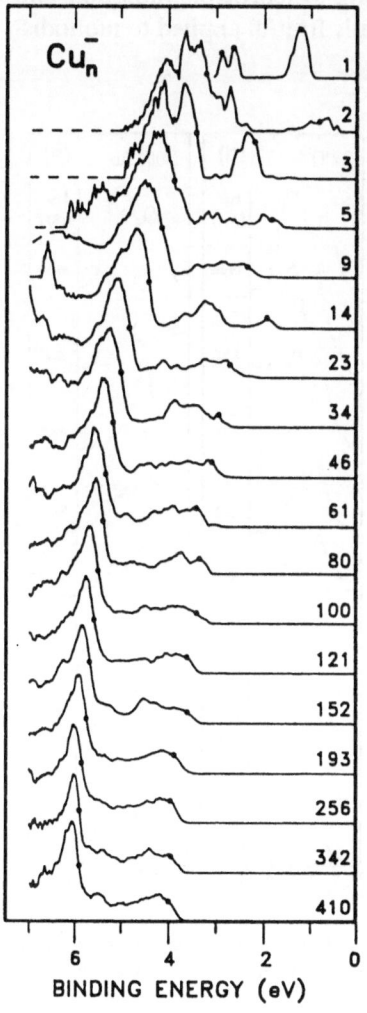

Fig. 9. Ultraviolet photoemission spectra of free, mass-selected, negatively charged Cu clusters, taken with 7.9 eV F_2 excimer laser radiation. (from [11])

the valence band shift together, but in the opposite sense from that seen in photoemission from supported clusters, which are neutral in the initial state. In particular, the top of the d-band and the Fermi cut-off show quite similar shifts in the larger clusters. Such monodisperse clusters, if collected on a suitable substrate, could even now be studied with synchrotron radiation to reveal the complete structure of the d-bands. Comparisons between spectra for monodisperse free and supported clusters could also give the first clear indication of the substrate interaction.

Figure 10 shows some very early data [34], which nevertheless suggest a future direction in cluster research. This is an elegant method of detecting the onset of delocalization, relying on the idea that in solids, the crystal momentum is a good quantum number. For clusters with this property, there will be dispersion, that is, the valence band spectra obtained with different photon energies will look different. This approach, using synchrotron radiation and a high-resolution spectrometer, should be especially fruitful applied to monodisperse clusters.

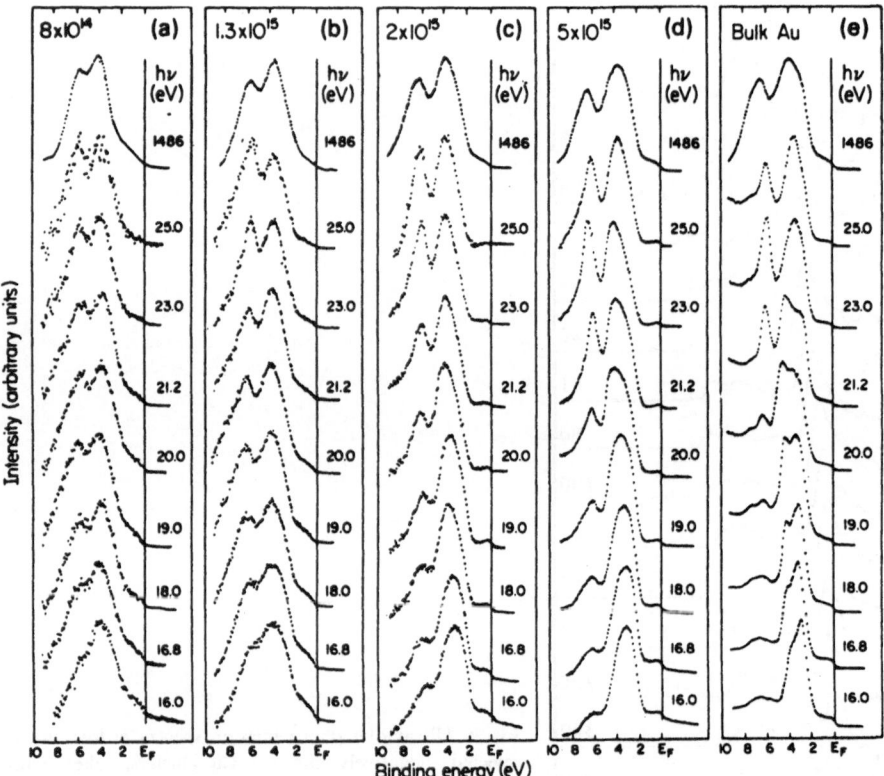

Fig. 10. Photoemission spectra of polydisperse Au clusters at four coverages and bulk Au, taken with photon energies ranging from 16 to 1486 eV. The signal from the carbon support has been subtracted. (from [30])

3.6.9   Conclusions

Photoemission studies of supported, polydisperse clusters have demonstrated the narrowing of the d-bands in noble and transition metal clusters due to the loss of long range periodicity. The transition from discrete molecular orbitals to electronic bands has eluded observation, even though the effects of the insulator-metal transition has been seen in core-level shifts. The major source of core-level shifts in the larger metallic clusters is the coulomb energy due to the charge left on the cluster by the ejection of a core electron. This effect depends only on the size of the cluster, and not on its electronic properties. These become important only in the smallest supported clusters typically studied in XPS.

The future clearly belongs to studies of mass-selected, monodisperse supported clusters with 2 to 200 atoms. The technological problems of preparing suitable samples are well defined, and have been overcome in a few experiments. In order to observe the details of the valence structure, future photoemission experiments will require the higher flux and resolution of the third-generation synchrotron sources.

Supported clusters are inherently more complex than free clusters. The contact with the substrate lowers the symmetry of the system, and creates a larger class of inequivalent atoms. However, the real motivation for the study of supported clusters lies in these complications, i.e., in the prospect of one day understanding the interaction with the substrate.

References

1. M.G. Mason: Phys. Rev. **B 27**, 748 (1983), and references therein
2. K. Siegbahn, C. Nordling, A. Fahlman, R. Nordberg, K. Hamrin, J. Hedman, G. Johansson, T. Bergmark, S.-E. Karlsson, I. Lindgren, B. Lindberg: Nova Acta Regiae Soc. Sci. Ups. Ser. IV, 20 (1967)
3. S.B. DiCenzo, S.D. Berry, E.H. Hartford, Jr.: Phys. Rev. **B 38**, 8465 (1988)
4. W. Eberhardt, P. Fayet, D.M. Cox, Z. Fu, A. Kaldor, R. Sherwood, D. Sondericker: Phys. Rev. Lett. **64**, 780 (1990)
5. J.F. Hamilton, P.C. Logel: Thin Solid Films **16**, 49 (1973); **23**, 89 (1974)
6. S.B. DiCenzo, G.K. Wertheim: Phys. Rev. **B 39**, 6792 (1989)
7. R. Unwin, A.M. Bradshaw: Chem. Phys. Lett. **58**, 58 (1978); Y. Takasu, R. Unwin, B. Tesche, A.M. Bradshaw, A.M. Grunze: Surf. Sci. **77**, 219 (1978)
8. G.K. Wertheim, S.B. DiCenzo, D.N.E. Buchanan: Phys. Rev. **B 33**, 5384 (1986)
9. P.H. Citrin, G.K. Wertheim, Y. Baer: Phys. Rev. Lett. **41**, 1425 (1978)
10. E. Wimmer, A.J. Freeman, H. Krakauer: Phys. Rev. **B 30**, 3113 (1984)
11. O. Cheshnovsky, K.J. Taylor, J. Conceicao, R.E. Smalley: Phys. Rev. Lett. **64**, 1785 (1990)
12. P.H. Citrin, G.K. Wertheim, Y. Baer: Phys. Rev. **B 27**, 3160 (1983)
13. G.K. Wertheim, S.B. DiCenzo, D.N.E. Buchanan, P.A. Bennett: Solid State Commun. **53**, 377 (1985)
14. J.M. Smith: AIAA J. **3**, 648 (1965)
15. D.M. Wood: Phys. Rev. Lett. **46**, 749 (1981)
16. G. Makov, A. Nitzan, L. Brus: J. Chem. Phys. **88**, 5076 (1988)

17. G.K. Wertheim, S.B. DiCenzo, S.E. Youngquist: Phys. Rev. Lett. **51**, 3210 (1983)
18. T.T.P. Cheung: Surf. Sci. **127**, L129 (1984); **140**, 151 (1984)
19. G.K. Wertheim, S.B. DiCenzo: Phys. Rev. **B 37**, 844 (1988)
20. G.K. Wertheim: Phys. Rev. **B 36**, 9559 (1987)
21. S. Kohiki, S. Ikeda: Phys. Rev. **B 34**, 3786 (1986)
22. B. Johansson, N. Mårtensson: Phys. Rev. **B 21**, 4427 (1980)
23. A.R. Miedema, P.F. de Châtel, F.R. de Boer: Physica **100B**, 1 (1980)
24. G.K. Wertheim: Z. Phys. B-Condensed Matter **66**, 53 (1987)
25. J. Colbert, A. Zangwill, M. Strongin, S. Krummacher: Phys. Rev. **B 27**, 1378 (1983)
26. C.D. Wagner: Faraday Discuss. Chem. Soc. **60**, 291 (1975)
27. G. Apai, J.F. Hamilton, J. Stöhr, A. Thompson: Phys. Rev. Lett. **43**, 165 (1079)
28. P.A. Montano, G.K. Shenoy, E.E. Alp, W. Schultze, J. Urban: Phys. Rev. Lett. **56**, 2076 (1986)
29. L.B. Hansen, P. Stoltze, J.K. Norskov, B.S. Clausen, W. Nieman: Phys. Rev. Lett. **64**, 3155 (1990)
30. R. Kubo, A. Kawabata, S. Kobayashi: Ann. Rev. Mater. Sci. **14**, 49 (1984)
31. K. Rademann, B. Kaiser, U. Even, F. Hensel: Phys. Rev. Lett. **59**, 2319 (1987)
32. C. Bréchignac, M. Broyer, Ph. Cahuzac, G. Delacretaz, P. Labastie, J.P. Wolf, L. Wöste: Phys. Rev. Lett. **60**, 275 (1988)
33. R. Kawai, J.H. Weare: Phys. Rev. Lett. **65**, 80 (1990)
34. S.T. Lee, G. Apai, M.G. Mason, R. Benbow, Z. Hurych: Phys. Rev. **B 23**, 505 (1981)

3.7 Nanocrystalline Materials

R. Birringer and *H. Gleiter*

3.7.1 Introduction

In Materials Science and Solid State Physics progress has been made in many cases by one of the following two approaches: either by developing and applying new methods of investigation or by preparing materials with novel structural features and/or properties. Transmission electron microscopy, Mössbauer spectroscopy, X-ray diffraction, are examples for the first approach. The discovery of metallic glasses, high temperature superconductors or quasi crystals represent developments resulting in materials with novel structural features and/or properties. In the case of crystalline materials the generation of compositionally, topologically and/or morphologically metastable solids was attempted in order to achieve materials with new and technologically attractive properties. Nanocrystalline materials seem to be suitable candidates for tailoring types and volume fractions of metastable elements in a solid.

3.7.2 Basic Ideas

Nanocrystalline materials (n-materials) are solids containing such a high density of defects that the number of atoms which are located in the cores of these defects becomes comparable ($\geq 10\%$) with the number of atoms located at regular lattice sites. Depending on the type of defects involved (grain boundaries, interphase boundaries, dislocations, internal surfaces etc.), different types of n-materials may be generated. For illustration, Fig. 1 shows the core structure of a dislocation and a grain boundary, respectively, on an atomistic scale. It may be noticed that in both cases the atomic density and special correlation between nearest neighbours is changed in comparison to the perfect lattice. It is the basic idea [2] of n-materials that new properties should emerge resulting from the large number density of atoms situated in highly defective core structures. But also properties which are sensitive to length scales should be influenced because in n-materials the succession of crystalline and defective regions is modulated on a length scale of typically a few nanometers, which is comparable to fundamental quantities such as the electron mean free path or the domain-wall

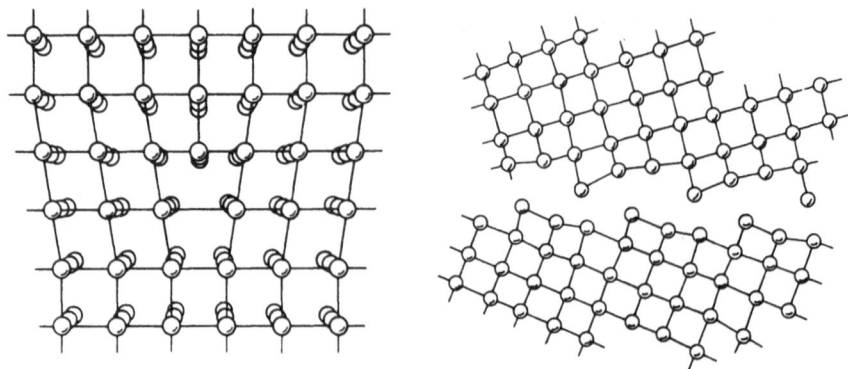

Fig. 1. Edge dislocation (to the left) in a simple cubic lattice and a 37° [1 0 0] tilt-grain boundary (to the right). In the case of the dislocation (1 dim) and the grain boundary (2 dim) [1] the atomic density is reduced and the coordination is changed (breaking of the lattice-symmetry) in comparison to the perfect lattice

thickness in systems showing order–disorder transitions. As a model system of a n-material we shall discuss polycrystals with an incorporated number density of grain boundaries of typically $10^{19}/cm^3$, requiring an average crystallite size of about 7 nm. In order to improve the understanding of the microscopic details of our model system it may be helpful to summarize some facts about grain boundaries.

Experimental and theoretical investigations on grain boundaries [3] in bicrystals and/or coarse grained polycrystals yield evidence that the atomic structure of the boundary core represents a two-dimensional atomic arrangement lying in the boundary plane that is located between two adjacent crystallites of the same kind which are differently oriented relative to each other; the thickness of the boundary core is several atomic layers ($\approx$ 1 nm). A section perpendicular to the boundary plane reveals that the boundary core may be conceived as being built up of different structural units (groups of atoms where a central atom is surrounded by 5, 6 or 7 nearest-neighbour atoms) in which each unit (e.g. the one with 5 atoms) may differ by variable distances between the central atom and the nearest-neighbour atoms. Depending on the boundary crystallography (i.e. the misorientation between adjacent crystallites, the boundary inclination and the translational position of both crystals relative to each other), different sequences of structural units repeat along the grain boundary core. The length of such a repeat-sequence is termed the periodicity length of a boundary which lies typically in the order of 10 interatomic spacings. Obviously the boundary core cannot be described as thin (disordered) glassy layer. The boundary core structure tends to relax in an atomic arrangement of minimum energy in the potential field of the adjacent crystals. As a consequence, the boundary core structure depends on the interatomic bonding forces and the boundary crystallography. Hence in a polycrystal with random crystallographical parameters a large variety of grain boundary core structures should exist. On an atomic scale the boundary core regions are characterized by a reduced atomic density (typically less than 20%) and interatomic spacings deviating from the nearest neighbour spacings in the perfect lattice. The physical reason for the reduced density and the non-lattice spacings between the atoms in the boundary core is the misfit between two crystal lattices of different orientations joined along a common interface. The misfit between the crystals not only reduces the atomic density in the interfacial region in comparison to the perfect lattice but also causes strain fields to extend from the boundary core region into the crystallites. Those strain fields additionally displace atoms from their ideal lattice sites. The amount of atomic displacements in both, core and lattice depends primarily on the interatomic potential. Hence, materials with different interatomic potentials are expected to exhibit different atomic structures in the grain boundaries (cores of defects) and different displacement fields extending into the crystallites.

Nevertheless, all of these materials have the following common microstructural feature. They consist of comparable volume fractions of defect cores on the one hand and crystal lattice regions (more or less strained) on the other hand. For illustration, Fig. 2 shows a hard sphere model of a two-dimensional n-material. It may be noticed that in the boundary core regions the atomic density and nearest neighbour coordination (non-lattice spacings) is changed in comparison to the perfect lattice.

A discussion of more sophisticated structural models of n-material [4] is omitted because of the present limitations in dealing with such systems theoretically. Furthermore, it is not clear at present to what extent the knowledge about grain boundaries in bicrystals or coarse grained polycrystals can be adopted in order to try to understand the structure and properties of n-materials. In n-materials, for example, the boundary length, the periodicity length and the

Fig. 2. Two-dimensional hard sphere model of a nanocrystalline material. For simplicity the nanocrystalline material is modeled to consist of perfect crystals of different crystallographic orientations (*black discs*) which are connected by grain boundaries (*white discs*)

crystallite diameter all are of the similar magnitude of roughly 10 lattice spacings, whereas in coarse grained materials the boundary length is typically 10^3 times more than the periodicity length. Hence it seems conceivable that a given crystallographic orientation between two adjacent crystals will give rise to different boundary structures in n-materials and coarse grained materials, respectively.

In most cases, n-materials are generated by configurationally freezing of metastable structures far from equilibrium. The study of such systems is complicated by the increase of parameters which are necessary, but not known ab initio, to specify the state of a sample under investigation. Hence many of the studies performed so far are hampered by a suitable specimen characterization. Therefore, this report and the several reviews [5–12] published on n-materials reflect the present state of a continuously developing understanding in this field. More recent results which may even qualify some views expressed in the present chapter, may be found in [57–60].

3.7.3 Preparation and Characterization

A widely applicable preparation technique for producing polycrystals with a crystal size of a few nm and randomly oriented crystallites is the following two-step procedure [13]: First, small crystallites are generated and accumulated; subsequently this assemblage of crystallites is compacted to a n-material. This technique allows the use of chemical (e.g. pyrolytic decomposition) as well as physical (e.g. inert-gas condensation) methods for generating small crystallites. The various methods for generating small crystallites are documented in several comprehensive reviews [14–16]. Other techniques by which a n-material may be generated in a single production step are high energy ball milling [17, 18], rapidly quenching of a melt with control of the nucleation parameters [19] or film techniques where the evaporated material condenses on a cold substrate [20]. Instead of discussing the details of the above mentioned techniques it seems preferable to summarize the principles of metastable structure synthesis reviewed by *Turnbull* in [21].

N-materials belong to the category of configurationally frozen metastable structures. The degrees of metastability may be characterized by structural variety and energy in excess of the stable equilibrium. The general procedure in metastable synthesis is to energize the material and then try by imposing constraints (e.g. quenching rates) to store as much as possible of the excess energy in a configurationally frozen state. For example, the volume density of the interfacial areas of a material which is phase separated by spinodal decomposition is comparable to that of n-materials. However, in the case of n-materials the interfaces are introduced by simultaneously transforming free surfaces into grain or interphase boundaries via the consolidation procedure resulting in a large excess energy per unit volume. Whereas the operation of the spinodal decomposition requires a small interfacial tension yielding a substantially smaller excess energy of such a material in comparison to n-materials. Obviously, the amount of energy in excess, the nature of the constraints and the path along which the energized material is configurationally frozen are crucial parameters in metastable synthesis. Table 1 shows a classification of metastable structures and the excess energies – in units of RT_m, where T_m is the average of the equilibrium melting temperatures, T_m, of the elementary constituents – approached by some synthetic members of each class [21]. In order to classify n-materials in the hierarchy of metastable structures, a preliminary result (see Sect. 3.7.5.5) concerning the excess energy of n-Cu is appended at Table 1. The value of the excess energy of 1.5 – in units of RT_m – for n-Cu should be compared with the excess energies of < 0.1 – in units of RT_m – of materials belonging to the category of morphological metastabilities. For comparison the enthalpy of fusion for Cu – in units of RT_m – is 1.16. Hence it may be suggested that n-materials are potential candidates for a new class of metastable materials with excess energies of an order of magnitude larger than the ones of the metastable materials produced so far. This also implies that for comparison of experimental data measured on different samples it is insufficient to give the crystal size only; chemistry, excess energy or related quantities have to be additionally known for comparative studies.

The experimental arrangement (Fig. 3) used most frequently so far for generating small isolated crystals is a modified gas condensation method [13]. The material (e.g. iron) is evaporated into an inert-gas atmosphere (e.g. helium, pressure about 1 kPa). As a result of interatomic collisions with the helium atoms, the evaporated iron atoms lose kinetic energy and condense in the form of small crystals. Via convective flow of the inert gas the as-produced crystallites are accumulated on a vertical cold finger in the form of a loose powder. Crystal

Table 1

Nature of meta- (or "in-") stabilities	Materials examples	Interfacial area/volume (cm^{-1})	Typical excess Energy (RT_m)
Compositional	Supersaturated solutions		< 1
Structural	Intermetallic cpds. Amorphous solids		< 0.5
Morphological	Microcrystalline compositionally modulated films	$10^4 - 10^7$	< 0.1
	Interphase dispersions		
	After D. Turnbull [21]		
	Nanocrystalline Cu (10 nm)	$\approx 10^7$	≈ 1.5

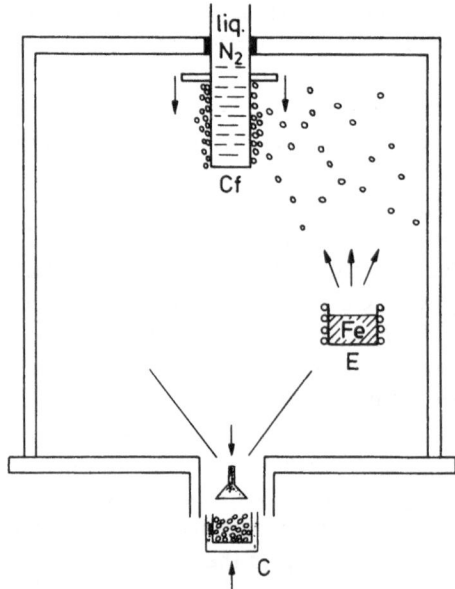

Fig. 3. Schematic representation of a gas-condensation chamber for the synthesis of nanocrystalline materials. The material evaporated from the source condenses in the gas and is transported by convective flow to the cold finger which is filled with liquid nitrogen. The powder of small crystals accumulating at the cold finger is subsequently scraped from the cold finger, collected and compacted in situ [13]

size and degree of aggregation and/or agglomeration of the powder depend on the material evaporated and the processing parameters as inert-gas pressure and evaporation rate etc. After restoring high vacuum (less than 10^{-6} Pa) the powder is stripped off from the cold finger and funnelled into a piston and anvil device where it is compacted (pressure up to 2 GPa) into a n-material.

The as-prepared samples are disc shaped (diameter 8 mm, thickness 0.1–0.3 mm) with a density varying between 70% and more than 90% of the crystalline density (depending on the material and the compaction procedure). Metallic impurities due to the evaporation – the scrapping – and the compaction procedure can be reduced to values in the order of 10^{-4} at. %. In n-Pd the concentration of helium was less than 50 ppm. The dominant impurity of metallic samples is oxygen and hydrogen. The concentrations of these contaminants depend on the affinity of the material to those elements and on the residual contamination of the vacuum system. For example n-Pd and n-Cu generated in high vacuum baked out at 150 °C have an oxygen content of about 1 at. % and a hydrogen content of about 2 at. %. Single-phase nanocrystalline materials exhibit crystal growth at elevated temperatures. For example the grain size of n-Fe or n-Pd remains stable up to 150 °C, n-Cu up to 100 °C, whereas n-Sn or n-Pb exhibit grain growth at ambient temperatures. Just as in conventional polycrystals, grain growth may be inhibited by second-phase particles and/or impurity drag. For example grain growth of n-Cu can be prevented by addition of a few per cent of tungsten oxide. Metallographic investigations of n-Cu revealed closed porosity in a few cases of up to 8 vol. %. In the case of n-Pd no pores could be detected.

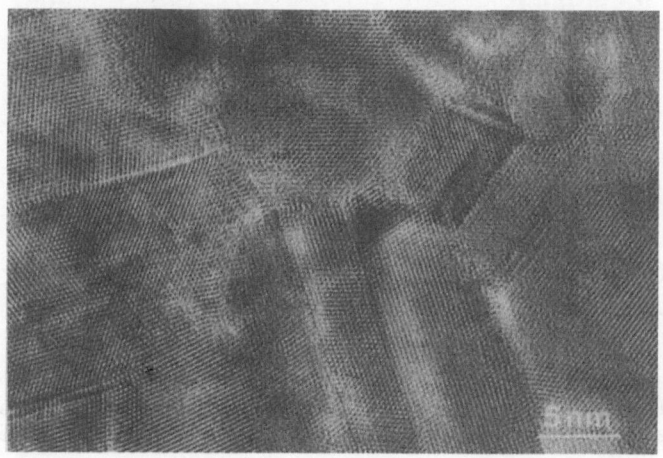

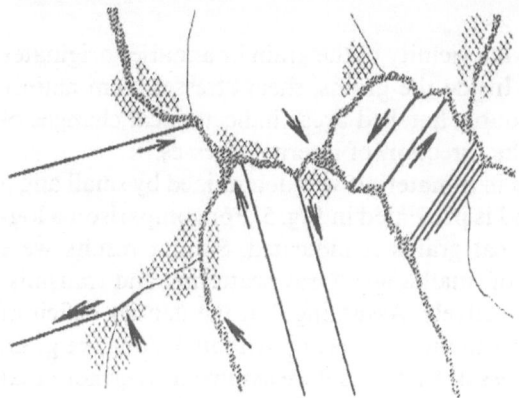

Fig. 4. *Top,* High resolution electron micrograph (HREM) of nanocrystalline palladium imaged at 400 kV. The average grain diameter was determined to 17.8 nm [22]. *Bottom,* Schematic drawing of the microstructure deduced from the HREM at left side. The grain boundary network is indicated by the *bold dotted contours.* The thickness of the grain boundaries was estimated to be more than 0.6 nm. The *double hatched areas* mark bended regions extending into the interior of the grains. The arrows symbolize the direction of internal stresses. The *straight lines* inside the grains indicate the presence of twin-boundaries which might be nucleated during the compaction of the nanocrystalline powder

3.7.4   Structural Studies

3.7.4.1   Microstructure

In Fig. 4 a high-resolution electron micrograph (HREM) through a thin section of n-Pd is shown [22]. An average crystal size of 17.8 nm was evaluated. The

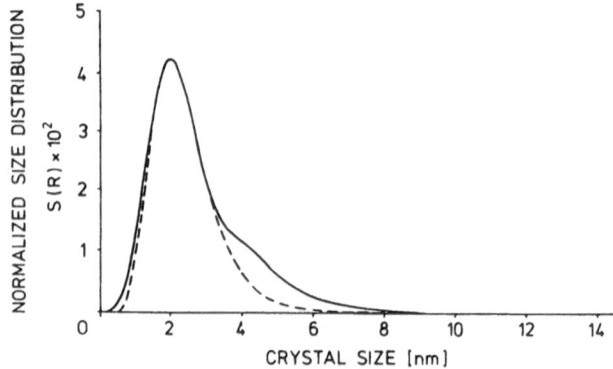

Fig. 5. Distribution of the crystal sizes in nanocrystalline Pd measured by small angle neutron scattering [23]. The *dashed line* represents the log-normal distribution approximated to the low-size regime of $S(R)$

variation of the contrast fringes in the vicinity of the grain boundaries originates from local atomic displacements. Inside the grains, shear stresses seem mainly relaxed by twin formation. The double hatched areas indicate local changes of orientation, the arrows indicate the direction of internal stresses.

The distribution of crystal sizes in n-materials was determined by small angle neutron scattering (SANS) [23] and is presented in Fig. 5. For comparison a log-normal size distribution of spherical grains is indicated. Similar results were found for Pd and TiO_2 by means of small angle X-ray scattering and transmission electron microscopy [7], respectively. Assuming that the density deficit of n-Pd (70–90% of the bulk density) is homogeneously distributed along the grain boundaries with an average thickness of 1 nm and if we assume furthermore that the crystals have the bulk density, SANS yields an average grain boundary density of about 60% of the crystalline density. The average grain boundary thickness of n-Pd was determined to be about 1 nm by means of hydrogen solubility measurements [24].

3.7.4.2 Atomic Structure

In order to study the atomic structure of n-materials, X-ray diffraction and extended X-ray absorption fine structure (EXAFS) were applied. In X-ray diffraction [25], structural information can be deduced by comparing the experimentally observed interference function with computed interference functions based on different structural models of a n-material. For computation (summing up of scattering amplitudes) the n-material was modeled by a crystalline component consisting of a randomly oriented array of 6 nm iron crystals and a grain boundary component , the density, thickness and degree of order (averaged over all boundaries of a specimen) of which could be varied. If

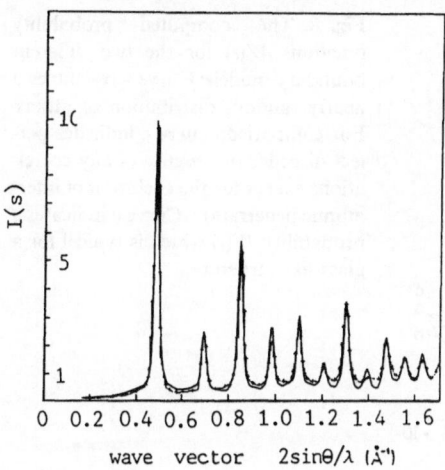

Fig. 6. Comparison of the measured and computed interference functions of nanocrystalline iron (6 nm crystal size) [25]. The model used for the computations assumes that the superposition of all different grain boundaries results in an "average boundary structure" which is characterized by a nearly random distribution of atoms (lack of short-range order). The average density in the boundaries was assumed to be 80% of the bulk density. A reduced density is a necessary prerequisite to get rid of the nearest neighbour correlations

the interfacial component was assumed to exhibit little short range order it was possible to reproduce the experimentally deduced interference function (Fig. 6). This was not the case for an interfacial component with a short-range ordered (glassy) structure (Fig. 7). The probability function $W(r)$ (which expresses the probability of finding the centre of a second atom at a distance r away from a specified central atom) of the interfacial component deduced from the computations underlying Fig. 6 is presented in Fig. 8 in comparison with the same function for an interfacial component with glassy (curve b) disorder. Curve c shows a probability function representing the maximum degree of disorder. As may be seen, the probability function reproducing the experimental data best (Fig. 6) resembles closely an "average boundary structure" with little short-range

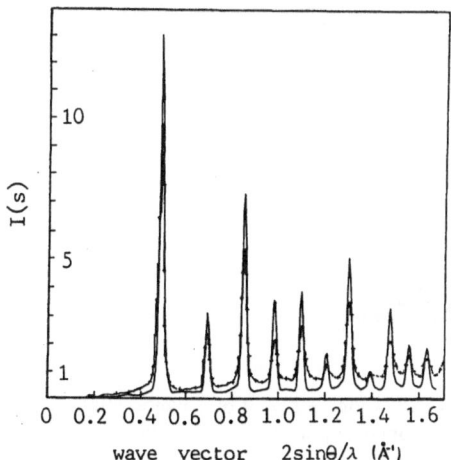

Fig. 7. Comparison of the measured and computed interference functions for nanocrystalline iron (6 nm crystal size). The model system assumed for the computations was an "average boundary structure" (see Fig. 7) which is similar to the short-range order measured in a $Fe_{80}B_{20}$ metallic glass. No density deficit was taken into account

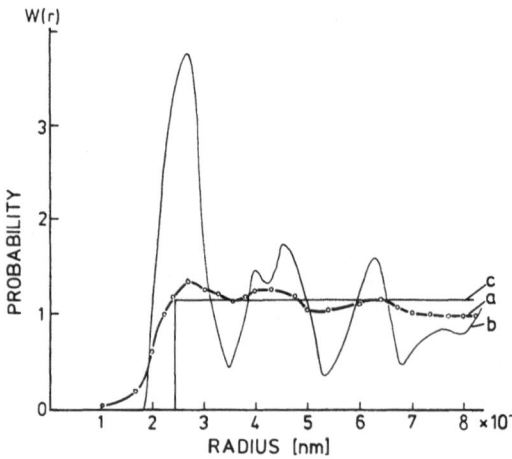

Fig. 8. The computed probability functions $W(r)$ for the two different boundary models: Curve a resembles a nearly random distribution of atoms. For comparison curve c indicates perfect disorder or absence of any correlations except for the exclusion of interatomic penetration. Curve b indicates a probability $W(r)$ which is typical for a glass-like structure

order suggesting that the individual grain boundaries are characterized by a variety of different structures the superposition of which can drastically reduce short-range order.

The oscillation of the extended X-ray absorption fine structures (EXAFS) yield information about the local atomic arrangement adjacent to the absorbing atom. EXAFS measurements [26] were carried out for nanocrystalline copper (10 nm) and conventional crystalline copper by normalizing both absorption spectra to the same number of absorbing atoms. The reduction by about 30% of the EXAFS-oscillations χk^3 – and their Fourier transformed FT (χk^3) respectively – of n-Cu (10 nm) in comparison with coarse-grained Cu (Fig. 9) indicates

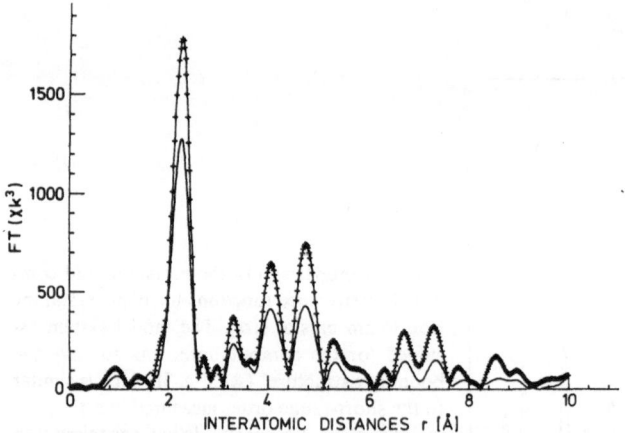

Fig. 9. Fourier transform of the weighted EXAFS oscillations $FT(\chi k^3)$ of a nanocrystalline Cu sample (10 nm, –) and the same function for crystalline copper (1 μm, + + + +) [26]. Both curves are normalized to the same number density of absorbing atoms

that the local atomic arrangements around absorbing atoms are disordered, in the sense that the number of nearest neighbour atoms is reduced and the atomic distance correlation is characterized by a wide spectrum of interatomic spacings. Hence the reduction of the oscillation scales with the reduction in crystal size, it may be suggested that this effect is due to the grain boundaries. In order to determine the effect of the grain boundaries on the intensity of the EXAFS-oscillations, the grain boundaries of a given n-material should be decorated (e.g. by grain boundary-diffusion) with chemically different atoms but similar in size. By measuring the EXAFS-spectra at the absorption edge of the decorating atoms it should be possible to probe the atomic arrangement in the grain boundaries. EXAFS experiments on n-Cu samples decorated with Bi or Co are in progress.

The spectroscopic methods discussed in the following were applied to get additional information on an atomic scale and in order to check whether or not the spectroscopic results agree with the structural concepts, as delineated by the methods based on diffraction.

The Mössbauer spectrum [27] of n-iron (Fig. 10) consists of two subspectra, one of which corresponds to crystalline α-iron. The second which is attributed to the interfacial component exhibits an enhanced hyperfine magnetic field H, a larger linewidth and an increased isomer shift IS. The enhancement of IS and H reflects a reduction of the electronic and atomic density in the interfacial component, in agreement with the X-ray results and high-pressure Mössbauer experiments. Compression of iron is known to result in a reduction of IS and H (*Pound* et al. 1961). If the temperature is increased, the larger value of H of the subspectrum 2 in Fig. 10 is followed by a faster decrease indicating a lower Curie temperature of the interfacial component in comparison to the α-iron lattice.

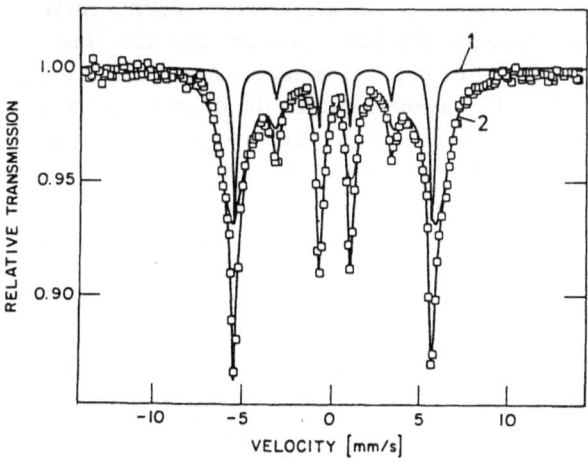

Fig. 10. Mössbauer spectrum of a nanocrystalline iron sample, measured at 77 K [27]. Two subspectra (*1*, sharp lines; *2*, broad lines) were used to fit the experimental data (*squares*)

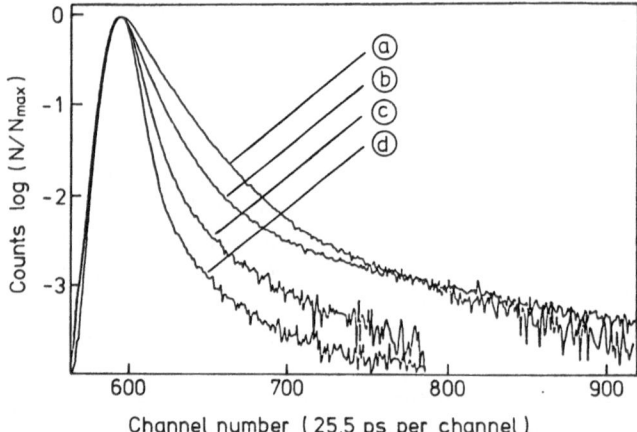

Fig. 11. Positron lifetime spectra on (a) uncompacted Fe powder, (b) a nanocrystalline Fe specimen (6 nm), (c) the amorphous $Fe_{85.2}B_{14.8}$ alloy, and (d) polycrystalline bulk iron, with the background of the spectra subtracted [28]

Similar behavior is known from observations on thin films. With increasing temperature, the relative intensity of subspectrum 2 decreased faster than the relative intensity of subspectrum 1, suggesting a smaller Debye–Waller factor for the interfacial component. In fact, a Debye temperature of 345 K was deduced for the n-iron which is to be compared with the bulk value of 467 K.

Positron lifetime spectroscopy [28] seems suitable for studying n-substances as it yields information about lattice defects as well as structural fluctuations in disordered materials in terms of the free volume associated with them. The different positron lifetime curves observed (Fig. 11) for crystalline, glassy and n-iron suggest three different atomic structures. The fact that the lifetime curve of the n-material is intermediate between the loose powder and the glassy material evidences a structure in which free volumes of larger sizes than in the glassy state are present. The same result was obtained for the other substances (palladium and copper) studied by positron annihilation. A compilation of positron lifetimes from studies on different n-materials is given in [29].

3.7.5 Properties

If the microstructure and the atomic structure of n-materials differ from the structures of glasses and crystals, the structure-dependent properties of n-materials are expected to deviate from the properties of the chemically identical substances in the glassy or crystalline state. In the following we shall present a selection of properties to give an overview on the different areas investigated so far.

3.7.5.1 Self-Diffusion

Measurements ([30], [31]) of the self-diffusivity D (using a ^{67}Cu tracer) in n-copper (8 nm crystal size, temperature range 293–393 K) revealed an enhancement of the self-diffusivity by a factor of about 10^{19} in comparison to lattice diffusion. This remarkable enhancement may be understood in terms of the high boundary density which provides a connective network of short-circuit diffusion paths. An enhancement of D in comparison to boundary diffusion by a factor of about 100 was noticed. The activation energy for grain boundary diffusion in n-Cu was found to be reduced by a factor of 1.7 in comparison to boundary diffusion in conventional polycrystalline materials. The differences may be explained to result from rapid diffusion along the connective network of boundary triple junctions, a different boundary structure in conventional polycrystals in comparison to nanocrystalline materials or surface diffusivity through internal surfaces.

3.7.5.2 Solute-Diffusion

The substitutional diffusion of silver in n-Cu [32] was investigated using electron beam microanalysis to determine the silver-concentration profile at various temperatures between 303 and 373 K in a plate shaped n-specimen the flat surface of which was coated by a thin Ag film. The activation enthalpies of 0.39 eV below 343 K and 0.63 eV above 353 K as given in Fig. 13 are to be compared with an activation enthalpy of 0.73 eV for grain boundary diffusion of

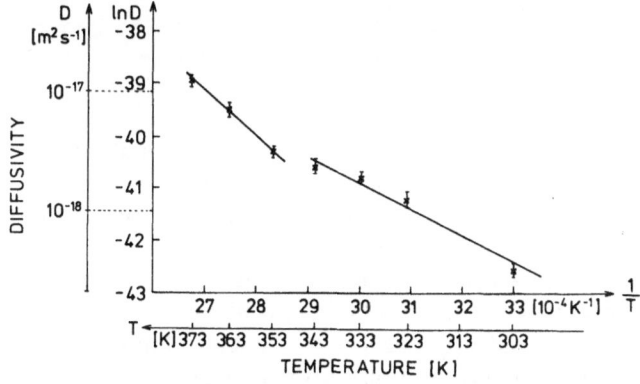

Fig. 12. Temperature dependence of the diffusion coefficients of Ag in nanocrystalline Cu. The vertical lines indicate the error bars of the measurements. The activation enthalpies and the pre-exponentials are 0.39 ± 0.1 eV and $1 \times 10^{-12} \pm 10^{-2}$ m^2/s, respectively, for temperatures below 343 K and 0.63 ± 0.1 eV and $3.8 \times 10^{-8} \pm 10^{-2}$ m^2/s, respectively, for temperatures above 353 K [32]

Ag in polycrystalline silver and 0.3–0.5 eV for Ag diffusion on different Cu surfaces. The two different slopes in the Arrhenius Plots (Fig. 12) are indicative of a reversible change-over between two different grain boundary diffusion mechanisms dominating at lower and higher temperatures, respectively.

For hydrogen as an interstitial solute atom, the diffusivity in n-Pd was noticed to become concentration dependent [33]. This behaviour was interpreted in terms of traps with different energies distributed along the network of grain boundaries in a n-material. The hydrogen diffuses by gradually filling first the deepest traps and subsequently the shallower ones. In other words, if the hydrogen concentration is low it diffuses slowly because the hydrogen atoms move between adjacent deep traps. However, in specimens with higher hydrogen contents, diffusion involves predominantly the motion of hydrogen between shallow traps resulting in an increase of the mobility with enhanced hydrogen concentrations yielding a diffusion coefficient exceeding that for diffusion in a single-crystal.

3.7.5.3   Enhanced Solute Solubility

The miscibility of a solute A is controlled by the chemical potential μ_A of A in a given solvent B. If the atomic structure of B is changed, the chemical potential and hence the solubility of A in B may be enhanced or reduced. As a consequence, the solute solubility of n-materials is expected to be different from that of single crystals or glasses of the same chemical composition. Two examples of this effect have been reported up to now: the enhanced solubility of hydrogen in n-palladium and the increase of the solubility of bismuth in copper from $< 10^{-4}\%$ (lattice solubility) to about 4% (solubility in n-copper) [31].

3.7.5.4   Specific Heat

The enhancement of the specific heat (Δc_p) of n-palladium (6 nm crystal size) in comparison to polycrystalline palladium is shown in Fig. 13 [34]. The enhancement varies between 29% (at 150 K) and 53% (at 300 K). For comparison, the c_p values of the polycrystalline palladium and glassy state ($Pd_{72}Si_{18}Fe_{10}$) differ by about 8% (only 4% of which was found to originate from glassy disorder). As palladium is paramagnetic, the enhanced c_p is likely to result from vibrational and/or configurational entropy effects (e.g., related to lattice vibration and/or variation of the defect concentration). If the enhancement of c_p is primarily due to the grain-boundary component, grain growth should reduce the specific heat of the n-materials. This was, in fact, observed when n-Pd samples were annealed at 750 K. In the case of 8 nm copper an enhancement of 8% at 150 K and 11% at 300 K was found. Similar results were found for Ru after 32 h of high energy ball milling [35].

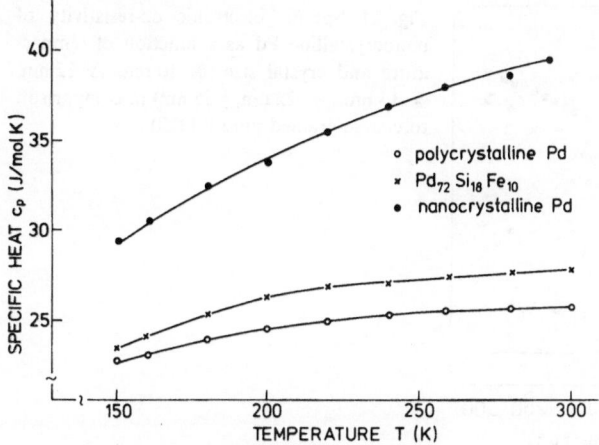

Fig. 13. Specific heat (c_p) of polycrystalline Pd in comparison with nanocrystalline Pd (10 nm) and a corresponding $Pd_{72}Si_{18}Fe_{10}$ metallic glass [11]

3.7.5.5   Entropy

The measured excess specific heat (Δc_p) may be used to compute the excess entropy (ΔS) of n-palladium relative to a palladium single crystal by integrating over $\Delta c_p/T$ [36]. At 300 K, ΔS of n-Pd is about 2.3×10^{-23} J/at. K ($\approx 1.7 k_B$). For most metals the entropy of melting is about $1 k_B$. In order to elucidate the degree of metastability of n-materials, electrochemical measurements yielding direct information on the Gibb's free energy are in progress. If the Gibb's free energy and the excess specific heat are known in a temperature-range where structural relaxations are absent, the excess energy and the configurational entropy can be calculated. A theoretical interpretation of the large enthalpies and entropies found in n-materials has been attempted in [37].

3.7.5.6   Electrical Resistivity

Studies of the electric resistivity of n-Cu, Pd and Fe samples have recently been performed ([38], [39]). Figure 14 shows the electric dc-resistivity of n-Pd specimens as a function of temperature for various grain sizes. Similar results were found for n-Cu and Fe. By deducing the temperature coefficient of the electric resistivity from Fig. 14 it was found that a reduction in the crystal size yields a decrease in the temperature coefficient. The observed effects may be understood by analogy to the treatment of the electric dc-resistivity of thin metallic films [40], [41]. If the crystal size is smaller than the electron mean free path, the resistivity is primarily governed by grain boundary scattering which was modeled in terms of the reflection and/or transmission of the conduction

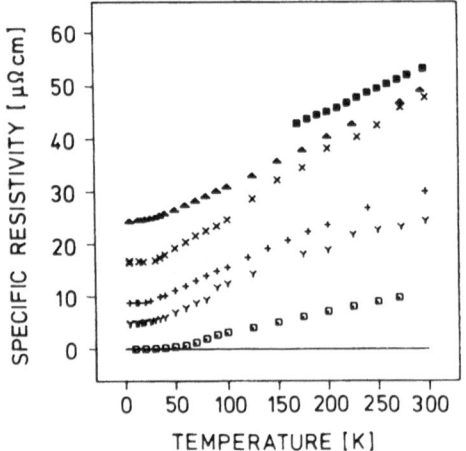

Fig. 14. Specific electronic dc-resistivity of nanocrystalline Pd as a function of temperature and crystal size (■ 10 nm, △ 12 nm, × 13 nm, + 22 nm, $\curlyvee$ 25 nm) in comparison to coarse-grained pure Pd (□)

electrons through the boundaries. Thus, an enhanced density of boundaries causes a decrease of the conductivity as well as of the temperature coefficient. For grain sizes larger than the electron mean free path, electron scattering inside the crystallites is the dominant scattering mechanism. In this scattering mode measured on annealed samples, enhanced values of the resistivity were observed in comparison to the resistivity of the bulk suggesting that a large density of point and/or line defects is incorporated in the crystal lattices.

3.7.5.7 Magnetic Properties

Measurements of the saturation magnetization (M_s) of n-iron (6 nm crystal size) revealed a reduction of M_s from 220 emu g^{-1} (α-iron) to about 130 emu g^{-1} [5]. For comparison, in metallic iron glasses (extrapolated to pure iron), M_s is only reduced to about 215 emu g^{-1}. A study of magnetic phase transitions has been carried out for n-erbium (10–70 nm grain size) [42]. The three normally observed magnetic transitions of erbium vanish and a new low-temperature transition to superparamagnetic behavior arises. On the other hand, for n-erbium with larger grain diameters, the normal magnetic transitions reappear but at different temperatures, while the low-temperature superparamagnetic behavior is retained. A reduction of the Curie temperature (T_c) of Ni by about 40 °C has been reported if the crystal size was lowered to about 70 nm. The observed reduction in T_c was attributed to a lower T_c of the grain boundary regions [43], [44]. The magnetic microstructure of n-Fe was found to differ from the well known microstructure of crystalline and amorphous iron which is characterized by ferromagnetic domains separated by domain walls. In n-iron no domain structure was revealed by applying the Bitter technique, Kerr microscopy and Lorentz-electron microscopy [45]. In order to elucidate the magnetic microstructure, neutron scattering and atomic force microscopy experiments are in progress. In a recent study on n-FeF$_2$ [46], it was found that the width of the paramagnetic to antiferromagnetic phase transition ($T_N = 77$ K) is spread over 12 K whereas conventional polycrystalline FeF$_2$ shows a width of the transition region of 2–3 K.

3.7.5.8 Mechanical Properties

The elastic constants of n-Pd were determined [47], [48] using different methods (elastic bending of a thin plate, measurements of the sound velocity, torsion pendulum) to be reduced by about 20% in comparison to polycrystalline

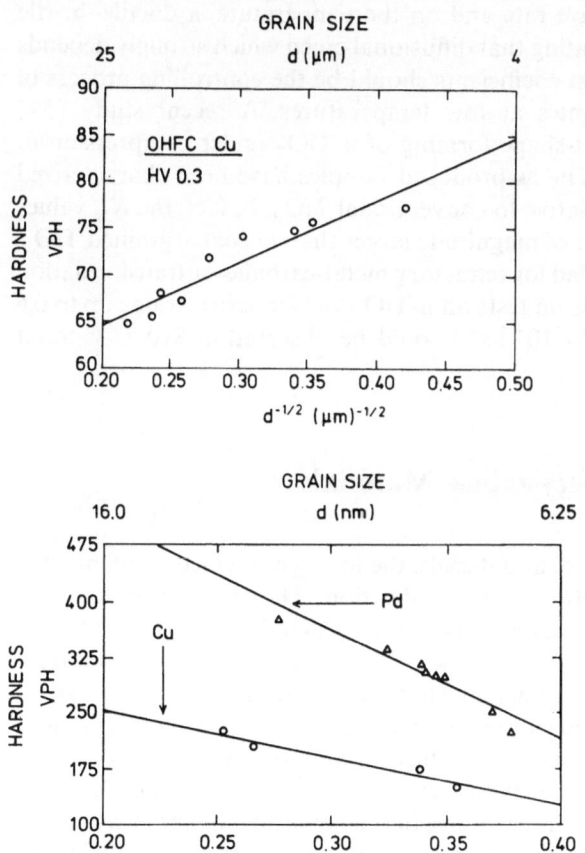

Fig. 15. *Top,* Variation of hardness of nanocrystalline Cu and Pd with a negative Hall–Petch slope as a function of grain size. *Bottom,* For comparison the grain size dependence of the hardness for coarse grained copper is shown. The positive slope of the data is typical for materials with grain sizes greater than about 1 μm

samples. The reduction in the elastic moduli is presumably due to the reduced density and the change of interatomic potentials in the boundary regions.

The plastic deformation [49], [50] of Pd and Cu has been studied in the grain size region from 25 µm to 6 nm by indentation measurements (Fig. 15). The results obtained show a deviation from the Hall–Petch relationship below a critical grain size, d_{crit}, whereas for grain sizes $d > d_{crit}$ the stress required to initiate plastic flow increases inversely proportional to $d^{1/2}$ as was predicted by the Hall–Petch equation. For $d < d_{crit}$ the resistance against plastic flow becomes smaller as the grain size is reduced. This effect was interpreted in terms of grain boundary sliding and diffusional creep. Stimulated by these experiments, it has been observed that conventionally brittle ceramics (TiO_2, CaF_2) become ductile, permitting large plastic deformations at low temperatures, if a ceramic material is generated in the nanocrystalline form [51]. Depending on the deformation rate and on the temperature, a ductile–brittle transition was observed indicating that diffusional creep which strongly depends on grain size and the diffusion coefficients should be the controlling process of plastic deformation of ceramics at low temperatures. A recent study [52] revealed the possibility of net-shape forming of n-TiO_2 under compression at temperatures below 800 °C. The as-produced samples have been characterized by an enhanced toughness relative to conventional TiO_2. In fact, the K_{Ic} values obtained were nearly an order of magnitude larger than in coarse grained TiO_2. Similar K_{Ic} values were reported for refractory metal-carbides or transformation toughened zirconia. Compression tests on n-TiO_2 yielded large strains up to 0.6 and strain rates as high as 8×10^{-5} s^{-1} could be observed at 810 °C without fracturing [53].

3.7.6   Multiphase Nanocrystalline Materials

In order to generate multiphase n-materials, the inert-gas condensation method may be used with a straightforward modification. The single evaporator as shown in Fig. 3 is replaced by several ones, from which different materials A, B and so on are evaporated. Under suitable preparation conditions a homogeneous mixture of crystals with different chemical compositions (A, B) can be collected on the cold finger and subsequently compacted into a "nanocrystalline alloy". Obviously, in these materials, alloying occurs on a scale of a few nanometers and hence deviates from a homogeneous solid solution. Nevertheless, these alloys may be of interest because chemically different atoms or molecules of the various crystallites are mixed on an atomic scale in the interfaces. Thus, nanocrystalline alloys allow the mixing of atoms or molecules irrespective of their bulk miscibility. In fact, in a recent study [54] of nanocrystalline Cu–Fe and Ag–Fe alloys such mixing effects were observed. The Ag–Fe alloy serves as model system because Ag–Fe are immiscible in the solid as well as in the molten state close to the melting point. The Mössbauer spectrum of a

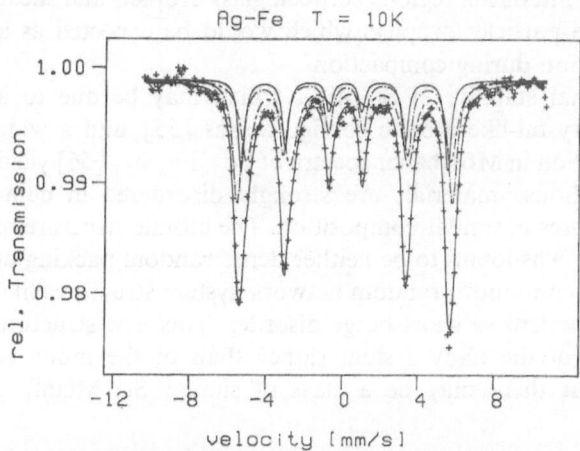

Fig. 16. Mössbauer spectrum of a nanocrystalline Fe–Ag alloy (30 at. % Fe; crystal size 8 nm) recorded at 10 K [17]. The spectrum consists of the following three components: (i) α-Fe (–), (ii) Fe-atoms dissolved in Ag (–––) and (iii) Ag atoms dissolved in Fe (–·–)

Ag–Fe alloy (Fig. 16) with 30 at. % Fe consists of the following three components: α-Fe crystal lattices, Fe-atoms dissolved in Ag and Ag-atoms dissolved in Fe. The dissolution of Ag in Fe and vice versa was interpreted as follows. In the strained lattice regions in the vicinity of the Ag–Fe interphase boundaries the solubility of Fe in Ag and the one of Ag in Fe is proposed to be enhanced as was demonstrated for the strained regions in the vicinity of precipitates. Furthermore, the segregation of Ag and Fe atoms along the grain boundaries may also contribute to the alloying of the different atoms.

3.7.7   Prospects

At present the mechanical properties of nanocrystalline ceramics seem to be attractive for technological applications; provided the economical production of large amounts of n-ceramics (typically kg/day) will be developed in the future. Tailoring of microstructure and chemistry by nanocrystalline alloys seems to be an upcoming promising discipline for the discovery of novel properties.

A new emerging field entitled "nanoamorphous" materials or nanometer-sized glasses deals with a novel mode of preparation of bulk amorphous solids by a procedure similar to the one described in Fig. 3. If the material evaporated has the ability to form a glass, the evaporation and subsequent rapid cooling in the He atmosphere may result in the formation of nanometer-sized glass droplets rather than crystallites. The compaction of the configurationally frozen and randomly oriented glass droplets may result in solids containing a high density of defects, due to the interfacial regions between glass droplets and shear bands in the interior of the particles/droplets which would be expected as a result of extensive deformation during compaction.

In fact, increased thermal stability of $Si_{75}Au_{25}$ which may be due to a lowered concentration of crystal-like atomic configurations [55], and a wide quadruple-splitting distribution in Mössbauer spectra of $Pd_{70}Fe_3Si_{27}$ [56] yield evidence that "nanoamorphous" materials are strongly disordered in comparison to conventional glasses of similar composition. The atomic structure of "nanoamorphous" $Si_{75}Au_{25}$ was found to be neither dense random packing of hard spheres structures nor continuous random network system structures like and to be characterized by extensive short-range disorder. This new structure seems to be characteristic of the alloy system rather than of the mode of preparation, suggesting that there may be a class of similar $Si_x \, Metal_{1-x}$ ($x \approx 0.25$) glasses.

Acknowledgements. The financial support of the Alcoa Foundation, the BMFT (Contract No. 523-4003 03 M 23-4) the DFG (G.W. Leibniz Programm) and the Fonds der Chemischen Industrie is gratefully acknowledged. The author would like to thank his colleagues at the University of Saarbrücken for stimulating discussions. The helpful suggestions of Prof. Dr. H. Gleiter, Drs. U. Herr, Th. Haubold, J. Weißmüller and A. Tschöpe are appreciated.

References

1. K.L. Merkle, J.F. Reddy, C.L. Wiley, D.J. Smith: J. de Phys. C5-**49**, 251 (1988)
2. H. Gleiter: Proc. Second Risø Int. Symp. on Metallurgy and Materials Science, ed. by N. Hansen, T. Leffers, H. Loholt, Risø National Laboratory, Roskilde, 1981, p. 15
3. H. Gleiter: Mat. Sci. and Eng. **52**, 91 (1982)
4. A. Holz, A. Patashinski: unpublished
5. R. Birringer, U. Herr, H. Gleiter: Suppl. Trans. Jpn. Inst. Met. **27**, 43 (1986)
6. H. Gleiter: *Proc. Ninth Int. Vac. Congress and Fifth Int. Conf. Sol. Surf.*, ed. by J.L. Segovia, Association Es-Espanola de Vacio Y Sus Aplicaiones, Madrid, 1983, p. 397
7. R.W. Siegel, H. Hahn: *Current Trends in Physics of Materials*, ed. by M. Yusouff, World Scientific Publ. Co., Singapore, 1987, p. 403
8. R. Birringer, H. Gleiter: *Advances in Materials Science, Encyclopedia of Mat. Sci. and Eng.*, ed. by R.W. Cahn, Pergamon Press, Oxford 1988, p. 339
9. H.E. Schaefer, R. Würschum, R. Birringer: J. Less-Common Metals **140**, 161 (1988)
10. R. Birringer: Mat. Sci. and Eng. A **117**, 33 (1989)
11. H. Gleiter: Europhysics News **20**, 130 (1989)
12. H. Gleiter: *Progress in Materials Science* 33(4), 223 Pergamon Press, Oxford (1989)
13. R. Birringer, H. Gleiter, H.-P. Klein, P. Marquardt: Phys. Lett., **102A**, 365 (1984)
14. W. Romanowski, S. Engels: *Hochdisperse Metalle*, Verlag Chemie, Weinheim 1982, p. 27
15. J.A.A.J. Perenboom, P. Wyder: Physics Reports **78**, No. 2, 173 (1981)
16. R.P. Andres, R.S. Averback, W.L. Brown, L.E. Brus, W.A. Goddard, A. Kaldor, S.G. Louie, M. Moscovits, P.S. Peercy, S.H. Riley, R.W. Siegel, F. Spaepen, Y. Wang: J. Mat. Res. **4**, 740 (1989)
17. E. Hellstern, H.J. Fecht, Z. Fu, W.L. Johnson: J. Appl. Phys. **65**, 305 (1989)

18. W. Schlump, H. Grewe: in *Proc. DGM Conf. "New Materials by Mechanical Alloying Techniques"*, ed. by E. Arzt, L. Schultz, Calw-Hirsau, Oct. 1988, DGM Verlag, Oberursel Germany (1989)
19. S.K. Menon, T.R. Jervis, M. Nastasi: Mat. Res. Soc. Symp. Proc. **80**, 369 (1987)
20. K.N. Tu, J.W. Mayer, L.C. Feldman: Thin Film Science, MacMillan Publ. Co. N.Y. 1992
21. D. Turnbull: Met. Trans. B Vol. **12 B**, 217 (1981)
22. W. Wunderlich, Y. Ishida, R. Maurer: Scripta Met. **24**, 403 (1990)
23. E. Jorra, H. Franz, J. Peisl, G. Wallner, W. Petry, R. Birringer, H. Gleiter, T. Haubold: Philos. Mag. **B 60**, 159 (1989)
24. T. Mütschele, R. Kirchheim: Scripta Met. **21**, 135 (1987)
25. X. Zhu, R. Birringer, U. Herr, H. Gleiter: Phys. Rev. B **35**, 9085 (1987)
26. T. Haubold, R. Birringer, B. Lengeler, H. Gleiter: Phys. Lett. A **135**, 461 (1989)
27. U. Herr, J. Jing, R. Birringer, U. Gonser, H. Gleiter: J. Appl. Phys. **50**, 472 (1987)
28. H.E. Schaefer, R. Würschum, M. Scheytt, R. Birringer, H. Gleiter: Mater. Sci. Forum, **15–18**, 955 (1987)
29. H.E. Schaefer, R. Würschum, R. Birringer, H. Gleiter: J. Less-Common Metals **140**, 161 (1988)
30. J. Horvath, R. Birringer, H. Gleiter: Solid State Comm. **62**, 391 (1987)
31. R. Birringer, H. Hahn, H. Höfler, J. Karch, H. Gleiter: Defect and Diffusion Forum **59**, 17 (1988)
32. S. Schumacher, R. Birringer, R. Strauss, H. Gleiter: Acta Metall. **37**, 2485 (1989)
33. R. Kirchheim, T. Mütschele, W. Kieninger, H. Gleiter, R. Birringer, T.D. Koblé: Mater. Sci. Eng. **99**, 457 (1988)
34. J. Rupp, R. Birringer: Phys. Rev. **B 36**, 7888 (1987)
35. E. Hellstern, H.J. Fecht, Z. Fu, W.L. Johnson: J. Appl. Phys. **65**, 305 (1989)
36. D. Korn, A. Morsch, R. Birringer, W. Arnold, H. Gleiter: J. de Physique **49**, C5-769 (1988)
37. H.J. Fecht: Acta Met. et Mat: **38**, 1927 (1990)
38. M. Kraack: Diploma Theses, University of the Saarland (1990)
39. H.J. Weber: Diploma Theses, University of Stuttgart (1989)
40. G. Reiss, J. Vancea, H. Hoffmann: Phys. Rev. Letters **56**, 2100 (1986)
41. H. Hoffmann, P. Kücher: Thin Solid Films, **146**, 155 (1987)
42. J.A. Cowen, B. Stolzmann, R.A. Averback, H. Hahn: J. Appl. Phys. **61**, 3317 (1987)
43. R.Z. Valiev, R.R. Mulyukov, Kh. Ya. Mulyukov, V.I. Novikov, L.I. Trusov: Pisma v szurnal tekhnicheskoi fiziki **15**, 78 (1989)
44. R.Z. Valiev, NATO ASI, Series E, **233**, 303 (1992) Eds. M. Nastasi, D.M. Parkin and H. Gleiter: Kluwer Acad. Publ. London 1992
45. W. Wagner, W. Petry, A. Geibel and H. Gleiter: J. Mat. Res. **6**, 2305 (1991)
46. J. Jiang, S. Ramasamy, R. Birringer, U. Gonser, H. Gleiter: Sol. State Comm. **80**, 525 (1991)
47. D. Korn, A. Morsch, R. Birringer, W. Arnold, H. Gleiter: J. de Physique **49**, C5-769 (1988)
48. M. Weller, J. Diehl, A. Seeger, H.E. Schaefer: to be sub. to Phil. Mag. A
49. A.H. Chokhsi, A. Rosen, J. Karch, H. Gleiter: Scr. Metall. **23**, 1679 (1989)
50. J. Karch: Diploma Thesis, University of the Saarland (1987)
51. J. Karch, R. Birringer, H. Gleiter: Nature **330**, 556 (1987)
52. J. Karch, R. Birringer: Ceramics International **16**, 291 (1990)
53. H. Hahn, R.S. Averback: J. Am. Cer. Soc. **74**, 2918 (1991)
54. U. Herr, J. Jing, U. Gonser, H. Gleiter: Solid State Comm. **76**, 197 (1990)
55. J. Weissmüller, R. Birringer, H. Gleiter: Phys. Letters A **145**, 130 (1990)
56. J. Jing, A. Krämer, R. Birringer, H. Gleiter, U. Gonser: J. of Non-Cryst. Solids **113**, 167 (1989)
57. R.W. Siegel: Mat. Sci. and Eng. A **168**, 189 (1993)
58. A.L. Greer: NATO ASI Series E **233**, 53 (1992) Eds.: M. Nastasi D.M. Parkin and H. Gleiter: Kluwer Acad. Publ. London
59. H. Gleiter: ibid, **233**, 3 (1992)
60. R. Birringer: Proc. NATO-ASI: Nanophase Materials Ed.: C. Haddjipanayis, Kluwer Acad. Press. (in press)

Subject Index

absorption, of large clusters 321
  optical 292, 295
  surface 22
  time resolved 293, 299, 302
  X-ray 286
adsorbate uptake 223
adsorption 281
adsorption site 229
aerosol 271
Ag 277, 292, 300, 304, 367, 371
aggregation, stochastic model for 364
air quality 284
Al 110
amorphous carbon 363
applications 284
Ar 19, 170, 174, 198
Ar_3^+ 125
Au 214, 277
Auger effect 198, 376
autoionisation 156

B 248
band structure 365, 369
bimetallic clusters 295
binding energy, electrostatic 56
bonding, hydrogen 46, 71
Born–Haber cycle 96, 203, 374
breakdown graph 176

caging 201
catalysis 305
CF_3Cl 37
CF_3Cl/Ar_n 22
$CH_3F–HCl$ 39
CH_3F/Ar_n 38
CH_3OH 89, 90
charge transfer 296
charging, diffusion 274
  multiple 183, 277
  photoelectric 275
chemi-ionisation 159
chemical amplification 306

chemical properties 298
chemistry, adsorbate decomposition 235
  beam techniqe 246, 257
  ion-molecule 242
  ions 241
  isomers 256
  kinetics 225
  neutral metals 221
  thermodynamics 227
chromophore 19, 31
church windows 2, 321
cluster, alkali-halide 150
  compaction of 389
  semiconductor 312
  transition metal 221
cluster, formation 107
cluster, hydrogen bonded 51
  multiply charged 183, 193
  solvated metal 7, 77, 290
  supported 361
  trimethylamine 102
  van der Waals 154
cluster growth 364
cluster matter 342
cluster reaction, enthalpy changes 111, 114
cluster reactions, entropy change 82, 116
  thermodynamics 82
clusters in polymers, Nafion 297
CO_2 124, 171, 214
$CO_3^-(H_2O)_n$ 122
coalescence dynamics 293
collective excitation 323
collision induced dissociation 249
collision rate 108
colloid, metal 294
  optical absorption of 263
combustion process 284
complex 78
  hydrogen bonded 38
complex formation 37
computer simulation 28
Coulomb repulsion 277

Coulomb shift   373
cross section, ionisation   158
$Cs(H_2O)_n$ 14
$Cs(NH_3)$ 14
Cu_n 109, 380

detection, velocity modulation   48
detector, bolometer   19
    condensation nucleus counter   274
development, photographic   307
    selective   306
dielectric function, nonlocal   192, 318, 323
dielectric screening   16
diffusion battery   273
dissociative electron attachment   142
Drude electron   331

electromagnetic response   270, 271
electron, trapped   135
electron affinity   136
    adiabatic   145
    dielectric sphere   146
    vertical   15, 145
electron diffraction   21
electron microscopy   389
electron storage   305
electron transfer   300
electron transfer process   97
electron–hole interaction   313
ellipsoidal particle   332
embedding medium   332
energy, appearence   185
    binding   176
    core level   367, 369
    Coulomb   370
    ionization   303
    surface   191
equivalent-cores approximation   374
evaporation, monomer   167
evaporative ensemble   81, 87, 90, 165, 168
exciton, Wannier   312
exciton annihilation   199

face-centered cubic   40
ferromagnetic properties   296
filling factor   264, 267
fission   191, 206
fission barrier   208
flow tube experiments   244
fragmentation, intramolecular   186
Franck–Condon principle   146, 155

Ge   203
grain boundary   384
Gspann parameter   168

H_2O 15, 56, 135
    protonated 21-mer   87
H_3O^+ 56
$H_3O^+(H_2O)_m$ 62
hydrocarbon, polycyclic aromatic   284

icosaeder   24, 36
interaction, environmental   290
interaction potential   28
interband transition   334
interfacial area   387
intermolecular energy transfer   164
internal rotation   61, 68, 69, 70
ion, transition metal   114
ion cyclotron resonance   245, 254
ion trap, octupol   52
ionisation, electron impact   154
    mechanisms   155
    Penning   97, 159
    resonant two photon   16
ionisation energy   10
    chemical measurement of   255
isomerisation, rate   30

jellium   110

kinetic energy release   99, 101, 104, 166, 170,
    213, 215
Kr   126
Kramers Kronig analysis   327

Landau damping   339
latent image   306
Lewis base   11
lifetime, radiative   316
ligand   112, 297

magic number   111, 172, 196
magnetic partial wave   334
MAMICS   49, 55, 73
matrix, annealed   40
matrix effect   269
metal-insulator transition   378
metastable dissociation   104, 154, 162, 165
mobility   36
mobility analyzer   273
model, liquid-drop   118

$Na(H_2O)_n$ 7
$Na(NH_3)_n$ 7

nanocrystalline material 384
nanocrystals, electrical resistivity 398
  magnetic properties 400
  mechanical properties 400
  specific heat 297
Ne 175
negatively charged cluster, He, Xe 138, 141
negatively charged clusters, H_2O, NH_3 140, 141
neutralization chamber 282
NH_3, negatively charged 15
NH_3 55, 56, 87, 88, 95, 102
NH_4^+ 55
$NH_4^+(NH_3)_3H_2O$ 66
$NH_4^+(NH_3)_n$ 58
non-retarded limit 263
nucleation 77, 117
  classical theory 118

odd–even effect 303
optical scattering 345
Oseen transition 343
oxido-reduction 298

paraxylene, Rydberg state 98
Pd 366, 368
phonon polariton band 322
photo-electron, angular distribution 149
photo-fragmentation 20
photoelectric yield 276, 279
photoemission, from rare gas matrix 376
PIE spectrum, $Na(H_2O)$ 10
PIPECO 160
plasmon, longitudinal 327
  multipolar 322
  surface 265, 269, 322, 323
plasmon polariton 324, 335, 346
polarisability, average 21
polariton 318, 326
potential, anisotropic 28
  Lennard–Jones 29
  parameters 29
  Rittner 151
predissociation 163
  vibrational 49
preexisting traps 143
proton affinity 102
proton transfer 46
pulse radiolysis 302

quantum clusters 41
quantum dots 312
quantum size effect 314
quasistatic approximation 325, 330

radiation damping 324
radiolysis, water 291
rate coefficient 86
reaction, charge-exchange 186
  S_N2 105
reaction, ligand exchange 105
reaction, dehydration 91
  internal 154, 159
  ion-molecule 87, 104, 160, 163, 242
  sequential 253, 254
redox properties 299
relaxation 332
  final-state 369
retardation 334
rotational tunneling 126

scaling law 191
segregation 401
self-diffusion 395
SF_6/Ar_n 21
SiF_4/Ar_n 25
SIFT 84, 244
simulation, molecular dynamics 28, 63
  Monte-Carlo 63
size effect 264
  extrinsic 323
  intrinsic 337
Sn 368, 373
solvated, fully 23
  loosely 23
solvated alkali, excited state 16
solvated electron 7, 134
solvation 47, 119, 137, 160
  structural change 101
solvation energy 313
solvation shell 61, 116
soot 285
source, ball milling 387
  corona discharge 51
  flowing afterglow 84
  gas aggregation 265, 387
  high pressure 84
  ion 51
  liquid metal 188
  pick-up 8, 20
spectroscopy, Auger 376
  consequence 48, 49, 53
  EXAFS 377, 391
  HEELS 322
  infrared 19, 55
  liquid phase 73, 394
  matrix isolation 19
  multiphoton 54
  photodetachment 145

spectroscopy, Auger (Contd.)
  photodissociation 122
  photoelectron 362
  positron life time 394
  vibrational 55, 56, 65
spectrum, coexpansion 26
  Fermi edge 371
  pick-up 26
spill-out effect 325, 340
stability gap 189
stretching mode, Na ··· NH_3 17
structure 231
structure effect 250
superfluidity, finite size 42
supported clusters, preparation of 305, 362
surface polarization 313
surface state 149
surface tension 207
surfactant 294, 297
synthesis of clusters 291

temperature 36, 52

theory, effective medium 325, 343
  extended Mie 344
  Gans–Happel 327
  Lorenz–Mie 314, 328, 334
  QET/RRK 99, 101, 161
  quantum path integral 148
thermionic emission 184
thermochemical properties 111
TOF spectrometer 79
transition, gas to liquid 71
transiton state 212
triple junction boundary 396

unimolecular dissociation 87, 99, 154

Voronoi tessellation 364

work function 276, 280

Xe_{55}^+ 210
Xe_n^- 141

Subject Index of Volume 52

activation barrier 306
adjacency matrix 187
alkali cluster 255
alkali halide cluster 357
aluminum cluster 83, 288
analytical cluster model 114
  quantum mechanical 116
  topological 116, 126
angular momentum 169
association reaction 174
Auger-process 88
autoionization resonance 89, 90, 94, 99, 100
autoionization spectrum 87

Balian and Bloch 159
band broadening 98
band formation 99, 297
band gap 95, 102, 103
benzene cluster 166, 175, 402
bond length 101, 310
bonding, covalent 91, 92
  hydrogen 396
  ionic 133
  metallic 86, 92, 105, 109
  van der Waals 86, 99, 105, 109, 396, 398
Born–Oppenheimer approximation 16
boron cluster 288
Buckminsterfullerene 11, 166, 333

caloric curve 200
carbazole 195
carbon cluster 331
catchment area 364
cesium oxide cluster 369
chain/ring structure 400
charged cluster 7, 8
charge fluctuation 105
charge localization 380, 390
chemical bonding 316
chemical potential 133
chemical probe 327
chemical reactivity 338

chemisorption 306
chemistry 10, 305
chromophore 245, 390
classical nucleation theory 213
classification of clusters 2
closed orbit 155
cluster beam 6
cluster formation 213
cluster geometry 38, 41
cluster growing 378
cluster, homogeneous 3
cluster ions, re-neutralised 237
CO-stretching mode 409, 410
coexistance region 202
cohesive force 5
collective resonance 285, 302
collision induced dissociation 182, 289, 310
commercial applications 328
compressibility 131
configuration interaction (CI) 17, 69
conformer, linear/cyclic 407
coordination number 57, 128
correlation diagram 192
correlation function 187, 188
Coulomb explosion imaging (CEI) 338

decay channel 7
decay rate 405
defect 192, 204
delocalization 91
density functional theory (DFT) 67, 70, 110
density of states (DOS) 102, 103, 104, 135,
   152, 165
deposited clusters 327
detection problems 229
diffusion constant 200
dimer ion 380, 390
divalent metal cluster 86, 110
divergent nozzle 212
donor state 371
doubly charged cluster 267
drift cell 249

effective-core potential   20
electron affinity   34, 36, 325, 336
electron correlation   21, 107
electron delocalization   105
electron diffraction   196, 231, 328
electron gas   80
electron localization   105
electron shell model   23
electronic density   34
electronic excitation   375
electronic structure   90
emission, infrared   181
energy, cohesive   92, 103, 271, 305
   dissociation   162, 266, 317, 320, 345
   electrostatic   358, 397
   exchange-correlation   71
   ionization   34, 87, 90, 93, 97, 110, 255,
      290, 317, 368
   promotion   317
entropy   192, 365
etching reaction   307
evaporation   175, 270, 293
evaporative cooling   179, 190
evaporative ensemble   179, 217, 219, 309
EXAFS   328
excited electronic state   183, 184
excited states, alkali   37
exciton   94, 95, 99, 376, 383
exciton splitting   402

fission   293
   nuclear   162
fluctuating state   202
fluorescence   137
fragmentation   196, 220, 235, 387, 397
fragmentation channel   34
Franck–Condon principle   375
free energy   192
freezing   202
fullerene   333, 339
   metal containing   348
fullerite   349

GaAs cluster   310
gallium clusters   161, 309
gap, HOMO-LUMO   342
geometry   21
giant atom   141
giant resonance   282, 352
Gutzwiller   159

Hartree–Fock   16, 69
   time dependent   29
heat capacity   180

helium   142
   three   142, 145
Hellman–Feynman theorem   21
Hubbard hamiltonian   91, 106, 107
Hückel-model   54
Hund's rule   149
hydrazine cluster   411
hydrogen cyanide   406

icosahedron   11, 198, 340, 378, 385
indium cluster   309
infinite linear chain   136
InP cluster   311
interband transition   324
internal energy distribution   179
interpolation formula   117
   asymptotic behaviour   133
   in other fields   135
   large N dependence   134
   theoretical background   133
   verification   119
ion cyclotron resonance (ICR)   175
ion-molecule reaction   172
ionization chromophore   386
ionization, electron impact   310
   resonant two photon   301
isomer   3, 184, 249, 319, 362, 407, 414
isomerization   167

Jahn–Teller effect   23, 56, 151, 324
jellium   10, 68, 82, 294

Kepler problem   155
kinetic energy distribution   175
kinetic shift   304
Knudsen cell   207, 212, 320

LCAO approximation   19
lifetime   304
Lindemann criterion   200
linear response theory   24
line shift   404, 409
linewidth   404
liquid drop model   154, 258, 274
liquid-like cluster   187, 193
lithium clusters   161
local density-functional theory   80, 296
local spin density (LSD)   72

magic fragment   293
magic number   8, 10, 23, 130, 141, 143, 149,
   154, 289, 317, 331, 385
magnetic bottle spectrometer   247, 298
magnetic moment   300

magnetic properties 249, 299, 322
magnetism 8
mass-spectrometer 239, 304
mean field 143, 146
melting 202, 203
metal clusters 119–126, 145, 249
metal-insulator transition 86
metallic properties 102
methanol 397, 409
microcrystal 367
Mie resonance 279
mode selective 407, 415
model, compound nucleus 168
  Debye 183, 191
  Einstein 191
  free electron 255, 293
  Gartenhaus–Schwartz 191
molecular cluster 128
molecular dynamics (MD) 6, 67, 73, 196
momentum gap law 405
momentum transfer 408
Monte Carlo (MC) 9, 196
Mott criterion 91
MRD-CI 21
multi-photon process 182, 321

neutral beam, mass-selected 232
nonrigidity parameter 191
non stationary property 136
normal mode 406
Nosé method 196
nuclear deformation 150
nuclear shell structure 146
nucleus 141, 142, 145

optical response 278
orbit, triangular/square 158, 159
organic superconductor 352
oxidation state 317

partition function 190
phase space theory 168
phosphorous cluster 79
photodepletion 37
photodetachment 37, 262
photodissociation 293
photoemission 8, 376
photoionization 88, 295, 318
plasmon 25, 26
polarizability 27, 34
  dynamic 24
  static 36, 281, 300
potential, Born-Mayer 9, 358
  harmonic oscillator 147, 151, 155, 307

intermolecular pair 128
  Lennard–Jones 9, 374, 377
  Nilsson 146
  square well 147
  Woods–Saxon 146
potential energy surface 131, 172, 305
pseudo Jahn–Teller effect 23
pump-probe experiment 249

quantum liquid 142
quantum shell 141
quasi atom 141

radiative emission 174
random phase approximation (RPA) 29
rare gas cluster 128, 141, 374
REMPI 244
rotation 153
RRK, RRKM model 162, 164, 217, 304, 322

saddle-point approximation 108, 109
scaling law, cluster formation 220
  electron affinity 299
  ionization energy 258. 261, 295, 318
scattering, inelastic electron 384
screening 91
selenium cluster 74, 77
self-trapping 375
semiconductor cluster 310
shape isomer 151
shell closing 33
shell model 9, 317
shell structure 141, 156, 157
silicon cluster 53,55
simulated annealing 68, 73
single particle energy 376
size inconsistency 21
size selection 408
skimmer 212
slave-boson 92, 101, 106, 107, 110
slit nozzle 212
smoke source 223
sodium-chloride cluster 363
soft mode 187
soft solids 188
solid-like clusters 187, 193
solvation 246
solvent 397
soot formation 343
source 207
  electro-spray 249
  gas aggregation 6, 223
  laser ablation 226, 289

source (Contd.)
  liquid-metal-ion 229
  pick-up 227, 246
  supersonic 208, 249
  surface erosion 225
spectral analysis 194
spectroscopy 37
  depletion 244
  double-resonance 415
  fluorescence 382
  infrared 245
  kinetic energy of product ions 176
  optical 243, 338, 382, 401
  photoabsorption 390
  photodissociation 302, 408
  photoelectron 247, 297, 316, 325, 336,
    389
  photofragmentation 346, 392
  photoionisation 387
  PIPECO 166
  TPEPICO 388
  two photon 311, 401, 414
  vibrational 366, 382, 404, 407, 414
  ZEKE 248
spillout, electronic 300
spinodal curve 204
spin relaxation 323
sputtering 7
stability 30
structure 3, 30, 33, 67
suboxide cluster 357

sulphur cluster 74, 75, 76
sum rule, Thomas–Reiche Kuhn 25
supercooled matter 202
supershell 153
supersonic jet 208
  pulsed 212
surface localization 127
surface melting 188
surface tension 133

Tamm–Dancoff method 29
temperature 8, 216, 218, 232, 249
  mean 198
  vibrational 196
tetramer ion core 391
thallium cluster 309
thermionic emission 347
tight-binding method 50, 90, 106
time scale 188, 194, 381, 389
transition metal cluster 315
transition state 164
trapped cluster 7

ultra-fast process 249
unimolecular dissociation 263, 344

vacuum level 377

water cluster 396
Whitten–Rabinovitch approximation 166

1   **Atomic Spectra and Radiative Transitions**
By I. I. Sobelman

2   **Surface Crystallography by LEED**
Theory, Computation and Structural
Results. By M. A. Van Hove, S. Y. Tong

3   **Advances in Laser Chemistry**
Editor: A. H. Zewail

4   **Picosecond Phenomena**
Editors: C. V. Shank, E. P. Ippen,
S. L. Shapiro

5   **Laser Spectroscopy**
Basic Concepts and Instrumentation
By W. Demtröder   3rd Printing

6   **Laser-Induced Processes in
Molecules**   Physics and Chemistry
Editors: K. L. Kompa, S. D. Smith

7   **Excitation of Atoms and Broadening
of Spectral Lines**   By I. I. Sobelman,
L. A. Vainshtein, E. A. Yukov

8   **Spin Exchange**
Principles and Applications in
Chemistry and Biology
By Yu. N. Molin, K. M. Salikhov,
K. I. Zamaraev

9   **Secondary Ion Mass Spectrometry
SIMS II**   Editors: A. Benninghoven,
C. A. Evans, Jr., R. A. Powell,
R. Shimizu, H. A. Storms

10   **Lasers and Chemical Change**
By A. Ben-Shaul, Y. Haas,
K. L. Kompa, R. D. Levine

11   **Liquid Crystals of One- and
Two-Dimensional Order**
Editors: W. Helfrich, G. Heppke

12   **Gasdynamic Laser**   By S. A. Losev

13   **Atomic Many-Body Theory**
By I. Lindgren, J. Morrison

14   **Picosecond Phenomena II**
Editors: R. M. Hochstrasser,
W. Kaiser, C. V. Shank

15   **Vibrational Spectroscopy of
Adsorbates**   Editor: R. F. Willis

16   **Spectroscopy of Molecular Excitions**
By V. L. Broude, E. I. Rashba,
E. F. Sheka

17   **Inelastic Particle-Surface Collisions**
Editors: E. Taglauer, W. Heiland

18   **Modelling of Chemical Reaction
Systems**   Editors: K. H. Ebert,
P. Deuflhard, W. Jäger

19   **Secondary Ion Mass Spectrometry
SIMS III**
Editors: A. Benninghoven, J. Giber,
J. László, M. Riedel, H. W. Werner

20   **Chemistry and Physics of Solid
Surfaces IV**   Editors: R. Vanselow, R. Howe

21   **Dynamics of Gas-Surface Interaction**
Editors: G. Benedek, U. Valbusa

22   **Nonlinear Laser Chemistry**
Multiple-Photon Excitation
By V. S. Letokhov

23   **Picosecond Phenomena III**
Editors: K. B. Eisenthal, R. M. Hochstrasser,
W. Kaiser, A. Laubereau

24   **Desorption Induced by Electronic
Transitions DIET I**   Editors: N. H. Tolk,
M. M. Traum, J. C. Tully, T. E. Madey

25   **Ion Formation from Organic Solids**
Editor: A. Benninghoven

26   **Semiclassical Theories of Molecular
Scattering**   By B. C. Eu

27   **EXAFS and Near Edge Structures**
Editors: A. Bianconi, L. Incoccia, S. Stipcich

28   **Atoms in Strong Light Fields**
By N. B. Delone, V. P. Krainov

29   **Gas Flow in Nozzles**
By U. G. Pirumov, G. S. Roslyakov

30   **Theory of Slow Atomic Collisions**
By E. E. Nikitin, S. Ya. Umanskii

31   **Reference Data on Atoms, Molecules,
and Ions**   By A. A. Radzig, B. M. Smirnov

32   **Adsorption Processes on Semiconductor
and Dielectric Surfaces I**
By V. F. Kiselev, O. V. Krylov

33   **Surface Studies with Lasers**
Editors: F. R. Aussenegg, A. Leitner,
M. E. Lippitsch

34   **Inert Gases**
Potentials, Dynamics, and Energy Transfer
in Doped Crystals.   Editor: M. L. Klein

35   **Chemistry and Physics of Solid
Surfaces V**   Editors: R. Vanselow, R. Howe

36   **Secondary Ion Mass Spectrometry,
SIMS IV**   Editors: A. Benninghoven,
J. Okano, R. Shimizu, H. W. Werner

37   **X-Ray Spectra and Chemical Binding**
By A. Meisel, G. Leonhardt, R. Szargan

38   **Ultrafast Phenomena IV**
By D. H. Auston, K. B. Eisenthal

39   **Laser Processing and Diagnostics**
Editor: D. Bäuerle

Springer Series in Chemical Physics

Editors: Vitalii I. Goldanskii   Fritz P. Schäfer   J. Peter Toennies

Managing Editor: H.K.V. Lotsch

Volume 40   **High-Resolution Spectroscopy of Transient Molecules**
By E. Hirota

Volume 41   **High Resolution Spectral Atlas of Nitrogen Dioxide 559–597 nm**
By K. Uehara and H. Sasada

Volume 42   **Antennas and Reaction Centers of Photosynthetic Bacteria**
Structure, Interactions, and Dynamics
Editor: M.E. Michel-Beyerle

Volume 43   **The Atom-Atom Potential Method.** Applications to Organic
Molecular Solids
By A.J. Pertsin and A.I. Kitaigorodsky

Volume 44   **Secondary Ion Mass Spectrometry SIMS V**
Editors: A. Benninghoven, R.J. Colton, D.S. Simons, and H.W. Werner

Volume 45   **Thermotropic Liquid Crystals, Fundamentals**
By G. Vertogen and W.H. de Jeu

Volume 46   **Ultrafast Phenomena V**
Editors: G.R. Fleming and A.E. Siegman

Volume 47   **Complex Chemical Reaction Systems**
Mathematical Modelling and Simulation
Editors: J. Warnatz and W. Jäger

Volume 48   **Ultrafast Phenomena VI**
Editors: T. Yajima, K. Yoshihara, C.B. Harris, and S. Shionoya

Volume 49   **Vibronic Interactions in Molecules and Crystals**
By I.B. Bersuker and V.Z. Polinger

Volume 50   **Molecular and Laser Spectroscopy**
By Zu-Geng Wang and Hui-Rong Xia

Volume 51   **Space-Time Organization in Marcomolecular Fluids**
Editors: F. Tanaka, M. Doi, and T. Ohta

Volume 52   **Clusters of Atoms and Molecules.** Theory, Experiment,
and Clusters of Atoms
Editor: H. Haberland

Volume 53   **Ultrafast Phenomena VII**
Editors: C.B. Harris, E.P. Ippen, G.A. Mourou, and A.H. Zewail

Volume 54   **Physics of Ion Impact Phenomena**
Editor: D. Mathur

Volume 55   **Ultrafast Phenomena VIII**
Editors: J.-L. Martin, A. Migus, G.A. Mourou, and A.H. Zewail

Volume 56   **Clusters of Atoms and Molecules II.** Solvation and Chemistry of
Free Clusters, and Embedded, Supported and Compressed Clusters
Editor: H. Haberland

Volume 1–39 are listed on the back inside cover